PENGUIN REFERENCE BOOKS

THE PENGUIN DICTIONARY OF SCIENCE

E. B. Uvarov was born in Russia in 1910, and educated at Haberdashers' Aske's Hampstead School and the Imperial College of Science and Technology. He graduated in chemistry in 1929 and spent two years in biochemical research. In 1932 he taught science at Bertrand Russell's School: this was followed by two years as biochemist and technical manager to a firm of food manufacturers. From 1935 to 1944 he was senior chemistry master at Dartington Hall and subsequently at Taunton School. For the following eleven years he was head of the Technical Information Bureau of Courtaulds before going into independent practice as a scientific literature consultant and translator.

Alan Isaacs was born in London in 1925 and educated at St Paul's School and the Imperial College of Science and Technology, where he graduated in 1946. He was then engaged in fundamental research into combustion problems associated with rocket propulsion and was awarded a Ph.D. in 1950. At the same time he was a part-time teacher of English and mathematics at the Polish University College in London. Dr Isaacs is the author of *Introducing Science* and *The Survival of God in the Scientific Age*. He has been science editor of the *Collins English Dictionary* since its first publication in 1979. As a full-time lexicographer his many reference books include the *Longman Dictionary of Physics* and the *Macmillan Encyclopedia*.

THE PENGUIN DICTIONARY OF
SCIENCE

E. B. UVAROV AND
ALAN ISAACS

Seventh Edition

PENGUIN BOOKS

PENGUIN BOOKS

Published by the Penguin Group
Penguin Books Ltd, 27 Wrights Lane, London W8 5TZ, England
Penguin Books USA Inc., 375 Hudson Street, New York, New York 10014, USA
Penguin Books Australia Ltd, Ringwood, Victoria, Australia
Penguin Books Canada Ltd, 10 Alcorn Avenue, Toronto, Ontario, Canada M4V 3B2
Penguin Books (NZ) Ltd, 182–190 Wairau Road, Auckland 10, New Zealand

Penguin Books Ltd, Registered Offices: Harmondsworth, Middlesex, England

First published 1943
Second edition 1951
Third edition 1964
Fourth edition 1972
Fifth edition 1979
Sixth edition 1986
Seventh edition 1993
3 5 7 9 10 8 6 4

Copyright 1943 by E. B. Uvarov and D. R. Chapman
New material for third, fourth and seventh editions copyright
© Alan Isaacs, 1964, 1972, 1993
New material for fifth and sixth editions
copyright © Alan Isaacs and E. B. Uvarov, 1979, 1986
All rights reserved

Printed in England by Clays Ltd, St Ives plc
Set in Times Roman

Foreword to the 1993 Edition

SINCE this dictionary was last revised in 1986 science has continued to advance – with a corresponding increase in its vocabulary. Many new words have therefore been added and many existing entries have been expanded and brought up to date. In this edition, as in previous editions, the general principle of selecting predominantly scientific, as opposed to technological, words has been maintained. Moreover, in this edition a wider coverage of the basic terms in biology has been added. Fuller treatment of words used in computers, electronics, physics, biology and botany will be found in the Penguin dictionaries covering these subjects.

The network of cross-references has been kept throughout this edition, and cross-references in the text are indicated by the use of italics. Italics have not been used for the elements, however, as the dictionary lists all the elements (including transuranic elements). Trade names are indicated by an asterisk.

Since E. B. Uvarov originated this dictionary in 1943 it has been through seven editions in English and over thirty reprintings; it has also been translated into eight foreign languages. This gratifying response seems to indicate that the book is serving a useful purpose.

A.I.

Abbreviations Used in the Text

At. No.	Atomic number
(astr.)	Astronomy; as used in astronomy
(bio.)	Biochemistry or biology; as used in biochemistry or biology
b.p.	Boiling point
(chem.)	Chemistry; as used in chemistry
conc.	Concentrated
f.p.	Freezing point
(math.)	Mathematics; as used in mathematics
m.p.	Melting point
(phot.)	Photography; as used in Photography
(phys.)	Physics; as used in physics
r.a.m.	Relative atomic mass
r.d	Relative density

Abbreviations for SI units are used throughout. A table of these abbreviations will be found on page 405.

A

AAS See *atomic absorption spectroscopy.*

ab- A prefix attached to the names of practical electric units (e.g. *ampere, volt*) to indicate the corresponding unit in the *electromagnetic system* (e.g. abampere, abvolt). See *absolute units.*

Abbe condenser An optical *condenser* used in *microscopes*, consisting of two or three *lenses* having a wide *aperture.* Named after Ernst Abbe (1840 – 1905).

aberration (astr.) A variation in the apparent position of a *star* or other heavenly body, due to the motion of the observer with the *Earth.*

aberration, chromatic The formation, by a *lens*, of an image with coloured fringes, due to the *refractive index* of *glass* being different for *light* of different *colours.* The light is thus dispersed (see *dispersion of light*) into a coloured band. The effect is corrected by the use of *achromatic lenses.*

aberration, spherical The distortion of the image produced by a *lens* or *mirror* due to different rays from any one point of the object making different angles with the line joining that point to the *optical centre* of the lens or mirror (see *mirrors, spherical*) and coming to a *focus* in slightly different positions.

abiogenesis The hypothetical process by which living *organisms* were formerly thought to have been created from non-living matter: spontaneous generation.

abrasive A substance used for rubbing or grinding down surfaces; e.g. *emery.*

abscissa of a point *P.* In *analytical geometry*, the portion of the *x* axis lying between the *origin* and a point where the line through *P* parallel to the *y* axis cuts the *x* axis. See Fig. 5 under *Cartesian coordinates.*

absolute Not relative; independent. E.g. *absolute zero* of temperature, as distinct from zero on an arbitrary scale such as the *Celsius temperature* scale.

absolute alcohol *Ethanol* containing not less than 99% pure ethanol by weight.

absolute configuration See *optical activity.*

absolute expansivity of a liquid. The true *expansivity*, not relative to the containing vessel. The absolute expansivity is equal to the sum of the relative or apparent expansivity of the liquid and the volume expansivity of the containing vessel.

absolute humidity The amount of *water vapour* present in the *atmosphere*, defined in terms of the number of *kilograms* (or grams) of water in one cubic *metre* of air. See also *relative humidity.*

absolute permittivity See *permittivity.*

absolute temperature See *thermodynamic temperature.*

absolute units 1. A system of electrical *units* based on the *c.g.s. system*; e.g. the *ab*volt which is 10^{-9} practical *volts.* **2.** Any system of units using the least possible number of *fundamental units.* See *SI units; coherent units.*

absolute value 1. Modulus. The square root of the sum of the squares of the real numbers in a complex number. For example, the complex number $x + iy$ has an absolute value, written $|z|$, equal to $\sqrt{(x^2 + y^2)}$. **2.** The positive real number equal to a given real number irrespective of its sign, i.e. $|r| = r = |-r|$.

absolute zero The lowest temperature theoretically possible; the zero of *thermodynamic temperature*. $0 \text{ K} = -273.15°\text{C}. = -459.67°\text{F}$. See also *zero-point energy*.

absorbed dose See *dose*.

absorptance α. The ratio of the radiant or luminous *flux* absorbed by a body to the flux falling on it. The absorptance of a *black body* is 1.

absorption coefficient α. **1.** The ratio of the sound energy absorbed by a material to the sound energy falling on it. **2.** See *linear absorption coefficient*; *linear attenuation coefficient*.

absorption edge The *wavelength* at which a discontinuity appears in the intensity of an *absorption spectrum*.

absorption of gases The *solution* of *gases* in *liquids*. It is sometimes also applied to the absorption of gases by *solids* when the gas permeates the whole body of the solid rather than its surface. Compare *adsorption*.

absorption of radiation *Radiant energy* is partly reflected, partly transmitted, and partly absorbed by a body on which it falls, the absorption being accompanied by a rise in *temperature* of the absorbing body. Dull black surfaces absorb the greatest proportion of the incident energy, and brightly polished (reflecting) surfaces the least. Surfaces that are the best absorbers are also the best radiators. See *absorptance*; *absorption coefficient*.

absorption spectrum A *spectrum* consisting of dark lines or bands obtained when the *light* from a source, itself giving a continuous spectrum, is passed through a *gas* into a *spectroscope*. The dark lines or bands will occur in some of the same positions as the coloured lines in that substance's *emission spectrum* and will be characteristic of the substance. When the absorbing medium is in the *solid* or *liquid state* the spectrum of the transmitted light shows broad dark regions, which are not resolvable into sharp lines. Characteristic *X-ray* and *ultraviolet* absorption spectra are also formed.

absorptivity of a surface. The fraction of the *radiant energy* incident on the surface that is absorbed. Now replaced by *absorptance*.

abundance 1. The ratio of the number of *atoms* of a particular *isotope* in a mixture of isotopes of an *element,* to the total number of atoms present. Sometimes expressed as a percentage, e.g. the abundance of U-235 in natural uranium is 0.71%. **2.** The ratio of the mass of an element in the earth's crust to the total mass of the earth's crust, usually expressed as a percentage. For example, the abundance of silicon in the earth's crust is 28%.

abyssal Denoting a zone of the ocean, or an organism that inhabits it, below about 2000 metres, where very little light penetrates.

a.c. See *alternating current*.

acceleration a. The rate of increase of *velocity* (v) or *speed*, i.e., $a = dv/dt = d^2s/dt^2$, where s is *displacement*. It is measured in m s^{-2}. See *motion, equations of*.

acceleration of free fall g. Acceleration due to gravity. The *acceleration* of a body falling freely in a *vacuum* in the Earth's *gravitational field*; it varies slightly in different localities as a result of variations in the distance from the centre of mass of the Earth. Standard accepted value = 9.806 65 m s^{-2} (32.174 ft s^{-2}).

accelerator (chem.) A substance that increases the rate of a *chemical reaction* (i.e. a *catalyst*), particularly in the manufacture of *vulcanized rubber.*

accelerator (phys.) A machine for increasing the *kinetic energy* of charged particles (e.g. *protons, electrons, nuclei*) by accelerating them in *electric fields.* In *electrostatic generators* (see also *Van de Graaff generator* and *tandem generator*) the acceleration is achieved directly by using a very high *potential difference.* In multiple accelerators a lower potential difference is used repeatedly to give the particle successive increments of energy. Multiple accelerators are classified as *linear accelerators* or cyclic accelerators. See *cyclotron*; *synchroton*; *synchrocyclotron*; *betatron*; *bevatron*; and *storage ring.*

accelerometer An instrument for measuring *acceleration*, especially the acceleration of an aircraft or *rocket.*

acceptor (chem.) A molecule, atom, or ion that accepts electrons in the formation of a coordinate bond (see *valence*).

acceptor (phys.) An impurity in a *semiconductor* that accepts *electrons* and therefore causes *hole* conduction.

access time The time taken by a *computer store* to provide information to the *C.P.U.* The access time for high-speed stores is of the order of nanoseconds: for *backing storage* it may be from 1 *milli*second to some minutes.

accommodation The process that enables the eye to form sharp images on the retina at a wide range of distances. It is achieved by automatic adjustments to the shape of the eye's *lens*, which alter its *focal length.*

accumulator Storage battery, secondary cell. A device for storing chemicals that can react together to produce an *electric current.* An electric current is passed between two plates in a *liquid*; this causes chemical changes (due to *electrolysis*) in the plates and the liquid. When the changes are complete, the accumulator is charged. When the charged plates are joined externally by a *conductor* of electricity, the chemical changes are reversed, a current flows through the conductor until the reversal is complete, and the accumulator is discharged. In the common lead accumulator, the liquid is *sulphuric acid* of relative density 1.20 to 1.28, the positive plate when charged is *lead(IV) oxide* (lead dioxide), PbO_2, and the negative plate is spongy lead. During discharge both plates tend to become lead(II) sulphate, $PbSO_4$, and the density of the acid solution falls. Discharge should not be continued beyond the point at which the relative density reaches 1.15, otherwise an insoluble sulphate of lead, not decomposed on re-charging, may be formed. When this occurs, the cell is said to be sulphated. Nickel-iron (Ni-Fe*) accumulators in which the negative plate is iron and the positive plate is *nickel oxide* are also widely used. In these cells the liquid is a 20% solution of *potassium hydroxide.*

The increasing interest in all-electric cars has stimulated development of accumulators in recent years. One of the most promising devices is the zinc-air accumulator, which derives its *energy* from the conversion of zinc to *zinc oxide.* The plates are made of zinc and oxygen is obtained from the air through a porous nickel *electrode,* the electrolyte is potassium hydroxide. The lead accumulator will provide some 8×10^4 *joules* per kg, whereas the zinc accumulator can provide 5 times this energy density. Even higher energy densities are obtainable from Na/S and Li/Cl accumulators but these require operating *temperatures* of 300–600°C. See also *fuel cells.*

acetal An *organic compound* of the general formula $RCH(OR')_2$, where R is hydrogen or an organic *radical,* and R′ is an organic radical. $CH_3CH(OC_2H_5)_2$, 1,1-diethoxyethane, was formerly known as acetal. It is a liquid, b.p. 104°C., used as a *solvent,* in perfumes, and in organic synthesis.

acetaldehyde See *ethanal.*

acetaldol See *aldols.*

acetamide See *ethanamide.*

acetanilide Antifebrin. See *phenylethanamide.*

acetate See *ethanoate.*

acetate plastics *Plastics* made from *cellulose ethanoate* (acetate). See also *rayon.*

acetic acid See *ethanoic acid.*

acetic anhydride See *ethanoic anhydride.*

acetic ether See *ethyl ethanoate.*

acetoin 3-hydroxy-2-butanone. $CH_3CH(OH)COCH_3$. A yellow *liquid,* b.p. 148°C., used in the manufacture of flavours.

acetolysis The conversion of a group of atoms in an organic compound to an *ethanoyl* (acetyl) group by reacting the compound with *glacial ethanoic acid.*

acetone See *propanone.*

acetonitrile Ethanenitrile, methyl cyanide. CH_3CN. A colourless poisonous *liquid,* b.p. 82°C., with an odour like *ethoxyethane* (ether). Used in organic synthesis and as a *solvent.*

acetophenone See *phenylethanone.*

acetyl See *ethanoyl.*

acetylation The introduction of an *ethanoyl* (acetyl) group, CH_3CO-, into an organic compound. Compare *acylation.*

acetylene See *ethyne.*

acetylide See *carbide.*

acetylsalicylic acid See *aspirin.*

achromatic lens A *lens* free from chromatic *aberration,* giving an image free from coloured fringes. It consists of a pair of lenses, one of *crown glass,* the other of *flint glass,* the latter correcting the *dispersion* caused by the former. Other pairs of glasses can also be used if their dispersions neutralize each other. In an apochromatic lens three types of glass are used.

acid A substance that liberates *hydrogen ions* in solution, reacts with a *base* to form a *salt* and water only, has a tendency to lose protons, and turns litmus red. The classical theory of acid relies on the equation:
$$HX \rightleftharpoons H^+ + X^-$$
where H^+ is the liberated hydrogen ion. In aqueous solution the hydrogen ion is solvated (see *solvation*) to form the *oxonium* ion, H_3O^+:
$$HX + H_2O \rightleftharpoons H_3O^+ + X^-.$$
Many acids are corrosive and have a sour taste. See also *strong acid*; *weak acid*; *Lewis acids and bases*; *Lowry–Brønsted theory.*

acid amides See *amides.*

acid dyes A group of *dyes,* nearly all *salts* of *organic acids* in which the *chromophore* is a negative *ion,* often an organic *sulphonate*; they are used chiefly for dyeing wool and natural silk from an acid dyebath. In metallized dyes, the negative ion contains a metal atom *chelate.*

acid halide Acyl halide. An organic compound with the general formula RCOX, where R is a hydrocarbon group and X is a *halogen* atom. They are obtained from carboxylic acids by replacing the hydroxyl group with a halogen atom. They are used in *halogenation*.

acidic Having the properties of an *acid*. Compare *alkaline*; *basic*.

acidic anhydride See *anhydride*.

acidic hydrogen That portion of the hydrogen in an *acid* that is replaceable by *metals* to form *salts*, i.e. the hydrogen atom in an acid that becomes a positive ion (H^+) when the acid dissociates.

acidimetry Determination of the amount of *acid* present in a *solution* by *titration*. See *volumetric analysis*.

acidity constant Acidity dissociation constant. The *dissociation constant* (K_a) when an *acid*, HX, dissociates into the ions H^+ and X^-, i.e.
$$K_a = [H^+][X^-]/[HX].$$
See also *basicity constant*.

acidolysis *Hydrolysis* by means of an *acid*.

acid radical A *molecule* of an *acid* without the *acidic hydrogen*. E.g., the *bivalent sulphate radical* $-SO_4$, from *sulphuric acid*, H_2SO_4, is present in all sulphates.

acid rain Rain that has become polluted by *sulphuric acid* and *nitric acid* as a result of absorption of *sulphur dioxide* (sulphur(IV) oxide) and *nitrogen oxides* in the atmosphere. The effects can include the destruction of crops, trees, and fish, as well as damage to buildings. The remedy is to control the pollution by the oxides, especially from industrial and vehicle emissions.

acid salt An *acid* in which only a part of the *acid hydrogen* has been replaced by a *metal*. E.g. *carbonic acid*, H_2CO_3, forms such acid salts as *sodium hydrogencarbonate*, $NaHCO_3$, which contain the ion HCO_3^-.

acid value of a fat or oil. A measure of the free *fatty acid* present; the number of *milligrams* of *potassium hydroxide* required to neutralize the free fatty acids in one gram of the substance.

aclinic line See *magnetic equator*.

acoustics 1. The study of *sound*. **2.** The characteristics of an auditorium that determines its ability to enable music and speech to be heard clearly within it.

acoustic spectrum The range of frequencies occurring in the sound emitted by a source.

acoustoelectronics The study and use of devices in which *electronic signals* are converted by *transducers* into surface acoustic waves and passed through tiny solid strips. As acoustic signals are propagated some 10^5 times more slowly than *electromagnetic waves*, this technique enables *delay lines* to be constructed that can be up to 50 times lighter than pure electronic devices.

acquired characteristic A physical characteristic of an individual that is acquired during its lifetime, such as the size of the muscles of an athlete. These characteristics are not passed on to their progeny as they are not genetically determined. See also *Lamarckism*.

acre British unit of area. 4840 square yards. 4046.86 square metres.

acriflavine, 3,6-diamino-10-methylacridinium chloride. $C_{14}H_{14}N_3Cl$. A yellow substance used as an *antiseptic*.

acrolein See *propenal*.

acrylaldehyde See *propenal*.

5

acrylate See *propenoate*.

acrylic acid See *propenoic acid*.

acrylic resins A class of *plastics* obtained by the *polymerization* of derivatives of *propenoic* (acrylic) *acid*. They are transparent, colourless, and *thermoplastic*. Examples are *Perspex** and Acrilan*.

acrylonitrile See *propenonitrile*.

ACTH Adrenocorticotrophic hormone. A *polypeptide hormone* secreted by the pituitary gland that controls the secretion of *corticosteroid hormones* by the adrenal glands.

actin One of the two major *proteins* of muscle *cells*. See *actomyosin*.

actinic radiation *Electromagnetic radiation* that can cause photochemical reactions, especially radiation that can be used as a source of illumination in photography. It includes *X-rays* and *infrared* and *ultraviolet* radiation, as well as light.

actinides Actinoids, actinons. The *elements* with *atomic numbers* from 89 (actinium) to 103; they are analogous to the *lanthanides*. See Appendix, Table 8. See also *transactinides*.

actinium Ac. Element. R.a.m. 227. At. No. 89. A *radioactive* substance, *half-life* 21.7 *years,* m.p. 1050°C., b.p. 3200°C. There are two natural isotopes, Ac-227 and Ac-228. Some 22 artificial isotopes are known.

actinium series See *radioactive series*.

actinometer Any instrument that measures the intensity of *electromagnetic radiation*, especially one that is based on fluorescence or the *photoelectric effect*.

actinon Actinium emanation. A gaseous *radioisotope* of radon, Rn-219, produced by the *disintegration* of actinium. It is now known as radon-219.

action potential The change in *electric potential* across the membrane of a *cell* when a nerve impulse is passing through it.

activated alumina *Aluminium oxide* which has been dehydrated in such a way that a porous structure of high surface area is obtained. Activated alumina has the power of adsorbing *water vapour* and certain gaseous *molecules*. It is used for drying air and other gases.

activated carbon Active charcoal. Carbon, especially *charcoal,* which has been treated to remove *hydrocarbons* and to increase its powers of *adsorption*. It is used in many industrial processes for recovering valuable materials from gaseous mixtures, as a deodorant, and in *gas masks*.

activation (phys.) The process of inducing *radioactivity*.

activation analysis A sensitive analytical technique that can be used to detect the presence of many elements in a sample weighing only milligrams by first activating it, usually by neutron bombardment (neutron activation analysis) in a *nuclear reactor*, and then examining the gamma-ray spectrum of the decay products to detect characteristic emission lines.

activation energy The energy that must be supplied to a system in a *metastable state* to make a particular process occur. It is usually applied to systems on the atomic scale and the process may be an atomic reaction, such as fission, or an emission event. It may also be applied to chemical reactions, in which it is the energy barrier that must be overcome to make the reaction occur.

active 1. Denoting an electronic component, such as a *transistor*, that is capable of amplification. **2.** See *satellites, artificial.* Compare *passive*.

active mass (chem.) In the law of *mass action*, the active mass is taken to mean the *molar concentration* of the substance under consideration.

active site The site on an *enzyme* at which the *substrate* molecule binds. The way in which the enzyme functions depends on the constituent *amino acids* and the active sites created by the arrangement of the *polypeptide* chains in space.

activity (chem.) A function used to calculate the *equilibrium constant* (K) for reactions involving nonideal gases and solutions. In the reaction $X \rightleftharpoons Y + Z$, K is given by $a_Y a_Z / a_X$, where a_Y is the activity of component Y, etc. For gases, the activity coefficient, γ, is given by a/p, where p is the pressure; for solutions it is aX, where X is the *mole fraction*. See also *fugacity*.

activity (radioactive) A. The number of *disintegrations* of a *radioactive* material per second. The *SI unit* is the *becquerel*. See also *specific activity* and *curie*.

actomyosin A complex of two *proteins,* actin and myosin, that is the major constituent of muscle. The contraction of muscles is due to the interaction of these two proteins, thin filaments of actin sliding between thicker filaments of myosin.

acute angle An angle of less than 90°.

acyclic Not cyclic; having an *open-chain* structure.

acyl The *univalent radical* RCO–, where R is an organic group; regarded as being derived from the corresponding *carboxylic acid*, RCOOH.

acylation The introduction of an *acyl group*, RCO–, into a compound.

acyl halide See *acid halide*.

adatom An adsorbed *atom*. See *adsorption*.

addition compound A chemical *compound* formed by the addition of an *atom* or group of atoms to a *molecule*. E.g. *phosgene,* $COCl_2$, is an addition compound of *carbon monoxide,* CO, and chlorine, Cl_2.

addition reaction A *chemical reaction* in which one or more of the *double bonds* or *triple bonds* in an *unsaturated compound* is converted to a single bond by the addition of other *atoms* or groups.

additive process The process of forming any *colour* by a mixture of red, green, and blue lights. The colours add together to form a new colour, the colour obtained depending on the proportions of each additive *primary colour*. Equal proportions give white light. Compare *subtractive process*.

adduct An *addition compound* formed by a reaction involving no *valence* changes.

adenine 6-aminopurine. $C_5H_3N_4NH_2$. A white crystalline *purine* base, m.p. 360–365°C., occurring in *nucleic acids,* which plays a part in the constitution of genetic information. It also occurs in *adenosine triphosphate*.

adenosine A *nucleoside* consisting of *adenine* linked to a D-*ribose* sugar molecule. The phosphate esters are biologically important. See *adenosine triphosphate*.

adenosine triphosphate ATP. $C_{10}H_{12}N_5O_3H_4P_3O_9$. A *nucleotide* of importance in the transfer of *energy* within living *cells.* One of the phosphate groups can be readily transferred to other substances, in the presence of the appropriate *enzymes,* and with it goes a considerable amount of stored energy. It is as a result of the transfer of these phosphate groups that energy is made available in cells for chemical synthesis, muscle contraction, etc. ATP that has lost one phosphate group becomes the diphosphate (ADP). Adenosine is a *nucleoside* consisting of *adenine* and D-ribofuranose.

adhesives Substances used for sticking surfaces together; e.g. *glues*, *cements*, etc. They may be based on animal substances, such as *collagen*, vegetable *gums*, or *synthetic resins*, such as *epoxy resins*.

adiabatic Taking place without *heat* entering or leaving the system.

adiabatic demagnetization A method of attaining temperatures in the region of *absolute zero* (down to 10^{-6} K) by magnetizing a *paramagnetic* salt, such as potassium chrome alum or gadolinium sulphate, and allowing it to demagnetize adiabatically. During magnetization, between the poles of an electromagnet, the heat produced is removed by helium; during the adiabatic demagnetization cooling to very low temperatures takes place as a result of the fact that the demagnetized state is less ordered and therefore requires more energy, which comes from the *internal energy* of the substance.

adipic acid See *hexanedioic acid*.

admittance *Y*. The reciprocal of *impedance*, measured in *siemens*.

adrenaline Epinephrine. 3,4-dihydroxy-α-(methylaminomethyl) benzyl alcohol, $C_9H_{13}NO_3$. A *hormone* produced by the *medulla* of the adrenal glands and synthetically. It is secreted in response to stress, preparing the body for emergencies by affecting blood vessels and *carbohydrate metabolism*, stimulating blood flow to vital organs and raising blood sugar levels.

adrenocorticotrophic hormone See *ACTH*.

adsorbate The substance that is adsorbed on a surface. See *adsorption*.

adsorbent A substance that adsorbs. *Silica gel* and many porous or powdered materials are effective adsorbents by virtue of their large *specific surface* in conjunction with their ability to form bonds with *adsorbates*. See *adsorption*.

adsorption The concentration of a substance on a surface; e.g. *molecules* of a *gas* or of a dissolved or suspended substance on the surface of a *solid*. In chemisorption a single layer of atoms or molecules of the adsorbed substance is held to the solid surface by *covalent bonds*. In physisorption, several layers of atoms or molecules are held by *Van der Waals forces*.

advanced gas-cooled reactor AGR. See *gas-cooled reactor*.

advection The process in which either *matter* or *energy* is transferred from one place to another by a horizontal stream of *gas*, as in wind systems.

aelotropic See *anisotropic*.

aerial (U.S.A., antenna) That part of a *radio* system from which *energy* is transmitted into, or received from, *space* (or the *atmosphere*).

aerobic In the presence of free oxygen. See *respiration*.

aerodynamics The study of the motion and control of solid bodies (e.g. aircraft, *rockets,* missiles, etc.) in *air*. The study of air or other *gases* in motion.

aerogenerator Wind generator, windmill. A device to extract usable energy from winds. While the old-fashioned windmill drove milling machinery, the modern aerogenerator drives an electrical generator. The power available from an aerogenerator is proportional to $\rho d^2 v^3$, where ρ is the air density, d is the blade diameter, and v is the wind speed.

aerolites *Meteorites*, especially those consisting of stony material rather than iron.

aero metal A casting *alloy* consisting chiefly of aluminium, zinc, and copper.

aerosol A dispersion of *solid* or *liquid* particles in a *gas*; e.g. *smoke*. Some substances are sold in the form of aerosol sprays in pressurized cans. These cans contain an inert propellant, liquefied under pressure, usually *chlorofluoro-carbons* (CFCs). These substances present an environmental hazard, however (see *ozone layer*), and their use is being discontinued.

aerospace The *Earth's atmosphere* and the *space* beyond.

AES See *atomic emission spectroscopy*.

aetiology (U.S.A., etiology) The science or philosophy of causation. It is used in medicine to mean the science of the causes of disease.

affinity (chem.) **1.** See *free energy*. **2.** A former name for chemical attraction; the *energy* binding *atoms* together.

aflatoxins Four related toxic compounds produced by the mould *Aspergillus flavus*. They occur in peanuts and cereals contaminated with the mould and can cause liver damage and cancer.

AFM See *atomic force microscope*.

afterburning 1. The *combustion* that results from the addition of *fuel* to the exhaust of a *jet engine* in order to increase *thrust* and reduce fuel consumption. **2.** The irregular burning of residual *propellant* in a *rocket* motor when the main combustion has finished.

after-damp A poisonous mixture of *gases,* containing *carbon monoxide,* formed by the explosion of *fire-damp* (*methane,* CH_4) in coal-mines.

after-glow A glow sometimes observed high in the western sky after sunset. It is caused by fine dust particles in the *upper atmosphere* scattering the *light* from the *Sun*.

after-heat *Heat* generated in a *nuclear reactor* after it has been shut down, by the *radioactive* substances formed in the *fuel elements*.

agar A *gelatin*-like material obtained from certain seaweeds; it is chemically related to the *carbohydrates*. A *solution* in hot water sets to a firm jelly, which is used as a base for *culture media* for growing *bacteria, fungi,* etc.

agate A very hard natural form of *silica*, used for knife-edges of *balances*, for mortars for grinding hard materials, and in ornaments.

aglycone A non-sugar component of a *glycoside*.

agonic line A line of zero *magnetic declination*.

AGR Advanced *gas-cooled reactor*.

air See *atmosphere*.

air equivalent The thickness of a layer of air at *s.t.p.* that causes the same amount of absorption of nuclear radiation as the substance being considered.

air thermometer See *gas thermometer*.

alabaster A natural opaque form of *hydrated calcium sulphate,* $CaSO_4.2H_2O$.

alanine A colourless crystalline *soluble amino acid.* See Appendix, Table 5.

albedo 1. The ratio of the *radiant flux* reflected by a surface to that falling on it. **2.** The probability that a *neutron* entering a material will be reflected back by that material through the surface by which it entered.

albumins A group of *soluble* globular *proteins* occurring in many animal *tissues* and *fluids*; e.g. egg-white (egg albumin, which is often called albumen), milk (lactalbumin), and *blood* (serum albumin).

albuminoids See *scleroproteins*.

alchemy The predecessor of scientific *chemistry.* An art by which its devotees sought, with the aid of a mixture of mysticism, *astrology,* practical chemistry, and quackery, to transmute *base metals* into gold, prolong human life, etc. It flourished from about A.D. 500 till the Middle Ages, when it gradually fell into disrepute.

alcoholates Metallic *salts* of *alcohols*, formed by replacement of hydrogen atoms in the *hydroxyl* groups of the latter by metals, e.g. sodium ethanolate (sodium ethoxide), C_2H_5ONa.

alcoholometry The determination of the proportion of *ethanol* in spirits and other *solutions*; it is usually performed by measuring the relative density of the liquid at a standard *temperature* by a specially graduated *hydrometer*.

alcohols A class of *organic compounds* derived from the *hydrocarbons,* one or more hydrogen *atoms* in *molecules* of the latter being replaced by *hydroxyl groups,* –OH. The names of alcohols are obtained from the hydrocarbons from which they are derived to which the suffix -ol is added. E.g. *ethanol* (ordinary 'alcohol') is C_2H_5OH, theoretically derived from *ethane,* C_2H_6. In primary alcohols, the carbon atom to which the hydroxyl group is attached has two hydrogen atoms attached to it, e.g. CH_3CH_2OH; in secondary alcohols this is reduced to one, e.g. $(CH_3)_2CHOH$; and in tertiary alcohols to none, e.g. $(CH_3)_3COH$. Alcohols that contain more than one hydroxyl group are called *polyhydric alcohols*. See also *diols*; *triols*.

aldehydes A class of *organic compounds* of the type R.CO.H where R is an *alkyl* or *aryl* group. The names of aldehydes are obtained from the corresponding primary alcohols, from which they are obtained by oxidation, with the suffix -al. E.g. *methanal* (formaldehyde) is obtained from *methanol* and *ethanal* (acetaldehyde) from *ethanol*.

aldols *Organic compounds* that contain both an *aldehyde* and an *alcohol*. 3-hydroxybutanal, $CH_3CH(OH)CH_2CHO$ (also called acetaldol or aldol), is an example. A thick oily liquid, b.p. 83°C, it is used in the vulcanization of rubber and in perfumes.

aldose A *monosaccharide* containing an aldehyde (*formyl*) group in the *molecule*.

algae A diverse group of simple plants that contain *chlorophyll* and are capable of *photosynthesis*. They exist in aquatic habitats and in moist situations on land. They are not now regarded as a single taxonomic group.

algebra The branch of mathematics dealing with the properties of, and relationships between, quantities by means of general symbols.

algebraic sum The total of a number of quantities of the same kind, with due regard to sign. Thus the algebraic sum of $3, -5$, and -2 is -4.

algin A loose term for *alginic acid* or its sodium *salt*.

alginic acid $(C_6H_8O_6)_n$. A complex *organic compound* related to the *carbohydrates,* found in certain seaweeds. It is used for preparing *emulsions* and as a thickening agent in the food industry; its *salts,* the alginates, can be made into textile fibres, which are *soluble* in *alkalis* and are used for special purposes.

algol Algorithmic language. A type of *computer* language, based on *Boolean algebra,* for expressing information in an algebraic notation.

algorithm Algorism (math.). A systematic mathematical procedure that enables a problem to be solved in a finite number of steps. Problems for which no algorithms exist require *heuristic* solutions.

alicyclic compound A type of *organic compound* that is essentially *aliphatic,* although it contains a *saturated* ring of carbon *atoms*.

alidade An instrument for measuring vertical heights and distances.

aliphatic compounds *Organic compounds* containing open chains of carbon *atoms* rather than the closed rings of carbon atoms of the *aromatic compounds*. They consist of the *alkanes*, *alkenes*, and *alkynes* as well as all their *derivatives* and *substitution products*. See also *alicyclic compound*.

aliquot part A divisor of a number or quantity that will give an *integer*. Thus 3 is an aliquot part of 6, but 5 is not.

alizarin 1,2-dihydroxyanthraquinone. $C_{14}H_6O_2(OH)_2$. An orange-red crystalline *solid*, m.p. 289°C. A colouring matter formerly extracted from the root of the madder plant, it is now made synthetically. Used in dyeing with the aid of *mordants*.

alkali A *soluble hydroxide* of a *metal*, particularly of one of the *alkali metals*, i.e. one that produces *hydroxide* ions. It is often applied to any substance that has an alkaline reaction (i.e turns *litmus* blue and neutralizes *acids*) in solution. See also *base*.

alkali metals The *univalent metals* lithium, sodium, potassium, rubidium, caesium, and francium belonging to Group 1A of the *periodic table*.

alkalimetry The determination of the amount of *alkali* present in a *solution*, by *titration*. See *volumetric analysis*.

alkaline Having the properties of an *alkali*; the opposite of *acidic*.

alkaline earth metals The *bivalent* group of *metals* comprising beryllium, magnesium, calcium, strontium, barium, and radium, belonging to Group 2A of the *periodic table*.

alkaloids A group of *basic* organic substances of plant origin, containing at least one nitrogen *atom* in a ring structure in the *molecule*. Many have important physiological actions and are used in medicine. E.g. *codeine, cocaine, nicotine, quinine, morphine*.

alkanal Any *aliphatic aldehyde*.

alkanes Paraffins. A *homologous series* of *saturated hydrocarbons* having the general formula C_nH_{2n+2}. Their systematic names end in -ane. They are chemically inert, stable, and flammable. The first four members of the series (*methane, ethane, propane, butane*) are gases at ordinary temperatures; the next eleven are liquids, and form the main constituents of *paraffin oil*; the higher members are solids. *Paraffin wax* consists mainly of higher alkanes.

alkanization The process of converting an *unsaturated hydrocarbon* into an *alkane*.

alkanol Any *aliphatic alcohol*.

alkenes Olefins. A *homologous series* of *unsaturated hydrocarbons* containing a double bond and having the general formula C_nH_{2n}. Their systematic names end in -ene. *Ethene* (ethylene), $CH_2:CH_2$, and *propene* (propylene), $CH_3CH:CH_2$ are the first members. Higher members have isomers depending on the position of the double bond. E.g. *butene*, C_4H_8, has isomers but-1-ene and but-2-ene.

alkoxide A salt-like compound formed by reacting an alcohol with sodium or potassium. They contain the ion $R–O^-$.

alkoxy *Univalent organic* groups having the formula RO–, where R is an *alkyl* group.

alkyd resins See *glyptal resins*.

alkyl *Univalent saturated hydrocarbon* groups having the general formula C_nH_{2n+1}, derived from *alkanes*. E.g. *methyl*, CH_3-; *ethyl*, C_2H_5-.

alkylarene An *arene* (e.g. *benzene*) with one or more hydrogen atoms in the molecule replaced by *alkyl* groups; e.g. ethylbenzene, $C_2H_5C_6H_5$.

alkylation The introduction of an *alkyl* group into a molecule; e.g. the addition of *alkanes* to *alkenes*.

alkyl halide See *haloalkane*.

alkynes Acetylenes. A *homologous series* of *unsaturated hydrocarbons* having the general formula C_nH_{2n-2} and containing a *triple bond* between two of the carbon atoms in the molecule. Their systematic names end in -yne, e.g. *ethyne* (acetylene) $CH{\equiv}CH$.

allele One of the alternative forms of a *gene*. In most organisms each parent contributes one allele of a gene; each allele occupies the same relative position on *homologous chromosomes* (one from each pair coming from the female parent and one from the male). See also *dominant*; *recessive*.

allo- Prefix meaning 'other', used in *chemistry* to denote a variation from the standard or normal form.

allobar A mixture of the *isotopes* of an element that does not occur naturally.

allochromy The emission of radiation by a surface at a *wavelength* that differs from that of the incident radiation. See *fluorescence*.

allomerism A similarity in the crystalline structure of substances of different chemical composition.

allomorphism A variability in the crystalline structure of certain substances. Allomorphs are different crystalline forms of the same *compound*.

allosterism The ability of a *protein* to shift reversibly between different stable *conformations*, thus affecting its ability to interact with other *molecules*. For example, an *enzyme* may be activated by binding a *ligand* to one part of its surface, so changing the shape of the *substrate* binding site. Allosteric proteins, e.g. *receptors* and *repressors*, are important in regulating *cell* functions.

allotropes Allotropic forms. See *allotropy*.

allotropy The existence of a chemical *element* in two or more forms differing in physical properties but giving rise to identical chemical *compounds*. E.g. sulphur exists in a number of different allotropic forms.

allowed bands See *energy bands*.

alloxan $(CO)_4(NH)_2$. A white crystalline *heterocyclic* compound, m.p. 170°C., derived from *uric acid* by treatment with dilute *nitric acid*. It destroys certain cells in the pancreas and is used to produce diabetes for experimental purposes.

alloy A composition of two or more *metals*; an alloy may be a *compound* of the metals, a *solid solution* of them, a heterogeneous *mixture*, or any combination of these. The term is sometimes extended to include non-metallic components; e.g. iron-carbon alloys.

alluvial Deposited by rivers.

allyl alcohol Prop-2-en-1-ol. $CH_2{:}CH.CH_2OH$. A colourless pungent *liquid alcohol*, b.p. 96.5°C., used in the manufacture of synthetic *resins* and pharmaceuticals.

allyl group The *univalent radical*, $CH_2{:}CH.CH_2-$, derived from *propene*.

allyl resins Synthetic *resins* formed by the *polymerization* of chemical *compounds* containing the *allyl group*.

Alnico* A series of *alloys* based on iron and containing nickel, aluminium, cobalt, and copper. They are used to make permanent magnets.

alpha decay A form of *radioactive decay* in which a nucleus spontaneously emits an *alpha particle*.

alpha-iron An allotropic (see *allotropy*) form of pure iron that exists up to 900°C.

alpha particle A helium *nucleus*; i.e. a close combination of two *neutrons* and two *protons* (see *atom, structure of*), and therefore positively charged. Alpha particles are emitted from the *nuclei* of certain *radioactive nuclides*. See *radioactivity*.

alpha radiation Streams of fast-moving *alpha particles*. Alpha radiation produces intense *ionization* in matter, is easily absorbed, travelling more than a few tenths of a millimetre only in gases. It produces *fluorescence* on a fluorescent screen.

altazimuth An instrument for the measurement of the *altitude* and *azimuth* of heavenly bodies.

alternating current a.c. A flow of *electric current* that, after reaching a maximum in one direction, decreases, finally reversing and reaching a maximum in the opposite direction, the *cycle* being repeated continuously. The number of such cycles per unit time is the *frequency* (f). If the unit of time is the second, the frequency is measured in *hertz*. The instantaneous value of an alternating current (I) is given by $I = I_0 \sin 2\pi f t$, where I_0 is the maximum value. See also *root mean square value of an alternating quantity*; *reactance*.

alternative energy sources See *renewable energy sources*.

alternator An electric *generator* for producing *alternating currents*.

altimeter An instrument used to measure height above sea-level. It usually consists of an *aneroid barometer* calibrated to read zero at sea-level and the height above sea-level in metres or feet. See *hydrostatic equation*.

altitude 1. Height. **2.** The altitude of a heavenly body is its *angular distance* from the horizon on the vertical circle passing through the body, the *zenith,* and the *nadir.* See Fig 2, under *azimuth*.

alum Potash alum. $K_2SO_4.Al_2(SO_4)_3.24H_2O$. Crystalline aluminium potassium sulphate. The *compound* occurs naturally and is used as a *mordant* in dyeing, for fireproofing, and other technical purposes. See also *alums*.

alumina See *aluminium oxide*.

aluminate A *salt* containing the aluminate ion $[Al(OH)_4]^-$. It is formed when *aluminium hydroxide* is dissolved in a solution of a strong *base*.

aluminium Al. Element. R.a.m. 26.98154. At. No. 13. A light white *metal*, r.d. 2.7, m.p. 659.70°C., b.p. 2467°C., ductile and malleable, and a good *conductor* of electricity. It occurs widely in nature in *clays* and is the third most abundant element in the Earth's crust (8%). It is extracted mainly from *bauxite* by *electrolysis* of a molten mixture of purified bauxite and *cryolite* (see *Hall-Héroult cell*). The metal and its *alloys* are used for aircraft, cooking utensils, electrical apparatus, and for many other purposes where its light weight is an advantage.

aluminium brass *Brass* containing small amounts of aluminium.

aluminium bronze An *alloy* of copper containing 4%–13% aluminium.

aluminium chloride $AlCl_3$. A white crystalline solid, which fumes in moist air and reacts violently with water. It is known as the *anhydrous salt* (m.p. 190°C. at 2.5 atm) and the hexahydrate, $AlCl_3.6H_2O$. It is used as a *catalyst* in the oil industry.

aluminium ethanoate Aluminium acetate. $Al(CH_3COO)_3$. A white *soluble amorphous* powder, used as an *astringent* and *antiseptic*. *Basic* aluminium ethanoate, $AlOH(CH_3COO)_2.xH_2O$, a white crystalline powder, is used as a waterproofing and fireproofing compound in the textile industry.

aluminium hydroxide $Al(OH)_3$. A white *insoluble amphoteric* powder used in the manufacture of *glass* and *ceramics,* and as an *antacid* in medicine.

aluminium oxide Alumina. Al_2O_3 A white crystalline substance, m.p. 2015°C., used in *cement,* as a *refractory,* and in the manufacture of aluminium. It occurs naturally as *corundum* and *emery* and in a hydrated form as *bauxite*. See also *activated alumina*.

aluminium sulphate $Al_2(SO_4)_3$. A white crystalline *soluble* substance, known as the *anhydrous salt* and the hydrate $Al_2(SO_4)_3.18H_2O$. It is used in purifying *water,* in the manufacture of *paper,* and in *fire extinguishers*.

aluminosilicates A large class of *minerals*, both natural and synthetic, containing aluminium and silicon combined with oxygen in their structure. It includes *clays*, *zeolites*, *micas*, and many other important mineral materials.

aluminothermic reduction High-temperature *reduction* of metal oxides to the corresponding metals by the *thermite* method.

alums Double *salts* of the general formula
$$M_2SO_4.R_2(SO_4)_3.24H_2O,$$
where M is a *univalent* ion such as sodium, potassium, or ammonium, and R is a *tervalent* ion, such as aluminium or chromium. See also *alum*.

alum-stone See *alunite*.

alunite Alum-stone. A natural *compound* of potassium and *aluminium sulphate* and *aluminium hydroxide,* $K_2SO_4.Al_2(SO_4)_3.4Al(OH)_3$. It is used as a source of *alum*.

AM See *amplitude modulation*.

amalgam An *alloy* of mercury.

amalgamation process for gold. Gold-bearing rock or sand, after crushing, is treated with mercury, which forms an *amalgam* on the surface of the gold. The amalgamated particles are allowed to stick to amalgamated copper plates, the rest of the ore being washed away; they are then removed, the mercury is distilled off in iron *retorts,* and the remaining gold purified by *cupellation*.

amatol An explosive mixture of 80% *ammonium nitrate* and 20% *T.N.T.*

amber Succinite. A fossil *resin,* derived from an extinct species of pine. Obtained from mines in East Prussia, and found on seashores, it is a yellow to brown solid, which contains *succinic acid,* and is used for ornamental purposes and as an electrical insulator.

ambergris A grey or black waxy material that occurs (probably as the result of disease) in the intestines of the sperm whale. It is used in perfumery.

americium Am. *Transuranic element.* At. No. 95. *Radioactive.* A member of the *actinide* series. The most stable *isotope,* americium-243, has a *half-life* of 7.95×10^3 years. R.d. 13.7, m.p. 995°C. Ten isotopes are known.

amethyst A violet variety of *quartz*; impure crystalline *silica,* SiO_2.

amidases *Enzymes* that control the *hydrolysis* of *amides*.

amides A group of *organic compounds* formed by replacing the hydrogen *atoms* of *ammonia,* NH_3, by *acyl radicals.* E.g. *ethanamide,* CH_3CONH_2. The general formula is $RCONH_2$, where $-CONH_2$ is the amide group.

Amidol* 2,4-diaminophenol dihydrochloride,
$$C_6H_3(OH)(NH_2)_2.2HCl;$$
used in *photography* as a developer.

aminases *Enzymes* that *catalyze* the *hydrolysis* of *amines.*

amination The introduction of an *amino group* into a compound.

amines Compounds formed by replacing one or more hydrogen *atoms* of *ammonia,* NH_3, by hydrocarbon *radicals* linked to nitrogen through carbon atoms. They are classified into primary amines of the type NH_2R; secondary amines, NHR_2; and tertiary amines, NR_3. See also *quaternary ammonium compounds.*

amine salt A salt formed by a reaction between an *acid* and an *amine.* For example, dimethyl amine, $(CH_3)_2NH$ reacts with hydrochloric acid, HCl, to form the ionic compound $[(CH_3)_2NH]^+Cl^-$. These compounds, especially when used medically, were formerly called hydrochlorides.

amino acid A *carboxylic acid* that contains the *amino group* $-NH_2$. These acids are the units that link together into *polypeptide* chains to form *proteins*; they are therefore of fundamental importance to life. Some twenty different amino acids occur in nature, nearly all of which have the general formula: $R-CH(NH_2)COOH$. See Appendix, Table 5. 'Essential' amino acids are those that an *organism* is unable to synthesize and therefore has to obtain from its environment. There are eight 'essential' amino acids for man.

amino group The *univalent* group $-NH_2$.

aminoplastic resins Synthetic *resins* derived from the reaction of *urea, melamine,* or allied amino compounds with *aldehydes.* They form the basis of *thermosetting* moulding materials.

amino sugar A sugar in which a hydroxyl group has been replaced by an amino group.

ammeter An instrument for the measurement of *electric current.* In moving-iron ammeters, a strip of soft iron is caused to move in the *magnetic field* set up by the current flowing through a coil; these instruments can measure both a.c. and d.c. The more accurate moving-coil instruments contain a permanent *magnet* between the poles of which is pivoted a coil carrying the current to be measured; they essentially measure only d.c., but can be used with a rectifier to measure a.c. In each type of instrument a pointer attached to the moving portion moves over a scale graduated in *amperes.* In thermoammeters, the current to be measured heats a resistance wire and can be used with a.c. or d.c. In one form, usually used for a.c., the wire heats up a thermocouple connected to a galvanometer, in the other, *hot-wire* instrument, the expansion of the wire causes a pointer to move over a scale.

ammines *Coordination compounds* containing *ammonia* molecules as *ligands*; *complex compounds* formed by ammonia with *salts* or *bases.*

ammonal A *mixture* of *ammonium nitrate,* NH_4NO_3, and aluminium. It is used as an *explosive.*

ammonia NH_3. A pungent-smelling very *soluble gas,* m.p. $-74°C$., b.p. $-30.9°C$., giving an *alkaline solution* containing *ammonium hydroxide,* NH_4OH. It is obtained synthetically from atmospheric nitrogen (see *Haber process*). It is used in the manufacture of *resins*, *explosives*, and *fertilizers*.

ammonia clock An atomic clock based on the vibrational frequency with which the nitrogen atom in the *ammonia* molecule passes through the plane of the three hydrogen atoms and back again. The vibration has a frequency of 23 870 *hertz* and a *quartz crystal* is used to supply ammonia gas with energy at this frequency. Because the ammonia will only absorb energy at this frequency, the ammonia can be used to regulate the frequency of the quartz oscillator, through a feedback circuit.

ammonium carbonate $(NH_4)_2CO_3$. A white *soluble* crystalline salt, that usually occurs as the monohydrate $(NH_4)_2CO_3.H_2O$. It decomposes slowly, yielding ammonia, carbon dioxide, and water. See also *sal volatile*.

ammonium chloride Sal ammoniac. NH_4Cl. A white *soluble* crystalline *salt*, used in *dry cells* and *Leclanché cells*.

ammonium ethanoate Ammonium acetate. CH_3COONH_4. A white *deliquescent* solid, m.p. 114°C., used as a meat preservative and in the manufacture of *dyes*.

ammonium hydroxide NH_4OH. A *compound* presumed to exist in *aqueous solutions* of *ammonia*; the name is often applied to the *solution*.

ammonium nitrate NH_4NO_3. A white *soluble* crystalline *salt*, m.p. 169.6°C., that decomposes on heating to form *dinitrogen oxide*, N_2O, and water. It is used in *explosives*, e.g. *ammonal, amatol.*

ammonium ion NH_4^+. A *univalent* ion that in *compounds* behaves similarly to an *alkali metal*, giving rise to ammonium *salts*.

ammonium sodium hydrogen orthophosphate Microcosmic salt. NH_4NaHPO_4. $4H_2O$. A white crystalline *soluble salt*, used as a *flux.*

ammonium sulphate $(NH_4)_2SO_4$. A white *soluble* crystalline *salt*, obtained as a by-product of *coal-gas* manufacture, now produced from *ammonia* and *sulphuric acid* and used as a *fertilizer.*

ammonium thiocyanate NH_4SCN. A colourless *soluble* crystalline substance, m.p. 149.6°C., used as a herbicide and in the textile industry.

ammonolysis A chemical reaction in which one group of an organic compound is converted to an *amine* group, by reacting the compound with ammonia.

amorphous Non-crystalline; having no definite form or shape. *Glass* is an amorphous solid (see *solid state*).

amount of substance *n.* A basic physical quantity that is proportional to the number of specified particles of a substance. The specified particle may be an *atom, molecule, ion, radical, electron,* etc., or any specified group of such particles. The constant of proportionality, the reciprocal of the *Avogadro constant,* is the same for all substances. The basic *SI unit* of amount of substance is the *mole.*

ampere A unit of *electric current* approximately equivalent to the flow of 6×10^{18} *electrons* per second. The absolute ampere, which is one-tenth of an abampere (see *ab-*), is equal to 1.000 165 International amperes. The International ampere was originally defined as the unvarying current that when passed through a *solution* of *silver nitrate,* deposits silver at the rate of 0.001 118 00 grams per second. The ampere was redefined in 1948 as the intensity of a constant current that, if maintained in two parallel, rectilinear *conductors* of infinite length, of negligible circular section and placed at a distance of one *metre* from one another in vacuo, will produce between the conductors a *force* equal to 2×10^{-7} *newton* per *metre* of length. The ampere so defined is the basic *SI unit* of current. Symbol A. Named after A. M. Ampère (1775–1836).

ampere-hour The practical unit of *electric charge*, the charge flowing per hour through a *conductor* when the current in it is one *ampere*. 3600 *coulombs.*

Ampere's law The strength of the *magnetic field* induced by a *current* flowing through a *conductor* is, at any point, directly proportional to the *product* of the current and the length of the conductor and inversely proportional to the *square* of the distance between the point and the conductor. The direction of the field is perpendicular to the *plane* joining the point and the conductor.

ampere-turns A unit of *magnetomotive force*. The product of the number of turns in a coil and the current in *amperes* that flows through it.

amphetamine $C_6H_5CH_2CHNH_2CH_3$. A *drug,* used in the form of the *sulphate,* that stimulates the central nervous system in cases of depression. Also known under the trade name, Benzedrine*.

amphiboles A group of complex *silicate minerals* that includes *hornblende* and *asbestos.*

amphichroic Amphichromatic. Giving one colour on reaction with an *acid* and another colour on reaction with a *base.*

amphipathic A *molecule* in which one end is *polar* and *soluble* in water (hydrophilic) and the other end non-polar and insoluble in water but soluble in *fats* and *fat solvents*. In water they form *micelles* and can form bimolecular layers, the non-polar ends associating in the middle, with the polar ends outwards. Amphipathic molecules, e.g. *phospholipids*, are important components of *cell membranes.*

amphiprotic Capable both of accepting and of yielding *protons* in solution; *amphoteric*. See *solvent.*

ampholyte An *amphoteric* substance, especially one that functions as an *electrolyte.*

amphoteric Chemically reacting as *acidic* to strong *bases* and as *basic* towards strong *acids*. E.g. the amphoteric *oxide, zinc oxide,* gives rise to zinc *salts* of strong *acids* and zincates of the *alkali metals.*

amplifier An *electronic* device that increases the strength of a signal fed into it, by obtaining *power* from a source other than the input signal.

amplitude (phys.) If any quantity is varying in an oscillatory manner about an equilibrium value, the maximum departure from that equilibrium value is called the amplitude; e.g. in the case of a *pendulum* the amplitude is half the length of the swing. For a *wave motion,* e.g. *electromagnetic waves* or *sound* waves, the square of the amplitude of the wave is proportional to the amount of *energy* carried by the wave.

amplitude modulation AM. One of the principal methods of transmitting information by *radio* waves. The *amplitude* of a *carrier wave* is modulated (see *modulation*) in accordance with the *frequency* of the signal to be transmitted.

AMU See *atomic mass unit.*

amyl The former name of the radical C_5H_{11}–; n-amyl compounds are now called *pentyl* compounds and isoamyl compounds are 3-methylbutyl compounds.

amyl acetate See *pentyl ethanoate.*

amyl alcohol See *pentanol.*

amylases A group of *enzymes* capable of splitting *starch* and *glycogen* into *sugars*. They are found in many plants and animals (e.g. the pancreatic juices of mammals).

amylopectin The principal component (about 80%) of most cereal starches (see *starch*). A *polysaccharide* whose molecules consist of long cross-linked chains of *glucose* units. It is insoluble in water. Compare *amylose*.

amyloplast See *plastid*.

amylose A water-soluble component (about 20%) of most cereal starches (see *starch*). A *polysaccharide* whose molecules consist of long unbranched chains of *glucose* units, structurally related to *cellulose*. Compare *amylopectin*.

amylum See *starch*.

anabolic steroid A *steroid* that promotes the growth of tissue, especially muscle. They include naturally occurring *androgens* and synthetic substances used medically to increase weight. The use of anabolic steroids by athletes to build muscles is forbidden by most athletic associations, as they cause liver damage.

anabolism Part of *metabolism*, comprising the building-up of complex substances from simpler material, with absorption and storage of *energy*.

anaerobic In the absence of free oxygen. See *respiration*.

anaesthetic A substance used in medicine to produce insensibility or loss of feeling.

analgesic A substance used in medicine to relieve pain.

analog computer A *computer* in which numerical magnitudes are represented by physical quantities, such as *electric current*, *voltage*, or *resistance*. See also *digital computer*.

analogous (bio.) Organs of living creatures that are similar in function but different in structure and development and are therefore considered to be of separate origin and not indicative of evolutionary relationship; e.g. the wings of an insect and of a bird. See *homologous*.

analysis (chem.) The process of determining the composition of a substance. See *activation, colorimetric, gravimetric, qualitative, quantitative, spectrographic,* and *volumetric analysis; chromatography*.

analytical geometry Coordinate geometry. A form of *geometry* based upon the use of *coordinates* to define positions in *space*. See *Cartesian coordinates* and *polar coordinates*.

anastigmatic lens A *lens* designed to correct *astigmatism*. In spectacles, such a lens has different radii of curvature in the vertical and horizontal planes.

anatase Crystalline natural *titanium dioxide*, TiO_2.

AND circuit See *logic*.

androgen A *steroid sex hormone*, such as *testosterone*, or other substance that promotes male secondary characteristics in vertebrates. Testosterone is produced by the testes.

anechoic Having a low degree of reverberation. An anechoic chamber is used in *acoustics* for experimental purposes. Its walls are covered by absorbent material to avoid reflections and also with small pyramids to avoid *standing waves*.

anemo- Prefix denoting the wind.

anemometer Instrument for measuring the speed of wind or any other moving gas.

aneroid Without liquid. The aneroid *barometer* is an instrument for measuring atmospheric *pressure*; it consists of an exhausted metal box with a thin corrugated metal lid. Variations in atmospheric pressure cause changes in the displacement of the lid; this displacement is magnified and made to actuate a pointer moving over a scale by means of a system of delicate levers.

anethole $CH_3CH:CHC_6H_4OCH_3$. A white crystalline powder, m.p. 22.5°C., used in perfumes, flavouring, and in medicine.

aneurine See *thiamine*.

angle The difference in direction between two intersecting lines or *planes*. It is measured in *degrees* or in *radians* (see *circular measure*).

angstrom Ångström unit. Å. 10^{-10} *metre*. A unit of length, formerly used for measurement of *wavelengths* of *light* and intra-molecular distances. 10 Å = 1 nanometre.

angular acceleration α. The rate of change of *angular velocity*.

angular displacement θ. The *angle* through which a point, line, or body has been rotated in a specified direction, about a specified *axis*.

angular distance The distance between two bodies, measured in terms of the *angle* subtended by them at the point of observation; it is used in *astronomy*.

angular frequency Pulsatance. The *frequency* of a periodic process expressed in *radians* per *second*; it is equal to 2π times the number of *cycles* per second.

angular momentum L. The product of *moment of inertia* and *angular velocity*. For the angular momentum of *elementary particles* see *spin*.

angular velocity ω. Rate of motion through an *angle* about an *axis*. It is measured in *degrees*, *radians*, or revolutions per unit time.

anhydride The anhydride of a substance is that which, when chemically combined with *water*, gives the substance. A *basic* anhydride is the *oxide* of a *metal* and forms a *base* with *water* (e.g. $Na_2O + H_2O = 2NaOH$): an *acidic* anhydride is the oxide of a non-metal and forms an *acid* with water (e.g. $SO_3 + H_2O = H_2SO_4$). In organic chemistry an anhydride is formed by the action of *dehydrating agents* on *carboxylic acids*, giving the anhydride group –CO–O–CO–: e.g. two molecules of *ethanoic acid* (CH_3COOH) on *dehydration* yield one molecule of *ethanoic anhydride* (CH_3CO–O–$COCH_3$).

anhydrite A naturally occurring form of *calcium sulphate*, CaSO4.

anhydrous Without *water*; it is often applied to *salts* without *water of crystallization*.

anilide An *organic compound* analogous to an *amide* but derived from an *aromatic amine*, especially from *phenylamine* (aniline).

aniline See *phenylamine*.

aniline dyes *Dyes* prepared or chemically derived from *phenylamine* (aniline).

animal charcoal Bone black, bone char. Material containing 10% carbon and 90% inorganic matter, chiefly *calcium phosphate*, $Ca_3(PO_4)_2$, obtained by charring bones and other animal substances. It is used as a decolorizing agent.

animal starch See *glycogen*.

anion A negatively charged *ion*; an ion that is attracted towards the *anode* in *electrolysis*. Compare *cation*.

anisaldehyde Aubepine, 4-methoxybenzenecarbaldehyde. A colourless oily *liq-

uid, the *para-* form of $CH_3OC_6H_4CHO$. B.p. 247°C., it is used in cosmetics and perfumes.

anisole See *methoxybenzene*.

anisometric Not *isometric*. Denoting *crystals* that have *axes* of different lengths.

anisotropic Aelotropic. Possessing different physical properties in different directions; e.g. certain *crystals* have a different *refractive index* in different directions. Compare *isotropic*.

anisyl alcohol Anisalcohol, methoxyphenylmethanol $CH_3OC_6H_4CH_2OH$. A colourless *liquid*, b.p. 258.8°C., used in perfumes.

annealing Very slow regulated cooling, especially of *metals,* to relieve *strains* set up during heating or other treatment. It has the effect of making the metal easier to work. It can be used with both *ferrous* and *non-ferrous metals*. Each metal requires a different temperature to which it must be heated and a different rate of cooling to achieve annealing.

annihilation The process that occurs when a particle comes sufficiently close to its corresponding *antiparticle*, to cause both to be destroyed, their total mass converting at least initially to energy, in accordance with the *mass-energy equation*. At low energies the collision between an *electron* and a *positron* usually results in the creation of two gamma-ray *photons*, which are emitted in opposite directions if the initial particles are close to being at rest. At higher energies the total annhilation energy can be enough to produce sprays of new *elementary particles*, such as *mesons*. When a *proton* and an antiproton collide, the annihilation occurs between the *quarks* and antiquarks that the larger particles contain, and again new particles emerge from the collision.

annual variation A very small regular variation that the *magnetic declination* undergoes in the course of a year.

annular Ringed. An annular space is the space between an inner and outer ring.

annular eclipse An *eclipse* of the *Sun* in which a ring of its surface is visible surrounding the darkened *Moon*.

annulus A plane figure consisting of the area between two concentric circles of different radii. Its area is $\pi(R^2 - r^2)$, where R and r are the two radii.

anode Positive *electrode*. In *electrolysis* the anode attracts the negative *ions* (*anions*). In a *primary cell* the anode is the electrode that becomes positively charged. See also *thermionic valve*.

anode sludge See *electrolytic refining*.

anodizing Producing an *oxide* coating on a metallic surface by making it the *anode* in an electrolytic bath (see *electrolysis*). It can be used as a decorative finish on a metal object by making the oxide coating absorb a coloured dye.

anolyte The *electrolyte* near the *anode* during *electrolysis*.

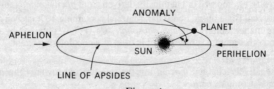

Figure 1.

anomaly (astr.) An angle used to describe the position of a *planet* in its *orbit*. The 'true anomaly' is the *angle* between the *perihelion,* the *Sun,* and the planet, in the direction of the planet's motion. See Fig. 1. The 'mean anomaly' is the angle between the perihelion, the Sun and a fictitious planet having the same *period* as the real planet, but assumed to be moving with a constant *velocity.*

antacid A pharmaceutical term for a substance, such as aluminium hydroxide, magnesium hydroxide, or calcium carbonate, that counteracts stomach acidity.

antenna See *aerial.*

anthracene $C_6H_4(CH)_2C_6H_4$. A white crystalline tricyclic *aromatic hydrocarbon* with a blue *fluorescence*; it is often yellowish due to impurities. M.p. 215.8°C. Obtained by the distillation of crude oil or from *coal-tar,* it is used in the manufacture of *dyes.*

anthracite A hard form of *coal,* containing more carbon (92–98%) and far less *hydrocarbons* than other forms. It is probably the oldest form of coal.

anthraquinone $C_6H_4(CO_2)_2C_6H_4$. The common isomer is anthracene-9,10-dione. A yellow *insoluble* powder, derived from *anthracene* and used as an *intermediate* in the manufacture of an important class of *vat dyes.*

anti- Prefix denoting opposite, against. E.g. *antichlor.*

antiaromatic See *pseudoaromatic.*

antibiotics Chemical substances produced by *microorganisms,* such as moulds and *bacteria,* which are capable of destroying bacteria and fungi or preventing their growth. They thus provide a means of controlling diseases caused by bacteria and fungi. Numerous antibiotics have been discovered, the first of which was *penicillin.* See *Aureomycin, Chloromycetin*; *erythromycin*; *nystatin*; *streptomycin*; *Terramycin*; *tetracyclines.*

antibody A *protein* produced by animal plasma *cells* (of the reticuloendothelial system) as a result of the presence of an *antigen.* Specific antigens stimulate the formation of specific antibodies. The function of the antibodies is to combine chemically with antigens and thereby to render them harmless to the *organism* that they are invading. As parasitic *organisms* and *viruses* produce, or are associated with, specific antigens, the consequent antibody formation provides a defence mechanism, called the immune response, against these invading parasites. Although antibodies themselves have a limited life, enduring immunity arises because specific plasma cells (B lymphocytes) that produce the antibodies persist in the bloodstream. Immunity to disease by inoculation is brought about by injecting antigens into the bloodstream with the object of stimulating the formation of antibodies. See also *vaccine.*

antichlor A substance used to remove chlorine from materials after *bleaching.* E.g. *sodium thiosulphate,* $Na_2S_2O_3$.

antidiuretic hormone See *vasopressin.*

antidote A remedy for a particular poison, which generally acts chemically upon the poison, by neutralizing it, making it *insoluble,* or otherwise rendering it harmless.

antifebrin See *phenylethanamide.*

antiferromagnetism A type of *magnetism* that occurs in certain inorganic compounds, such as MnO, MnS, and FeO. These materials have a low *susceptibility,* which increases with temperature up to the *Néel temperature,* above which the susceptibility falls and the material becomes *paramagnetic.* The phenomenon arises in substances in which interaction between neighbouring *atoms* leads to an *antiparallel* arrangement of magnetic *dipole moments*.

antifreeze A substance added to the cooling water in radiators of *internal-combustion engines* in order to lower the *freezing point* of the water. *Ethanediol* (ethylene glycol), $CH_2OH.CH_2OH$, is frequently used.

antigen A *protein* or *carbohydrate* that is foreign to an *organism* and capable of stimulating an immune response and the formation of *antibodies*.

antigorite See *serpentine*.

antihistamines A group of *drugs* that counteract the effect of *histamine* in the body and are therefore used in the treatment of allergic reactions.

antiknock agent See *knocking*.

antilogarithm Antilog. The number represented by a *logarithm*.

antimatter Hypothetical *matter* composed of *antiparticles*. Antihydrogen, for example, would consist of an anti*proton* and an orbital *positron*. While theoretically possible, the existence of antimatter in the *Universe* has never been detected. Contact between antimatter and matter would result in the *annihilation* of both.

antimony Sb. (Stibium.) Element. R.a.m. 121.75. At. No. 51. A brittle crystalline silvery-white *metal*, r.d. 6.69, m.p. 630°C., b.p. 1380°C., that expands on solidifying. It occurs as the *element, oxide,* and *sulphide* (*stibnite*, Sb_2S_3). It is extracted from its ores by roasting the ore and reducing with carbon. Antimony is used in *type metal* and other *alloys*.

antimony hydride See *stibine*.

antimony pentasulphide Antimony(V) sulphide. Sb_2S_5. A yellow *insoluble* powder, used as a *pigment* and in the *vulcanization* of *rubber*.

antimony potassium tartrate Tartar emetic. Potassium antimonyl tartrate. $2K(SbO)C_4H_4O_6.H_2O$. A white *soluble* poisonous powder, used as an emetic and as a *mordant*.

antimony sulphate $Sb_2(SO_4)_3$. A white crystalline *insoluble* solid, used in *explosives*.

antimony trisulphide Antimony(III) sulphide, stibnite. Sb_2S_3. A black or red *insoluble* crystalline solid, m.p. 550°C., used as a *pigment* and also in fireworks and *matches*.

antinodes Points of maximum displacement in a series of *standing waves*. Two similar and equal *wave motions* travelling with equal *velocities* in opposite directions along a straight line give rise to antinodes and *nodes* alternately along the line. The antinodes are separated from their adjacent nodes by a distance corresponding to a quarter of the *wavelength* of the wave motions.

antioxidants Agents added to certain materials, such as *rubber*, *plastics*, *paints*, and *oils*, to prevent the harmful effects to the materials of oxidation.

antiparallel vectors Vectors that have parallel lines of action but act in opposite directions.

antiparticle An *elementary particle* that has the same mass as another particle but an equal and opposite value of some other charge-like property. The antiparticle of the negatively charged *electron* is the positively charged *positron*. The antiproton has a negative charge equal in magnitude to the *proton's* positive charge. The antineutron has an opposite *magnetic moment*, relative to its spin, to the neutron. Particles with no charge-like properties, such as the *photon* and neutral *mesons*, are their own antiparticles. See also *antimatter*.

antipyretic Febrifuge. A substance used medically to lower the body *temperature*.

antiseptic A substance that destroys disease-causing *microorganisms* but does not harm body cells or tissues. Compare *disinfectant*.

antisquawk agents Substances added to lubricating oils to suppress noise in the operation of automatic clutches, etc.

apatite A natural *phosphate* and *fluoride* of calcium, $CaF_2.3Ca_3(PO_4)_2$, that is, used in the manufacture of *fertilizers*. The enamel of teeth consists largely of apatite, hence the importance of fluorides in the drinking water of children. Other forms of apatite exist.

aperture Opening; in optical instruments, the size of the opening admitting *light* to the instrument. In spherical *mirrors* or *lenses,* the diameter of the reflecting or refracting surface. The ratio of this diameter to the *focal length* is called the relative aperture, the reciprocal of which is the focal ratio. In photography the focal ratio of a lens is called its f-number.

aperture synthesis The use of two small *aerials* in a *radio telescope* to synthesize a large *aperture*. This principle can be used both with *parabolic reflectors* and *radio interferometers,* but it usually best employed in conjunction with an *unfilled aperture.*

aphelion The time, or point, in a *planet's orbit* at which it is furthest from the *Sun*. The opposite of *perihelion*. See Fig. 1 under *anomaly*.

aplanatic If any reflecting or refracting surface produces a point image at *B* of a point object at *A* irrespective of the angle at which the rays fall on the surface from *A,* then that surface is said to be aplanatic with respect to *A* and *B*.

apocynthion The time, or point, in the orbit of a lunar satellite at which it is farthest from the *Moon's* surface. The opposite of *pericynthion.*

apogee The *Moon* or any other Earth *satellite* is said to be in apogee when it is at its greatest distance from the Earth. The opposite of *perigee*.

apomorphine $C_{17}H_{17}NO_2$. A crystalline *alkaloid*, derived from *morphine*, used in the form of its hydrochloride as an emetic.

apothecaries' fluid measure
    1 minim = 0.0591 cc (about 1 drop)
    60 minims = 1 fluid drachm = 3.55 cc
    8 fl dr = 1 fluid ounce = 28.41 cc
    20 fl oz = 1 pint = 568 cc
These measures have now been replaced by *metric units*.

apothecaries' weights See *Troy weight.*

apothem A *perpendicular* from the centre of a regular *polygon* to one of its sides.

apparent depth The depth of a *liquid* viewed from above appears to be less than the true depth, owing to the *refraction* of *light*. The ratio of the true depth to the apparent depth is equal to the *refractive index* of the liquid.

apparent expansivity Relative expansion of a *liquid*. See *expansivity*.

Appleton layer See *ionosphere*.

aprotic Unable to either accept or donate *protons*. See *solvent*.

apsis (plural **apsides**) One of the extremities of the major *axis* of the *orbit* of a *planet* or *comet*. See *perihelion* and *aphelion*. The 'line of apsides' joins one apsis to the other. See Fig 1 under *anomaly*.

aq Symbol denoting *water*; e.g. H_2SO_4.aq. is *aqueous sulphuric acid*.

aqua fortis Concentrated *nitric acid*, HNO_3.

aquamarine A bluish form of *beryl*.

aqua regia A mixture of concentrated *nitric* and *hydrochloric acids* (1 to 4 by volume). A highly corrosive *liquid* that dissolves gold and attacks many substances unaffected by other reagents. It turns orange-yellow owing to the formation of nitrosyl chloride, $NOCl$, and free chlorine.

aqueous Watery. Denoting *solutions* in which *water* is the *solvent*.

arabinose Pectinose. $C_5H_{10}O_5$. A white *soluble* crystalline solid, m.p. 164.5°C., obtained from *gums* or synthetically from *glucose*, used as a *culture medium* in *bacteriology*.

arachidic acid Eicosanoic acid. $CH_3(CH_2)_{18}COOH$. A white crystalline *insoluble* solid, m.p. 76.3°C., obtained from peanut oil and used in lubricants, plastics, and *waxes*.

arc, electric A highly luminous discharge, accompanied by a *temperature* of over 3000°C, which is produced when an *electric current* flows through a gap between two *electrodes*. The current being carried by the *vapour* of the electrode; e.g. the common carbon arc is formed between two carbon rods, and constitutes a very bright source of *light*. In the same way metallic arcs are formed between two similar metallic surfaces.

Archimedes' principle The apparent loss in *weight* of a body totally or partially immersed in a *liquid* is equal to the weight of the liquid displaced. See *buoyancy*. Named after the Greek mathematician (287–212 B.C.).

arc lamp A technical application of the electric *arc* to produce a very bright *light*. The *carbon arc* lamp consists of an electric arc between two carbon *electrodes*, with suitable automatic mechanism for striking the arc and drawing the carbons closer together as they are vaporized away. The mercury arc lamp is important for laboratory use.

arc of circle See *circle*.

arc sin, tan, cos See *inverse trigonometrical functions*.

are Metric unit of *area* equal to 1 square dekametre, 100 square *metres*, or 119.60 square yards.

area, Imperial units
    1 square inch = 6.4516 square cm
    144 sq ins = 1 sq foot = 929 sq cm
    9 sq ft = 1 sq yard
    30¼ sq yds = 1 sq pole
    40 sq pls = 1 rood
    484 sq yds = 1 sq chain
    4 roods = 4840 sq yds = 1 acre
    640 acres = 1 sq mile
  See also Appendix, Table 1.

area, metric units
    1 sq centimetre = 0.155 sq inch
    10 000 sq cm = 1 centare = 1 sq metre
    100 sq m = 1 are

100 ares = 1 hectare = 2.47105 acres
100 hectares = 1 sq kilometre
See also Appendix, Table 1.

arene An *aromatic hydrocarbon*.

Argand diagram 1. The representation of a *complex number*, $z = x + iy$, as the point (x,y) in *Cartesian coordinates*, using the horizontal (x-axis) to represent the real part of the number and the vertical (y-axis) to represent the imaginary part of the number. In *polar coordinates*, the point is represented by (r, θ), where θ is the argument of the complex number and r is its modulus. **2.** A *vector* diagram showing the magnitude and phase angle of any vector with respect to another. Named after J. R. Argand (1768–1822).

argentic Containing silver in its +2 *oxidation* state, e.g. argentic oxide is silver(II) oxide, AgO.

argentiferous Silver-bearing.

argentite Silver glance. Natural silver sulphide, Ag_2S. An important *ore* of silver.

argentous Containing silver in its +1 *oxidation* state, e.g. argentous oxide is silver(I) oxide, Ag_2O.

arginine An essential *amino acid*. See Appendix, Table 5.

argol Tartar. A reddish-brown crystalline deposit consisting mainly of *potassium hydrogen tartrate*, which separates in wine-vats.

argon Ar. Element. R.a.m. 39.948. At. No. 18. A *noble gas*, m.p. –189°C., b.p. –185°C., that occurs in the air (0.9%). It is used for filling electric lamps and in fluorescent tubes at a pressure of about 3 mm of mercury (400 N m^{-2}). See also *potassium-argon dating*.

argument (math.) **1.** An independent *variable* that forms part of a *function*. **2.** See *Argand diagram*.

arithmetic mean Arithmetic average. The sum of a set of n numbers divided by n. E.g. the arithmetic mean of 4, 3, and 8 is 5.

arithmetical series A *series* of quantities in which each term differs from the preceding by a constant common difference. For an arithmetic series in which the first term is a, the common difference d, the number of terms n, the last term L, and the sum of n terms S,

$$S = n[2a + (n-1)d]/2$$
$$S = n(a + L)/2$$
$$L = a + (n-1)d.$$

armature The coil or coils, usually rotating, of a *dynamo* or *electric motor*. Also, more widely, an armature is any part of an electric apparatus or machine in which a *voltage* is induced by a *magnetic field*, e.g. in electromagnetic loud-speakers, *relays*, etc.

aromatic (chem.) A compound that contains a *benzene ring* or has similar properties to benzene. Although the molecules contain *double bonds*, they tend to undergo *substitution* rather than *addition reactions*. The classical formula devised by Kekulé for the benzene ring, with alternating double bonds and single bonds, is now regarded as a simplification. In fact, the bonds are all of equal length and the properties result from the π-orbital electrons being delocalized over the whole ring. See also *aromaticity*.

aromaticity The degree to which a cyclic organic compound or *ion* with *double bonds* in the ring exhibits the high stability and specific reactivity (i.e. tendency to undergo substitution rather than addition reactions) characteristic of *aromatic* compounds. It is exhibited by such compounds as *pyridine, quinoline,* and *thiophene.*

arsenate An arsenate(V) is a *salt* or *ester* of *arsenic*(V) *acid.* An arsenate(III) is a *salt* or *ester* of *arsenic*(III) *acid* (formerly called arsenious acid). An arsenate(V) was formerly called an arsenate; an arsenate(III) was formerly called an arsenite.

arsenic As. Element. R.a.m. 74.9216. At. No. 33. It exists in three *allotropic* forms: ordinary grey metallic arsenic, r.d. 5.727, black arsenic, r.d. 4.5, and yellow arsenic, r.d. 2.0. It occurs combined with sulphur as *realgar,* As_4S_4, *orpiment,* As_2S_3; with oxygen as *white arsenic,* As_2O_3; with some *metals* and as the *element.* Arsenic is used in *semiconductors* and in *alloys. Compounds* are very poisonous and are used in medicine and for destroying pests.

arsenic acid Arsenic(V) acid (formerly arsenic acid), H_3AsO_4, is a white *soluble* crystalline powder, m.p. 35.5°C., used to manufacture *arsenate(V)* salts. Arsenic(III) acid (formerly arsenious acid), H_3AsO_3, is a solution of *arsenic(III) oxide* in water.

arsenical pyrites See *mispickel.*

arsenic(III) oxide White arsenic, arsenic trioxide, arsenious oxide, arsenious anhydride. As_2O_3. A white *amorphous* powder used in the manufacture of *pigments* and formerly as an *insecticide.*

arsenic(V) oxide Arsenic oxide. As_2O_5. A white *amorphous deliquescent* solid that decomposes at 315°C. It loses oxygen on heating to give *arsenic(III) oxide* and in solution in water yields *arsenic(V) acid.*

arsenic trisulphide Arsenic(III) sulphide, *orpiment.* As_2S_3. A yellow *soluble* solid, m.p. 300°C., used as a *pigment.*

arsenious acid See *arsenic acid.*

arsenite See *arsenate.*

arsine Hydrogen arsenide. AsH_3. An intensely poisonous colourless *gas.* It is used to dope microelectronic components with arsenic.

artificial radioactivity See *induced radioactivity.*

aryl An *organic univalent* group derived from an *arene;* e.g. *phenyl,* C_6H_5-, derived from *benzene.*

asbestos A variety of fibrous *silicate* minerals, mainly calcium magnesium silicate. It is used as a heat-insulating material and for fire-proof fabrics. Inhalation of the fibres can be extremely dangerous, causing asbestosis.

ascorbic acid Vitamin C. See *vitamins.*

aseptic Free from microorganisms, especially *bacteria* and their spores.

ash The incombustible residue left after the complete *combustion* of any substance. It consists of the non-*volatile, inorganic* constituents of the substance.

asparagine A white crystalline *soluble amino acid* occurring in some leguminous plants. See Appendix, Table 5.

aspartic acid Asparaginic acid, aminosuccinic acid. A white crystalline *amino acid* found in sugar beet and some other plants. See Appendix, Table 5.

asphalt A black semi-solid sticky substance composed of *bitumen* with mineral matter. It consists mainly of complex *hydrocarbons,* and occurs naturally in asphalt lakes or in deposits mixed with *sandstone* and *limestone.* It is made artificially by adding mineral matter to bitumen and is used in road-making and building.

aspirator An apparatus for drawing a current of air or other *gas* through a *liquid*.

aspirin Acetylsalicylic acid. $CH_3COOC_6H_4COOH$. A white solid, m.p. 133°C., used in medicine as an *antipyretic* and *analgesic*. It functions by inhibiting the formation of *prostaglandins*. It also reduces the agglutination of *blood platelets* and is therefore given to patients with *cardiovascular* problems.

assaying Analysing for one constituent of a *mixture*, particularly the estimation of *metals* in *ores*. See also *bioassay*.

association (chem.) Under certain conditions, e.g. in *solution*, the *molecules* of some substances associate into groups of several molecules, thus causing the substance to have an abnormally high *relative molecular mass*. Molecules of different substances may also associate, usually being held together by *hydrogen bonds*. See *water*.

associative law An algebraic law stating that the value of an expression does not depend on the way the terms in the expression are grouped. For example, $a + b + c = a + (b + c) = (a + b) + c$ or $(ab) \times c = a \times (bc)$, i.e. addition and multiplication are associative, while subtraction and division are not. Compare *commutative law*; *distributive law*.

astatic coils An arrangement of wire-wound coils used in sensitive electrical instruments; the coils are arranged to give zero *resultant* external *magnetic field* when an *electric current* passes through them, and to have zero *electromotive force* induced in them by an external magnetic field.

astatic galvanometer A type of moving-magnet *galvonometer*, in which two equal small magnets are arranged parallel but in opposition at the centres of two oppositely wound coils, the system being suspended by a fine torsion fibre. Since the resulting *magnetic moment* is zero, the Earth's *magnetic field* exerts no controlling *torque* on the moving system. Instead, the restoring torque is supplied by the suspending fibre and is made very small by using a fine *quartz* fibre; the sensitivity of the galvanometer is thus very large.

astatic pair of magnets An arrangement of *magnets* used in *astatic galvanometers*.

astatine At. Element. At. No. 85. The last member of the *halogen* group and the only one without a stable *isotope*. The most stable isotope, astatine-210, has a *half-life* of only 8.3 hours.

asteroids Planetoids, minor *planets*. A belt of small bodies rotating round the *Sun* in orbits between those of *Mars* and *Jupiter*. The largest, Ceres, has a diameter of 685 km, but most are much smaller. It is thought that there are many thousands of these bodies, but only about 200 have a diameter in excess of 100 km.

astigmatism A defect of *lenses* (including the eye) caused by the curvature being different in two mutually perpendicular *planes*; thus *rays* in one plane may be in focus while those in the other are out of focus, producing distortion. Astigmatism of the eye is corrected by the use of *anastigmatic lenses*.

astringent A substance that by contracting body *tissues*, veins etc, reduces the discharge of mucus or *blood*.

astrocompass An instrument for determining direction relative to the *stars*. Unaffected by the errors to which *magnetic* or *gyro compasses* are subject, it is used to determine the errors of such instruments.

astrolabe An instrument used by early astronomers to measure the *altitude* of heavenly bodies. The simplest form consists of a graduated circular ring with a movable sighting arm. It has now been replaced by the *sextant*.

astrology The ancient art or pseudo-science of predicting the course of human destinies by indications derived from the positions and movements of the heavenly bodies.

astrometric binary A *binary star* in which one component is too faint to be observed with an optical *telescope*, its presence being deduced from the *perturbations* in the motions of the other. Compare *spectroscopic binary*; *visual binary*.

astrometry The branch of *astronomy* concerned with measurements of the positions of celestial bodies on the *celestial sphere*.

astronautics The scientific study of travel outside the *Earth's atmosphere*.

astronomical telescope See *telescope*.

astronomical unit The mean distance from the centre of the *Earth* to the centre of the *Sun*. 1.495×10^{11} *metres*, approximately 92.9×10^{6} miles.

astronomy The scientific study of the heavenly bodies, their motions, relative positions, and nature. Its main branches are *astrometry*, *celestial mechanics*, and *astrophysics*. See also *radio astronomy* and *cosmology*.

astrophysics The branch of *astronomy* concerned with the physical properties and evolution of celestial bodies, and the interaction between *matter* and *energy* within them (and in the *space* between them). See *cosmology*.

asymmetric Not possessing *symmetry*.

asymmetric carbon atom A carbon *atom* in a *molecule* of an *organic compound* with four different atoms or groups attached to its four *valences*. Such a grouping permits of two different arrangements in space, leading to the existence of optical isomers. See *optical activity*, *stereoisomerism*.

asymptote A line approaching a curve, but never reaching it within a finite distance.

asymptotic freedom The theory that the forces between *quarks* become weaker as the particles come closer together and vanish when the distance becomes zero. This theory is a consequence of some *gauge theories*.

atactic polymer A *polymer* in which the groups attached to the main chain are not arranged regularly. In isotactic polymers the same irregularity is repeated along the chain, whereas in syndiotactic polymers there are *asymmetric carbon atoms* in the chain and successive groups lie on alternate sides of the chain. Compare *tactic polymer*.

athermancy The property of being opaque to *radiant heat*; i.e. of absorbing heat radiations.

atherodyde Athodyd. See *ram jet*.

atmolysis The separation of a *mixture* of *gases* through the walls of a porous vessel by taking advantage of the different rates of *diffusion* of the constituents.

atmometer Evaporometer. An instrument for measuring the rate of *evaporation* of *water*.

atmosphere The gaseous envelope surrounding the *Earth* (or other heavenly body). The composition of the Earth's atmosphere varies very slightly in different localities and according to altitude. Volume composition of dry air at sea-level (average values) are: nitrogen, 78.08%; oxygen, 20.95%; argon, 0.93%; *carbon dioxide*, 0.03%; neon, 0.0018%; helium, 0.0005%; krypton, 0.0001%; xenon, 0.00001%. Air generally contains, in addition, *water vapour*, *hydrocarbons*, *hydrogen peroxide*, sulphur *compounds*, and dust particles in small and very variable amounts. See also *upper atmosphere*.

atmosphere A unit of *pressure*. The pressure that will support a column of mercury 760 mm high (29.92 inches) at 0°C., sea-level and latitude 45°. 1 normal atmosphere = 101 325 *pascals* = 14.72 lb/sq in (approx). Atmospheric pressure fluctuates about this value from day to day.

atmospherics Electrical discharges that take place in the atmosphere, causing crackling sounds in *radio* receivers.

atom The smallest portion of an *element* that can take part in a *chemical reaction*. See *atom, structure of*; *atomic theory*.

atom, structure of The *atom* consists of a positively charged central core, the *nucleus*, surrounded by one or more negatively charged planetary *electrons*. The openness of atomic structure is indicated by the following approximate dimensions:

Effective radius of atom	10^{-10} m
Effective radius of nucleus	10^{-14} m
Effective radius of electron	10^{-15} m

Almost all the *mass* of the atom resides in the nucleus, which is composed of two different types of stable particle of almost equal mass, the *proton*, which is positively charged, and the *neutron*, which is electrically neutral. The mass of the electron is 1/1836th of that of the proton, and although its charge is opposite in sign, it is numerically equal to that of the proton. The number of planetary electrons in the electrically neutral atom is therefore equal to the number of protons in the nucleus. The chemical behaviour of an atom is determined by its number of planetary electrons (characterized by the *atomic number*), chemical combination between atoms taking place by the transfer or sharing of outer electrons between combining atoms. See *valence*.

According to the *Bohr theory*, the planetary electrons of an atom were to be thought of as moving in well defined *orbits* about the nucleus, corresponding to specific *energy levels* – the emission or absorption of a *photon* of *electromagnetic radiation* occurring when an electron made a *quantum* jump from one permitted orbit, or energy level, to another (see *quantum numbers*). In the more modern *wave mechanics* the electrons are regarded as having a dual wave particle existence, which is expressed mathematically by a *wave function*. The precise position of the electron in the Bohr model of the atom is therefore replaced in the wave mechanical model by a *probability* that a particular planetary electron, visualized as a particle, may be found at a particular point in the path of a wave. Thus, in this model the atom is visualized as a central nucleus surrounded by a distribution of probabilities that individual electrons will exist at certain points at certain instants of time.

Atoms of an *element* that have the same number of protons, p, in their nuclei, but a different number of neutrons, n, are called *isotopes* of that element. When a particular isotope is being considered the following notation is used: to the chemical *symbol* of the element, the *mass number* $(n + p)$ of the isotope is added as a superscript. The atomic number of the element may also be added as a subscript; e.g. $^1_1$H, $^{12}_6$C, $^{197}_{79}$Au, are the most abundant isotopes of hydrogen, carbon, and gold.

atomic absorption spectroscopy AAS. A form of *spectrographic analysis* in which a sample of material to be analysed is vaporized and the nonexcited atoms in the sample absorb electromagnetic radiation at specific wavelengths. These wavelengths can be used to identify the atoms present.

atomic bomb See *nuclear weapons*.

atomic clock A very accurate form of clock in which the basis of the time scale is derived from the vibrations of *atoms* or *molecules*. See *caesium clock*; *ammonia clock*; *International Atomic Time*.

atomic constants See Appendix, Table 2.

atomic emission spectroscopy AES. A form of *spectrographic analysis* in which a sample of material to be analysed is vaporized; the atoms present emit electromagnetic radiation as specific wavelengths, which can be used to identify them.

atomic energy See *nuclear energy*.

atomic force microscope AFM. A *microscope* in which a tiny chip of diamond is held in a spring-loaded device that keeps it in contact with the surface of the sample. The diamond is raised and lowered so that the tracking force between the diamond and the surface is kept constant by a *computer*, which generates a contour map of the surface, in which individual molecules can be resolved. Unlike the *scanning tunnelling microscope*, which relies on electrical forces, the AFM relies on mechanical forces and can therefore be used on materials that are nonconducting.

atomic heat The numerical product of the *relative atomic mass* and the *specific heat capacity* of an *element*, i.e. what is now called the *molar heat capacity*. *Dulong and Petit's law* states that the atomic heat of all *solid* elements is approx 25 joules per *mole* per degree, i.e. it is approximately equal to $3R$, where R is the *gas constant*. The law is obeyed by many elements at ordinary *temperatures*, but at lower temperatures the atomic heat of all elements falls below this value, tending to zero as *absolute zero* of temperature is approached.

atomicity The number of atoms in a molecule. E.g. water has an atomicity of 3.

atomic mass The *mass* of an *isotope* of an *element* measured in *atomic mass units*.

atomic mass unit Dalton. AMU. A unit used for expressing the masses of individual atoms of *elements*, approximately equal to 1.66×10^{-27} kg. It was formerly defined so that the atoms of the most abundant isotope of oxygen, O-16, had a mass of 16 atomic mass units. In 1961 the 'unified atomic mass unit' was defined as 1/12 of the mass of an *atom* of C-12, and was adopted by the International Union of Pure and Applied Physics and the International Union of Pure and Applied Chemistry.

atomic number Proton number. Z. The number of *electrons* orbiting the *nucleus* of the neutral *atom* of an *element*, or the number of *protons* in the nucleus. See *atom, structure of* and Appendix, Table 3.

atomic orbital See *orbital*.

atomic physics The study of the physics of the *atom*, its structure, energy, and physical properties. See also *nuclear physics*.

atomic pile The original name for a *nuclear reactor*.

atomic theory An hypothesis as to the structure of *matter*, foreshadowed by Democritus, put forward as a formal explanation of chemical facts and laws by Dalton in the beginning of the nineteenth century. It assumes that matter is made up of small indivisible particles called *atoms*; the atoms of any one *element* are identical in all respects, but differ from those of other elements at least in *mass*. Chemical *compounds* are formed by the union of atoms of different elements in simple numerical proportions. Modern views on the structure of the atom (see *atom, structure of*) diverge considerably from Dalton's hypothesis, but it is still of value in affording a simple explanation of the laws of *chemical combination*.

atomic volume The *relative atomic mass* of an *element* divided by its *density*.

atomic weight See *relative atomic mass*.

atom smasher A popular name for an *accelerator*.

ATP See *adenosine triphosphate*.

atropine $C_{17}H_{23}NO_3$. A colourless crystalline *insoluble alkaloid*, m.p. 115°C. It is extremely poisonous, has a powerful effect upon the nervous system, and is used in medicine to dilate the pupil of the eye. It occurs in the deadly nightshade and henbane.

attenuation (phys.) The loss of *power* suffered by *radiation* as it passes through *matter*.

atto- Prefix denoting one million million millionth; 10^{-18}. Symbol a, e.g. am = 10^{-18} *metre*.

aubepine See *anisaldehyde*.

audibility, limits of The limits of *frequency* of *sound* waves that are audible as sound to the human ear. The lowest is about 30 *hertz*, corresponding to a very deep vibrating rumble, and the highest in the region of 20 000 *hertz*, corresponding to a shrill hiss.

audiofrequency A *frequency* between 30 and about 20 000 *hertz*, which in the case of *sound* waves would be audible.

audiometer An instrument for measuring the level of human hearing.

Auer metal A *pyrophoric alloy* of 65% *misch metal* (a mixture of cerium and other metals) and 35% iron. It is used as 'flint' in lighters.

Auger effect The emission of an *electron* (Auger electron) by an *atom*, without the emission of *X*- or *γ-radiation*, as a result of a change from an excited state (see *excitation*) to a lower energy state. Named after Pierre Auger (born 1899).

Aureomycin* Chlortetracycline. $C_{22}H_{23}N_2O_8Cl$. A broad-spectrum *antibiotic* used against many *organisms* that are resistant to *penicillin*; it is also used to stimulate growth of animals.

auric Containing gold in its higher (+3) *oxidation state*, such as auric chloride, i.e. *gold(III) chloride*, $AuCl_3$.

auric chloride See *gold(III) chloride*.

auriferous Gold-bearing.

aurora borealis Northern lights. A display of coloured light streamers and glows, mainly red and green, visible in the regions of the North and South Poles. It is caused by the interaction between charged particles emanating from the *Sun* and atomic oxygen and other molecules in the *upper atmosphere*; it is most prominent when large *sunspots* are observed. In southern latitudes the effect is called the aurora australis. See *solar wind*.

aurous Containing gold in its lower (+1) *oxidation state*, such as aurous chloride, i.e. gold(I) chloride, AuCl.

austenite A *solid solution* of carbon or of iron carbide in the *gamma* form of iron; it is normally stable only at high *temperatures*, but may be preserved at normal temperatures by certain alloying *elements* or by rapid cooling.

autocatalysis *Catalysis* in which the catalyst is produced during the course of the *reaction* that is being catalysed.

autoclave A thick-walled vessel with a tightly fitting lid, in which substances may be heated under pressure to above their *boiling points*. It is used in the manufacture of chemicals, for sterilizing medical instruments, etc., and in cooking.

autolysis The self-destruction of biological *cells*, tissues, etc. after death, as a result of the action of their own *enzymes*.

automation The application of mechanical, or more commonly *electronic* or computerized, techniques to minimize the use of manpower in any process.

autoradiograph An image obtained by placing a thin biological or other specimen, containing a *radioactive isotope*, in contact with a photographic plate, exposing for a suitable period and *developing*. The image shows the distribution of the radioactive *element* in the specimen.

autosome Any *chromosome* other than a sex chromosome.

auxin A type of *plant hormone* (known as a growth substance) that promotes the elongation and growth of plant *cells* and stimulates rooting; e.g. *indole-3-acetic acid*.

auxochrome A group of atoms, such as $-NH_2$ or $-OH$, that intensifies the colour of a *dye* by reacting with the *chromophore*.

avalanche (phys.) A *shower* of ionized particles caused by a single *ionization* in which the ion and electron produced are accelerated in an *electric field* resulting in further (secondary) ionizations.

average See *arithmetic mean, geometric mean*.

Avogadro constant Avogadro's number. L or N_A. The number of atoms or molecules in a mole of a substance: it has the value $6.022\ 1367 \times 10^{23}$.

Avogadro's Law Avogadro's hypothesis. Equal *volumes* of all *gases* contain equal numbers of *molecules* under the same conditions of *temperature* and *pressure*. It is true only for *perfect gases*. Named after Count Amadeo Avogadro (1776–1856).

avoirdupois weights A system of weights used in the English-speaking countries. See *weight, British units of*.

axis 1. An imaginary line about which a given body or system is considered to rotate. **2.** One of two or three reference lines in a system of *Cartesian coordinates*.

axis of mirror See *mirrors, spherical*.

axis of symmetry A line about which a given figure is symmetrical; e.g. the diameter of a *circle*.

axon A long nerve fibre that carries impulses away from the body of a *neurone*.

azeotrope Constant-boiling mixture. A mixture of two or more liquids that distils at a certain constant *temperature* and has a constant composition at a given *pressure*. Its boiling point may be a maximum or a minimum relative to the original components.

CELESTIAL SPHERE

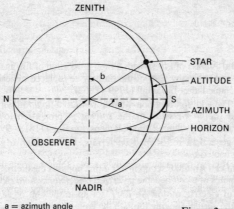

a = azimuth angle
b = zenith angle

Figure 2.

azide A *compound* containing the *univalent* azido group, $-N_3$, or ion, N_3^-, e.g. *sodium azide*, NaN_3.

azimuth (astr.) The *angular distance* from the north or south point of the horizon to the foot of the vertical circle through a heavenly body. The azimuth of a horizontal direction is its deviation from the north or south. See Fig. 2.

azimuthal quantum number See *quantum number*.

azines *Organic* derivatives of *hydrazine*, of the general formula

$$RR'C=N-N=CRR'$$

where R and R′ are *univalent organic radicals*. The suffix -azine is also used in systematic naming of six-membered *unsaturated heterocyclic compounds* containing nitrogen in the ring. Such compounds are sometimes described as azines.

azino The *quadrivalent* group $=N.N=$

azo compound A compound containing an azo group, $-N=N-$, attached to two carbon atoms $(-CN=NC-)$. *Aromatic* azo compounds are usually prepared by *azo coupling*.

azo coupling The formation of an *azo compound* by the reaction of an *aromatic diazo compound* with a suitable *nucleophilic reagent*, such as an *amine* or a *phenol*.

azo dyes *Azo compounds* used as dyes of many application classes (*acid, direct, disperse,* azoic, etc.).

azoic dyes Insoluble *azo dyes* that are formed within the fibre by the *azo coupling* of a *diazo compound* with a suitable *azo-coupling* component, often a *naphthol* derivative.

azoimide See *hydrogen azide*.

azote Former name for *nitrogen*.

azurite Natural *basic* copper(II) carbonate, $2CuCO_3.Cu(OH)_2$. Having an intense blue colour it is used as a gemstone.

B

Babbitt metal A class of *alloys* with a high proportion of tin, and small amounts of copper and antimony. Part of the tin may be replaced by lead. It is used for bearings. Named after I. Babbitt (1799–1862).

Babo's law The addition of a non-volatile *solid* to a *liquid* in which it is *soluble* lowers the *vapour pressure* of the *solvent* in proportion to the amount of substance dissolved. Named after Clemens von Babo (1818–99).

bacillus In general, a rod-shaped *bacterium*. In particular, a genus of spore-producing bacteria.

back EMF of cell An *EMF* set up in a *cell* that opposes the normal EMF. It occurs when the poles of a *cell* become polarized (see *polarization, electrolytic*).

back EMF of electric motor An *EMF* set up in the coil of an *electric motor*, opposing the current flowing through the coil, when the *armature* rotates.

background (phys.) The counting rate of a *counter tube* caused by sources other than the one being measured. It is due primarily to natural *radioactivity* in the soil, and *cosmic rays*. See also *microwave background*.

backing store *Computer memories* with a capacity to store enormous quantities of information, but with an *access time* much greater than the main store. The commonest types are *magnetic tape* decks, fixed magnetic disk stores, and exchangeable magnetic disk stores.

bactericide A substance that kills *bacteria*.

bacteriology The study of *bacteria*.

bacteriophage Phage. A *virus* that requires a *bacterium* in which to replicate.

bacterium A cellular *microorganism* that is single celled, *haploid*, has a characteristic cell wall, and has no nuclear *membrane*. Most bacteria reproduce by *binary fission*. Bacteria are the causes of many diseases, most of which can now be treated by the use of *antibiotics*. However, bacteria also perform an indispensable function in nature by bringing about the decay of plant and animal debris in the *soil*; they also perform an essential function in the guts of many animals, providing a source of nutrients for the host and a buffer against infection. Bacteria are broadly classified by their shape into four main groups: the spherical or *coccus* form, the spiral-shaped organism called a *spirillum*, the rod-shaped or *bacillus* type, and the filamentous or *mycelial* type.

Bakelite* Trade name for various synthetic *resins* of which *phenol-formaldehyde resins* are amongst the most widely known. Named after Leo Hendrick Baekeland (1863–1944).

baking powder A *mixture* that produces *carbon dioxide* gas, CO_2, on wetting or heating, thus causing the formation of bubbles in dough and making it 'rise'. It usually contains *sodium hydrogencarbonate*, $NaHCO_3$, and *tartaric acid* or *cream of tartar*.

baking soda See *sodium hydrogencarbonate*.

balance An apparatus for weighing. A beam balance consists of a lever with two equal arms, with a pan suspended from the end of each arm. Masses placed in the pans are subject to pulls of *gravity;* when these *forces* are equal, as indicated by the beam being horizontal, the masses themselves must be equal. Sensitive balances have beam and pans poised on knife-edges of *agate* resting on agate surfaces. An accurate balance will weigh to the nearest 10^{-5} g. In the substitution balance, weights are removed from a single arm to bring the single pan back into balance with a fixed counter-weight. These can be accurate to 10^{-9} g. In electronic balances, mass is measured by the displacement of the pan support, which generates a current proportional to the displacement. This current is used to generate a magnetic force that makes the support return to its equilibrium position. The magnitude of the force required, translated into units of mass, is given by a digital display.

balanced reaction See *chemical equilibrium.*

balata A natural *rubber*-like material very similar to *gutta-percha.*

ballistic galvanometer An instrument for measuring electric *charge* by detecting a momentary current. Any *galvanometer* may be used ballistically provided that its period of oscillation is long compared with the time during which the current flows. It is usually a moving-coil instrument with a heavy coil and little damping. The initial deflection is proportional to the charge passed.

ballistic missile A ground-to-ground missile with a parabolic flight path. A missile that is propelled and guided only during the initial phase of its flight.

ballistic pendulum A device for measuring the *velocity* of a *projectile,* such as a bullet. It consists of a large *mass* freely suspended from a horizontal beam and a means of measuring the displacement of the mass when it is struck by the projectile. The displacement of the mass is a function of the projectile's velocity.

ballistics The study of the flight path of *projectiles.*

Balmer series A series of sharp distinct lines in the visible *spectrum* of hydrogen, the *wavelengths, λ,* of which may be represented by the formula:
$$1/\lambda = R(1/2^2 - 1/n^2);$$
$n = 3, 4, 5,$ etc., R is a constant known as the *Rydberg constant,* which has the value $1.096\ 77 \times 10^7$ m^{-1}. Named after J. J. Balmer (1825–98).

band spectrum An *emission* or *absorption spectrum* consisting of a number of fluted bands each having one sharp edge. Each band is composed of a large number of closely spaced lines. Band spectra arise from *molecules.*

band theory See *energy bands.*

bandwidth The range of *frequencies* within which the performance of a *circuit,* receiver, or *amplifier* does not differ from its maximum value by a specified amount. The bandwidth of a *radio* emission is the width of the *frequency band* that carries a specified proportion (usually 99%) of the total *power* radiated.

bar A unit of *pressure* in the *c.g.s. system;* a pressure of 10^6 *dynes* per sq cm. It is equivalent to a pressure of 0.986 923 *atmosphere* (approx. 750 mm Hg). 1 bar = 10^5 *pascals* and the commonly used millibar is equal to 100 Pa.

barbitone Barbital, 5,5-diethylbarbituric acid. $CO(HNCO)_2(C_2H_5)_2$. A crystalline substance derived from *barbituric acid,* m.p. 191°C., used in the form of its sodium *salt* as a *hypnotic.*

barbiturates A class of *organic compounds* derived from *barbituric acid.* Many of these compounds have a powerful soporific effect. They were formerly used extensively in sleeping tablets, but as an overdose could be fatal they have been largely replaced by safer substances.

barbituric acid Malonylurea. $CO(NH.CO)_2CH_2$. A white crystalline powder, m.p. 248°C., used in the synthesis of drugs and *plastics*.

Barff process Prevention of rusting of iron by the action of *steam* upon the surface of the red-hot metal, resulting in a surface coating of black oxide of iron, Fe_3O_4.

barium Ba. Element. R.a.m. 137.33. At. No. 56. A silvery-white soft *metal*, which tarnishes readily in air. R.d. 3.5, m.p. 725°C., b.p. 1640°C. It occurs as *barytes*, $BaSO_4$, and as *witherite*, $BaCO_3$. *Compounds* resemble those of calcium but are poisonous. Compounds are used in the manufacture of *paints, glass*, and fireworks.

barium carbonate Witherite. $BaCO_3$. A heavy white poisonous *insoluble* powder, used in rat poisons, certain optical glasses, and various industries.

barium hydroxide Caustic baryta. $Ba(OH)_2.8H_2O$. A white poisonous crystalline solid, m.p. 76°C., used for recovering sugar from waste molasses, for refining *vegetable oils*, and in *glass* manufacture.

barium oxide Baryta. BaO. A white crystalline powder, m.p. 1923°C., used as a dehydrating agent and in the manufacture of *glass*.

barium peroxide BaO_2. A white *insoluble* powder, m.p. 450°C., used as a bleaching agent.

barium sulphate Blanc fixe. $BaSO_4$. A white crystalline *insoluble* powder, m.p. 1580°C., used as a *pigment* and, because it is *opaque* to *X-rays*, as the basis of 'barium meal' in X-ray diagnosis.

barium titanate $BaTiO_3$. A crystalline substance with good *ferroelectric* and *piezoelectric* properties, used in *transducers*.

Barkhausen effect The effect observed when a *ferromagnetic substance* is magnetized by a slowly increasing *magnetic field*; the magnetization does not take place continuously, but in a series of small steps. The effect is due to orientation of *magnetic domains* present in the substance. Named after H. Barkhausen (1881–1956).

barn A unit of area for measuring the *cross-section* of *nuclei*. 1 barn equals 10^{-28} sq metre.

barograph An instrument used in *meteorology* for recording on paper the variations in atmospheric pressure over a period of time.

barometer An instrument for measuring atmospheric pressure. A mercury barometer consists of a long tube (about 800 mm) closed at the upper end, filled with mercury, and inverted in a vessel containing mercury; the vertical height of the mercury column that the atmospheric pressure is able at any time to support being taken as the atmospheric pressure at that time. At standard atmospheric pressure the height of the column is 760 mm, irrespective of the diameter of the tube. See also *aneroid* barometer, *Fortin barometer*.

barrel A unit of volume used in the oil industry and the chemical industry. 1 barrel equals 35 gallons (Imperial) or 45 gallons (U.S.).

barrier-layer rectifier A *rectifier* that consists of a *semiconductor* between rectifying and non-rectifying metal *electrodes*.

barycentre *Centre of mass*: particularly the centre of mass of the *Earth/Moon* system.

barye A unit of pressure in the *c.g.s. system*, equal to one *dyne* per sq cm or 0.1 *pascal*.

baryon A collective name for *hadrons* that consist of three *quarks* bound together (see *elementary particles*). They are either *nucleons* or other particles that have a nucleon in their decay products. The number of baryons minus the number of corresponding anti-baryons taking part in a process is called the baryon number–a quantity that appears to be conserved in all processes. All baryons have *spin* $\frac{1}{2}$.

baryta See *barium hydroxide*; *barium oxide*.

barytes Heavy spar. Natural *barium sulphate*, $BaSO_4$.

basalt A rock of volcanic origin, chemically resembling *feldspar*.

base (chem.) A substance that liberates *hydroxide ions* in solution, reacts with an *acid* to form a *salt* and water only, has a tendency to accept *protons*, and turns *litmus* blue. Bases include *oxides* and *hydroxides* of metals and compounds that release hydroxide ions in aqueous solution, such as *ammonia*. See also *alkali*; *organic base*; *Lewis acids and bases*; *Lowry–Brønsted theory*.

base (math.) **1.** The horizontal line upon which a geometric figure stands. **2.** The number that is a starting point for a numerical or logarithmic system. E.g. the *binary notation* is a numerical system to the base 2; common *logarithms* are to the base 10.

base (phys.) The part of a *transistor* that separates the *emitter* from the *collector*.

base exchange Cation exchange. See *ion exchange*.

base metals. Metals that corrode, tarnish, or oxidize on exposure to air, moisture, or heat. Compare *noble metals*.

base pairing The association of pairs of nitrogenous *bases* in the two *polynucleotide* strands of *deoxyribonucleic acid* (DNA), or of a strand of *ribonucleic acid* (RNA) being synthesized on a strand of DNA (see *transcription*). The position of *hydrogen bonds* between the bases stabilizes *guanine* (G) only with *cytosine* (C) and *adenine* (A) only with *thymine* (T) (or with *uracil* (U) in RNA). Thus the sequence CGAT in one strand of DNA will be complementary to the sequence GCTA in the other strand, or to GCUA in RNA.

base unit A *unit* that is not defined in terms of other units but in terms of some arbitrarily fixed physical parameter. For instance, the *metre* is defined as the distance travelled by light in a vacuum in a specified time. Speed, however, is a *derived* unit, based on the metre and the *second*. In *SI units* there are seven base units.

basic (chem.) Having the properties of a *base*; opposite to *acidic*; reacting chemically with *acids* to form *salts*. See also *alkali*.

basicity constant Base dissociation constant. The *dissociation constant* (K_b) when a *base* MOH dissociates into the *ions* M^+ and OH^-, i.e.

$$K_b = [M^+][OH^-]/[MOH]$$

Compare *acidity constant*.

basic-oxygen process B.O.P. A method of making *steel* that has largely replaced the *Bessemer process*. It originated in the Linnz-Donnewitz (L-D) process. Scrap iron (up to 30% of the charge) and molten *pig iron* are inserted into a tilting furnace and converted to steel by blowing high-pressure oxygen through a water-cooled lance onto the surface of the metal.

basic salt A *salt* formed by the partial neutralization of a *base*; it consists of the normal salt combined with a definite molecular proportion of the base. E.g. basic *lead carbonate* or lead(II) carbonate hydroxide, $2PbCO_3.Pb(OH)_2$ (formerly known as white lead).

basic slag An impure *mixture* of tetracalcium phosphate, $Ca_4P_2O_9$, *calcium silicate*, $CaSiO_3$, *lime*, CaO, and *iron(III) oxide*, Fe_2O_3. A by-product of iron smelting in a *blast furnace*. Its high phosphorus content makes it a valuable *fertilizer*.

bastnasite A rare yellow-brown *mineral* consisting of a mixed *carbonate* and *fluoride* of lanthanum and several other lanthanides, i.e. $LaFCO_3$. It is used as a source of lanthanides, including praseodymium.

bath salts The main constituent is generally sodium sesquicarbonate, Na_2CO_3. $NaHCO_3.2H_2O$, or some other *soluble* sodium *salt* to soften the *water*. See *hard water*.

bathymetry Measurement of depth, especially of the sea.

battery A number of *primary* or *secondary cells* arranged in series or parallel. In series, they give a multiple of the *EMF* of the cell; in parallel, they give the same EMF as the cell, but have a greater capacity, i.e. a given current can be supplied for a longer period; it is usual to give the capacity of a battery in ampere hours. The common 'dry batteries' usually consist of *Leclanché cells*.

Baumé scale A scale of *relative density* (specific gravity) of *liquids*. Named after A. Baumé (1728–1804).

$$\text{Degrees Baumé} = 144.3 \text{ (r.d. } -1)/\text{r.d.}$$

bauxite Natural *hydrated aluminium oxide*, $Al_2O_3.xH_2O$. The most important *ore* of aluminium.

bauxite cement Ciment fondu. A rapid-hardening *cement* consisting mainly of calcium aluminate; made from *bauxite* and *lime* in an electric furnace.

BCS theory See *superconductivity*.

beam (phys.) *Radiation* travelling in a particular direction.

beam hole A hole made in the shield, and usually through the *reflector*, of a *nuclear reactor* to permit the escape of a beam of radiation, particularly *neutrons*, for experimental purposes.

beam riding A method of *rocket* guidance in which the missile steers itself along the axis of a *beam* of *radiation*, usually a conically scanned *radar* beam.

beam transmission *Radio* transmission in which the *electromagnetic waves* are sent in a particular direction in a *beam* instead of being radiated in all directions.

bearing (math.) The direction of a point B from a fixed point A in terms of the angle the line AB makes with the line running due North and South through A (e.g. 20° East of North) or in terms of the angle the line AB makes with the line running due North through A, considered in a clockwise direction.

beat frequency See *beats*.

beats (phys.) A periodic increase and decrease in loudness heard when two notes of nearly the same *frequency* are sounded simultaneously. It is caused by *interference* of *sound* waves. If the two notes have frequencies f_1 and f_2, there will also be a beating sound heard, the amplitude of which will vary between the sum of the amplitudes of the two notes and the difference between them. The frequency of this note, called the 'beat frequency', will be $(f_2 - f_1)$.

Beaufort scale A numerical scale for the estimation of wind force, based on its effect on common objects. Named after Admiral Sir F. Beaufort (1774–1857). The scale for various wind forces is given in the table.

Beaufort number	Description of wind	Wind speed metres per sec.
0	Calm	< 0.3
1	Light air	0.3–1.5
2	Light breeze	1.6–3.3
3	Gentle breeze	3.4–5.4
4	Moderate breeze	5.5–7.9
5	Fresh breeze	8.0–10.7
6	Strong breeze	10.8–13.8
7	Near gale	13.9–17.1
8	Gale	17.2–20.7
9	Strong gale	20.8–24.4
10	Storm	24.5–28.4
11	Violent storm	28.5–32.6
12	Hurricane	⩾ 32.7

Beckmann thermometer A sensitive *thermometer* for measuring small differences or changes in *temperature*. The quantity of mercury in the bulb can be varied by causing it to overflow into a reservoir at the top, thus enabling the thermometer to be used over various ranges of temperature. The scale covers 6–7 degrees and is graduated to 0.01 degree. Named after E. O. Beckmann (1853–1923).

becquerel The derived *SI unit* of *activity* (radioactive). The activity of a *radionuclide* that decays at an average rate of one spontaneous nuclear transition per second. 1 curie = 3.7×10^{10} Bq. Symbol Bq. Named after Antoine Henri Becquerel (1852–1908).

beeswax A whitish *wax* consisting of a *mixture* of *compounds*, secreted by bees for the purpose of building their honeycombs. It is used in polishes and cosmetics.

beet sugar See *sucrose*.

behenic acid See *docosanoic acid*.

bel Ten *decibels*.

bell, electric A simple device making use of the magnetic effect of an *electric current*. Closing the switch (see Fig. 3) causes a current to flow through a small *electromagnet*. This then attracts a piece of soft iron attached to a hammer, causing the latter to strike the gong of the bell. The movement of the iron breaks the circuit; the current ceases to flow through the electromagnet, and the iron and attached hammer spring back into their original position, thus closing the circuit again; this process continues as long as the switch is closed.

bell metal An *alloy* of copper (60%–85%) and tin, sometimes including zinc or lead. It is the type of *bronze* used to cast bells.

bending moment The bending moment about any point in a loaded beam is the *algebraic sum* of the *moments* of all the vertical *forces* to one side of that point.

beneficiation The separation of *ores* into valuable components (concentrates) and wastes (gangue). It can be achieved in various ways; e.g. by crushing, magnetic separation, *flotation*, etc.

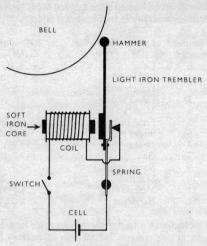

Figure 3.

benthos *Organisms* that exist on the bottom of the sea or other aqueous environment. Compare *pelagic*.

bentonite A clay-like material similar to *fuller's earth*.

benzaldehyde See *benzenecarbaldehyde*.

Benzedrine* See *amphetamine*.

benzene Benzol. C_6H_6. A colourless *liquid aromatic hydrocarbon* found in *coal-tar* but now made from *petroleum*, b.p. $80.1°C$. It is used as a *solvent* and in the manufacture of numerous *organic compounds*. See also *benzene ring*; *aromaticity*.

benzenecarbaldehyde Benzaldehyde. C_6H_5CHO. A colourless oily *liquid*, b.p. $178.1°C$. with a smell of almonds, in the kernels of which it occurs. It is used as a *solvent*, in the manufacture of *dyes*, and in perfumes and flavours.

benzenecarbonitrile See *benzonitrile*.

benzenecarbonyl Benzoyl. The univalent group C_6H_5CO-. E.g. benzenecarbonyl chloride (benzoylchloride), C_6H_5COCl.

benzenecarboxylate Benzoate. A *salt* or *ester* of *benzenecarboxylic acid* (benzoic acid).

benzenecarboxylic acid Benzoic acid, C_6H_5COOH. A white crystalline powder, m.p. $122°C$. It is used as a food preservative because it inhibits the growth of yeasts and moulds. It is also used for this purpose in the form of its sodium salt, which is highly water soluble.

benzene-1,3-diol See *resorcinol*.

benzene-1,4-diol Hydroquinone. Quinol. $C_6H_4(OH)_2$. A white crystalline substance, m.p. $170°C$. It can be reversibly oxidized to cyclohexadiene-1,4-dione (quinone) and is used as a *reducing agent*, as an antioxidant, and in photographic developing.

benzene hexachloride BHC. Hexachlorocyclohexane. $C_6H_6Cl_6$. A crystalline substance, m.p. 59°C., used as a pesticide. It has undesirable environmental effects.

benzene ring In the *benzene* molecule the six carbon atoms are joined in a hexagon known as the benzene ring, generally represented by the Kekulé formula in which hydrogen and carbon have their usual *valences* of one and four respectively, and the carbon atoms are linked by alternating single and double *chemical bonds*. Derivatives are formed by substitution of the hydrogen atoms, positions being indicated by numbering the ring as shown.

usually abbreviated to:

Although this has been shown to be incorrect as a representation of the actual state of a benzene molecule, this 'classical' formula with alternating double bonds but with a geometrically correct arrangement of atoms can be used as one of a number of reference formulae with different classical bond arrangements, the most important of which is also a Kekulé formula with the double bonds in the other three alternate positions. These reference formulae of nonexistent forms of benzene, known as *resonance* (mesomeric) structures, can be used in the description of the actual, highly stable, state of the benzene molecule, treated as a resonance hybrid of all the contributing structures (see also *orbital*). This quantum-mechanical resonance is general for all molecules; thus, H–Cl and H^+Cl^- are resonance forms of the hydrogen chloride molecule. It is distinct from an equilibrium between actually existing interconvertible forms, as in the case of *tautomerism*.

In *structural formulae* the benzene molecule is usually represented as shown. See also *aromatic*; *aromaticity*.

benzenesulphonic acid $C_6H_5SO_3H$. A crystalline *soluble* solid, m.p. 52.5°C., used in organic synthesis and as a *catalyst*.

benzene-1,2,3-triol See *pyrogallol*.

benzidine See *biphenyl-4,4'-diamine*.

benzine Petroleum benzin, petroleum ether, solvent naphtha. A mixture of *hydro-carbons* (mainly *alkanes*) obtained from petroleum; it boils between 35°C. and 80°C. and is used as a solvent. Because of possible confusion with *benzene*, the word 'benzine' should be avoided in scientific writing.

benzoate See *benzenecarboxylate*.

benzoic acid See *benzenecarboxylic acid*.

benzoin 1. See *2-hydroxy-1,2-diphenylethanone*. 2. Gum Benjamin. A natural brown *aromatic resin* obtained from certain trees (Styrax benzoin), used in incense and in the manufacture of cosmetics and perfumes.

benzol See *benzene*.

benzonitrile Benzenecarbonitrile. Phenyl cyanide. C_6H_5CN. A colourless poisonous *liquid*, b.p. 190.7°C., used in organic synthesis.

benzophenone See *diphenylmethanone*.

benzopyrene $C_{20}H_{12}$. A yellow crystalline *polycyclic hydrocarbon*, m.p. 179°C., found in small quantities in *coal-tar*. It is a *carcinogen* and is one of the most harmful constituents of tobacco smoke.

benzoyl See *benzenecarbonyl*.

benzoyl peroxide See *di(benzenecarbonyl) peroxide*.

benzyl The *univalent group* $C_6H_5.CH_2-$.

benzyl alcohol See *phenylmethanol*.

benzyl cellulose A *benzyl ether* of *cellulose*, possessing good electrical insulating properties and forming the basis of a *plastic* material.

benzylidene The *bivalent group* $C_6H_5CH=$.

benzylidene chloride Benzal chloride. $C_6H_5CHCl_2$. A colourless oily *liquid*, b.p. 205.2°C., used in the manufacture of *dyes*.

benzylidyne The *trivalent group* $C_6H_5C\equiv$.

berberine $C_{20}H_{19}NO_5$. A *soluble* crystalline *alkaloid*, m.p. 145°C., used in the form of its *sulphate* or *hydrochloride* in medicine.

Bergius process A process for the manufacture of *oil* from *coal*. Coal, made into a paste with heavy oil, is heated with hydrogen under a pressure of 250 *atmospheres* to a *temperature* of 450°–470°C., in the presence of a *catalyst* (originally iron(III) oxide, but later other substances were used). The carbon of the coal reacts with the hydrogen to give a mixture of various *hydrocarbons*. Named after F. Bergius (1884–1949).

berkelium Bk. *Transuranic element*. At. No. 97. A member of the *actinide* series. Most stable *isotope*, berkelium-247, has a *half-life* of about 1400 years.

Bernoulli's theorem At any point in a tube through which a *liquid* is flowing, the sum of pressure energy, *potential energy*, and *kinetic energy* is constant. Named after Daniel Bernoulli (1700–1782).

Berthollide compounds Non-stoichiometric compounds. Chemical *compounds* the composition of which does not conform to a simple ratio of *atoms* in the *molecule*.

beryl Natural beryllium silicate, $3BeO.Al_2O_3.6SiO_2$. It is the main *ore* of beryllium. The green variety is highly prized as emerald.

beryllate A compound of beryllium and oxygen formed in solution when the *metal*, *oxide*, or *hydroxide* dissolve in a strong *alkali*. $BeO_2{}^{2-}$ is the beryllate *ion*, which may occur with the hydroxy ion, $Be(OH)_4{}^{2-}$.

beryllia See *beryllium oxide*.

beryllium Glucinum. Be. Element. R.a.m. 9.01218, At. No. 4. A hard white *metal*, r.d. 1.85, m.p. 1285°C, b.p. 2970°C. It occurs as *beryl*, from which it is obtained by *electrolysis*. It is used for light, corrosion-resisting *alloys* some of which are used as *moderators* in *nuclear reactors*. See also *beryllium oxide*.

beryllium oxide Beryllia. BeO. An *insoluble* crystalline compound, m.p. 2550°C., which occurs naturally in *beryl*. It is an *amphoteric oxide* that forms *beryllates* with *alkalis* and is used in *refractories*, *nuclear reactors*, and *semiconductors*.

Bessemer process A process for making *steel* from *cast iron*. Molten iron from the *blast furnace* is run into the Bessemer converter, a large egg-shaped vessel with holes below. Through these, air is blown into the molten metal, and the carbon is oxidized. The requisite amount of *spiegel* is then added to introduce the correct amount of carbon for the type of steel required. In some converters, instead of air a mixture of oxygen and steam is blown into the molten metal to avoid the absorption of nitrogen by the steel. This is known as the VLN (very low nitrogen) process. Named after H. Bessemer (1813–98). See also *basic-oxygen process*.

beta decay A *radioactive* disintegration of an unstable *nucleus* in which a *neutron* changes to a *proton* with the emission of an *electron* and an antineutrino or in which a proton changes to a neutron with the emission of a *positron* and a *neutrino*. Thus a beta decay involves unit change of *atomic number* but no change of mass number. All neutrons outside a nucleus are unstable to beta decay, which is a form of *weak interaction*.

beta emitter An unstable *nucleus* that emits *electrons* as a result of a *beta decay*.

beta-iron An allotropic (see *allotropy*) form of pure iron, stable between 768°C. and 910°C.; it is similar to *alpha-iron* except that it is non-*magnetic*.

beta particle An *electron* or *positron* emitted by a *radioactive nucleus*. See *beta decay*.

beta radiation A stream of *beta particles*; they usually possess greater penetrating power than *alpha particles* and are emitted with speeds in some cases exceeding 98% of the *speed* of light.

betatron A cyclic *accelerator* for accelerating a continuous beam of *electrons* to high speeds by means of the *electric field* produced by a changing magnetic flux. The electrons move in stable circular orbits in an evacuated *torus*-shaped chamber. By allowing the fast electrons to strike a metal target a continuous source of *gamma rays* with energies up to 340 MeV can be produced.

BeV See *GeV*.

BHC See *benzene hexachloride*.

BHT See *Ionol*.

bi- 1. Prefix denoting two, e.g. **binomial. 2.** Prefix formerly used in chemical nomenclature to indicate an *acid salt* of a *dibasic acid*. See *bicarbonate*.

bicarbonate *An acid salt* of *carbonic acid*, H_2CO_3; carbonic acid in which half the *acidic hydrogen* has been replaced by a *metal*. E.g. sodium bicarbonate, $NaHCO_3$. However the use of 'bi-' as a prefix in such *compounds* has now been abandoned and the correct name for this substance is sodium *hydrogencarbonate*.

biconcave Denoting a *lens* that is *concave* on both sides. See Fig. 24 under *lens*.

biconvex Denoting a *lens* that is *convex* on both sides. See Fig. 24 under *lens*.

big-bang theory Superdense theory. The theory in *cosmology* that the *Universe* has evolved from one 'superdense' agglomeration of *matter* that suffered a cataclysmic explosion. The observed *expansion of the Universe* is regarded as a result of this explosion, the *galaxies* flying apart like fragments from an exploding bomb. This hypothesis, which presupposes a finite beginning and probably a finite end to the history of the Universe, is in opposition to the *steady-state theory*. At present the evidence of the *microwave background* favours the big-bang theory and the observed *abundance* of helium in the Universe.

bile An alkaline secretion of the liver of vertebrates important in the digestion of *fats*. It consists of *cholesterol*, bile salts (salts of cholic acid), and bile pigment (degradation products of *haemoglobin*).

billion One thousand million, 10^9. Formerly, in the UK one billion meant 10^{12}, but this usage has now almost entirely given way to the US usage, 10^9.

bimetallic strip A strip composed of two different *metals* welded together in such a way that a rise of *temperature* will cause it to buckle as a result of unequal expansion. Used in *thermostats*.

bimorph cell Two plates of *piezoelectric* material joined together so that they bend in proportion to an applied *potential difference*.

binary acid Hydracid. An *acid* in which the *acidic hydrogen* atom is bound to an atom other than oxygen, e.g. *hydrochloric acid*. Compare *oxo acid*.

binary compound A chemical *compound* of two *elements* only. The traditional names for these compounds are denoted by the suffix -ide; e.g. *calcium dicarbide*, CaC_2.

binary fission A form of asexual reproduction in which the *nucleus* of a *cell* divides into two parts, followed by a similar division of the *cytoplasm*. Most bacteria reproduce in this way.

Decimal system	Binary system
1	0001
2	0010
3	0011
4	0100
5	0101
6	0110
7	0111
8	1000
9	1001
10	1010

binary notation Binary number system. A system of numbers that has only two different *digits*, usually 0 and 1. There are several ways of representing numbers in the binary notation; one common method is given in the table. Because it has only two digits, which can be represented by an electric current switched on or switched off, this notation is used in *computers*.

binary stars Two *stars* gravitationally attracted to each other, so that they revolve around their common *centre of mass*, thus forming a *double star*. See also *astrometric binary*; *spectroscopic binary*; *visual binary*.

binding energy (phys.) The *energy* that must be supplied to a *nucleus* in order to cause it to decompose into its constituent *neutrons* and *protons*. The binding energy of a neutron or a proton is the energy required to remove a neutron or a proton from a nucleus. Because the nucleus provides a stable ordered environment, nuclear nucleons are in a lower energy state than free nucleons. This is the energy that must be supplied to remove them from the nucleus.

binocular Any optical instrument designed for the simultaneous use of both eyes; e.g. *binocular field-glasses*; binocular microscope (see *microscope*).

binocular field glasses A pair of astronomical refracting *telescopes* arranged for *binocular* vision. Good quality field glasses contain a pair of *prisms* in each telescope to increase their effective length and give upright *images*. Field glasses are specified by two numbers, the first of which gives the angular *magnification*, while the second gives the diameter of the *objective* in millimetres. Cheaper field glasses, such as opera glasses, consist of a pair of *Galilean telescopes*.

binomial A mathematical expression consisting of the sum or difference of two terms; e.g. $a^2 - 3b$.

binomial nomenclature (bio.) The method of naming living organisms introduced by Linnaeus (Carl Linné; 1707 – 78) in the mid-eighteenth century. Every plant or animal has two Latin names: a generic name designating its *genus*, and a specific name indicating the *species*; e.g. *Felis tigris,* the tiger.

binomial theorem The expansion of

$$(x + y)^n = x^n + nx^{n-1}y + n(n - 1)x^{n-2}y^2/2! + \ldots + y^n,$$

n being a positive *integer*. In general, for n not a positive integer, the following expression is valid if the numerical value of x is less than unity:

$$(1 + x)^n = 1 + nx + n(n - 1)x^2/2! + \ldots \text{to } \infty.$$

bioassay A quantitative estimation of the effect a substance has on a living organism.

biochemical oxygen demand See *BOD*.

biochemistry The *chemistry* of living organisms. It is important in agriculture and medicine, forming the basis of such life sciences as *physiology*, *pharmacology*, *genetics*, nutrition, etc.

biodegradation The chemical *degradation* by living organisms of substances introduced into the environment. Substances that are biodegradable, such as sewage, cause no permenent damage if correctly treated. However, substances that are nonbiodegradable, such as some insecticides and some heavy metals, can cause serious problems by accumulating in the environment.

bioenergetics The flow of energy in a living organism, including the intake from food, sunlight, etc.; the amount used in growth and maintenance of tissues, maintaining temperature, movement, etc.; and the amount lost through wastes, *respiration*, *transpiration*, etc.

bioengineering The application of engineering principles and techniques to living organisms. It is largely concerned with the design of replacement body parts, such as limbs, heart valves, etc.

biofuel See *biomass energy*.

biogas See *biomass energy*.

biogenesis The biological doctrine that only life begets life, as opposed to the unsubstantiated theory that animate matter may still be spontaneously generated from inanimate matter. See *abiogenesis*.

biological clock The hypothetical mechanism that enables certain physiological processes to occur in an organism at regular intervals, which appear to be independent of periodic events in the environment.

biological control The control of a pest by means of a biological technique rather than a chemical pesticide. E.g. by using predators or parasites of the pest or reducing the population of an insect pest by introducing large numbers of sterile males into the population.

biology The science of life, the main branches of which are *botany*, *zoology*, *cytology*, *histology*, *morphology*, *physiology*, *embryology*, *ecology*, *genetics*, and *microbiology*, although in recent years there has been a tendency to reduce the compartmentalization of the subject. Related subjects are *biochemistry*, *biophysics*, and *biometry*.

bioluminescence A form of *luminescence* occurring in living creatures, such as fire flies, glow worms, etc. The light is emitted when the substance luciferin is oxidized in the presence of the *enzyme* luciferase.

biomass The mass of living matter in a population of particular organisms in a particular area.

biomass energy A *renewable energy source* that makes use of such biofuels as methane (biogas) generated by sewage, farm, industrial, or household organic waste materials. Other biofuels include trees grown in so-called 'energy forests' or other plants, such as sugar cane, grown for their energy potential. It is estimated that by the end of the first quarter of the 21st century some 15–20% of the UK's energy needs could be met by biomass energy and specific methods are being actively researched. Biomass energy relies on combustion and therefore produces carbon dioxide; its use would not, therefore, alleviate the *greenhouse effect*.

biometry The application of mathematical and statistical methods to the study of *biology*.

biophysics The application of *physics* to the study of *biology*.

biopoiesis The evolution of living matter from self-replicating but nonliving molecules. This is the process by which life on Earth is believed to have originated.

biosphere See *ecosphere*.

biosynthesis The *synthesis* of chemical compounds by living organisms.

biotechnology The technological techniques that make use of biological processes in the industrial production of such substances as cheese, *antibiotics*, beer, etc.

biotin $C_{10}H_{16}O_3N_2S$. A crystalline substance, m.p. 230°C.; a *vitamin* of the B complex, widely distributed in nearly all living *cells* in very small quantities. It is a *coenzyme* needed in the incorporation of *carbon dioxide* into various compounds. It is produced by intestinal *bacteria* in animals, and also occurs in cereals, vegetables, milk, and liver.

biotite A black or green mineral consisting of magnesium iron potassium aluminium silicate with the general formula $K(Mg,Fe)_3(Al,Fe)Si_3O_{10}(OH)_2$. It is a member of the *mica* group.

biotype A group of individual *organisms* having the same genetic characteristics.

biphenyl Diphenyl. $C_6H_5C_6H_5$. An *insoluble* colourless powder, m.p. 70°C., used in organic synthesis and in the manufacture of *dyes*. See also *polychlorinated biphenyl*.

biphenyl-4,4′-diamine Benzidine. $NH_2C_6H_4C_6H_4NH_2$. An *aromatic* base, m.p. 128°C., of importance in the dyestuff industry.

bipolar transistor See *transistor*.

biprism An optical device for obtaining *interference* fringes; it consists of two acute-angled *prisms* placed base to base.

birefringence See *double refraction*.

Birkeland and Eyde process A process for the fixation of atmospheric nitrogen (see *fixation of nitrogen*). Nitrogen and oxygen from the atmosphere are made to combine to form *nitrogen oxides* by the action of an electric *arc*. It is still sometimes used where plentiful hydroelectricity is available, but is becoming obsolete. Named after Kristian Birkland (1867–1917) and Samuel Eyde (1866–1940).

bisection Division into two equal parts.

bisector A straight line that divides another line or angle into two equal parts.

bismuth Bi. Element. R.a.m. 208.9804. At. No. 83. A white *metal* with a reddish tinge, r.d. 9.78, m.p. 271°C., b.p. 1560°C. It is a brittle, rather poor *conductor* of *heat* and *electricity*, that expands on solidifying. It occurs as the metal, or as the *oxide*, Bi_2O_3 (biomite) and is extracted by roasting the *ore* and heating with *coal*. It is used in casting *alloys* of low *melting point* (see *Rose's metal*, *Wood's metal*), *catalysts*, and *nuclear reactors*; *compounds* are used in medicine.

bismuth(III) chloride oxide Bismuth oxychloride. BiOCl. A white crystalline *insoluble* powder, used in the manufacture of *pigments* and artificial pearls.

bismuth nitrate $Bi(NO_3)_3.5H_2O$. A colourless *deliquescent* crystalline substance that, with a large excess of water, forms bismuth(III) nitrate oxide, $BiONO_3.H_2O$, a crystalline substance, m.p. 105°C., used in medicine.

bistable circuit See *flip-flop*.

bisulphate See *hydrogensulphate*.

bisulphite See *hydrogensulphite*.

bit A unit of information in *information theory*. The amount of information required to specify one of two alternatives, e.g. to distinguish between 1 and 0 in the *binary notation* as used in *computers*. It is also used as a unit of capacity in a *store*. See also *byte*, *character*, and *word*.

bittern (chem.) The *mother-liquor* remaining after the crystallization of common *salt*, NaCl, from *sea-water*. It is a source of *compounds* of magnesium, bromine, and iodine.

bitumen Various *mixtures* of *hydrocarbons*, more particularly solid or tarry mixtures, *soluble* in *carbon disulphide*, obtained from *petroleum*.

bituminous Containing, or yielding upon *distillation*, bitumen or *tar*.

bituminous coal *Coal* with a higher carbon content than *lignite*; it is the form of coal most widely used as a fuel.

biuret Carbamoylurea. $NH_2CONH.CONH_2.H_2O$. An *insoluble* crystalline substance formed from *urea*. See *biuret reaction*.

biuret reaction A *chemical reaction* in which an *alkaline solution* of *biuret* gives a purple colour on the addition of 1% *copper(II) sulphate* solution in the presence of *peptide* bonds. It is used as a biochemical test for *protein* and *urea*.

bivalent Divalent. Having a *valence* of two.

black ash Impure *sodium carbonate* obtained in the *Leblanc process*.

black body A hypothetical body that has an *absorptance* and an *emissivity* of 1, i.e. absorbs all the *radiation* falling on it. No such body can exist, but a small hole in a furnace wall approaches it.

black body radiation Full or complete *radiation*; radiation of all *frequencies* that would be emitted by a *black body*, which absorbs all radiations falling upon it. As the *absorptance* of a black body is one, the radiation that it emits is a function of *temperature* only (see *Stefan's Law*). The *energy* distribution has a maximum at certain frequencies, which depends on the temperature (see *Wien displacement*).

blackdamp *Carbon dioxide* (in coal mines).

black hole A hypothetical region of space possessing a *gravitational field* so intense that no matter or radiation can escape from it. Such regions are believed to form as a result of the *gravitational collapse* of a *star* after it has used up all its nuclear fuel (see *stellar evolution*). Smaller stars create *supernova* explosions when they die, leaving *neutron stars*: it is the more massive stars that are believed to create black holes.

The boundary of the black hole is thought to be a sphere (called the event horizon) with a radius (called the Schwartzchild radius) $2GM/c^2$, where M is the mass of the region, G is the *gravitational constant*, and c is the speed of light.

The problem of detecting black holes is that, being unable to emit or reflect radiation, they are invisible. However, it is thought that some *X-ray binary stars* exist in which one member of the pair is a black hole.

blacklead Plumbago, graphite. A natural crystalline form of carbon. A soft grey-black *solid*; used for making vessels to resist high *temperatures*, in pencils, and as a lubricant.

blanc fixe Artificial *barium sulphate*, $BaSO_4$. It is used as an *extender* in the *paint* industry.

blanket (phys.) A layer of *fertile* material surrounding the core of a *nuclear reactor* to act as a *reflector*, or for the purpose of breeding new fuel. See *breeder reactor*.

blast furnace A furnace for the smelting of iron from iron oxide *ores* (see *haematite*, *magnetite*). It is constructed of *refractory* bricks covered with *steel* plates and charged from above with a mixture of the ore, *limestone* ($CaCO_3$), and *coke*. The coke is ignited at the bottom of the furnace by a blast of hot air; the *carbon monoxide* so produced reduces the iron oxide to iron, while the heat of the action decomposes the limestone into *carbon dioxide* and *lime*, CaO. The overall reaction is:

$$Fe_3O_4 + CaCO_3 + 2CO + 2H_2 \rightarrow 3Fe + CaO + 3CO_2 + 2H_2O.$$

The lime combines with the sand and other impurities in the ore to form a molten *slag*. The molten iron and the slag are tapped off at the bottom of the furnace. The resulting *pig-iron* or *cast iron* contains up to 4.5% carbon. In some modern processes the carbon monoxide and hydrogen are added separately, so that the reduction can occur at a lower temperature.

blasting gelatin Jelly-like mixture of *gun-cotton* with *nitroglycerin*. A very powerful *explosive*.

blastula A hollow ball of *cells* that forms in the very early embryonic development of animals.

bleaching Removing the colour from coloured materials by chemically changing the dyestuffs into colourless substances. *Bleaching powder* and other *oxidizing agents*, or *sulphur dioxide* and other *reducing agents* are often used.

bleaching powder Chloride of lime. A whitish powder, consisting mainly of calcium chlorate(I), $CaOCl_2$, (formerly called calcium oxychloride) with water; prepared by the action of chlorine on *calcium hydroxide*, $Ca(OH)_2$. The action of *dilute acids* liberates chlorine, which acts as an *oxidizing agent* and so bleaches the material.

blende Natural *zinc sulphide*, ZnS.

blink microscope Blink comparator. An instrument for examining photographs of the sky taken in rapid succession to each other. Minor *planets* and *stars* with large *proper motions*, or rapid changes of *luminosity*, are thereby made conspicuous.

blood A *liquid* that circulates throughout the body of the higher animals, transporting oxygen, *hormones*, and *cell* foods to all the component cells of the body, and removing their excretions. It is also a defence system, providing protection against infection. Blood consists of a liquid, *blood plasma*, in which *blood cells* and *blood platelets* are suspended. The average human male has about 11 pints (6.2 litres) of blood in his body.

blood cells Blood corpuscles, haemocytes. There are two types of blood cell: red cells (*erythrocytes*) and white cells (*leucocytes*). The function of the red cells is to transport oxygen throughout the body, by way of the *haemoglobin* that they contain. The function of the white cells is to combat infection. See also *blood platelets*.

blood plasma *Blood* from which all *blood cells* and *blood platelets* have been removed. Plasma is 90% water, in which the principal *solutes* are *proteins*, *salts*, *sugar*, and *urea*. *Hormones*, *vitamins*, and excretions are also present in the plasma.

blood platelets Thrombocytes. Small membrane-bounded coin-shaped particles that circulate in the *blood*. If a blood vessel should break, the platelets clump together to form a plug to stop the bleeding. Platelets contain substantial quantities of ATP, and it is the diphosphate that causes the agglutination; they also release *serotonin*, which causes the blood vessels to constrict and bleeding to be reduced. Human blood contains about 250 000 platelets per cubic millimetre.

blown oil A thickened *oil* made by blowing air through a natural vegetable or animal oil.

blowpipe A device for producing a jet of *flame* by forcing a flammable *gas* mixed with air or oxygen through a nozzle at high pressure.

blue vitriol Bluestone. See *copper(II) sulphate*.

board of trade unit B.O.T. unit. A former name for the *kilowatt-hour*. The energy obtained when a *power* of 1 *kilowatt* is maintained for 1 hour.

boart See *bort*.

BOD Biochemical (or biological) oxygen demand. A measure of the content of organic matter in water and wastes. It is the amount of oxygen (mg of O_2 per cubic decimetre of water) when a sample containing a known mass of oxygen in solution is kept at 20°C. for five days. The oxygen is consumed by microorganisms that feed on the organic matter in the sample. A high BOD implies a high organic matter, suggesting a high degree of *pollution*.

body-centred Denoting a crystal in which there is a *lattice* point at the centre of the body of the crystal as well as at the corners. It is said to be 'face-centred' when there is a lattice point at the centre of each face. See Fig. 4.

bog iron ore An impure form of *hydrated iron(III) oxide*, $Fe_2O_3.xH_2O$, found in bogs and marshes.

Bohr theory A theory of the *atom* put forward by Niels Bohr (1885–1962) in 1913 to explain the *line spectrum* observed for hydrogen (see *Balmer series*). It is based on three postulates: 1. The *electrons* rotate in certain *orbits* round the *nucleus* of the atom without radiating *energy* in the form of *electromagnetic waves*. 2. These orbits are such that the *angular momentum* of the electron about the nucleus is an *integral* multiple of $h/2\pi$, where $h =$ the *Planck constant*. 3. Emission or absorption of radiation occurs when an electron jumps from one of these so-called *stationary states* of energy E_1 to another of energy E_2, the *frequency*, v, of the emitted (or absorbed) *light* being given by $E_1 - E_2 = hv$. If E_1 is greater than E_2, light is emitted; conversely, light is absorbed. See *quantum mechanics*. This theory has now been superseded by the application of *wave mechanics*, which has shown that for the hydrogen atom spectrum, Bohr's theory is a very good approximation. Wave mechanics has the advantage of requiring no *ad hoc* assumptions and can deal more effectively with the problem of atoms with two or more electrons. (See also *atom, structure of*.)

boiled oil *Linseed oil* boiled with, or containing, a drying agent, such as *lead(II) oxide* (lead monoxide), PbO. It is mainly used in *paints*.

boiling Ebullition. The state of a *liquid* at its *boiling point* when the equilibrium *vapour pressure* of the liquid is equal to the external pressure to which the liquid is subject, and the liquid is freely converted into *vapour*.

boiling point B.p. The *temperature* at which the equilibrium *vapour pressure* of a *liquid* is equal to the external pressure; the temperature at which the liquid boils freely under that pressure. Boiling points are normally quoted for standard atmospheric pressure, i.e. 760 mm of mercury (101 325 Pa).

boiling water reactor BWR. A *nuclear reactor* in which *water* is used as *coolant* and *moderator*. *Steam* is thus produced in the reactor under pressure, and can be used to drive a *turbine*. This avoids the use of *heat exchangers* but additional equipment is required to prevent the escape of *radioactive* gases from the *turbogenerator*. Compare *pressurized water reactor*.

bolide A large bright *meteor*; some of these objects explode on entering the *Earth's atmosphere*.

bolometer An extremely sensitive instrument for measuring *heat radiations*. Early models consisted of two very thin, blackened platinum gratings, forming two arms of a *Wheatstone bridge* circuit. Radiant heat is allowed to fall upon one grating while the other is shaded from it; the change in resistance of the exposed grating causes a deflection of the needle of a *galvanometer* in the circuit. More modern instruments use *semiconductors* instead of platinum gratings.

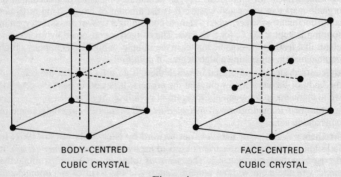

BODY-CENTRED
CUBIC CRYSTAL

FACE-CENTRED
CUBIC CRYSTAL

Figure 4.

Boltzmann constant $k = R/L = 1.380\,658 \times 10^{-23}$ *joule* per *kelvin*, where R = the *gas constant* and L = *Avogadro constant*. Named after L. Boltzmann (1844–1906).

bomb calorimeter A strong metal vessel used for measuring *heats of reaction*, especially *heats of combustion*; e.g. for determining the *calorific value* of a *fuel*. To do this, a known *mass* of the substance under test is burnt in the vessel, and by measuring the quantity of heat produced, the calorific value is calculated.

bond See *chemical bond*; *hydrogen bond*.

bond energy The *energy* characterizing a *chemical bond* between two *atoms*. It is measured by the energy required to separate the two atoms.

bond length The distance between the *nuclei* of two *atoms* joined by a *chemical bond*.

bone ash *Ash* obtained by heating bones in air. It consists mainly of *calcium phosphate(V)*, $Ca_3(PO_4)_2$.

bone black See *animal charcoal*.

bone char See *animal charcoal*.

bone oil Dippel's oil. A product obtained by the *destructive distillation* of bones. It is a dark oily evil-smelling liquid used as a source of *pyridine*.

Boolean algebra A branch of symbolic logic used in *computers*. Logical operations are performed by operators, such as 'and', 'or', 'not-and', in a way that is analogous to mathematical signs. Named after George Boole (1815–64).

booster See *rocket*.

boracic acid See *boric acid*.

boranes *Hydrides* of *boron*, having the general formula B_nH_{n+2}; the simplest is diborane, B_2H_6. Many may be made by the action of *acid* on magnesium boride (MgB_2).

borate Any of a number of *ionic compounds* containing negative *ions* consisting of boron and oxygen, such as calcium borate, CaB_2O_4 or *disodium tetraborate*, $Na_2B_4O_7.10H_2O$.

borax See *disodium tetraborate*.

borax bead test A chemical test for the presence of certain *metals*. A bead of *borax* fused in a wire loop will react chemically with the *salts* of a number of metals, often producing colours that help to identify the metal; e.g. manganese *compounds* give a violet bead, cobalt a deep blue.

Bordeaux mixture A mixture of *copper(II) sulphate*, $CuSO_4$, *calcium oxide*, CaO, and *water*. Used for spraying plants as a *fungicide* for plant diseases.

boric acid Orthoboric acid, boracic acid, trioxoboric(III) acid. H_3BO_3. A white crystalline *soluble solid*, m.p. 169°C. It occurs naturally in volcanic regions and is manufactured from *borax*. It used as a mild *antiseptic*, in detergents and glazes, and in various other industries. On heating it forms tetraboric acid (pyroboric acid), $H_2B_4O_7$.

boric oxide Boric anhydride. B_2O_3. An *oxide* that exists either as a transparent crystalline substance, m.p. 460°C., or a transparent *amorphous glass*. It is used in the manufacture of special glasses.

boride A *binary compound* with boron. Some metal borides, such as zinc boride, ZnB_2, and titanium boride, TiB_2, are used as *refractories*.

borneol Bornyl alcohol. $C_{10}H_{17}OH$. A white *optically active translucent* solid, m.p. 210.5°C., used in the manufacture of synthetic *camphor* and in perfumes.

bornite An *ore* of copper consisting of a copper iron sulphide, Cu_5FeS_4.

bornyl ethanoate $C_{10}H_{17}COOCH_3$. A colourless *liquid*, b.p. 223°C., with a *camphor*-like odour. It is used in the manufacture of perfumes and as a *plasticizer*.

boron B. Element. R.a.m. 10.811. At. No. 5. A brown *amorphous* powder, r.d. 2.37, or yellow *crystals*, r.d. 2.34; m.p. 2300°C., b.p. 2550°C. It occurs as *borax* and *boric acid*. It is used for hardening *steel* and for producing *enamels* and *glasses*. As boron-10 (present to an extent of 18–20% in natural boron) absorbs slow *neutrons*, it is used in steel *alloys* for making *control rods* in *nuclear reactors*.

boron carbide B_4C. A very hard black crystalline substance, m.p. 2350°C., used as an *abrasive* and in control rods in *nuclear reactors*.

boron chamber An *ionization chamber* lined with boron or boron *compounds* or filled with boron trifluoride *gas*. It is used in *boron counter tubes* to count *slow neutrons*.

boron counter tubes A *proportional counter tube* containing a *boron chamber* used for counting *neutrons*. The counting pulse results from particles emitted when *neutrons* react with the boron-10 *isotope*.

borosilicates A group of silicates in which the SiO_4 units are linked with BO_3 units in a variety of structures. The borosilicates glasses, such as *Pyrex**, have a smaller plastic range than soda *glasses* and are resistant to chemical attack.

bort Boart. Impure or discoloured *diamond*; useless as a gem, it is as hard as pure diamond and is used for drills, cutting tools, etc.

Bosch process An industrial process for the manufacture of hydrogen. *Water gas*, a *mixture* of *carbon monoxide* and hydrogen, is mixed with *steam* and passed over a heated *catalyst*. The steam reacts chemically with the carbon monoxide to give *carbon dioxide*, CO_2, and hydrogen. The CO_2 is then removed by dissolving it in water under pressure. The process was formerly used to produce hydrogen for the *Haber process*. The Bosch process has largely been replaced by *steam reforming* of *natural gas*. Named after C. Bosch (1874–1940).

Bose–Einstein statistics The branch of *quantum statistics* used with systems of identical particles having the property that the *wave function* remains unchanged if any two particles are interchanged. See *bosons*. Named after S. N. Bose (1894–1974) and Albert Einstein (1879–1955).

bosons *Elementary particles*, such as *photons* and *mesons*, that conform to *Bose-Einstein statistics*; their numbers are not conserved in particle interactions. Bosons have *integral spin* (0, 1, 2). Compare *fermions*.

botany The scientific study of plants.

bottom A type of *quark*. See *elementary particles*.

boundary layer The layer of *fluid* closest to a body over which the fluid is flowing; owing to the *force* of *adhesion* between the body and the fluid the boundary layer has a reduced rate of flow.

Bourdon gauge A *pressure gauge* that depends on the tendency of a partly flattened curved tube to straighten out when under internal pressure. One end of the tube, which is usually C-shaped, is closed while the other is connected to the source of pressure; the movement of the closed end as the tube tends to straighten under pressure is transmitted by a simple mechanism to a needle moving round a dial or to a digital display.

Boyle's law At a constant *temperature*, the *volume* of a given quantity of *gas* is inversely proportional to the *pressure* upon the gas; i.e pV = constant. It is only true for an ideal *gas*. Named after Robert Boyle (1627–91). See also *gas laws*.

Bragg's law When a *beam* of *X-rays*, of *wavelength* λ, strikes a *crystal* surface, the maximum intensity of the reflected ray occurs when $sin\ \theta = n\lambda/2d$. Where d is the distance separating the layers of the *atoms* or *ions* in the crystal, θ is the *complement* of the angle of *incidence*, and n is an *integer*. Named after W. L. Bragg (1890–1971).

brake horsepower The *horsepower* of an engine measured by the degree of resistance offered by a brake; it represents the useful horsepower that the engine can develop.

branched chain A chain of carbon atoms in an organic molecule, in which the main chain has one or more branches.

branching (phys.) The occurrence of more than one *radioactive disintegration* scheme for a particular *nuclide*.

brass A large class of *alloys*, consisting principally of copper and zinc. Small quantities of aluminium, iron, nickel, tin, or lead are also often added.

breeder reactor A *nuclear reactor* that produces the same kind of *fissile* material as it burns. E.g. a reactor using plutonium as a fuel can produce more plutonium than it uses by conversion of uranium-238.

Bremsstrahlung (German, meaning 'brake radiation') *X-rays* emitted when a charged particle, such as an *electron*, is rapidly slowed down by an *electric field*, as when an electron strikes a positively charged *nucleus*; it results from the direct conversion of *kinetic energy* into *electromagnetic radiation*.

brewing The making of beer. *Malt* is ground and mixed with water. In the resulting 'mash', chemical changes take place, the chief of which is the conversion of *starch* into *maltose*, forming a sweetish liquid known as wort. This is boiled with the addition of hops. After cooling and removal of *solids*, *yeast* is added and *fermentation* occurs.

Brewster's law The tangent of the angle of *polarization* of light (Brewster angle) is numerically equal to the *refractive index* of the reflecting medium when the polarization is a maximum. Named after David Brewster (1781–1868).

brimstone Sulphur fused into blocks or rolls.

Brinell test A test for the hardness of *metals*. A ball of chrome *steel*, or other hard material, of standard size, is pressed by a heavy load into the surface of the metal, and the diameter of the depression is measured. The Brinell number is the ratio of the load in *kilograms* to the *area* of the depression in square millimetres. Named after J. A. Brinell (1849–1925).

Britannia metal An *alloy* of variable composition, containing 80%–90% tin, with some antimony and copper, and sometimes also zinc and lead.

British thermal unit The Imperial unit of heat, originally defined as the quantity of *heat* required to raise the *temperature* of 1 lb of *water* through 1° Fahrenheit; it is now defined as 1055.06 *joules*.

bromate A *salt* or *ester* of *bromic acid*.

bromic acids Bromic(I) acid, hypobromous acid, HBrO, is a yellow liquid that is a weak acid but a strong *oxidizing agent*. Bromic(V) acid, $HBrO_3$, formed by the action of *sulphuric acid* on barium bromate, is a strong acid and is also used as an oxidizing agent.

bromide A *salt* of *hydrobromic acid*, HBr; a *binary compound* with bromine. 'Bromide' of pharmacy is *potassium bromide*, KBr.

bromide paper Photographic paper containing *silver bromide*, AgBr.

bromination A reaction in which one or more bromine atoms are substituted for hydrogen atoms in an organic molecule.

bromine Br. Element. R.a.m. 79.909. At. No. 35. A dark red fuming *liquid* with a choking, irritating smell, r.d. 3.12, m.p. $-7.2°$C., b.p. $58.8°$C. It occurs as magnesium bromide, $MgBr_2$, in *bittern* from *sea-water*, in the *Stassfurt deposits*, in marine plants and animals, and in some inland lakes. It is used as a *disinfectant* and in the manufacture of some *organic compounds*. Compounds are used in *photography* and medicine.

bromoform See *tribromomethane*.

bronze 1. A class of *alloys* of copper and tin (1–30%). Other elements are sometimes added, especially phosphorus (phosphor bronze). Bronzes are widely used for casting. See also *bell metal*; *gun metal*. 2. A copper alloy containing no tin, e.g. *aluminium bronze* is an alloy of copper and aluminium.

Brownian movement Erratic random movements performed by microscopic particles in a *disperse phase*; e.g. particles in suspension in a *liquid*, or *smoke* particles in air. It is caused by the continuous irregular bombardment of the particles by the *molecules* of the surrounding medium. Named after Robert Brown (1773–1858).

brucite A mineral consisting of *magnesium hydroxide*.

brush discharge An electric discharge from sharp points on a *conductor*. The surface density (i.e. quantity of electric charge per unit area) is greatest at sharp points; the high charge at such points causes a displacement of the charge on the air particles near the points, and hence an attraction to the points. On reaching the points, the particles acquire some of the charge on the points and are repelled. This causes a stream of charged air particles to leave the vicinity of the points.

bubble chamber An instrument for making the tracks of ionizing particles visible as a row of bubbles in a *liquid*. The liquid is maintained under pressure, at a temperature close to the boiling point for that pressure. Immediately before the passage of the particles the pressure is reduced so that the liquid is now *superheated* and the trails of ionized atoms left by the ionizing particles act as centres for the formation of small *vapour* bubbles. These can be photographed to give a record of the tracks of the particles.

Büchner funnel A funnel, usually of *porcelain*, with a flat circular base perforated with small holes. It is used for filtering by suction. Named after E. Büchner (1860–1917).

buffer solution A *solution* the *hydrogen ion concentration* of which, and hence the acidity or alkalinity, is practically unchanged by dilution. It also resists a change of *pH* on the addition of *acid* or *alkali*. An acid buffer consists of a weak acid and a *salt* of the acid (e.g. *carbonic acid* and *sodium hydrogencarbonate*); when acid is added the hydrogen *ions* produced react with the negative ions from the salt and when a base is added the *hydroxide* ions react with the undissociated molecules of the acid to form water and the negative ions of the salt. Thus there is no great change in either the hydrogen ion or hydroxide ion concentrations. A basic buffer consists of a weak base and a salt of the base with similar reactions taking place.

bulk density The *density* of a powder or of a porous or granular substance, calculated for unit volume of the substance including the pores or spaces between the grains; it is generally less than the true density of the material.

bulk modulus *Elastic modulus* applied to a body having uniform *stress* distributed over the whole of its surface. Its value is given by the expression pV/v where p = intensity of stress, V = original *volume* of the body, and v = change in volume.

Bunsen burner A gas burner used in laboratories. It consists of a metal tube with an adjustable air-valve for burning a mixture of gas and air. Named after R. W. Bunsen (1811 – 99).

Bunsen cell A *primary cell* in which the *anode* consists of zinc and is immersed in dilute *sulphuric acid*, and the *cathode* consists of carbon immersed in concentrated *nitric acid*. It gives an EMF of 1.9 volts.

buoyancy The upward thrust exerted upon a body immersed in a *fluid*; it is equal to the *weight* of the fluid displaced. (See *Archimedes' principle*). Thus a body weighs less when weighed in water, the apparent loss in weight being equal to the weight of the water displaced. For accurate weighing of bodies in air, a small allowance has to be made to correct for the buoyancy of the body.

burette A graduated glass tube with a tap, for measuring the volume of *liquid* run out from it. It is used for *titration* in *volumetric analysis*.

burning See *combustion*.

burnt alum A white porous mass of *anhydrous* potassium aluminium sulphate, $K_2SO_4.Al_2(SO_4)_3$, obtained by heating *alum*.

butadiene Buta-1,3-diene. $CH_2:CH.CH:CH_2$. A *gas*, b.p. $-4.5°C.$, used in the manufacture of synthetic *rubbers*. See *styrene-butadiene rubber*, *nitrile rubber*, and *stereo-regular rubbers*.

butanal Butyraldehyde. $CH_3(CH_2)_2CHO$. A colourless flammable *liquid*, b.p. $75.7°C.$, used in the *plastics* and *rubber* industries.

butane C_4H_{10}. A *hydrocarbon* of the *alkane series*. A *gas* at ordinary *temperatures*, b.p. $-0.5°C.$, it is used in the manufacture of *synthetic rubber* and as a *fuel* (e.g. in cylinders under pressure under the trade name Butagas*). It is isomeric with 2-methylpropane, $CH_3CH(CH_3)CH_3$, which was formerly called isobutane.

butanedioic acid Succinic acid. $(CH_2COOH)_2$. A white crystalline organic *dibasic acid*, m.p. $185°C.$, used in the manufacture of *dyes*, lacquers, etc.

butanedione Diacetyl. $CH_3COCOCH_3$. A yellow *liquid*, b.p. $89°C.$, that occurs in butter, and is used as a flavour.

butanoic acid. Butyric acid. C_3H_7COOH. A *liquid* with a rancid odour, b.p. $163.5°C.$, which occurs in rancid butter. It is used in the form of its *esters* as a flavouring.

butanol Butyl alcohol. C_4H_9OH. A *liquid* that exists in two *isomeric* forms. Butan-1-ol, $CH_3CH_2CH_2CH_2OH$, has a b.p. $117.5°C.$ and is used as a *solvent*. Butan-2-ol, $CH_3CH(OH)C_2H_5$, b.p. $100°C.$, is also used as a solvent.

butanone Ethyl methyl ketone. $C_2H_5COCH_3$. A flammable *liquid*, b.p. $77.6°C.$, used as a *solvent* and in the manufacture of *plastics*.

butenedioic acid HOOCCH:CHCOOH. Two compounds that exhibit *cis-trans* *isomerism* (see illustration at this entry). The cis form, maleic acid, converts to the more stable trans form, fumaric acid, at 120°C. The cis form can eliminate water on heating to form maleic anhydride. Maleic acid is used in the manufacture of synthetic resins and dyes and as a preservative. Fumaric acid is used in making *baking powders*.

butenoic acid See *crotonic acid*.

butterfly effect An effect that sometimes occurs in systems with many variables in which it is difficult to specify the initial parameters with accuracy (see *chaos theory*). The butterfly effect takes its name from such a system in which the flapping of a butterfly's wings can so distort the meteorological dynamics in one part of the world that a tornado results from it in another.

butter of antimony Antimony(III) chloride, antimony trichloride. $SbCl_3$. A white crystalline substance, m.p. 73°C.

butter of zinc See *zinc chloride*.

butyl The *univalent alkyl group* C_4H_9-.

butyl rubber A synthetic *rubber*; a copolymer (see *polymerization*) of 2-methylpropene (iso-butylene) and sufficient methylbuta-1,3-diene (*isoprene*) (2%–3%) to enable vulcanization to be effected. Owing to its low *permeability* to *gases*, butyl rubber has been used in the manufacture of tyre inner tubes.

butyraldehyde See *butanal*.

butyric acid See *butanoic acid*.

butyryl The *univalent group* $CH_3(CH_2)_2CO-$.

BWR See *boiling water reactor*.

bypass capacitor Bypass condenser. A *capacitor* that provides a path of low *impedance* over a certain range of *frequencies*.

by-product A substance obtained incidentally during the manufacture of some other substance. It may be as important as the manufactured substance itself.

byte A single unit of information handled by a *computer*; usually 8 *bits*.

C

cacodyl The former name for the dimethylarsino group, $(CH_3)_2As-$, derived from *arsine*.

cadmium Cd. Element. R.a.m. 112.41. At. No. 48. A soft silvery-white *metal*; r.d. 8.65, m.p. 320.9°C., b.p. 765°C. It occurs together with zinc and as greenockite (see *cadmium sulphide*). It is used in the manufacture of *fusible alloys* and for *electroplating*. As cadmium is a good absorber of *neutrons* it is used in the manufacture of *control rods* for *nuclear reactors*.

cadmium cell Standard *primary cell*. See *Weston cell*.

cadmium sulphide CdS. A yellow *insoluble* powder, used as a *pigment*, known as 'cadmium yellow', and in photoconductive cells (see *photoelectric cell*). In the impure natural form it is known as 'greenockite'.

caesium Cesium. Cs. Element. R.a.m. 132.905. At. No. 55. A highly reactive silvery-white *metal* resembling sodium in its physical and chemical properties; r.d. 1.87, m.p. 28.4°C., b.p. 678°C. The natural *isotope*, caesium-133 is stable but there are 15 *radioisotopes*. *Compounds* are very rare. It is used in *photoelectric cells* and as a *catalyst*.

caesium clock A device used in the *SI unit* definition of the *second*. It is based on the energy difference between two states of the caesium-133 nucleus in a magnetic field. This energy difference corresponds to a *frequency* of 9 192 631 770 *hertz*. A beam of caesium atoms is split into the two components by a non-uniform magnetic field. Nuclei in the lower state are irradiated in a cavity by radio-frequency radiation at the difference frequency. Some are excited to the higher frequency by absorbing this radiation. By reanalysing the mixture of atoms and using a feedback system, the r-f oscillator can be locked to the difference frequency with an accuracy of one part in 10^{13}. It thus constitutes an extremely accurate clock.

caffeine Theine. $C_8H_{10}O_2N_4$. A white crystalline *purine*, m.p. 237°C., that occurs in coffee-beans (0.75–1.5% by weight), tea-leaves (0.3–0.7%), and other plant material. It has a powerful action on the heart and is used in medicine.

cage compound See *clathrate*.

calamine 1. A zinc mineral consisting of *zinc carbonate*, $ZnCO_3$. 2. In US usage, a zinc mineral consisting of *zinc silicate*, $Zn_2SiO_4.H_2O$ (or $2ZnO.SiO_2.H_2O$). 3. A skin preparation consisting of *zinc oxide* with ½% of *iron(III) oxide*.

calciferol Ergocalciferol, vitamin D. See *vitamins*.

calcination 1. The conversion of *metals* into their *oxides* by heating in air. 2. The deposition of *calcium carbonate* by *hard water*.

calcite Calcspar. Natural crystalline *calcium carbonate*, $CaCO_3$.

calcium Ca. Element. R.a.m. 40.078. At. No. 20. A soft white *metal* that tarnishes rapidly in air; r.d. 1.55, m.p. 840°C., b.p. 1484°C. *Compounds* are very abundant, widely distributed, and essential to life. It occurs as *calcium carbonate*, $CaCO_3$ (*limestone*, marble, and *chalk*) and *calcium sulphate*, $CaSO_4$ (*gypsum*, *anhydrite*); it is an essential constituent of bones and teeth. Compounds are of great industrial importance; e.g. *lime*.

calcium carbide See *calcium dicarbide*.

calcium carbonate $CaCO_3$. A white *insoluble* solid; it occurs naturally as *chalk*, *limestone*, marble, and *calcite*. It is used in the manufacture of *lime* and *cement*. See also *Solvay process*.

calcium chloride $CaCl_2$. A white *deliquescent* substance, m.p. 772°C., obtained by reacting *calcium carbonate* with *hydrochloric acid*. It is used as a drying agent, refrigerant, and preservative.

calcium cyanamide Cyanamide, nitrolime. $CaCN_2$. A black crystalline powder made by heating *calcium dicarbide*, CaC_2, in nitrogen at 1000°C. It is used as a *fertilizer* and converted by *water* in the soil into *ammonia* and *calcium carbonate*. It is used in the manufacture of some *plastics*.

calcium cyclamate $(C_6H_{11}NHSO_3)_2Ca.2H_2O$. A white crystalline *soluble* powder, formerly used as a sweetening agent in soft drinks, but its excessive consumption has been shown to be undesirable and it has therefore been banned.

calcium dicarbide Calcium carbide. CaC_2. A greyish solid, colourless when pure; it is prepared by heating *calcium oxide* with carbon in an electric furnace. It reacts with *water* to give *ethyne*.

calcium fluoride See *fluorspar*.

calcium hydroxide Slaked lime. $Ca(OH)_2$. A white crystalline powder, obtained by the action of *water* on *calcium oxide*, used in *mortars*, plaster, and *cement*.

calcium nitrate $Ca(NO_3)_2$. A white *deliquescent* solid, m.p. 561°C., used in the manufacture of *fertilizers*, fireworks, *matches*, and *explosives*.

calcium oxide Quicklime. CaO. A white solid, m.p. 2580°C., made by heating *calcium carbonate* (*limestone*) in lime-kilns. It combines with *water* to form *calcium hydroxide* (slaked lime); it is used in *cements* and *mortars* and in the manufacture of calcium *compounds*.

calcium phosphate There are several *phosphates* of calcium that occur in rocks and animal bones. Calcium phosphate(V), formerly called tricalcium diorthophosphate, $Ca_3(PO_4)_2$, is a white *amorphous* powder, m.p. 1670°C. (see *bone ash*). It is converted to the more *soluble* calcium dihydrogen phosphate(V), $Ca(H_2PO_4)_2.H_2O$, a *deliquescent* crystalline substance, which is the main constituent of *superphosphate*. See also *octacalcium phosphate*.

calcium silicates A range of compounds, including native *minerals*, composed of *calcium oxide* (CaO) and *silica* (SiO_2) in various molecular ratios; e.g. calcium metasilicate, $CaSiO_3$, and calcium orthosilicate, Ca_2SiO_4. Various calcium silicate phases are formed in *glass* and *cement* during the manufacture of these materials. See *silicates*.

calcium sulphate $CaSO_4$. A white *salt* that is slightly soluble in water. It exists in a number of crystalline forms, including anhydrite ($CaSO_4$) and gypsum ($CaSO_4.2H_2O$). The latter is converted to *plaster of Paris* (calcium sulphate hemihydrate) on heating. $CaSO_4$ is widely used in the ceramic, paper, and paint industries.

calcium sulphide CaS. A colourless crystalline substance, having an odour of bad eggs, used in the manufacture of luminous *paints* and in cosmetics.

calculus A powerful method of solving numerous mathematical problems. It is divided into two main parts, *differential calculus* and *integral calculus*. It was developed independently by Isaac Newton (1642–1727) and Gottfried Leibniz (1641–1716).

calibration The *graduation* of an instrument to enable measurements in definite *units* to be made with it; thus the arbitrary scale of a *galvanometer* may be calibrated in *amperes*, thereby converting the instrument into an *ammeter* for measuring *electric current*.

caliche Impure natural *sodium nitrate* $NaNO_3$, found in Chile.

californium Cf. *Transuranic element*. At. No. 98. The most stable *isotope*, californium-251, has a *half-life* of 700 years. It is useful as a *neutron* source in *activation analysis*.

callipers Calipers. An instrument for measuring the distance between two points, especially on a curved surface; e.g. for measuring the internal and external diameters of tubes.

calomel See *mercury(I) chloride*.

calomel electrode A *half cell* consisting of a mercury electrode covered with *calomel* (mercury(I) chloride) and a solution of mercury in *potassium chloride*. It is used as a standard electrode, its potential being 0.2415 volt at 25°C. with respect to a *hydrogen electrode*.

calorie Unit of quantity of *heat*. The quantity of heat required to raise the *temperature* of 1 g of *water* through 1°C. The 15° calorie is defined as the amount of heat required to raise the temperature of 1 g of water from 14.5°C. to 15.5°C. This calorie is equal to 4.1855 *joules*. The International Table Calorie is defined as 4.1868 joules. The joule is the *SI unit* of *heat* and has largely replaced the calorie.

calorie, large Kilogram-calorie. 1000 *calories*. Written Calorie or kcalorie, it is still used for quoting energy values of foods but is becoming obsolete.

calorific value The quantity of *heat* produced by a given *mass* of a *fuel* on complete *combustion*. It is expressed in *joules* per *kilogram* (*SI units*) or frequently megajoules per kilogram. Calories are still occasionally used. Calorific values are determined by the *bomb calorimeter*.

calorimeter 1. An instrument for determining quantities of *heat* evolved, absorbed, or transferred. In its simplest form it consists of an open cylindrical vessel of copper or other substance of known *heat capacity*. See also *bomb calorimeter*. **2.** In *nuclear physics* and *particle physics*, a detector designed to absorb particles or *gamma rays* in order to measure their total *energy*.

Calvin cycle A chain of reactions important in *photosynthesis*, by which *adenosine triphosphate* (ATP) and reduced NADP (see *nicotinamide adenine dinucleotide*) are used to reduce carbon dioxide to *carbohydrate*. The cycle occurs in *chloroplasts* and does not require light. Named after Melvin Calvin (born 1911).

calx 1. The powdery *oxide* of a *metal* formed when an *ore* or a *mineral* is roasted. **2.** Quicklime (see *calcium oxide*).

camera, photographic A device for obtaining photographs or exposing cinematic film, either coloured or black and white. A camera consists essentially of a *light*-proof box with a *lens* at one end and a light-sensitive film or plate at the other. An 'exposure' is made by opening a 'shutter' over the lens for a predetermined period during which an image of the object to be photographed is thrown upon the light-sensitive film. Focusing is carried out by varying the distance of the lens from the film by a suitable device. The amount of light that enters the camera is determined by the amount of light available, the *aperture* of the lens (see *f number*), and the shutter speed. In the simplest cameras the shutter speed and aperture are fixed, so that satisfactory photographs for a given film speed can only be obtained in bright sunlight. In more expensive cameras the aperture can be controlled by a variable *iris* and several separate shutter speeds are provided. In some modern cameras the iris or the shutter speed is controlled by the current from a built-in *photoelectric cell* (*exposure meter*), which measures the light available. Thus for a given film speed and aperture (or shutter speed) the camera automatically takes a correctly exposed photograph. In cinematic cameras the opening of the shutter is mechanically synchronized with the passage of the film through the camera so that, at normal speeds, between 16 and 24 frames are exposed every second. See also *photography*.

camera, television The part of a *television* system that converts optical images into electrical signals. It consists of an optical *lens* system similar to that used in a photographic *camera*, the image from which is projected into a 'camera tube'. The camera tube comprises a *photosensitive mosaic* that is scanned by an *electron* beam housed in an evacuated glass tube. The output signals of the camera tube are usually pre-amplified within the body of the camera.

camphor $C_{10}H_{16}O$. A white crystalline *solid* with a characteristic smell, m.p. 179°C. It occurs in the camphor tree and is used in the manufacture of *celluloid* and in other industries.

Canada balsam A yellowish *liquid* derived from fir trees, with similar optical properties to that of *glass*. Used for mounting microscopic slides and as an *adhesive* for optical instruments.

canal rays Positively charged *ions* produced during an *electric discharge* in gases. They are accelerated to the *cathode* by the applied *potential difference* and allowed to pass through canals bored in the cathode.

candela New candle. The *SI unit* of *luminous intensity*. The luminous intensity in a given direction of a source that emits *monochromatic* radiation of *frequency* 540×10^{12}Hz and has a *radiant intensity* in that direction of 1/683 watt per steradian. Symbol cd.

candlepower The *luminous intensity* of a light source in a given direction expressed in terms of the *candela*. Formerly expressed in terms of the *international candle*.

candle wax The wax used to make candles, it is usually either *paraffin wax* or *stearine*.

cane sugar Sucrose, saccharose. $C_{12}H_{22}O_{11}$. A *disaccharide* obtained from the sugar-cane. It is chemically identical with *beet sugar*.

Cannizzaro reaction A reaction in which two *molecules* of an *aldehyde* yield a *carboxylic acid* and an *alcohol*, i.e.

$$2RCHO + H_2O \rightarrow RCOOH + RCH_2OH$$

It occurs in the presence of strong *bases* with aldehydes that do not have hydrogen atoms on the carbon atom next to the *carbonyl* group. If there is a hydrogen atom in this position, an *aldol* is formed. Named after S. Cannizzaro (1826–1910).

canonical form See *resonance* hybrid.

Canton's phosphorus Impure *calcium sulphide*, CaS, having the property of *phosphorescence* after exposure to *light*. It is used in luminous *paints*.

caoutchouc Raw *rubber*.

capacitance Electrical capacity. *C*. The property of a system of electrical *conductors* and *insulators* that enables it to store *electric charge* when a *potential difference* exists between the conductors. It is measured by the charge that must be communicated to such a system to raise its potential by one unit; i.e. $C = Q/V$, where Q is the charge stored on one conductor and V is the potential difference between the conductors. The *SI unit* of capacitance is the *farad*.

capacitor Electrical condenser. A system of electrical *conductors* and *insulators*, the principal characteristic of which is its *capacitance*. The simplest form consists of two parallel *metal* plates separated by a layer of air or some other insulating material, such as *mica* (see *dielectric*). The capacitance, *C*, of such a parallel plate capacitor is given by:

$$C = A\varepsilon/d$$

where ε is the *permittivity*, in *farad* per *metre*, *A* the area of plate, and *d* the distance between them. See also *electrolytic capacitor*.

capacitor microphone Condenser microphone. A *microphone* consisting essentially of an electrical *capacitor*, one plate of which is fixed and the other plate forms the diaphragm upon which the *sound* waves fall. The vibrations of the diaphragm vary the *capacitance* of the capacitor, which in turn alters the *potential* across a high *resistance*. This varying potential is then amplified in the normal way.

capillary action Capillarity. A general term for phenomena observed in *liquids* due to inter-molecular attraction at the liquid boundary; e.g. the rise or depression of liquids in narrow tubes, the formation of films, drops, bubbles, etc. See also *surface tension*.

capillary tube A tube of small internal diameter.

capric acid See *decanoic acid*.

caproic acid See *hexanoic acid*.

caprylic acid See *octanoic acid*.

capture A process by which an atomic or nuclear system acquires an additional particle, e.g. the capture of *electrons* by *ions* or of *neutrons* by *nuclei*. 'Radiative capture' is a nuclear capture process that results in the emission of *gamma rays* only. See also *K-capture*.

caramel (chem.) A brown substance of complex composition, formed by heating *sugar*.

carat 1. A measure of weight of *diamonds* and other gems; formerly 3.17 grains (0.2053 g), now standardized as the international carat, 0.200 g. **2.** A measure of *fineness* of gold, expressed as parts of gold in 24 parts of the *alloy*. Thus, 24 carat gold is pure gold, 18 carat gold contains 18 parts in 24 or has a fineness of 750.

carbamide See *urea*.

carbamoyl The *univalent radical* NH_2CO-.

carbanion A negative *ion* containing a carbon atom. It has the structure R_3C^-, where R is an organic group. They occur in some organic reactions as *intermediates*. Compare *carbocation*.

carbene A transient organic group of the type R_2C:, in which the carbon atom has two *electrons* that do not form bonds. Methylene, H_2C:, is the simplest example.

carbide 1. True carbides contain the C^{4-} *ion*, e.g. Al_4C_3, and yield *methane* on *hydrolysis*. **2.** Dicarbides, such as *calcium dicarbide*, contain the ion C_2^{2-}. They yield *ethyne* on hydrolysis. They were formerly called acetylides. **3.** *Interstitial* carbides are formed by transition metals, with the carbon atoms occupying interstices in the metal *lattice*. These are usually hard materials with metallic conductivity.

carbocation A positive *ion* containing a carbon atom. It has the structure R_3C^+, where R is an organic group. They occur in some organic reactions as *intermediates*. The former name was 'carbonium ion'. Compare *carbanion*.

carbocyclic compounds A class of organic compounds containing closed rings of carbon atoms in their molecules. It includes *alicyclic* (e.g. *cycloalkanes*) and *aromatic* (e.g. *benzene*) compounds.

carbohydrases *Enzymes* that hydrolyse (see *hydrolysis*) *carbohydrates*; e.g. *amylase*, *lactase*, and *maltase*.

carbohydrates A large group of *organic compounds* composed of carbon, hydrogen, and oxygen only, with the general formula $C_x(H_2O)_y$. They include *monosaccharides*, *disaccharides* (both *sugars*), and *polysaccharides* (*starch*, *glycogen*, and *cellulose*). Carbohydrates play an essential part in the *metabolism* of all living *organisms*, *glycogen* (animal starch) being the principal form in which *energy* is stored in animals and cellulose being the principal structural material of plants.

carbolic acid See *phenol*.

carbon C. Element. R.a.m. 12.011, At. No. 6, m.p. 3550°C., b.p. 4289°C. It occurs in several allotropic forms (see *allotropy*) including *diamond* (r.d. 3.51) and *graphite* (r.d. 2.25); and as *amorphous* carbon (r.d. 1.8–2.1) in the forms of *lamp-black*, *gas carbon*, etc. *Compounds* occur as the metallic *carbonates*, *carbon dioxide* in the air, and an enormous number of *organic compounds*. Owing to its *valence* of four, carbon *atoms* are able to unite with each other to form the very large *molecules* upon which life is based. See *carbon cycle* (bio.). Animals obtain their energy by the *oxidation* of carbon compounds eaten as food. See also *radiocarbon dating*.

carbonado A black, discoloured, or impure variety of *diamond*, useless as a gem but very hard and used for drills, etc.

carbonate A *salt* of *carbonic acid*, H_2CO_3.

carbonation Treatment with *carbon dioxide*, usually for the formation of *carbonates*.

carbon black A finely divided soot-like form of *carbon*, produced by *pyrolysis* or by incomplete *combustion* from carbon-rich materials, such as *mineral oils*, *ethyne*, or *natural gas*. It is used mainly as a reinforcing *pigment* in rubber, and also as a black pigment in inks, plastics, etc.

carbon cycle (bio.) The circulation of carbon *atoms* between living *organisms* and the *atmosphere*. Carbon dioxide is built into complex carbon *compounds* by plants during *photosynthesis*; animals obtain their carbon atoms by feeding on plants or other animals; during *respiration*, and by decay after death, some of this carbon is returned to the atmosphere in the form of carbon dioxide.

carbon cycle (phys.) A cycle of six consecutive *nuclear reactions* resulting in the formation of a *helium* nucleus from four *protons*. The carbon nuclei with which the cycle starts are reformed at the end and therefore act as a *catalyst*. The *energy* liberated by the carbon cycle is thought to be the main source of energy in a large class of *stars*.

carbon dating See *radiocarbon dating*.

carbon dioxide Carbonic acid gas. CO_2. A colourless *gas* with a faint tingling smell and taste. It occurs in the *atmosphere* as a result of the *oxidation* of carbon and carbon *compounds*. Atmospheric carbon dioxide is the source of carbon for plants (see *photosynthesis* and *carbon cycle* (bio.). It forms a solid at $-78.5°C$. at atmospheric pressure, and is used as a *refrigerant* in this form as *dry ice*, for the reservation of frozen foods, etc. As carbon dioxide gas is heavier than air and does not support *combustion*, it is used in *fire extinguishers*. See also *greenhouse effect*.

carbon disulphide Carbon bisulphide. CS_2. A colourless flammable liquid, b.p. $46°C$., with a high *refractive index*. It is made by heating sulphur with carbon or with *methane* at high temperatures. It is used as a *solvent* in various industrial processes, in manufacture of viscose *rayon*, and as a *pesticide*. Due to its high flammability and toxicity its use is declining.

carbon fibre A material consisting of black silky threads of pure carbon that can be made stronger and stiffer than any other material of the same *weight*. Typical fibres are about 7 μm in diameter and have a *tensile strength* of up to 220 000 kg per square cm. They are made by heat-treating organic textile fibres in such a way that the side chains are stripped off, leaving only the carbon backbone. This backbone is subjected to further mechanical and heat treatment so that the crystallites are pulled into orientation along the axis of the fibre. They are used to reinforce a matrix of *resin*, *ceramic*, or *metal* with up to 600 000 fibres per square centimetre of cross-section and in this form make a valuable constructional material where strength is required at high *temperatures*, such as in components for jet engines and *rockets*.

carbonic acid H_2CO_3. A very weak *acid* probably formed in small amounts when *carbon dioxide* dissolves in *water*. It is never obtained pure as it breaks up almost completely into carbon dioxide and water when obtained in a chemical reaction. It gives rise to two series of *salts*, the *carbonates* and *hydrogencarbonates* (bicarbonates).

carbonium ion See *carbocation*.

carbonization See *destructive distillation*.

carbon monoxide CO. A colourless, almost odourless *gas* that is very poisonous when breathed, as it combines with the *haemoglobin* of the *blood* to form bright red carboxyhaemoglobin. This is chemically stable, and thus the haemoglobin is no longer available to carry oxygen. It burns with a bright blue *flame* to form *carbon dioxide*. It is formed during the incomplete *combustion* of *coke*, *charcoal*, and other carbonaceous *fuels*; it occurs in in the exhaust fumes of motor engines. It also occurs in cigarette smoke, which is a source of inhaled carbon monoxide. It is used in the *Mond process* for nickel and in organic synthesis.

carbon tetrachloride See *tetrachloromethane*.

carbonyl The *divalent* group =CO, characteristic of *aldehydes*, *ketones*, and *carboxylic acids*. Inorganic carbonyls are *coordination compounds* of *metals* and *carbon monoxide*, e.g. *nickel carbonyl*.

carbonyl chloride See *phosgene*.

carborundum See *silicon carbide*.

carboxyl group The *univalent* group, –COOH, characteristic of the organic *carboxylic acids*.

carboxylic acids *Organic acids* containing one or more *carboxyl groups* in the molecule; e.g. *ethanoic acid*, CH_3COOH. They are *weak acids* that form *salts* with *bases* and *esters* with *alcohols*. In systematic nomenclature they end in -oic. See also *dicarboxylic acid*; *fatty acid*; *tricarboxylic acid*.

carburettor A device in the *internal-combustion* petrol engine for mixing air with *petrol vapour* prior to explosion.

carbylamine See *isonitrile*.

carcinogen A substance or radiation capable of producing cancer (carcinoma).

cardiovascular Denoting the heart and system of blood vessels of a mammal that supply blood to the body through the arteries and the return to the heart through the veins.

carnallite Natural potassium magnesium chloride, $KCl.MgCl_2.6H_2O$, found in the *Stassfurt deposits*. An important source of potassium *salts*.

carnosine $C_9H_{14}N_4O_2$. An *optically active* crystalline *dipeptide*, m.p. 260°C., found in muscle *tissue*.

carnotite An *ore* of uranium consisting of uranium potassium *vanadate* of variable composition.

Carnot's cycle An ideal reversible cycle of operations for the working substance of a *heat engine*. The four steps in the cycle are: (a) *isothermal* expansion, the substance taking in *heat* and doing *work*; (b) *adiabatic* expansion, without heat change, external work done; (c) isothermal compression, heat given out, work done on the substance by external forces; (d) adiabatic compression, no heat change, work done on the substance. See also *Carnot principle*. Named after N. L. S. Carnot (1796–1832).

Carnot's principle The *efficiency* of any reversible *heat engine* depends only on the *temperature* range through which it works and not upon the properties of any material substance. If all the heat is taken up at a *thermodynamic temperature* T_1 and all given out at a thermodynamic temperature T_2 (as in *Carnot's cycle*), the efficiency is $(T_1 – T_2)/T_1$.

Caro's acid See *sulphuric acids*.

carotene $C_{40}H_{56}$. A yellow *unsaturated hydrocarbon* present in carrots and butter. It is converted into vitamin A (see *vitamins*) in the animal *organism*. Carotene acts as a photosynthetic pigment (see *photosynthesis*) in certain plant *cells*.

carrier (bio.) **1.** Membrane carrier. A *molecule* or group of molecules, usually *protein*, that transports a different kind of molecule from one side of a cell *membrane* to the other. **2.** An individual whose genetic make-up includes a recessive *gene* for a defective condition, which is then masked by a normal dominant gene. Such an individual does not suffer from the defect but may pass it on to its progeny. **3.** An individual whose body contains a pathogenic organism; although the individual may suffer no ill-effects from the organism it can transmit it to other individuals who may.

carrier (chem.) **1.** A substance assisting a *chemical reaction* by combining with part or all of the *molecule* of one of the reacting substances to form a *compound* that is then easily decomposed again by the other reacting substance; the carrier is thus left unchanged. See *catalyst*. **2.** An inactive substance used to transport a *radioisotope* in *radioactive tracing*. A radioisotope is said to be 'carrier-free' if it can be used without a carrier. **3.** Sometimes called carrier gas. The gas used to carry the sample through the column in gas *chromatography*.

carrier (phys.) Charge carrier. The particles that carry the *charge* when an electric current flows. In a *metal* they are the *free electrons*, in a *semiconductor* they are electrons or *holes*. In an *electrolyte* they are *ions*, and in a gas they are ions and electrons.

carrier wave A continuous *electromagnetic radiation*, of constant *amplitude* and *frequency*, emitted by a *radio* transmitter. By *modulation* of the carrier wave, oscillating *electric currents* caused by *sounds* at the transmitting end are conveyed by it to the receiver.

carron oil A mixture of *vegetable oil* (olive or cotton-seed) with *lime-water*. Used as an application for burns.

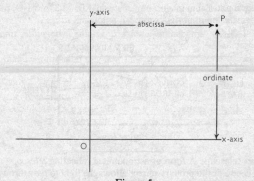

Figure 5.

Cartesian coordinates A system for locating a point, P, in a *plane* by specifying its distance from two *axes* at right angles to each other, which intersect at a point O, called the *origin*. The distance from the horizontal or *x*-axis is called the *ordinate* of P; the distance from the *y*-axis is called the *abscissa*. See Fig. 5. The system may also be used to locate a point in space by using a third, *z*-axis. Named after R. Descartes (1596–1650). Compare *polar coordinates*.

cartography The scientific study of maps and *map projections*.

carvacrol $(CH_3)_2CH.C_6H_3CH_3OH$. A colourless oily *liquid*, b.p. 237.7°C., with a mint-like odour. Used as a *disinfectant*, and in perfume.

carvone Carvol. $C_{10}H_{14}O$. An optically active liquid *ketone* related to the *terpenes*, b.p. 231°C., found in *essential oils* and used in flavours and perfumes.

cascade liquefier An apparatus used for liquefying air, oxygen, etc. A *gas* cannot be liquefied until it is brought to a *temperature* below its *critical temperature*. In the cascade liquefier the critical temperature of the gas is reached step by step, using a series of gases having successively lower *boiling points*. The first of these, which can be liquefied by compression at ordinary temperatures, is allowed to evaporate under reduced *pressure*; this produces a temperature below the critical temperature of the second gas, which can then be liquefied. This is similarly allowed to evaporate, and the step is repeated until finally the desired liquefaction is reached.

cascade process A process used in the separation of *isotopes*. It consists of a series of stages connected so that the separation produced by one stage is multiplied in subsequent stages. In a 'simple cascade' the enriched fraction is fed to the succeeding stage and the depleted fraction to the preceeding stage.

cascade shower See *shower*.

casein The main *protein* of milk. A pale yellow *solid* obtained from milk by the addition of *acid* ('acid casein'), by controlled souring ('self-soured casein'), or by curdling with *rennet* ('rennet casein'). It is used in paper-coating, *paints*, *adhesives*, *plastics*, and for making artificial textile fibres.

caseinogen British term for *casein* before precipitation. The American terms are casein before precipitation, and paracasein after.

CASSEGRAINIAN TELESCOPE

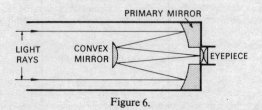

Figure 6.

Cassegrainian telescope A form of astronomical reflecting *telescope* in which a hole in the centre of the primary mirror allows the light to pass through it to the *eyepiece* or the photographic plate. See Fig. 6.

cassiopeium See *lutetium*.

cassiterite SnO_2. Natural *tin(IV) oxide*. It is the principal *ore* of tin.

cast iron Pig-iron. An impure, brittle form of iron, as produced in the *blast furnace*. It contains from 2%–4.5% carbon in the form of *cementite* and usually also some manganese, phosphorus, silicon, and sulphur. It may be cast into specific objects or converted into *steel* or *wrought iron*.

castor oil A *vegetable oil* extracted from the seed of the castor plant, consisting of *glyceryl esters* of *fatty acids*; the predominant acid (about 85%) being *ricinoleic acid*, $C_{17}H_{32}(OH).COOH$. It is used in the *paint* and varnish industry as well as medically as a laxative.

catabolism Katabolism. The part of *metabolism* dealing with the chemical *decomposition* of complex substances into simple ones, often with a release of *energy*.

catalase An *enzyme* that decomposes *hydrogen peroxide*.

catalysis The alteration of the rate at which a *chemical reaction* proceeds, by the introduction of a substance (*catalyst*) that remains unchanged at the end of the reaction. Small quantities of the catalyst are usually sufficient to bring the action about or to increase its rate substantially, usually by providing a different *pathway* for the reaction in which the *activation energy* is lower.

catalyst A substance that alters the rate at which a *chemical reaction* occurs, but is itself unchanged at the end of the reaction. Catalysts are widely used in the chemical industry; *metals* in a finely divided state, and *oxides* of metals, are frequently used. The *enzymes* are organic catalysts produced by living *cells*.

catalytic cracking The use of a *catalyst* to bring about the *cracking* of high boiling mineral *oils*.

cataphoresis See *electrophoresis*.

catechol See *1,2-dihydroxybenzene*.

catecholamines A group of *amines* derived from catechol (see *1,2-dihydroxybenzene*) that includes the *hormones adrenaline* and *noradrenaline* and the *neurotransmitter dopamine*.

catenary A curve formed by a chain or string hanging from two fixed points. Its equation is $y = k \cosh x/k$, where k is the distance between the *vertex* of the curve and the *origin*.

catenation The process of chain formation in molecules (particularly carbon chains in organic molecules).

catenoid The surface generated by rotating a *catenary* about its vertical axis.

cathetometer A *telescope* mounted on a graduated vertical pillar along which it can move. The instrument is used for measuring lengths and displacements at a distance of a few feet.

cathode Negative *electrode*. The negatively charged *conductor* in *electrolysis* and in *thermionic valves*. See *discharge in gases*. In a *primary cell* or *accumulator* the cathode is the pole from which *electrons* emerge during discharge.

cathode-ray oscilloscope CRO. An instrument based upon a *cathode-ray tube*, which provides a visible image of one or more rapidly varying electrical quantities. If an internal *time base* is fed to the horizontal plates and the signal is fed to the vertical plates, the image appearing on the screen of the CRO will be a graph of the signal against time.

cathode rays A stream of *electrons* emitted from the negatively charged *electrode* or *cathode* when an electric discharge takes place in a vacuum tube, i.e. a tube containing a gas at very low pressure. See *discharge in gases*.

cathode-ray tube CRT. A vacuum tube that allows the direct observation of the behaviour of *cathode rays*. It consists essentially of an *electron gun* producing a beam of *electrons* that, after passing between horizontal and vertical deflection plates or coils, falls upon a luminescent screen: the position of the beam can be observed by the *luminescence* produced upon the screen. *Electric potentials* applied to the deflection plates are used to control the position of the beam, and its movement across the screen, in any desired manner. It is used as the picture tube in *television* receivers, in *cathode-ray oscilloscopes*, and *radar* viewers.

catholyte The *electrolyte* near the *cathode* during *electrolysis*.

cation A positively charged *ion*; an ion that, during *electrolysis*, is attracted towards the negatively charged *cathode*. Compare *anion*.

CAT scanner See *tomography*.

causality The relating of causes to the effects that they produce. Many contemporary physicists believe that no coherent causal description can be given of events that occur on the sub-atomic scale.

caustic Corrosive towards organic matter (but not applied to *acids*). E.g. *caustic soda*.

caustic (phys.) Parallel rays of *light* falling on a *concave* spherical mirror do not form a point image at the *focus* (see *mirrors, spherical*). Instead, there is a region of maximum concentration of the rays forming a curve or surface of revolution, called a caustic, the apex or cusp of which is at the focus of the mirror. A similar caustic occurs in the image formed by a *convex lens* receiving parallel light. Such a curve may be seen on the surface of a liquid in a cup, formed by the reflection of light upon the curved wall of the cup.

caustic alkali *Sodium* or *potassium hydroxide*.

caustic potash See *potassium hydroxide*.

caustic soda See *sodium hydroxide*.

cavitation The formation of cavities in *fluids* when the pressure drops as a result of high *velocity*, in accordance with *Bernouilli's theorem*. These vapour-filled cavities collapse when they are carried to regions of higher pressure and the resulting impact pressure can cause pitting of such parts as propellers.

celestial equator (astr.) The circle in which the *plane* of the Earth's *equator* meets the *celestial sphere*.

celestial mechanics The branch of *astronomy* concerned with the motions of celestial bodies or systems under the influence of *gravitational fields*.

celestial sphere (astr.) The imaginary *sphere* to the inner surface of which the heavenly bodies appear to be attached; the observer is situated at the centre of the sphere, enabling it to be used to locate a celestial body with reference to the observer. See Fig. 2 under *azimuth*.

celestine Natural crystalline strontium sulphate, $SrSO_4$, mined as a source of strontium.

cell (bio.). The unit of *life*. All living *organisms* are composed of discrete, membrane-bounded units, which usually comprise two distinct forms of *protoplasm*: the *nucleus* and the *cytoplasm*. The former contains the *nucleic acids* responsible for organizing the synthesis of the cell's *enzymes* and for controlling the characteristics of its progeny, while the latter contains the enzyme systems that control the cell's *metabolism* and manufacture its constituents. Many *microorganisms* (e.g. *bacteria*, protozoa, etc.) consist of only one cell, whereas a man consists of some million million cells. Some organisms, such as bacteria and blue-green algae, do not have separate nuclei. See also *wall of a cell*.

cell (phys.) A voltaic cell is a device for producing an *electric current* by chemical action. See *accumulator, primary cell*. An electrolytic cell is one in which *electrolysis* takes place as a result of an *electric current* being passed through an *electrolyte*. See *half cell*.

celluloid A *thermoplastic* material made from *cellulose nitrate* and *camphor*. Because it is highly flammable it is no longer widely used.

cellulose A *polysaccharide* that occurs widely in nature in fibrous form as the structural material in the cell walls of plants. Its *macromolecules* consist of long unbranched chains of *glucose* units. It is obtained from wood pulp, cotton, and other plant sources; it is used in the manufacture of *paper, rayon, plastics*, and *explosives*.

cellulose ethanoate Cellulose acetate. An *ester* obtained by the action of *ethanoic anhydride* on *cellulose* (e.g. wood pulp). It is a white *solid*, used in the manufacture of *rayon* and *plastics*.

cellulose nitrate Nitrocellulose, gun-cotton. *Nitric acid ester* of *cellulose*. A range of *compounds* formed by treatment of cellulose with a mixture of nitric and *sulphuric acids;* its properties depend on the extent to which the *hydroxyl groups* of the cellulose are esterified (see *esterification*). It is used in the manufacture of *plastics*, lacquers, and *explosives*.

Celsius temperature *Temperature* measured on a scale originally devised by Anders Celsius (1701–44) in which the *melting point* of ice was 0° and the *boiling point* of water was 100°. This definition has been superseded by the *International Practical Temperature Scale* of 1968, which is expressed in both *kelvins* and degrees Celsius. The unit for both means of expressing temperature is the kelvin, and temperature differences may be expressed in kelvins even when using Celsius temperatures. The relation between the Kelvin temperature (T) and the Celsius temperature (t) is given by: $T = t + 273.15$. This temperature scale was formerly called the 'centigrade scale' but this name was dropped in 1948 to avoid confusion with the European unit of angle, the grade.

celtium See *hafnium*.

cement 1. Any bonding material. **2.** Portland cement and allied cements are made from materials containing *lime*, *alumina*, and *silica* (e.g. *limestone* and *clay*), which are heated strongly in a kiln to form clinker (consisting mainly of calcium *silicates* and *aluminates*). The finely ground clinker undergoes complex *hydration* processes when mixed with water, setting and hardening to a stone-like material.

cementation 1. An early process for *steel* manufacture. Bars of *wrought iron* were heated for several days in *charcoal* at red heat. **2.** Contact precipitation of a *metal* from a solution of a *compound* of that metal by a more *electropositive* metal.

cementite Iron carbide. Fe_3C. A hard, brittle *compound* that is responsible for the brittleness of *cast iron* and is present in *steel*.

centi- Prefix denoting one hundredth; 10^{-2}. Symbol c, e.g. cm = 0.01 metre.

centigrade temperature See *Celsius temperature*.

central processing unit Central processor. See *CPU*.

centre of curvature of a lens or spherical mirror The centre of the sphere of which the surface of a *lens* or *mirror* forms a part. See also *radius of curvature*.

centre of gravity The fixed point in a body through which the *resultant force* of *gravity* always passes, irrespective of the position of the body. This is identical to the *centre of mass* in a uniform *gravitational field*.

centre of mass The point at which the *mass* of a body may be considered to be concentrated. The point from which the sum of the *moments of inertia* of all the component particles of a body is zero.

centrifugal force See *centripetal force*.

centrifuge An apparatus for separating particles from a *suspension*. Balanced tubes containing the suspension are attached to the opposite ends of arms rotating rapidly about a central point; the suspended particles are forced outwards, and collect at the bottoms of the tubes. See also *ultracentrifuge*.

centripetal force A *force* that causes a body to move in a circular path. For example, if a body is attached to a string and swung in a horizontal circle, there will be a continuous change in the body's *velocity*, even though its *speed* may remain unchanged. This change in velocity results from the change in the body's direction; it will create a centripetal *acceleration* (an acceleration towards the centre of the circle) equal to v^2/r, where v is the body's velocity and r the length of the string. The magnitude of the centripetal force, i.e. the tension in the string, is then mv^2/r, where m is the *mass* of the body.

In the case of a satellite orbiting the *Earth*, the centripetal force is the *gravitational force* between the bodies and therefore:

$$GmM/r^2 = mv^2/r,$$

where G is the *gravitational constant* and M is the mass of the Earth. Until recently it was conventional to assume that the centripetal force was always balanced by an equal and opposite force called the centrifugal force. In this convention the centrifugal force was said to balance the gravitational force when the body is in stable orbit. However, this can cause confusion as the centrifugal force is a fictitious force, although in some cases the concept can be useful.

centroid The *centre of mass* of a uniform surface or a body with uniform density. If the surface or body is not uniform the centroid is not coincident with the centre of mass and its position would have to be determined by experiment or by integration.

ceramic Pertaining to products or industries involving the use of *clay* or other *silicates*.

cerargyrite See *horn silver*.

Cerenkov (Cherenkov) radiation *Light* emitted when charged particles pass through a transparent medium at a *speed* greater than the *speed* of light in that medium. The effect is utilized in *particle detectors* (Cerenkov detectors), which can reveal different types of particles as a result of their different speeds. The particles pass through a liquid and the Cerenkov radiation is registered by a *photomultiplier*. Named after P. A. Cerenkov (1904–90).

ceresin Hard, brittle *paraffin wax* with a *melting point* in the range of 70°–100°C. It is used as a substitute for *beeswax* in *paints* and polishes.

ceric Containing the cerium(IV) *ion*.

cerium Ce. Element. R.a.m. 140.12. At. No. 58. A steel-grey soft *metal*; r.d. 6.7, m.p. 798°C., b.p. 3433°C. It occurs in several rare minerals, e.g. *monazite* sand, and is used in *pyrophoric alloys* (e.g. *misch metal*) for lighter 'flints'; *compounds* are used in the manufacture of certain glasses.

cerium dioxide Ceria. CeO_2. A white crystalline powder, m.p. 2600°C., used in *glass* polishing.

cermet Ceramet. Abbreviation of CER(A)mic and METal. A very hard mixture of a *ceramic* substance and *sintered* metal, used where resistance to high *temperature*, *corrosion*, and abrasion is required.

CERN The European Laboratory for Particle Physics, which was formerly known as the Organisation (but previously Conseil) européenne pour la Recherche nucléaire. It has several large accelerators at its Geneva headquarters, including the Large Electron–Positron Collider (LEP) in which 50GeV beams are made to collide.

cerous Containing the cerium(III) *ion*.

cerussite Natural lead carbonate, $PbCO_3$, often formed by the weathering of *galena*. It is used as an ore of lead.

cetane $C_{16}H_{34}$. See *hexadecane*.

cetane number A measure of the ignition characteristics of a diesel fuel by comparison in a standard diesel engine with a range of mixtures, in which *cetane* is given a value of 100 and α-methylnaphthalene is 0.

cetyl alcohol See *hexadecanol*.

CFC See *chlorofluorocarbon*.

c.g.s. system Centimetre-gram-second system. A system of physical *units* derived from the centimetre, *gram* mass and the *second*. E.g. speed in c.g.s. units is measured in centimetres per second. It has been replaced for scientific purposes by the *SI units*.

chabasite A natural *zeolite*, calcium aluminium silicate. See *ion exchange*.

chain reaction Any self-sustaining molecular or *nuclear reaction*, the products of which contribute to the propagation of the reaction. In a *nuclear fission* chain reaction one nuclear transformation is capable of initiating a chain of similar transformations. For example, when nuclear fission occurs in a uranium-235 *nucleus*, between 2 and 3 *neutrons* are emitted, each of which is capable of causing the fission of further uranium-235 nuclei. The chain reaction so created is the basis of *nuclear weapons* and the *nuclear reactor*. If the average number of *transformations* directly caused by one transformation is less than one, the reaction is said to be convergent or *subcritical*; if it is equal to one, the reaction is self-sustained or critical; if it exceeds one, the reaction is divergent or *supercritical*.

In chemical reaction, chain reactions usually involve such intermediates as free radicals. *Combustion* is an example of a series of chemical chain reactions.

chalcedony A variety of natural impure *silica*, SiO_2, that has a fibrous structure and a waxy lustre. It is used for ornaments.

chalcocite Copper glance. Natural copper sulphide, Cu_2S. It occurs in veins with other copper ores.

chalcogens The elements of group VIA of the *periodic table*: oxygen, sulphur, selenium, tellurium, and polonium.

chalcopyrite Copper pyrites. A natural sulphide of copper and iron, $(Cu,Fe)S_2$; the most abundant ore of copper.

chalk Natural *calcium carbonate*, $CaCO_3$, formed from the shells of minute marine *organisms*. Blackboard chalk sticks are *calcium sulphate*, $CaSO_4$.

chalybeate Chalybite. Natural iron(II) carbonate, $FeCO_3$.

Chandrasekhar limit The maximum mass that a non-rotating *white dwarf* can have without further collapsing into a *neutron star* or a *black hole*. The value is about 1.4 solar masses, which is raised considerably if the star has a rotating core.

change of phase Change of state. The conversion of a substance from one of the *physical states* of *matter* (*solid*, *liquid*, or *gas*) into another. E.g. the melting of *ice*. See *latent heat*.

channel 1. In *telecommunications*, a path for the transmission of electrical signals, often specified by its *frequency band*. **2.** In *information theory*, a path or route along which information may flow or be stored. **3.** In a field-effect *transistor*, the region between the *source* and the *drain*; its *conductivity* is controlled by the *voltage* applied to the *gate*.

channel capacity The number of signals per second that can be transmitted through a *channel*. Also, in *information theory*, the hypothetical limiting rate at which information could be communicated by a given channel, with the frequency of errors tending to zero.

chaos theory The theory that deals with unpredictable random behaviour in a system that should obey deterministic laws. This usually occurs if the system is unusually sensitive to variations in the initial conditions or if its behaviour is determined by a large number of independent variables. Such conditions can occur in *meteorology*, in which the laws controlling the weather are more accessible than the exact *parameters* to insert in them (see *butterfly effect*). Other fields in which chaos theory is used include *turbulent* fluid flow, reaction *kinetics*, and many situations in *astronomy* and economics.

character A unit of information as handled by *computers*, usually six *bits*.

characteristic (math.) The *integral* or whole-number part of a *logarithm*.

charcoal Various forms of generally impure *carbon*; it is generally made by heating vegetable or animal substances with exclusion of air. Many forms are very porous and adsorb various materials readily. See *activated carbon*.

charge See *electric charge*.

charge carrier See *carrier*.

charge conjugation The operation that changes the *wave function* describing an *elementary particle* into that for an *antiparticle*, and vice versa. *Strong interactions* and *electromagnetic interactions* are symmetric with respect to this operation; that is, the overall wave function does not change sign. In *weak interactions*, however, the wave function does change sign under charge conjugation. It is restored in most cases by combining the operations of charge conjugation (C) and *parity* (P) in the combined CP operation.

charge density The *electric charge* on unit surface area of a body (surface charge density) or on unit volume of a body or medium.

Charles' law At constant *pressure* an *ideal gas* expands by 1/273 of its *volume* at $0°C$., for each $1°C$. rise in *temperature*; the volume of a fixed mass of gas at constant pressure is proportional to the *thermodynamic temperature*. Named after J. A. C. Charles (1746 – 1823). See also *gas laws*.

charm A property of matter postulated to account for the characteristics of the *psi particle* (discovered in 1974). According to this hypothesis a *quark* (and its antiquark) exists having the property called charm. The psi particle itself is not charmed as it consists only of a charmed quark and its antiquark, which give zero charm. However, other charmed *hadrons* exist (see *elementary particles*). Charm is conserved in *strong interactions* and *electromagnetic interactions* but not in *weak interactions*.

cheddite Class of *explosives* containing *sodium* or *potassium chlorate* with dinitrotoluene and other organic substances.

chelation The formation of a closed ring of *atoms* by the attachment of *compounds* or *radicals* to a central *polyvalent metal ion* (occasionally non-metallic); it is usually due to the sharing of a *lone pair of electrons*, from oxygen or nitrogen atoms in the compounds or radicals, with the central ion, e.g. two *molecules* of ethane-1,2-diamine ($NH_2CH_2CH_2NH_2$) form a 'chelate ring' with a copper(II) ion as shown in the diagram.

Chelating agents are used for 'locking up' (sequestering) unwanted metal ions; for instance they are added to shampoos with the object of softening the water by locking up iron, calcium, and magnesium ions. When used for this purpose they are called sequestering agents. Many tests for identifying metal ions depend on the formation of coloured *insoluble* chelates. *Chlorophyll* and *haemoglobin* are naturally occurring chelate compounds in which the central ions are magnesium and iron respectively.

$$
\begin{array}{ccc}
CH_2-NH_2 & & NH_2-CH_2 \\
| & \searrow \; ^{++} \; \swarrow & | \\
& Cu & \\
| & \nearrow \quad \nwarrow & | \\
CH_2-NH_2 & & NH_2-CH_2
\end{array}
$$

chemical affinity See *affinity* and *free energy*.

chemical bond The force holding the *atoms* of a *molecule* or a crystal in place. In ionic (electrovalent) bonds the transfer of an electron between the bonding atoms results in an *electrostatic* attraction between them. This attraction is the chemical bond. For example, a sodium atom has one electron in its outer shell, if it donates this electron to a chlorine atom, with seven outer electrons, both species will have the stable eight-electron configuration of a *noble gas* outer shell. However, the sodium atom will have become a positive *ion*, Na^+, while the chlorine atom will have become the negative ion, Cl^-. The electrostatic force between these ions will hold them together in a molecule of *sodium chloride*, or a crystal lattice of sodium chloride if there are many of them.

In covalent bonds, electrons are shared rather than transferred. For example, in the *water* molecule, H_2O, the oxygen atom, with six outer electrons, shares the single electrons from each of the two hydrogen atoms. Thus, the oxygen atom ends up with the stable noble gas configuration of neon, while the hydrogen atoms, which each share their own electron with an electron from the oxygen atom, have the stable helium noble gas electronic structure. See also *orbital*.

If one of the atoms in a covalent bond supplies both the electrons, this is called a coordinate, dative, or *semipolar bond*. See *ligand*.

In practice, many bonds have both ionic and covalent components. For example, in hydrogen chloride the bond is primarily covalent, however because chlorine is more electronegative than hydrogen, the molecule is polarized with a negative charge on the chlorine, which has the effect of an ionic bond. See also *hydrogen bond*; *metallic crystals*.

chemical change A change in a substance involving an alteration in its chemical composition, due to an increase, decrease, or rearrangement of *atoms* within its *molecules*. See *equation, chemical*; *molecule*.

chemical combination, laws of Three laws defining the ways in which chemical *compounds* are formed:

Law of constant composition. A definite chemical compound always contains the same *elements* chemically combined in the same proportions by *mass*.

Law of multiple proportions. When two elements unite in more than one proportion, for a fixed *mass* of one element there is always a simple relationship with the mass of the other element present.

Law of combining masses (also called the law of reciprocal proportions, law of equivalents). Elements combine in the ratio of their combining weights or *chemical equivalents*; or in some simple multiple or sub-multiple of that ratio.

chemical dating Any technique used to determine the age of a mineral, fossil, etc., by a chemical means. For example, fluoride *ions* in ground water slowly replace the phosphate in buried bones, enabling a measurement of the proportion of fluorine present to give an estimate of the time that the bone has been buried. Compare *radiometric dating*. See *dating*.

chemical energy That part of the *energy* stored within an *atom* or *molecule* that can be released by a *chemical reaction*.

chemical engineering The design, operation, and manufacture of plant or machinery used in industrial chemical processes.

chemical equilibrium Many *chemical reactions* do not go to completion; in such cases a state of equilibrium or balance is reached when the original substances are reacting at the same rate as the new substances are reacting with each other to form the original substances. Thus, if two substances *A* and *B* react to form *C* and *D*, the state at equilibrium is denoted by the balanced equation

$$A + B \rightleftharpoons C + D.$$

If one of the substances is removed, the system readjusts the equilibrium; thus, if *C* is constantly removed as soon as formed, more *A* and *B* react until the action is completed. An equilibrium reaction that could thus be made to complete itself in either direction is termed a *reversible reaction*. E.g. if *steam* is passed over red-hot iron, iron oxide and hydrogen are formed, the latter being constantly removed by more steam which passes through; the reaction thus goes to completion according to the equation

$$4H_2O + 3Fe = Fe_3O_4 + 4H_2.$$

If, however, hydrogen is passed over red-hot iron oxide, the reverse action takes place:

$$Fe_3O_4 + 4H_2 = 4H_2O + 3Fe.$$

If the reaction is allowed to proceed in an enclosed space, a state of equilibrium is reached, all four substances being present. See also *equilibrium constant*.

chemical equivalents Combining weights. The combining proportions of substances by *mass*, relative to hydrogen as a standard. The equivalent of an *element* is the number of grams of that element which will combine with or replace 1 g of hydrogen or 8 g of oxygen. The gram-equivalent, or equivalent weight, is the equivalent expressed in grams. The equivalent weight or gram-equivalent as a unit quantity of substance in chemical calculations has been replaced in *SI units* by the *mole*.

chemical potential μ. A thermodynamic concept equal to the change in Gibbs *free energy* (G) of a component of a mixture for a particular change in the *amount of substance* (n) of that component, when the pressure, temperature, and amounts of substance of the other components remain constant, i.e. $\mu = \partial G/\partial n$. See also *fugacity*.

chemical reaction The interaction of two or more substances, resulting in *chemical changes* in them.

chemical shift A change in the *wavelength* at which *electromagnetic radiation* is absorbed or omitted in a process as a result of a change in the energy levels of the *electrons* in an *atom's* inner shell or of its *nucleus*.

chemiluminescence Cold *flame*. The evolution of *light* accompanied by some *heat* during a *chemical reaction*. See *luminescence*.

chemisorption See *adsorption*.

chemistry The study of the composition of substances, and of their effects upon one another. The main branches are *inorganic chemistry*, *organic chemistry*, and *physical chemistry*. See also *biochemistry*; *geochemistry*.

chemotherapy The treatment of disease by chemical substances that are toxic to the causative *microorganisms* or directly attack *neoplastic* growths.

chemurgy The study of chemical industrial processes based on organic substances of agricultural origin.

chert A natural form of silica, SiO_2, resembling flint.

Chile saltpetre Impure *sodium nitrate*, $NaNO_3$. It occurs in huge deposits in Chile and is used as a source of nitrogen in agriculture.

china clay Kaolin. A pure natural form of *hydrated* aluminium silicate, $Al_2Si_2O_5(OH)_4$. On heating, it loses *water* and changes its chemical composition. It is used for making *porcelain*.

Chinese white *Zinc oxide*, ZnO.

chip Silicon chip. A crystal of a silicon *semiconductor* fabricated to carry out a number of electronic functions in an *integrated circuit*. Single crystals are grown as rods up to 15 cm in diameter, which are then sliced into thin wafers. Each wafer is divided into many chips, each of which is typically a few millimetres square.

chirality The concept of 'handedness' (right- or left-handedness) applied to *stereoisomerism*. A geometrical figure representing the configuration of a *molecule* in space is said to have chirality if it cannot be made to coincide with its image in a plane mirror.

Chiron A minor planet, discovered in 1977 by Charles Kowal, that revolves around the Sun between the orbits of Saturn and Uranus.

chitin A *glycosaminoglycan* that forms microfibrils similar to *cellulose*. It forms an essential part of the exoskeletons of crustaceans and insects. It is also found in some fungi.

chlor(o)acetic acids See *chloroethanoic acids*.

chloracne A disfiguring skin disease that is caused by certain chlorinated aromatic hydrocarbons. It can result from contact, ingestion, or inhalation of the chemicals.

chloral See *trichloroethanal*.

chloral hydrate See *2,2,2-trichloroethanediol*.

chloranil $C_6Cl_4O_2$. A yellow *insoluble* crystalline substance, m.p. 290°C., used as a *fungicide* and in the manufacture of *dyes*.

chlorargyrite See *horn silver*.

chlorate A salt of a *chloric acid*. It usually refers to salts containing the ClO_3^-, chlorate(V), ion. Other chlorates contain the ions ClO^- (chlorate(I) or hypochlorite), ClO_2^- (chlorate(III) or *chlorite*), or ClO_4^- (chlorate(VII) or *perchlorate*).

chloric acids Any of four *oxoacids*. The most common is chloric(V) acid, $HClO_3$, an unstable liquid prepared by the action of *sulphuric acid* on barium chlorate. It is a strong *acid* and *oxidizing agent*. Chloric(I) acid, or hypochlorous acid, HOCl, is stable only in solution and is prepared by the reaction of chlorine on mercury(II) oxide. It is a weak acid but is used as a *bleaching* agent. Chloric(III) acid, or chlorous acid, $HClO_2$, is pale yellow and known only in solution. It is formed by mixing chlorine dioxide with water. It is a weak acid and oxidizing agent. Chloric(VII) acid, or perchloric acid, $HClO_4$, is an unstable liquid that explodes at 90°C. It is a strong acid and oxidizing agent.

chloride A *salt* of *hydrochloric acid*, HCl. See also *halide*.

chloride of lime Calcium chlorate(I), $CaOCl_2$. See *bleaching powder*.

chlorination 1. The introduction of a chlorine atom into a compound by *substitution* or by an *addition reaction*. **2.** The treatment of drinking water with chlorine or a chlorine compound, such as *sodium hypochlorite* or *bleaching powder*.

chlorine Cl. Element. R.a.m. 35.453. At. No. 17. A greenish-yellow poisonous *gas*, m.p. $-100.98°C$., b.p. $-34.6°C$., with a choking irritating smell. *Compounds* occur as common salt (*sodium chloride*), NaCl, in *sea-water* and as *rock salt*; and as *chlorides* of other *metals* (e.g. *carnallite*). Manufactured almost entirely by the *electrolysis* of sodium chloride by the *Down's process* or of brine. Used in the manufacture of *bleaching powder*, *disinfectants*, *hydrochloric acid* and many organic compounds. It is also used in the *chlorination* of drinking-water.

chlorite 1. A *salt* of *chloric(III)* or chlorous *acid*. **2.** A group of *mineral silicates* of aluminium, iron, and magnesium.

chlorobenzene Phenyl chloride. C_6H_5Cl. A colourless flammable *liquid*, b.p. $132°C$., used as a *solvent* and in the synthesis of *drugs*. It is made by the *Raschig process*.

2-chlorobuta-1,3-diene Chloroprene. $CH_2:CH.CCl:CH_2$. A colourless liquid, b.p. $59.4°C$., used in the manufacture of *neoprene* synthetic *rubber*.

chloroethane Ethyl chloride. C_2H_5Cl. A colourless poisonous gas, used as a refrigerant and alkylating agent and in the manufacture of *lead tetraethyl*.

chloroethanoic acids Chloroacetic acids. Three substituted *ethanoic acids*. Monochloroethanoic acid, $CH_2ClCOOH$, is a crystalline solid, m.p. $63°C$. Dichloroethanoic acid, $CHCl_2COOH$, is a colourless liquid, m.p. $10°C$., b.p. $192-3°C$. Trichloroethanoic acid, CCl_3COOH, is a *deliquescent* solid, m.p. $56.3°C$. They are all stronger *acids* than ethanoic acid and are used in the manufacture of *dyes* and as wart removers.

chloroethene Vinyl chloride. $CH_2:CHCl$. A colourless gas, b.p. $-13.4°C$., that polymerizes to form polyvinylchloride (PVC). It is a powerful *carcinogen*.

chlorofluorocarbon CFC. An alkane in which all the hydrogen atoms have been substituted by chlorine and fluorine atoms, e.g. dichlorodifluoromethane (CCl_2F_2), which is sold under the trade name Freon*-12. These substances are inert and stable at high temperatures. They were formerly widely used as *refrigerants* and *propellants* in *aerosol* cans but because they are known to diffuse into the *upper atmosphere* causing the *ozone layer* to break down, their use is now being discontinued. See also *fluorocarbons*.

chloroform See *trichloromethane*.

chlorohydrins *Organic compounds* containing a *chlorine* atom and a *hydroxyl* group attached to adjacent *carbon* atoms in a *hydrocarbon* molecule; they are formed by *addition* of *chloric(I) acid* at the *double bond* to *alkenes*.

chloromethane Methyl chloride. CH_3Cl. A colourless poisonous *gas*, b.p. $-24°C$., used as a *refrigerant*, local anaesthetic, and as a methylating agent.

Chloromycetin* Chloramphenicol. $C_{11}H_{12}Cl_2N_2O_5$. A colourless crystalline *antibiotic*, active against certain *bacteria*.

chlorophenol ClC_6H_4OH. A substituted *phenol* that exists in three *isomeric* forms. The *ortho-* form has m.p. $8.7°C$. and b.p. $175°C$., the *meta-* form has m.p. $32.8°C$., and the *para-* form has m.p. $43°C$. All forms are used in the manufacture of *dyes*.

chlorophyll A green plant *pigment*, which absorbs *energy* from sunlight, enabling the plants to build up *carbohydrates* from atmospheric *carbon dioxide* and *water* by *photosynthesis*. It consists of a mixture of two *lipid* pigments, chlorophyll-a ($C_{55}H_{72}O_5N_4Mg$) and cholorophyll-b ($C_{55}H_{70}O_6N_4Mg$).

chloropicrin See *trichloronitromethane*.

chloroplast A *membrane*-bounded *organelle* containing *chlorophyll*; the site on which *photosynthesis* takes place in a living *cell*.

chloroplatinic acid Platinum chloride solution. $H_2PtCl_6.6H_2O$. A brown *hygroscopic soluble* substance, m.p. 60°C., used in platinizing *glass* and *ceramics*.

chloroprene See *2-chlorobuta-1,3-diene*.

chlorous acid See *chloric acids*.

choke Choking coil. A coil of low *resistance* and high *inductance* used in electrical circuits to pass *direct currents* while suppressing *alternating currents*, e.g. to smooth the output of a *rectifier*.

choke-damp See *after-damp*.

cholecalciferol Vitamin D_3. See *vitamins*.

cholesteric crystals *Liquid crystals* in which the molecules are arranged in layers, with their axes parallel and in the planes of the layers. See also *smectic crystals*; *nematic crystals*.

cholesterol $C_{27}H_{45}OH$. A white waxy *sterol* that is an important constituent of animal cell membranes. Its excessive production in man is suspected of being a contributory cause of coronary thrombosis.

choline $OH.C_2H_4N(CH_3)_3OH$. An *organic base* that is a constituent of some *fats* and of egg yolk in the form of *phospholipids*. It is a member of the *vitamin* B complex.

chondrite A type of stony meteorite (see *meteor*) that contains the small round masses of *olivine* or *pyroxene* known as chondrules.

chord (math.) A straight line joining two points on a curve. See *circle*.

chromate A *salt* of *chromic acid*.

chromatic aberration See *aberration, chromatic*.

chromaticity A trichromatic method of specifying objectively the colour quality of a visual stimulus that is independent of its *luminance* and that depends on the amounts of three reference stimuli, X, Y, and Z, needed to match the light being specified. The colour quality is expressed by three chromaticity coordinates, x, y, and z, which are defined by $x = X/(X + Y + Z)$, $y = Y/(X + Y + Z)$, and $z = Z/(X + Y + Z)$.

chromatids The two identical strands into which a *chromosome* splits during *cell* reproduction.

chromatography A method of chemical analysis in which a mobile phase, carrying the mixture to be analysed, is caused to move in contact with a selectively absorbent stationary phase. The mobile phase may be a solution of a mixture of compounds in a suitably inert solvent or it may be a mixture of compounds in a vapour diluted with an inert carrier gas. The stationary phase may be an absorbent (active) solid or a liquid supported on an absorbent solid: it is characterized by its ability to retain the components of the mixture to different degrees. During the progress of the mobile phase in contact with the stationary phase, the components of the mixture become separated and can be identified; in some cases they can be determined quantitatively.

When the mobile phase is a gas and the stationary phase is a liquid on a solid support the process is known as 'gas-liquid chromatography' (see *gas chromatography*, to which it is often shortened). This is one of the most powerful methods of analysis. When the stationary phase is an active solid, the process is known as 'gas-solid chromatography'.

When the mobile phase is a liquid, it can be applied to a column of the active solid (see *column chromatography*) or to a thin layer of the solid on a plate (see *thin-layer chromatography*). Filter paper can also be used as the stationary phase (see *paper chromatography*). The last two processes provide particularly useful methods of chemical investigation.

chromatophore (bio.) **1.** In animals, a *cell* containing pigment, especially cells that, by moving the pigment, cause colour changes in the skin (e.g. in chameleons or prawns). **2.** In plants, a *plastid* containing pigment. **3.** In *bacteria*, a *membrane*-bounded body containing pigment.

chromatron Chromoscope. A type of *cathode ray tube* that has four screens; used as a colour picture-tube in *television*.

chrome alum See *potassium chromium sulphate*.

chrome iron ore Chrome ironstone, chromite, ferrous chromite. $FeO.Cr_2O_3$. A source of chromium *metal* and its *compounds*.

chrome red *Basic* lead chromate, $PbO.PbCrO_4$. Used as a *pigment* in *paints*.

chrome yellow Lead chromate, $PbCrO_4$. Used as a *pigment*.

chromic Containing chromium in +3 or +6 *oxidation states*, e.g. chromic oxide, chromium(VI) oxide, CrO_3.

chromic acid H_2CrO_4. A hypothetical *acid* known only in *solution* or in the form of its *salts*, the *chromates*.

chromite See *chrome iron ore*.

chromium Cr. Element. R.a.m. 51.9961. At. No. 24. A hard white *metal* resembling *iron*; r.d. 7.18, m.p. 1900°C., b.p. 2640°C. It occurs as *chrome iron ore* and is extracted by reducing the *oxide* with aluminium (see *Goldschmidt process*) or by reducing the ore in a *blast furnace* with carbon or silicon to form *ferrochrome*, which is used in many iron-containing *alloys*. Pure chromium is also produced by *electrolysis* of chromium compounds. It is used in the manufacture of *stainless steel* and for *chromium plating*.

chromium oxides Four oxides of chromium are known. Chromium(II) oxide, CrO, is a black insoluble powder made by oxidizing a chromium *amalgam* with air. Chromium(III) oxide, Cr_2O_3, is a green insoluble salt made by heating the metal in a stream of oxygen. An *amphoteric* oxide, it is used as a *pigment*. Chromium(IV) oxide, or chromium dioxide, CrO_2, is an unstable black insoluble solid made by heating chromium(VI) oxide at 450°C. under pressure. Chromium(VI) oxide, or chromium trioxide, CrO_3, is a red crystalline compound made by the action of sulphuric acid on sodium dichromate. It is a powerful *oxidizing agent*.

chromium plating The deposition of a thin resistant film of chromium *metal* by *electrolysis* from a bath containing a *solution* of *chromic acid*.

chromium steel *Steel* containing varying amounts of chromium, usually 8–25%; it is strong and tough, and used for tools, etc.

chromophore Any chemical group, such as the azo group, that causes an *azo compound* to have a distinctive colour.

chromosomes Thread-like bodies that occur in the *nuclei* of living *cells* and are handed on from cell to daughter cell at *mitosis* and from generation to generation by way of the *gametes*. They carry the information (genetic information) that determines the inherited characteristics of an organism (see *genetic code*; *protein synthesis*). Chromosomes consist of *deoxyribonucleic acid* and *proteins* (*nucleoprotein*) and each chromosome can be regarded as comprising a number of *genes*. Chromosomes occur in pairs in *somatic* cells, each species being characterized by the different number of chromosomes that its cells contain (humans have 46 chromosomes per cell). See also *sex chromosome*.

chromosphere The layer of the *Sun's* atmosphere surrounding the *photosphere*, which is visible during a total *eclipse*. The chromosphere is several thousand miles thick and has an estimated average temperature of 20 000 K.

chromous Containing chromium in its +2 *oxidation state*, e.g. chromous chloride, chromium(II) chloride, $CrCl_2$.

chromyl The *bivalent* group $CrO_2=$, containing chromium in its +6 *oxidation state*, e.g. in chromyl chloride, CrO_2Cl_2.

chronograph An accurate time-recording instrument.

chronometer An accurate clock, especially one used on a ship in navigation.

chronon A hypothetical particle of time defined as the ratio of the diameter of an *electron* to the *speed of light*: i.e. the time taken for light to traverse an electron. Approximately 10^{-24} *second*.

Chronotron* A device that measures the time between two events, by measuring the positions on a transmission line of pulses initiated by the events.

chrysotile See *serpentine*.

cilia (singular **cilium**) Thread-like projections from a living cell; their back-and-forth whip-like motion moves the fluid medium past the cell.

ciment fondu See *bauxite cement*.

cinchonidine $C_{19}H_{22}N_2O$. A white crystalline *alkaloid*, m.p. 207.2°C., used as a substitute for *quinine*. One of its *isomers*, cinchonine, m.p. 265°C., is also used for this purpose.

cineole $C_{10}H_{18}O$. A colourless oily *liquid terpene*, b.p. 176.4°C., with an odour of *camphor*. It is found in certain *essential oils* and used in perfumes and medicine.

cinnabar Natural *mercury(II) sulphide*, HgS. A bright red crystalline *solid*, r.d. 8.1. It is the principal *ore* of mercury.

cinnamic acid 3-phenylpropenoic acid. $C_6H_5CH:CHCOOH$. A white crystalline *insoluble* substance the *cis*-form of which has a m.p. 42°C., and the *trans-form* has a m.p. 133°C. Used in perfumes.

cinnamyl group The *univalent* group $C_6H_5CH:CH.CH_2-$, derived from *cinnamic acid*.

circadian Diurnal. Occurring every 24 hours, used especially of a physiological process, such as sleep.

circle (math.) A plane figure contained by a line, called the circumference, which is everywhere equidistant from a fixed point within it, called the centre. The distance from the centre to the circumference is the radius; a straight line joining any two points on the circumference is a *chord*; a chord passing through the centre, equal in length to twice the radius, is a diameter; any portion of the circumference is an arc; a portion cut off by a chord is a segment; a portion cut off by two radii is a sector. The ratio of the circumference to the diameter,

79

denoted by π ('pi')=3.141 59...(approx. 22/7). Length of circumference = $2\pi r$; area = πr^2, where r = radius. The equation of a circle in *Cartesian coordinates* is $x^2 + y^2 = r^2$.

circuit, electrical The complete path traversed by an *electric current*.

circularly polarized light *Light* that can be resolved into two vibrations lying in *planes* at right angles, of equal *amplitude* and *frequency* and differing in *phase* by 90°. The electric *vector* of the wave describes, at any point in the path of the wave, a *circle* about the direction of propagation of the light as axis. See also *polarization of light*.

circular measure The measurement of *angles* in *radians*.

circular mil A unit of area. The area of a *circle* whose diameter is 0.001 inch, i.e. 0.785×10^{-6} sq in. It was formerly used in measuring the cross-section of fine wire.

circumference See *circle*.

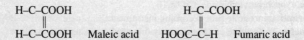

H–C–COOH ‖ H–C–COOH   Maleic acid   H–C–COOH ‖ HOOC–C–H   Fumaric acid

cis-trans isomerism A form of *isomerism* associated with *compounds* in which functional groups may be differently positioned with respect to a *double bond*, central *atom*, or ring. Like groups in such compounds may be either on the same side of the plane of the double bond, central atom, or ring (*cis*-form) or on opposite sides (*trans*-form). E.g. *maleic acid* (*cis*-butenedioic acid) and *fumaric acid* (*trans*-butenedioic acid) are respectively *cis*- and *trans*-forms. See also *stereoisomerism*.

cistron The functional unit of genetic information; the length of *deoxyribonucleic acid* in a *chromosome* that determines one *protein* (e.g. one *enzyme*) in a cell or organism. It was originally defined by the way in which an abnormal (mutant) gene in one chromosome may be compensated for by a normal gene either in the same chromosome (cis-configuration) or its pair (trans-configuration).

citrate A *salt* or *ester* of *citric acid*.

citric acid $C_3H_5O(COOH)_3$. A white crystalline *soluble* organic *tribasic acid*, m.p. 153°C. It has a sour taste, and occurs as the free acid in lemons (6%) and other sour fruits. It is used in the preparation of effervescent salts and as a food flavouring.

citric-acid cycle Krebs cycle. A complex cycle of *enzyme*-controlled biochemical reactions, which occur within living *cells*, as a result of which *pyruvic acid* is broken down into *carbon dioxide* with the release of *energy* (see *mitochondria*; *electron transport chain*). The citric-acid cycle is a most important clearing-house of metabolic intermediates (see *metabolism*), since it deals with the final stages of the *oxidation* of *carbohydrates* and *fats* and is also involved in the synthesis of some *amino acids*.

citronellal $C_9H_{17}CHO$. A colourless *liquid aldehyde* existing in several *isomeric* forms, b.p. 205–8°C., with a lemon-like odour. It is used as a flavouring and in the manufacture of perfume.

citronellol $C_9H_{17}CH_2OH$. A colourless *liquid alcohol* existing in several *isomeric* forms, b.p. 110°C., used in the manufacture of perfumes.

cladding (phys.) The covering of a *fuel element* in a *nuclear reactor* by a thin layer of another *metal*, to prevent corrosion by the *coolant* and the escape of *fission products*.

cladistics A method of classifying plants and animals into clades, i.e. groups that share a common ancestor. It assumes that a new species arises by splitting off from a common ancestor, rather than by gradual evolution.

Claisen condensation A reaction between two *molecules* of an *ester*, catalysed by *sodium ethoxide*, in which a keto ester is formed:

$$2CH_3COOR \rightarrow ROH + CH_3COCH_2COOR.$$

Clark cell A *primary cell*, formerly used as a standard of *EMF*, that gives 1.4345 *volts* at 15°C. It consists of a zinc *amalgam anode* and a mercury *cathode*, both immersed in a *saturated solution* of *zinc sulphate*.

class (bio.) A group (*taxon*) of similar *orders* (see *classification*).

classical physics *Physics* prior to the *quantum theory* (or in some senses prior to the theory of *relativity*).

classification (bio.) A system for putting into order the huge variety of living organisms, by grouping together those with similar characteristics. Each such group is a *taxon*. In the Linnaean system of classification commonly used in *biology*, the taxa are arranged in the hierarchy: *species*, *genus*, *family*, *order*, *class*, *phylum*, kingdom. For example, humans are the species *sapiens* of the genus *Homo*, in the family Hominidae, in the order Primates, of the class Mammalia in the phylum Chordata of the kingdom Animalia. See also *binomial nomenclature*.

clathrate compounds Cage compounds. Chemical *compounds* formed not by *chemical bonds* but by 'molecular imprisonment', the combined *molecules* being held together mechanically by virtue of their configuration in space.

Claude process A process for producing *liquid air*, based on the cooling that results from the adiabatic expansion of a gas that is performing external work. Air under pressure is divided into two separate channels. The first channel leads to a compressor, where the air performs external work by driving the compressor. The cool air so produced is used to reduce the temperature of the compressed air from the second channel in a counter-current *heat exchanger*.

clays Finely-divided *rock* materials whose component *minerals* are various *silicates*, mainly of *magnesium* and *aluminium*.

cleavage (bio.) A series of *cell* divisions that transforms a single fertilized egg cell into a *blastula*.

cleavage (chem.) The manner of breaking of a crystalline substance, so that more or less smooth surfaces are formed.

clinical thermometer See *thermometer, clinical*.

clone A group of organisms that have arisen from a single individual by asexual reproduction. All members of the group are genetically identical to each other and to the parent organism. Plants propagated by cuttings are clones, as are organisms produced from *somatic* cells.

cloud chamber (phys.) Wilson cloud chamber. An apparatus for making the tracks of ionizing particles visible as a row of droplets. It consists of a chamber filled with a saturated alcoholic *vapour* and fitted with a piston to enable the vapour to be expanded adiabatically. This causes sudden cooling and supersaturation of the vapour. In this state, a beam of particles of *ionizing radiation* passing through the chamber creates a stream of *ions* along its path. The vapour forms liquid droplets on the ions, thus producing a visible track.

cloud point The temperature at which a *homogeneous liquid* becomes cloudy or turbid, owing to separation into two *phases*, when cooled under specified conditions.

Clusius column A device for separating gaseous *isotopes*. It consists of a high column with a central heated wire. As a result of *thermal diffusion* the lighter isotopes collect at the top of the tube. Typically, this fraction is subjected to further passes through the column for greater enrichment.

cluster (astr.) An aggregation of *stars* that move together. A *globular cluster* is an aggregation of stars in a roughly spherical arrangement.

cluster compound A compound in which groups of *transition metal* atoms are held together by bonds between the metal atoms. If the atoms are all of the same element this is an isopoly compound; in heteropoly compounds the atoms are of different metals.

coagulation of proteins When *solutions* of water-soluble *proteins* (*albumins*) are heated, the protein becomes 'denatured' at a definite *temperature*; it then becomes *insoluble* and either remains in *suspension* or is precipitated as a clot or curd. Other types of proteins, e.g. *globulins*, may be denatured and coagulated by *heat*, or by the addition of *acids* or *alkalis*. A denatured protein cannot be easily reconverted into the original *compound* (see *denature*).

coal A combustible material, occurring in large underground deposits, consisting of carbon and various carbon *compounds*. It formed by the *decomposition* of vegetable matter during periods of many millions of years. 'Humic coal' is derived from plants, while 'sapropelic coal' is derived from algae, spores, and fragments of plant material. The process of forming coal (coalification) occurs with increased pressure and temperature, the percentage of carbon increasing with time (as the oxygen and volatile matter decreases). During this process the coal changes from *peat* to *lignite* to *bituminous coal* to *anthracite*.

coal-gas Fuel *gas* manufactured by the *destructive distillation* of *coal* in closed iron *retorts*. Its composition by *volume* is usually: hydrogen 50%, *methane* 30%, *carbon monoxide*, 8%, other *hydrocarbons* 4%, nitrogen, *carbon dioxide*, and oxygen 8%. Widely used as an energy source in the late 19th and first half of the 20th century, it has now been largely replaced by *natural gas*, which is much less toxic as it contains no carbon monoxide. See also *SNG*.

coal-gas by-products Amongst the valuable substances obtained during the manufacture of *coal-gas* were *coke*, *coal-tar*, *ammonia*, *sulphuric acid*, and *pitch*. Since coal-gas has now largely been replaced by *natural gas*, these substances are now largely obtained by other means.

coal-tar A thick black oily *liquid* formerly obtained as a by-product of *coal-gas* manufacture. It is now made during the manufacture of *coke* (from coal) for steel making. *Distillation* and purification yields, amongst other valuable products: *benzene*, C_6H_6; *methylbenzene*, $C_6H_5CH_3$; *dimethylbenzene*, $C_6H_4(CH_3)_2$; *phenol*, C_6H_5OH; *naphthalene*, $C_{10}H_8$; *cresol*, $CH_3C_6H_4OH$, and *anthracene*, $C_{14}H_{10}$. *Pitch* is left as a residue.

coaxial Having a common axis. Coaxial cable consists of a central conducting wire and a concentric cylindrical conductor, the space between the two being filled with a *dielectric*, such as *polythene*. The outer conductor is normally connected to earth. Its main use is to transmit high-frequency power or signals from one place to another with minimum energy loss and with minimum *interference*.

cobalt Co. Element. R.a.m. 58.933. At. No. 27. A hard silvery-white magnetic *metal* resembling iron. R.d. 8.9, m.p. 1495°C., b.p. 2870°C. It occurs combined with sulphur and with arsenic and is extracted by converting the *ore* into the *oxide* and reducing with aluminium, or with carbon in an electric furnace. The metal is used in many *alloys*; compounds are used to produce a blue colour in *glass* and *ceramics*.

cobaltic Containing cobalt in its +3 *oxidation state*, e.g. cobaltic chloride, cobalt(III) chloride, $CoCl_3$.

cobaltous Containing cobalt in its +2 *oxidation state*, e.g. cobaltous chloride, cobalt(II) chloride, $CoCl_2$.

cobalt steel *Steel* containing cobalt (usually 5–12%), and often other *metals* such as tungsten, chromium, and vanadium. The addition of cobalt results in greater hardness and brittleness, improves the cutting power of *high-speed-steel* tools, and alters the magnetic properties.

cocaine $C_{17}H_{21}O_4N$. A white crystalline *alkaloid* that occurs in the coca plant, m.p. 98°C. It is used as a local *anaesthetic* and is a dangerous habit-forming drug.

coccus A globular or spherical-shaped *bacterium*.

cochineal A natural red dyestuff obtained from the dried body of the Coccus cacti insect.

Cockcroft-Walton generator or accelerator A high voltage *direct current accelerator* used for accelerating nuclear particles (particularly *protons*). The DC voltage is obtained by multiplying a low AC voltage by an arrangement of *rectifiers* and *capacitors*.

codeine $C_{18}H_{21}O_3N$. A white crystalline *alkaloid*, m.p. 158°C., obtained by methylation of *morphine*. Used in medicine (often in the form of its *phosphate*) as an *analgesic*, *hypnotic*, and in the treatment of coughs.

codon See *genetic code*.

coefficient (math.) A number or other known *factor* written in front of an algebraic expression. E.g. in the expression $3x^4$, 3 is the coefficient of x^4.

coefficient (phys.) A *factor* or multiplier that measures some specified property of a given substance, and is constant for that substance under given conditions. E.g. coefficient of friction. (See *friction*, *coefficients of*.)

coelostat A device used in conjunction with an astronomical *telescope* to follow the path of a celestial body and reflect its *light* into the telescope. It consists essentially of two mirrors, one movable and one fixed.

coenzyme A nonprotein organic molecule that plays an essential part in some reactions catalysed by *enzymes*, it often acts as a temporary *carrier* of an intermediate product of the reaction. See also *cofactor*.

coercive force The strength of the *magnetic field* to which a *ferromagnetic substance* undergoing an *hysteresis cycle* must be subjected in order to reduce the *flux density* to zero. If the substance is magnetized to saturation during the cycle, the coercive force is called the *coercivity*. See Fig. 21 under *hysteresis cycle*.

cofactor A nonprotein required to make an enzyme an efficient *catalyst*. They may be *coenzymes* (which are organic molecules) or inorganic *ions*.

Coffey still An apparatus for the *fractional distillation* of *solutions* of *ethanol* as obtained by *fermentation* on an industrial scale; the product is known as *rectified spirit*.

coherent Denoting a beam of *light*, or other *electromagnetic radiation*, in which the waves are in *phase* or have a constant phase relationship. See *laser*.

coherent units A system of *units* in which the *quotient* or *product* of any two quantities has a unit equal to the quotient or product of the units of these quantities. E.g. when the unit of length is divided by the unit of time, the unit of *velocity* results. The basic units of a coherent system are arbitrarily defined physical quantities. All other units are obtained from these basic units by defining relations and are called 'derived units'. The coherent units now in scientific use are the *SI units*.

coinage metals The *metals* copper, silver, gold.

coincidence circuit Coincidence gate. An *electronic* circuit that produces an output only when two or more input signals arrive simultaneously, or within a specified time interval. A coincidence counter incorporates a counter circuit.

coke A greyish porous brittle *solid* containing about 80% carbon. It was formerly obtained as a residue in the manufacture of *coal-gas* ('gas coke'); it is now made specially in coke ovens, in which the *coal* is treated at lower *temperatures* than in gas manufacture. It is used in *blast furnaces* and other metallurgical processes as well as a smokeless domestic fuel.

colchicine $C_{22}H_{25}NO_6$. A yellow crystalline *alkaloid*, m.p. 156°C., obtained from the autumn crocus, that interferes with the process of *mitosis* in such a way that it causes a doubling of the number of *chromosomes* in a *cell*. Used as an artificial method of obtaining new agricultural and horticultural varieties and in the treatment of gout.

colcothar Rouge, red *iron(III) oxide*, Fe_2O_3. It is used as a *pigment* and for polishing.

cold emission The emission of *electrons* from a solid surface without the use of heat (as in *thermal emission*), i.e. by *field emission* or *secondary emission*.

cold fusion See *nuclear fusion*.

collagen A *protein* that is the major fibrous constituent of skin, tendon, ligament, and bone: it is, therefore, probably the most abundant protein in the animal kingdom. Collagen owes its unique properties not only to its chemical composition, but also to the physical arrangement of its individual *molecules*. The basic molecular *polypeptide* chain forms a left-handed *helix*, and three such helices are wrapped around each other to form a right-handed super-helix. On boiling with *water* collagen gives rise to *gelatin*.

collargol A powder containing *protein* material and finely divided silver; with *water* it forms a *colloidal solution* of silver.

collector The *electrode* in a *transistor* through which a primary flow of *carriers* leaves the inter-electrode region.

collider See *storage ring*.

colligative properties Those properties of a substance (e.g. a *solution*) that depend only on the *concentration* of particles (*molecules* or *ions*) present and not upon their nature; e.g. *osmotic pressure*.

collimator 1. A tube containing a *convex achromatic lens* at one end and an adjustable slit at the other, the slit being at the focus of the lens. *Light* rays entering the slit thus leave the collimator as a parallel *beam*. **2.** An arrangement of absorbers for limiting a beam of *radiation* to the required dimensions and angular spread in *radiology*. **3.** A small fixed *telescope* attached to a larger one for the purpose of accurately setting the line of sight of the larger instrument.

collision density The number of collisions per unit volume per unit time that a given *neutron* flux makes when passing through *matter*.

collodion A *solution* of *cellulose nitrate* in a mixture of *ethanol* and *ethoxyethane*.

colloid A substance present in *solution* in the *colloidal state*.

colloidal metals *Colloidal solutions* or *suspensions* of *metals,* the metal being distributed in the form of very small electrically charged particles. They are prepared by striking an electric *arc* between poles made of the metal, under water or by the chemical *reduction* of a *solution* of a *salt* of the metal. They are used in medicine.

colloidal solution Sol. A *solution* in which the *solute* is present in the *colloidal state*. Common examples include solutions of *starch*, *albumin*, *colloidal metals*, etc. The *solvent* is termed the *dispersion medium* or continuous phase and the dissolved substance the *disperse phase*. Several types of colloidal solution arc possible, depending upon whether the dispersion medium and the disperse phase are respectively *liquid* and *solid* (suspensoid sols), liquid and liquid (emulsoid sols), *gas* and solid, etc. If the disperse phase, when removed from solution by *evaporation* or coagulation, returns to the colloidal state on merely mixing with the dispersion medium, it is termed a reversible or *lyophilic colloid*, and the solution a reversible sol. If the disperse phase does not return to the colloidal state on simple mixing, it is termed an irreversible or *lyophobic colloid*.

colloidal state A system of particles in a *dispersion medium* (or continuous phase), with properties distinct from those of a true *solution* because of the larger size of the particles. The presence of these particles, which are approximately 10^{-4} to 10^{-6} mm across, can often be detected by means of the *ultramicroscope*. As a result of the grouping of the *molecules*, a *solute* in the colloidal state cannot pass through a suitable *semipermeable membrane* and gives rise to negligible *osmotic pressure*, *depression of freezing point*, and *elevation of boiling point* effects. The molecular groups or particles of the solute carry a resultant *electric charge*, generally of the same sign for all the particles.

cologarithm The *logarithm* of the *reciprocal* of a number, expressed with a positive *mantissa*.

colophony See *rosin*.

colorimeter Apparatus used in *colorimetric analysis* for comparing intensities of *colour*. See also *tintometer*.

colorimetric analysis A form of *quantitative analysis* in which the quanitity of a substance is estimated by comparing the intensity of *colour* produced by it with specific *reagents*, with the intensity of colour produced by a standard amount of the substance.

colour The visual sensation resulting from the impact of *light* of a particular *wavelength* on the *cones* of the retina of the eye (see *photopic vision*). Coloured light has three characteristics: *hue*, which is determined by its wavelength;

saturation, the extent to which a colour departs from white; and *luminosity*, a measure of its brightness (for a light or other emitting source). If the source is a pigment, dye, etc., that reflects rather than emits light, this last characteristic is called *lightness*. Coloured lights mix to form a different colour by an *additive process*. Pigments, dyes, etc., mix by a *subtractive process*. See *colour vision*; *spectrum colours*.

colour blindness The condition that results if the cones in the retina of the eye are absent or not functioning. Because this occurs as a result of a recessive *gene* on the X-chromosome (see *sex chromosome*), the condition is rare in women, who may be carriers, and not uncommon in men. See also *dichromatism*.

colour charge See *elementary particles*; *quantum chromodynamics*.

colour photography A system for forming coloured images on film or paper by photographic methods. The most common process is based on a subtractive reversal system using a film with three layers of *emulsion*, each of which is sensitive to one of the *primary colours* red, green, and blue. When an *exposure* is made the blue components of the object photographed will be stored as a latent image in the blue-sensitive emulsion, likewise the other two emulsions. The final image, after development, is formed from the correct proportions of the *complementary colours* cyan, magenta, and yellow.

colour temperature The *temperature* of a full radiator (see *black body radiation*) that would emit visible *radiation* of the same spectral distribution as the radiation from the *light* source under consideration.

colourtron A type of *cathode-ray tube*, used as a colour picture-tube in *television*, that has three *electron guns*, one for each *primary colour*.

colour vision *White light*, such as daylight, consists of a mixture of *electromagnetic radiations* of various *wavelengths* (see *spectrum colours*). A surface that reflects all of these will appear white; some surfaces, however, have the property of absorbing some of the radiations they receive, and reflecting the rest. Thus, a surface that absorbs all light radiations excepting those corresponding to green, will appear green by reflecting only those radiations. In the cases of colour seen by transmitted light, as in coloured *glass*, the glass absorbs all the radiations except those that are visible and pass through. See *colour*; *photopic vision*; *pigment colour*; *surface colour*.

columbium Cb. See *niobium*.

column chromatography A form of *chromatography* in which the mobile phase is liquid and the stationary phase is *activated alumina*, or a similar substance, contained in a vertical glass column. The mixture is introduced at the top of the column and washed through the stationary phase by a solvent. The components of the mixture are selectively adsorbed, forming coloured bands down the length of the column (if the components are coloured). The technique is used in laboratory preparations as well as in analysis, the eluate (see *elution*) being separated into fractions.

colza oil Rapeseed oil. Yellow oil obtained from the seeds of various Brassica plants. Used as an edible oil, illuminant, lubricant, and in the *quenching of steel*. 'Mineral colza' oil is a mixture of *paraffin hydrocarbons* with a boiling range of $250°$–$350°$C.

coma 1. The nebulous patch of *light* that surrounds the *nucleus* of a *comet*. **2.** An error of a *lens* or spherical *mirror* that causes a blurred *comet*-like image.

combination, laws of chemical See *chemical combination, laws of.*

combination (math.) A selection of a specified number of different objects from some larger specified number. The number of combinations of *r* different objects selected from *n* objects (i.e. the number of combinations of *n* objects taken *r* at a time) is denoted by the expression ^nC_r and is equal to $n!/r!(n-r)!$. See also *factorial* and *permutation*.

combustion Burning. A *chemical reaction*, or complex of chemical reactions, in which a substance combines with oxygen producing *heat, light,* and *flame*. The combustion reactions that supply most of the *energy* required by human civilization involve the oxidation of *fossil fuels* in which carbon is converted into *carbon dioxide* and hydrogen is converted into *water (steam)*. These reactions can be summarized by:

$$C + O_2 \rightarrow CO_2$$
$$2H_2 + O_2 \rightarrow 2H_2O$$

However, these *equations* conceal a number of free-radical *chain reactions* that proceed very rapidly.

comet A heavenly body, moving under the attraction of the *Sun* in an eccentric orbit. It consists of a hazy gaseous cloud (see *coma*) containing a brighter nucleus and a fainter tail. The nucleus is thought to consist of ice and dust particles. A large comet might have a nucleus 10^4 m in diameter with a coma 10^8 m in diameter, and a tail 10^{10} m in length. Their orbital *periods* range from 150 to 10^5 years. See *Halley's comet.*

command guidance A method of missile or *rocket* guidance in which computed information is transmitted to the missile and causes it to follow a directed flight path.

comma of Pythagoras See *temperament.*

common logarithm See *logarithm.*

communication satellite See *satellite, artificial.*

commutative law The law of algebra in which the order of the terms is not important. For example, $a + b = b + a$ and $ab = ba$ are commutative equations. Subtraction and division are not commutative. Compare *associative law; distributive law.*

commutator A device for connecting the *armature* of an *electric motor* or *generator* to an external circuit; it is also used in the *dynamo* to convert the *alternating current* into a direct one if required. It consists of a cylindrical assembly of insulated *conductors* each of which is connected to sections of the winding. Spring-mounted carbon brushes make contact with the conductors and thus carry the current to external circuits.

compass, magnetic In its simplest form a compass consists of a magnetized needle pivoted at its centre so that it is free to move in a horizontal plane. The effect of the Earth's *magnetic field* is to cause the needle to set along the *magnetic meridian*. The needle is usually placed at the centre of a circular scale marked with the points of the compass. As such a compass is also affected by magnetic fields other than that of the Earth, for navigation the *gyrocompass* is used.

complementarity A term introduced into *quantum theory* by Niels Bohr (1885–1962), implying that evidence relating to atomic systems that has been obtained under different experimental conditions cannot necessarily be comprehended by one single model. Thus, for example, the wave model of the *electron* is complementary to the particle model.

complementary (bio.) See *base pairing*.

complementary angles *Angles* together totalling 90° or one right angle.

complementary colours Pairs of coloured lights that, when combined, give the effect of white light. See *colour vision*.

complementary DNA cDNA. A form of *deoxyribonucleic acid* synthesized *in vitro* from *ribonucleic acid* (RNA) using the enzyme *reverse transcriptase*. The cDNA has a base sequence complementary to the RNA template. It is used in cloning *genes*.

complete radiation See *black body radiation*.

complex (chem.) Complex compound. The term originally derives from the recognition that compounds, which can exist as separate entities, may combine together by the formation of bonds (usually *coordinate bonds*) between atoms of the two components. The product is a complex compound, but the term now covers all analogous *coordination compounds*. Thus, a compound may form a *derivative* (*salt*) with a metal, but may also contain atoms that can coordinate with the metal in the product, so that the latter becomes a complex compound. See *complexons*; *porphyrins*; *chelation*.

complex number A complex number consists of two parts, 'real' and 'imaginary', and can be expressed in the form $x + iy$, where both x and y are real quantities and i is the square root of -1, i.e. $i^2 = -1$. The real part of the complex number is 'x' and the imaginary part 'iy'. Such numbers obey the ordinary laws of *algebra* except that in *equations* containing them the real and imaginary parts are equated separately. See *Argand diagram*.

complexometric analysis A method of chemical analysis based on *titration* of metal ions in solution with chelating agents (see *chelation*), such as *EDTA* or other *complexons*.

complexon(e)s Complex-forming or chelating agents (see *chelation*) used in *complexometric analysis*; e.g. *EDTA* and similar compounds.

component (chem.) The least number of substances from which every *phase* of a system may be constituted. E.g. each of the phases *ice*, *water*, *water vapour* in equilibrium is composed of one component, H_2O. In a mixture of *ethanol* and water there is one phase but two components. See *phase rule*.

component forces and velocities Two or more *forces* or *velocities* that produce the same effect upon a body as a single force or velocity, known as the *resultant*.

compound (chem.) A substance consisting of two or more *elements* chemically united in definite proportions by *weight*. The formation of a compound involves a *chemical reaction* and the elements cannot be separated by purely physical means.

compound, interstitial A *compound* of a *metal* and certain *metalloids* in which the metalloid *atoms* occupy the interstices between the atoms of the metal *lattice*.

compound microscope See *microscope, compound*.

compressibility The coefficient of compressibility (*isothermal*) of a substance is given by $\kappa = -1/V.\partial V/\partial p$, where ∂V is the change in the *volume* V of the substance resulting from a change of *pressure* ∂p, the *temperature* remaining constant. It is the *reciprocal* of *bulk modulus* and is measured in square *metres* per *newton*.

compression ratio The ratio of the volume formed in the cylinder and combustion chamber of an internal-combustion engine when the piston is at its lowest point (outer dead-centre) to the volume at the end of the compression stroke. In petrol engines the compression ratio is 8.5–9 to 1 (now tending to the lower end to avoid *knocking* with unleaded petrol) and 12–18 to 1 in *Diesel engines*.

compressive stress An axial *force* per unit area that tends to compress a body to which it is applied. Compare *tensile stress*. See also *stress*.

Compton effect The reduction in the *energy* of a *gamma-ray* or *X-ray photon*, as a result of its interaction with a free *electron*. Part of the photon's energy is transferred to the electron (Compton or recoil electron) and part is redirected as a photon of reduced energy (Compton scatter). In the 'inverse Compton effect', low-energy photons gain energy when they are scattered by free electrons having much higher energy. These electrons, consequently, lose energy. Named after Arthur H. Compton (1892–1962).

computer An *electronic* device that can accept data, apply a series of logical processes to it, which are incorporated in its *programs*, and supply the results of these processes as information, either on a *visual-display unit* (VDU), on a magnetic tape, or printed on paper. Computers are used to perform complex series of mathematical calculations at very great speed; this makes them of great use for a variety of purposes, such as routine office calculations, control of industrial processes, and the control of spacecraft flight paths. Their ability to perform these operations depends not only on their mathematical capabilities, but also on their ability to store information and retrieve specified *bits* of information very rapidly in the appropriate circumstances. The two main types of computer are: the *analog computer* in which numbers are represented by magnitudes of such physical quantities as *voltages*, mechanical movements, etc., and the *digital computer* in which numbers are expressed directly as *digits*, usually in the *binary notation*. This latter type is more versatile and most modern computers are digital. The main components of the hardware of a computer (the electronic equipment) are the input devices (keyboard, tape reader, disc reader, etc.), the central processing unit (see *CPU*), the *memory*, and the output devices. See also *microcomputer*; *minicomputer*; *peripherals*.

concave Curving inwards; thus, a concave (or bi-concave) *lens* is thinner at the centre than at the edges. See Fig. 24 under *lens*.

concavo-convex Denoting a *lens* that curves inwards on one side and outwards on the other. See Fig. 24 under *lens*.

concentrated (chem.) Denoting a *reagent* containing the minimum of *water* or other *solvent*; the opposite of *dilute*.

concentration The quantity of a substance present in a given space or defined quantity of another substance. The concentration (symbol c) of a solution is the amount of substance dissolved in unit volume of solution (measured in mol dm^{-3}). The mass concentration (symbol ρ) is the mass of solute per unit volume of solvent (kg dm^{-3}). The molal concentration or molality (symbol m) is the amount of substance of solute per unit mass of solvent (mol kg^{-1}). See also *molarity*; *solubility*.

concentration cell A *primary cell* whose *EMF* is due to a difference in *concentration* between different parts of the *electrolyte*.

concentric Having the same centre. E.g. two *coaxial* tubes would appear, in cross-section, as two concentric *circles*.

conchoidal fracture A type of break or fracture characteristic of an *amorphous solid*; an irregular break with a curved face exhibiting concentric rings.

concrete A building material composed of stone, *sand*, *cement*, and *water*. Reinforced concrete has *steel* rods or meshes imbedded in it to increase its *tensile strength*. In prestressed concrete the concrete is maintained in compression by stretching the reinforcing rods and keeping them in tension after the concrete has set.

condensation (chem.) A chemical change in which two or more *molecules* react with the elimination of *water* or of some other simple substance. E.g. *ethanoic anhydride*, $(CH_3CO)_2O$, may be regarded as a condensation product of *ethanoic acid*, CH_3COOH, a molecule of the anhydride being formed when two molecules of the acid combine with the elimination of one molecule of water. See also *polymerization*.

condensation of vapour The change of *vapour* into *liquid*, which takes place when the *pressure* of the vapour becomes equal to the *vapour pressure* of the liquid at that *temperature*. It is accompanied by the evolution of the *latent heat*.

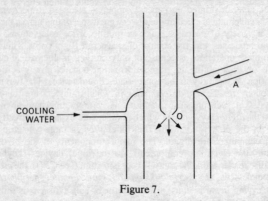

Figure 7.

condensation pump Diffusion pump. An apparatus used to obtain high vacua, i.e. *pressures* of the order of 10^{-7} pascal. Mercury or oil *vapour* issuing as a jet through the orifice O exhausts the system attached to the tube A. *Gas molecules* in A diffuse through the layer of mercury vapour around the orifice and are carried down with the vapour stream by molecular bombardment. The mercury vapour is cooled in a cold trap causing it to condense, so preventing it from diffusing back into the system that is being exhausted. See Fig. 7.

condensed-matter physics See *solid-state physics*.

condenser (chem.) An apparatus for converting *vapour* into *liquid* by means of cooling. The Liebig condenser consists of a tube along which the vapour passes and is cooled, usually by cold water flowing in the opposite direction through an outer jacket surrounding the tube. It is used for laboratory distillations.

condenser, electrical See *capacitor*.

condenser, optical A device used in optical instruments to converge rays of *light*; e.g. in the *microscope* a condenser *lens* is used to converge upon the object to be viewed. It often consists of two plano-convex lenses, with the convex faces facing each other.

condenser microphone See *capacitor microphone*.

conductance *G*. **1.** The conductance of a *direct current* circuit is the *reciprocal* of its *resistance*. The conductance of an *alternating current* circuit is its resistance divided by the square of its *impedance*. The *SI unit* is the *siemens*, formerly called the *mho* or reciprocal ohm. **2.** See *heat transfer coefficient*.

conductiometric titration A *titration* in which the electrical *conductivity* of the reacting chemicals is monitored as one reactant is added. The *equivalence point* is denoted by a sharp change in conductivity.

conduction, electrical The passage of an *electrical current* through a conductor as a result of the influence of an *electric field* on the charge *carriers*.

conduction, thermal The transmission of *heat* from places of higher to places of lower *temperature* in a substance, by the interaction of *atoms* or *molecules* possessing greater *kinetic energy* with those possessing less. In *gases* the heat energy is transmitted by collision of the gaseous molecules, those possessing the greater kinetic energy imparting, on collision, some of their energy to molecules having less. Conduction in *liquids* is mainly due to the same process. In solid electrical *conductors*, the chief contribution to thermal conduction arises from a similar process taking place between the free *electrons* and *ions* present. The interaction of the molecules responsible for thermal conduction in solid electrical *insulators* arises from the elastic binding *forces* between the molecules, which are effectively fixed in space.

conduction band The range of energies (see *energy bands*) in a *semiconductor* corresponding to states in which the *electrons* can be made to flow by an applied *electric field*.

conductivity, electrical *σ*. The *reciprocal* of the *resistivity* or specific resistance of a *conductor*. It is measured in *siemens* per *metre*.

conductivity, thermal *λ*. The rate of transfer of *heat* along a body by *conduction*. For a cube of side *l*, the energy *E*, transferred in unit time *t*, is given by:
$$E = t\lambda l(T_2 - T_1),$$
where *λ* is the conductivity and T_2 and T_1 are the temperatures of a pair of opposite faces. *λ* is measured in $J \ s^{-1} \ m^{-1} \ K^{-1}$. See also *heat transfer coefficient*.

conductivity water Water that has been repeatedly distilled in vacuo to bring its conductivity down to about $4 \times 10^{-6} \ S \ m^{-1}$, which is about one twentieth that of ordinary *distilled water*. The limit of the conductivity of water results from the ionization: $H_2O \rightleftharpoons H^+ + OH^-$.

conductor, electrical A body capable of carrying an *electric current*; a body that, if given an *electric charge*, will distribute that charge over itself. In metals the abundance of *free electrons* makes them good conductors.

conductor, thermal A body that will permit *heat* to flow through it by *conduction*.

Condy's fluid A *solution* of sodium or calcium (or sometimes aluminium) permanganate (manganate(VII)), $NaMnO_4$ or $Ca(MnO_4)_2$. It is used as a *disinfectant*.

cone (math.) A solid figure traced by a straight line passing through a fixed point, the *vertex*, and moving along a fixed *circle*. For a cone of vertical height *h*, slant height s, and radius of base r, the *volume* is given by $V = \pi r^2 h/3$, and the area of the curved surface $A = \pi rs$.

cone (optics) A photosensitive *cell* in *retina* of vertebrate eyes. They function in bright light and are sensitive to colour. They are concentrated in the fovea and are not present in the margins of the retina. Compare *rod*. See *photopic vision*.

confinement See *containment*.

conformation theory The principle that the three-dimensional structure of a *molecule* enables its stability and reactivity to be predicted. The theory pays special attention to the conformation of substituted hydrogen *atoms* in *organic* compounds; the axial (vertical) or equatorial (horizontal) disposition of *substituents* has been shown to be of great importance in predicting physical and chemical properties. The conformation of a *polypeptide* in water is of great importance in biological contexts. It may be stabilized by *hydrogen bonds*, by *hydrophobic* interactions of its constituent *amino acids*, by *salt* linkages, and by *covalent* bonds, such as disulphide bridges. The conformation is essential to its function and changes in its conformation may affect its activity (see *enzyme*; *allosterism*; *receptor*).

congeners *Elements* belonging to the same *group* in the *periodic table*.

congruent figures Geometrical figures equal in all respects.

CONIC SECTIONS

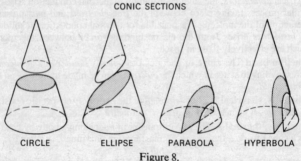

CIRCLE   ELLIPSE   PARABOLA   HYPERBOLA

Figure 8.

conic sections Curves obtained by the intersection of a *plane* with a *cone*; they include the *circle*, *ellipse*, *parabola*, and *hyperbola*. See Fig. 8.

coniine $C_8H_{17}N$. A poisonous *liquid alkaloid*, b.p. 166-8°C., which is the active constituent of hemlock.

conjugated double bond In an *unsaturated organic compound* two *double bonds* separated by a single bond are said to be conjugated, e.g. *buta-1,3-diene*, $CH_2:CH.CH:CH_2$. See *delocalization*.

conjugate points of a lens or mirror. Points on either side of the *lens* or mirror, such that an object placed at either will produce an image at the other.

conjunction (astr.) A *planet* (or other heavenly body) is said to be in superior conjunction when it is in a straight line with the *Sun* and the *Earth*; a planet with its orbit inside that of the Earth is in inferior conjunction when it is between the Sun and the Earth and in line with them.

conservation laws Laws stating that in a particular system the total amount of a physical quantity remains unchanged. Some of these laws appear to apply to the whole *Universe*, provided the Universe can be regarded as a closed system. See *conservation of charge*; *conservation of mass and energy*; *conservation of momentum*.

conservation of charge The principle that the total *electric charge* associated with a system remains constant: that electric charge can be neither created nor destroyed.

conservation of mass and energy A principle, resulting from Einstein's special theory of *relativity*, that combines the separate laws of the conservation of *energy* and of *mass*. The law of the conservation of energy states that in any system energy cannot be created or destroyed, and the law of the conservation of mass (or *matter*) states that in any system matter cannot be created or destroyed. As Einstein showed that the mass of a body is a measure of its energy content, according to $E = mc^2$, where c is the *speed of light*, the laws of conservation of mass and energy merge with each other.

conservation of momentum The principle that the total *momentum* of two colliding bodies before impact is equal to their total momentum after impact. When *velocities* comparable to the speed of *light* are being considered, the variation of *mass* with *velocity* (see *relativity, theory of*) must be taken into account, and the expression for the momentum becomes:
$$\text{Momentum} = mv = m_0 v (1 - v^2/c^2)^{-1/2}$$
where m_0 = *rest mass*, v = velocity of the body, and c = speed of light.

conservation of resources An awareness that the Earth's natural resources need to be used prudently as they are not unlimited. This applies particularly to *energy* deriving from *fossil fuels* and *nuclear* fuels (see *renewable energy sources*) and to deforestation, overfishing, and over-utilization of the land. Conservation also includes an awareness of the dangers caused to the environment by *pollution* and international agreement on restrictions to prevent the extinction of *endangered species*.

consolute temperature Critical solution temperature. The temperature at which two partially miscible liquids become completely miscible.

constant (math., phys.) Any quantity that does not vary; e.g. π ('pi'), the ratio of the circumference to the diameter of any *circle*. See also *fundamental constants*.

constantan An *alloy* of copper containing 10%–55% nickel; as its electrical *resistance* does not vary with *temperature* it is used in electrical equipment.

constant boiling mixture See *azeotropic mixture*.

constant composition, law of See *chemical combination, laws of*.

contact angle At a solid-liquid interface, the *angle* included between the tangent plane to the surface of a *liquid* and the tangent plane to the surface of a *solid* at any point along their line of contact.

contact potential difference If two dissimilar *metals*, a and b, are in contact (see Fig. 9), then a *potential difference* exists between point A, just outside *conductor a*, and a point B, just outside conductor b. This is the contact potential difference of the two conductors.

contact process An industrial process for the manufacture of *sulphuric acid*, H_2SO_4. *Sulphur dioxide*, SO_2, is made to combine with oxygen by passing over a heated *catalyst*, formerly platinum or *platinized asbestos*, but now more

93

usually a vanadium catalyst. The *sulphur trioxide*, SO_3, formed is combined with *water* to give sulphuric acid.

Figure 9.

containment Confinement. In a controlled *thermonuclear reaction*, the process of preventing the *plasma* from coming into contact with the walls of the containing vessel by means of *magnetic fields*. The approximate period for which the *ions* remain trapped by the containing field is referred to as the 'containment time'.

continental drift The theory that all the continents on the Earth's surface were once a single land mass, which subsequently split up. Evidence includes the apparent fit of South America into Africa, the distribution of rocks, and more recently the theory of *plate tectonics*. It is postulated that some 135 million years ago the single land mass, named Pangaea, split into the northerly Laurasia and the southerly Gondwanaland. 65 million years ago these two continents split again, Laurasia into North America and Europe-Asia, and Gondwanaland into South America and Africa.

continuous phase See *colloidal solution*; *colloidal state*.

continuous spectrum See *spectrum*.

continuous wave CW. *Radio* or *radar* transmissions generated continuously and not in short *pulses*.

continuum A continuous series of component parts passing into one another; e.g. the three *space* dimensions and the time dimension are considered to form a four-dimensional continuum.

control grid An *electrode* placed between the *cathode* and the *anode* of a *thermionic valve* or *cathode-ray tube* for controlling the flow of *electrons* through the valve or tube.

controlled thermonuclear reaction CTR. See *thermonuclear reaction*.

control rods Part of the control system of a *nuclear reactor* that directly affects the rate of reaction therein. Usually a number of rods or tubes, which can be moved up or down their axes, they are made of steel or aluminium containing boron, cadmium, or some other strong absorber of *neutrons*. When the rods are lowered into the core, the rods absorb some of the neutrons and thus bring the *chain reaction* to a halt.

convection Transference of *heat* through a *liquid* or *gas* by the actual movement of the *fluid*. Portions in contact with the source of heat become hotter, expand, become less dense, and rise; their place is taken by colder portions, thus setting up convection currents. If these currents occur spontaneously the process is called natural convection. If they have to be made to occur by a pump, fan, or similar device, the process is called forced convection.

convergence 1. The process of coming to a point. See *converging lens*. **2.** The process of tending to approach a finite limiting value. See *convergent series*.

convergent series A *series* in which the sum tends to a finite value or zero. Compare *divergent series*.

converging lens A *lens* capable of bringing to a point a *beam of light* passing through it; a *convex* lens. See Fig. 25 under *lens*.

converse The transposition of a statement consisting of a fact or datum and a consequent conclusion. Thus the converse of the proposition 'equal chords of a *circle* are equidistant from the centre' is 'chords that are equidistant from the centre of a circle are equal'. The converse of a true statement is not necessarily true.

conversion The process in a *nuclear reactor* as a result of which *fertile* material is transformed into *fissile* material, e.g. the conversion of thorium-232 into uranium-233. The 'conversion factor' is the number of fissile *atoms* produced from the fertile material per fissile atom destroyed in the fuel.

conversion electron An *orbital electron* ejected from an *atom* as a result of the *energy* it acquires from a transition of the *nucleus* from one energy state to another in the absence of *gamma-ray* emission. See also *internal conversion*.

converter 1. An electrical machine for converting *alternating current* to *direct current* or vice versa. **2.** The *retort* used in the *Bessemer process* or some similar steel-making process.

converter reactor A *nuclear reactor* that produces *fissile* material from *fertile* material by *conversion*.

convex Curving outwards; e.g. a convex *lens* is one thicker at the centre than at the edges. See Fig. 24 under *lens*.

coolant A *fluid* used for cooling, usually extracting *heat* from one source and transferring it to another. In a *nuclear reactor* the coolant transfers the heat from the *nuclear reaction* to the steam-raising plant.

Cooper pairs See *superconductivity*.

coordinate bond Dative Bond, semipolar bond. See *chemical bond*.

coordinate geometry See *analytical geometry*.

coordinates Magnitudes used to define the position of a point or line within a fixed frame of reference. See *Cartesian coordinates*; *polar coordinates*.

coordination compound A compound in which the *molecule* or a component *ion* of the molecule contains a central atom surrounded by atoms or groups of atoms (called *ligands*) attached to the central atom by a number of coordinate bonds (see *chemical bond*) in excess of the *stoichiometric* valence of the central atom. Thus, a *ferricyanide* is a coordination compound: in its *anion*, the hexacyano-ferrate(III) ion, $[Fe(CN)_6]^{3-}$, the central iron atom, which has a valence of three, is attached to six CN^- ions.

coordination number 1. The number of *ions* which surround a given ion in a *crystal lattice*. **2.** In the molecule of a *coordination compound*, the number of atoms, ions, molecules, or groups directly linked to the central atom or ion.

copal A natural *resin* obtained from certain trees. It is used in varnishes.

coplanar (math.) In the same *plane*.

copolymerization See *polymerization*.

copper Cu. Element. R.a.m. 63.546. At. No. 29. A red malleable and ductile *metal*, m.p. 1084°C., b.p. 2582°C., r.d. 8.93; after silver, it is the best *conductor* of electricity. It is unaffected by *water* or *steam*. It occurs as the free metal, and as cuprite or ruby *ore*, Cu_2O; *chalcocite*, Cu_2S; and *chalcopyrite*, $CuFeS_2$. Copper is extracted from *sulphide* ores by alternate roasting and fusing with sand, thus removing iron and *volatile* impurities, and leaving a mixture of *copper(I) oxide* and sulphide. This is then heated in a *reverberatory furnace*, giving impure copper, which is then refined by various methods (often by *electrolysis*). It is used for steam boilers, electrical wire and apparatus, in *electrotyping*, and in numerous *alloys*, e.g. *bronze, brass, speculum metal, gun metal, bell metal, Dutch metal, manganin, constantan, nickel silver, German silver,* etc.

copperas See *iron(II) sulphate.*

copper chlorides Copper(II) chloride (cupric chloride), $CuCl_2$, is a brownish powder, m.p. 620°C., obtained by passing chlorine over heated copper. The dihydrate is blue-green. Copper(I) chloride (cuprous chloride), $CuCl$, m.p. 430°C., is obtained by boiling copper(II) chloride with copper and *hydrochloric acid.* It is used as a *catalyst.*

copper glance See *chalcocite.*

copper oxides Copper(I) oxide (cuprous oxide or red copper oxide), Cu_2O, is a red *insoluble* powder, m.p. 1235°C., which occurs naturally as the mineral *cuprite.* It is used to colour glass red and as a *fungicide.* Copper(II) oxide (cupric oxide), CuO, is a black insoluble substance, m.p. 1326°C.

copper pyrites See *chalcopyrite.*

copper(II) sulphate Cupric sulphate, blue vitriol. $CuSO_4.5H_2O$. A blue crystalline *soluble salt,* obtained by reacting copper(II) oxide with dilute sulphuric acid. It is used as a *mordant, insecticide,* and *fungicide* (see *Bordeaux mixture*). It is also used in *electroplating* and to preserve wood.

coral Deposits of impure *calcium carbonate,* $CaCO_3$, formed of the hard skeletons of various marine *organisms.*

cordite An *explosive* prepared from *cellulose nitrate* and *nitroglycerine.*

core 1. A magnetic material that is used to increase the inductance of a coil through which it passes. It may be laminated or made of compressed ferromagnetic particles. **2.** The central part of a *nuclear reactor* that contains the fissile material and in which the nuclear reaction takes place. **3.** The devices, *semiconductors, ferrite* rings, etc., that constitute the memory of a computer. **4.** See *Earth.*

Coriolis force A fictitious force used to simplify calculations involving rotating systems, such as the movement of air on the surface of the Earth. For example, to an observer on a rotating disc, a particle moving in a straight line from the centre of the disc to its circumference would appear to be moving in a curved path. The Coriolis force is the fictitious force required to account for the tangential acceleration. Named after Gaspard de Coriolis (1792–1843).

corona A white irregular halo surrounding the *Sun,* which is visible during a total *eclipse.*

corona discharge A luminous discharge that appears round the surface of a *conductor* due to *ionization* of the air (or other *gas* surrounding it), caused by the *voltage* gradient exceeding a critical value, but not being sufficient to cause sparking.

corpuscle See *blood cell.*

corpuscular theory The theory that *light* consists of minute corpuscles in rapid motion. The original corpuscular theory was abandoned in the middle of the nineteenth century in favour of the *wave theory of light,* first put forward by

Huygens in 1678. Later research has shown that light phenomena must be interpreted in terms of *photons* (see *photoelectric effect*) and waves, as the two descriptions are merely two different ways of viewing one and the same reality. See *complementarity*.

corrosion Surface chemical action, especially on *metals*, by the action of moisture, air, or chemicals. See also *rust*.

corrosive sublimate See *mercury(II) chloride*.

cortex See *medulla*.

corticosteroid Any of a group of *hormones* produced by the cortex of the adrenal glands. They are all *steroids* and fall into two classes: glucocorticoids (such as *cortisone*), which are used as anti-inflammatory drugs and which regulate the body's use of *carbohydrates*, *fats*, and *proteins*, and mineralocorticoids (such as aldosterone), which regulate the *salt* and *water* balance.

cortisone 17-hydroxy-11-dehydrocorticosterone. $C_{21}H_{28}O_5$. A crystalline *corticosteroid hormone*, m.p. 215°C., secreted by the cortex of the adrenal gland. It reduces local inflammation and is used in the treatment of rheumatic and other conditions.

corubin Crystalline *aluminium oxide*, Al_2O_3, obtained as a *by-product* of the *aluminothermic reduction*.

corundum Natural *aluminium oxide*, Al_2O_3. A crystalline substance nearly as hard as *diamond*, used as an *abrasive*.

cosecant See *trigonometrical ratios*.

cosine See *trigonometrical ratios*.

cosine rule In any triangle,
$$c^2 = a^2 + b^2 - 2ab\cos\theta,$$
where θ is the angle between the sides a and b and opposite the side c.

cosmic dust Small particles of *matter*, probably ranging in size from one hundredth to one ten-thousandth of a *millimetre*, distributed throughout *space*.

cosmic rays Very energetic *radiation* falling upon the Earth from outer *space*, and consisting chiefly, if not entirely, of charged particles. The majority of these primary rays are most probably *protons*, although *electrons* and *alpha particles* are also present. There is also evidence that a small component (about 2%) of the primary radiation consists of heavy atomic *nuclei*. The primary particles, when incident upon our atmosphere, cause several secondary processes. Proton-*neutron* collisions in the top tenth of the atmosphere give rise to *mesons*. The charged mesons soon decay into *muons* while the neutral mesons decay generally to *photons*. The photons generate electrons and *positrons*; the muons, although unstable, have their *lifetimes* extended by *time dilation*, which at their high speeds enables many of them to survive to reach ground level and penetrate underground. *Neutrinos* are also produced in the decay of mesons and muons. Energies as high as 10^{20} *electronvolts* have been observed with cosmic ray particles. The origin of cosmic rays is not known with certainty although some appear to emanate from the *Sun*. See also *east-west asymmetry*.

cosmic ray shower Cascade shower. See *shower*.

cosmogony The science of the nature of the heavenly bodies, with particular reference to the formation of *planets*, *stars*, and *galaxies*.

cosmology The science of the nature, origin, and history of the *Universe*. A more general and widely used term than *cosmogony* when referring to the Universe as a whole. See *steady state theory, superdense theory.*

cotangent See *trigonometrical ratios.*

Cottrell precipitator A device used for removal of dust particles from gases by *electrostatic precipitation.*

COUDÉ TELESCOPE

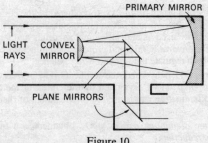

Figure 10.

coudé system A form of astronomical reflecting *telescope* in which *light* from the primary *mirror* is reflected back along the axis of the telescope by means of a system of mirrors as shown in Fig. 10. This system may also be adapted for use with a refracting telescope, enabling the telescope to rotate without rotating the base.

coulomb The derived *SI unit* of *electric charge*, defined as the quantity of electricity transferred by 1 *ampere* in one second. 10^{-1} *electromagnetic unit*; 3×10^9 *electrostatic units*. Symbol C. Named after Charles Augustin Coulomb (1736–1806).

coulomb explosion imaging A technique used to investigate molecular configuration. A beam of the neutral molecules to be investigated is accelerated and passed through a very thin metal foil, which scatters the electrons in the atoms of molecules, leaving only the atomic nuclei. This occurs in such a short period that the nuclei retain their configuration before exploding away from each other as a result of like electric charges. When this has happened the nuclei impinge on a detector, which records their speed and direction, enabling their original configuration to be calculated.

Coulomb scattering The *scattering* of sub-atomic particles caused by the *electrostatic* (coulomb) *field* surrounding an atomic *nucleus.*

Coulomb's law The *force* of attraction or repulsion between two charged bodies (assuming that the charges behave as though they were concentrated at a point) is proportional to the magnitude of the charges and inversely proportional to the square of the distance between them. In *SI units*, the equation is written $F = q_1q_2/4\pi\varepsilon_0d^2$, where F is the force in newtons, q_1 and q_2 are the charges in coulombs, d is the distance between them in metres, and ε_0 is the *electric constant.*

coulometer Coulombmeter. See *voltameter.*

coumarin $C_9H_6O_2$. A white crystalline substance, m.p. 71°C., with an odour of vanilla. Used as a flavour and in perfume.

coumarone Benzofuran. $\overline{C_6H_4OCH{:}CH}$. A *liquid* derived from *coal-tar*, b.p. 173°C., that polymerizes into a synthetic *resin*; used in the *paint* and varnish industry.

counter tube A device for counting individual ionizing events. It usually consists of a detector combined with a *scaler*. See *Cerenkov radiation*; *crystal counter*; *Geiger counter*; *proportional counter*; *scintillation counter*.

couple (phys.) Two equal and opposite parallel, but not colinear, *forces* acting upon a body. The *moment* of a couple is the product of either force and the perpendicular distance between the line of action of the forces.

coupling reaction See *azo coupling*.

covalent bond See *chemical bond*; *orbital*.

covalent crystal A *crystal* in which the *atoms* are held in the *lattice* by covalent bonds (see *chemical bond*). Typical examples are *diamond* and silicon(IV) oxide. See also *semiconductors*.

covalent radius The effective radius of an *atom* in a compound held together by covalent bonds (see *chemical bond*). In a *diatomic molecule* it will be half the distance between the atomic nuclei. With more complicated molecules this is not always the case as the ionic contribution to the bonding has to be taken into account.

CP See *parity*.

CPT See *parity*.

CPU Central processing unit or central processor. The central *electronic* unit in a *computer* that processes input information, and information from the *memory*, and produces the output information. The CPU and the memory store form the central part of the computer. The devices connected to them, known as *peripherals*, include the *backing store* and the input and output equipment. In recent years many computing systems have relied on some of the CPU functions being undertaken in various independent units; in these systems the CPU is usually called the processor.

cracking (chem.) Pyrolysis. The *decomposition* of a chemical substance by raising its *temperature*; especially the conversion of *mineral oils* of high *boiling point* into more *volatile* oils suitable for petrol engines, by 'cracking' the larger *molecules* of the heavy oils into smaller ones. In catalytic cracking the decomposition takes place in the presence of a *catalyst*, usually at a lower temperature.

cream of tartar See *potassium hydrogentartrate*. $C_4O_6H_5K$.

creatinine $C_4H_7N_3O$. A white crystalline substance, derived from the *amino acid* creatine, $C_3H_8N_3COOH$, found in urine, blood, etc.

creep A permanent change in the physical dimensions of a *metal* caused by the application of a continuous *stress*. This occurs between the *elastic limit* of the material and its *yield point*.

creosote A *distillation* product obtained from *coal-tar* or from the tar obtained by the *destructive distillation* of wood. An oily, transparent *liquid* containing *phenol* and *cresol*, it is used for preserving timber.

cresol See *methylphenol*.

cristobalite A mineral consisting of silicon(IV) oxide, SiO_2.

crith The *mass* of 1 litre of hydrogen at 0°C., and a pressure of 760 mm; approximately 0.09 g.

critical angle (phys.) The least angle of *incidence* at which *total internal reflection* occurs in a medium. When a ray of *light* passing from a denser to a less dense medium, e.g. *glass* to air, meets the surface, a portion of the light does not emerge, but is internally reflected. As the angle of incidence increases, the intensity of the internally reflected beam also increases until an angle is reached when the whole beam is thrown back, total internal reflection taking place.

critical damping A measuring instrument is said to be critically damped when it takes up its *equilibrium* deflection in the shortest possible time, the oscillations of the indicator (needle) about the equilibrium position being quickly damped out. *Galvanometers* are normally used critically damped.

critical mass The minimum *mass* of *fissile* material required in a *nuclear reactor* or a *nuclear weapon* to sustain a *chain reaction*. In a mass of fissile material below the critical mass, too many *neutrons* escape without causing fissions for the *chain reaction* to occur.

critical potential The minimum *energy* required to raise the *energy level* of an *orbital electron* (see *excitation*) or to remove it from the *atom*. See *ionization potential*; *radiation potential*.

critical pressure The *pressure* of the *saturated vapour* of a substance at the *critical temperature*.

critical reaction See *chain reaction*.

critical state Critical point. The state of a substance when its liquid and gaseous *phases* have the same *density*, at the same *temperature* and *pressure*.

critical temperature The *temperature* above which a gas cannot be liquefied by an increase of *pressure* alone.

critical velocity The *velocity* at which the flow of a *liquid* ceases to be *streamline* and becomes *turbulent*.

critical volume The *volume* occupied by unit mass of a substance at its *critical temperature* and *critical pressure*.

CRO See *cathode-ray oscilloscope*.

cross-linkage (chem.) The joining of polymer *molecules* (see *polymerization*) to each other by *chemical bonds*. A polymer may be imagined, in the simplest case, to consist of very long chain-like *molecules*; cross-linkage would have the effect of joining adjacent chains by lateral links.

cross-section In *nuclear physics* the cross-section represents the effective area that has to be attributed to a particular *atom*, *nucleus*, or *elementary particle* to account geometrically for its interaction with an incident *beam* of *radiation*. The 'total' (or 'collision') cross-section, which accounts for all interactions, is subdivided into the 'elastic cross-section' and the 'inelastic cross-section'. The elastic cross-section accounts for all elastic *scattering* in which the incident radiation suffers no loss of *energy* to the atom or nuclei. The inelastic cross-section accounts for all other interactions and may be further subdivided to account for specific interactions, e.g. 'capture cross-section', 'fission cross-section', 'ionization cross-section', etc.

crotonic acid Butenoic acid. $CH_3CH{:}CHCOOH$. A colourless crystalline *soluble* substance that exists in two *isomeric* forms. *Trans*-but-2-enoic acid has a m.p. of

71.6°C. and is used in organic synthesis. *Cis*-but-2-enoic acid (isocrotonic acid) has a m.p. of 14.5°C.

crown glass A variety of *glass* containing potassium or barium in place of sodium; it is less fusible than ordinary *soda glass*, and is used in optical instruments.

crucible A vessel of heat-resisting material used for containing high temperature *chemical reactions*.

crude oil See *petroleum*.

crust See *Earth*.

cryogen See *freezing mixture*.

cryogenic pump A *vacuum pump* in which gases are condensed on surfaces cooled to 4 K by means of liquid helium. Such a device, used with a *condensation pump*, enables pressures down to 10^{-13} pascal to be obtained.

cryogenics The study of materials and phenomena at *temperatures* close to *absolute zero*. By means of *adiabatic demagnetization* temperatures of 10^{-6} K can be reached.

cryohydrates Crystalline substances, containing the *solute* with a definite molecular proportion of *water*, that crystallize out from *solutions* cooled below the *freezing point* of pure water.

cryolite Natural sodium aluminium fluoride, Na_3AlF_6. Used in the manufacture of aluminium.

cryometer A *thermometer* especially designed for measuring low *temperatures*. A nuclear-resonance thermometer can measure temperatures as low as 3×10^{-7} K.

cryophorus An apparatus used to demonstrate the cooling effect of *evaporation*.

cryoscopic method Freezing-point method. The determination of the *relative molecular mass* of a dissolved substance by noting the *depression of freezing point* produced by a known *concentration* of it in *solvent*.

cryostat A vessel in which a specified low *temperature* may be maintained.

cryotron A switch based on *superconductivity*. The simplest form consists of a coil of wire of one superconducting material wound round a length of wire of another superconductor, all immersed in a bath of liquid helium. A control current passed through the coil produces a *magnetic field* strong enough to destroy the superconductivity of the central wire but not of the coil. Thus the current in the coil controls the *resistance* of the wire, switching it from zero to a finite value.

crystal A substance that has solidified in a definite geometrical form. Most *solid* substances, when pure, are obtainable in a definite crystalline form. Solids that do not form crystals are said to be *amorphous*. Crystals are classified according to the structure of their *lattices* (see *crystal systems*), or according to the type of *bond* that holds them together, i.e. *ionic crystals*, *covalent crystals*, or *metallic crystals*.

crystal counter A *counter tube* that depends upon a *crystal* in which the electrical *conductivity* is momentarily increased by an ionizing event.

crystal detector See *detector*. A fine wire ('cat's whisker') in contact with a crystal of *galena* (PbS) or other suitable *semiconductor*. This arrangement is a good conductor of *electric current* in one direction, and suppresses most of the flow of current in the other direction. It was used in crystal radio sets, before the advent of thermionic valves or *transistors*.

crystal lattice See *lattice*.

crystalline Consisting of *crystals* or having a regular structure of *atoms*, *ions*, or *molecules* characteristic of crystals (as in *metals*).

crystallite A small *crystal*, as in a microcrystalline substance.

crystallization The formation of *crystals* in a *liquid* or *gas*.

crystallography The study of the geometrical form of *crystals*. See *X-ray crystallography*.

crystalloids Substances that, in *solution*, are able to pass through a *semipermeable membrane*; substances that do not usually form *colloidal solutions*.

crystal microphone A microphone in which the sound waves to be amplified or transmitted vibrate a *piezoelectric crystal*, which generates a varying EMF.

crystal oscillator A source of electrical oscillations of very constant *frequency* determined by the physical characteristics of a *piezoelectric crystal*. The crystal, usually *quartz*, either produces the oscillation in a tuned circuit or is coupled to a tuned circuit to control its frequency. See *quartz clock*.

crystal pick-up A pick-up in a record player in which the varying EMF is produced by a *piezoelectric crystal* as a result of the vibrations obtained from the undulations in the groove of the record.

crystal rectifier A *semiconductor diode* used as a *rectifier*, usually in a manner similar to a *diode* valve.

crystal systems The seven classes into which crystals are divided: *cubic*, tetragonal, orthorhombic, hexagonal, trigonal, monoclinic, and triclinic. The definition of each class depends on the relative lengths of the sides of the *unit cell* and the angles between them.

CS $C_6H_4ClCH:C(CN)_2$. A white powder, m.p. 52°C., used as a tear gas and 'harassing agent' in crowd control. It causes tears, salivation, choking, and painful breathing.

CT scanner See *tomography*.

cube 1. A regular hexahedron; a regular solid figure with six square faces. **2.** The third power of a number. E.g. 8 is the cube of 2, 2^3.

cube root $\sqrt[3]{}$. The cube root of a number is the quantity that when raised to the third power gives that number. Thus 2 is the cube root of 8.

cubic crystal A *crystal system* in which the *unit cell* is a *cube*. In the simple cubic system there are *lattice points* at the cube's eight corners. See also *body-centred*.

culture medium A nutrient preparation used for growing and cultivating *cells*, *tissues*, or *microorganisms* for experimental purposes.

cumene (1-methylethyl)benzene. $C_6H_5CH(CH_3)_2$. A colourless liquid *aromatic hydrocarbon*, b.p. 152°C. It occurs in *petroleum* and is used as an *intermediate* in *organic synthesis*.

cumene process A method of making *phenol* by passing *benzene* and *propene* vapours over a catalyst (usually phosphoric acid) at 250°C. The resulting *cumene*, $C_6H_5CH(CH_3)_2$, can be oxidized to form the peroxide, $C_6H_5C(CH_3)_2O_2H$, which reacts with dilute acid to give phenol.

cupellation The separation of silver, gold, and other *noble metals* from impurities that are oxidized by hot air. The impure metal is placed in a cupel, a flat dish made of porous *refractory* material, and a blast of hot air is directed upon it in a special furnace. The impurities are oxidized by the air and are partly swept away by the blast and partly absorbed by the cupel.

cuprammonium ion The bivalent tetraamminecopper(II) cation, $[Cu(NH_3)_4]^{2+}$, formed by copper with *ammonia ligands*. Cuprammonium solution was used in the manufacture of *rayon* by the now obsolete cuprammonium process. See *Schweitzer's reagent*.

cupric Containing copper in its +2 *oxidation state*, e.g. cupric chloride, copper(II) chloride, $CuCl_2$.

cupric oxide See *copper oxides*.

cupric sulphate See *copper sulphate*.

cuprite See *copper oxides*.

cupronickel An *alloy* of copper and nickel used in coinage.

cuprous Containing copper in its +1 *oxidation state*, e.g. cuprous chloride, copper(I) chloride, $CuCl$.

cuprous oxide See *copper oxides*.

curare A very poisonous material, containing certain *alkaloids*. Obtained from various South American trees.

curie A former measure of the *activity* of a *radioactive* substance (see *radioactivity*). Originally defined as the quantity of *radon* in radioactive equilibrium with 1 g of radium, it was then extended to cover all radioactive substances and defined as that quantity of a radioactive isotope that decays at the rate of 3.7×10^{10} *disintegrations* per second. It has now been replaced by the *becquerel*. Named after Pierre Curie (1859–1906).

Curie point Curie temperature. The temperature for a given *ferromagnetic* substance above which it becomes merely *paramagnetic*. Named after Pierre Curie.

Curie's law The *magnetic susceptibility* (χ) of a *paramagnetic* substance is inversely proportional to the *thermodynamic temperature*. The constant of proportionality (C) in the equation $\chi = C/T$, is called the Curie constant. Named after Pierre Curie.

Curie-Weiss law A modification of *Curie's law*, of greater applicability, stating that $\chi = C/(T - \theta)$, where θ is the Weiss constant, which is characteristic of individual substances. Named after Pierre Curie and Pierre-Ernest Weiss (1865–1940).

curium Cm. *Transuranic element*, At. No. 96. A radioactive *actinide* whose most stable *isotope*, curium-247, has a *half-life* of 1.6×10^7 years.

current See *electric current*.

current balance An instrument for the determination of an *electric current* by measuring the *force* that the current produces between conductors. A common type consists of two similar coils attached to the extremities of a balance arm. Above and below each of these coils is a fixed coil. The six coils are connected in series in such a way that when the current is allowed to pass through them, the beam experiences maximum *torque*. The beam is restored to its horizontal *equilibrium* position by means of a known torque supplied by a rider sliding along the arm. From the known torque and the geometry of the system the current can be calculated.

current density 1. The current flowing through a conductor, *plasma*, etc., per unit cross-sectional area. It is usually expressed in amperes per square metre. **2.** (in *electrolysis*) The current flowing through an *electrolyte* per unit area of electrode.

cursor An indicating device, such as a spot of light, on a *VDU*.

cyanamide NH_2CN. A colourless crystalline *unstable* substance, m.p. 44°C. See also *calcium cyanamide*.

cyanate Fulminate. A *salt* or *ester* of *cyanic acid* (fulminic acid).

cyanic acid HOCN. An explosive *liquid*, with the structure $H-O-C\equiv N$, also known as fulminic acid. Its *salts*, the *cyanates* (or fulminates) are also explosive. Cyanic acid is isomeric with *isocyanic acid*, with the structure $H-N=C=O$.

cyanide A *salt* of *hydrocyanic acid*, HCN. All cyanides are intensely poisonous.

cyanide process The extraction of gold from its *ores* by dissolving the gold in a *solution* of potassium cyanide, KCN, reducing the resulting potassium auro-cyanide, $KAu(CN)_2$, with zinc, filtering off, melting down, and cupelling (see *cupellation*) the metal.

cyanine dyes A group of *dyes* containing a chain of carbon atoms with *conjugated double bonds* forming a bridge between two *heterocyclic nuclei*. They are used for photographic *sensitization*.

cyanite Al_2SiO_5. A blue mineral consisting of aluminium silicate, used as a *refractory*.

cyanocobalamin Vitamin B_{12}. See *vitamins*.

cyanogen C_2N_2. A colourless, very poisonous *gas* with a smell of bitter almonds. In its chemical properties it resembles the *halogens*, forming *cyanides* analogous to the *chlorides*. It is prepared by heating mercury(II) cyanide.

cyano group The *univalent group* –CN.

cyanoguanidine See *dicyandiamide*.

cyanuric acid Tricyanic acid. $C_3H_3O_3N_3.2H_2O$. A white crystalline *soluble* substance, used in organic synthesis. It consists of a six-membered ring of alternating –NH– and –CO– groups.

cybernetics The theory of communication and control mechanisms in living beings and machines.

cybotaxis The tendency of *molecules* in *liquids* to form regularly arranged groups, resembling *crystals*. See *liquid crystals*.

cyclamate See *sodium cyclamate*, *calcium cyclamate*.

cycle (phys.) Any series of changes or operations performed by or on a system, which brings it back to its original state. The time taken to complete a cycle is the *period*. The number of cycles in unit time is the *frequency* (measured in *hertz*).

cyclic (chem.) Having a ring structure. See *carbocyclic* and *heterocyclic compounds*.

cyclic figure (math.) A figure through all the vertices or corners of which a *circle* may be drawn; a figure inscribed in a circle.

cyclic quadrilateral A four-sided plane *rectilinear* figure through the vertices of which a *circle* may be drawn. The pairs of opposite angles are *supplementary* (i.e. total 180°).

cyclization The conversion of an open chain molecule into a *cyclic* compound.

cycloalkane A *saturated cyclic* compound with the general formula C_nH_{2n}. They are chemically similar to the *alkanes* but are less reactive. See *cyclohexane*; *cyclopentane*; *cyclopropane*.

cyclohexane C_6H_{12}. A colourless flammable *liquid cycloalkane*, b.p. 81°C., consisting of a six-membered ring. It is used as a *solvent* and in the manufacture of *plastics*.

cyclohexanol $C_6H_{11}OH$. A crystalline *soluble* substance, m.p. 25.1°C., b.p. 161.1°C., used as a *solvent*.

cycloid A figure traced out in space by a point on the circumference of a *circle*, which rolls without slipping along a fixed straight line.

cyclonite Hexogen, R.D.X.*. $(CH_2N.NO_2)_3$. A very powerful *explosive* made from *hexamine*.

cyclopentane C_5H_{10}. A colourless *liquid cycloalkane*, b.p. 49.2°C., obtained from *petroleum* and used as a *solvent*.

cyclopropane C_3H_6. A colourless flammable *cycloalkane gas*, used as an *anaesthetic*.

cyclotron An *accelerator* for imparting to charged particles of atomic magnitudes *energies* of several million *electronvolts*. The *ions* or charged particles are caused to traverse a spiral path between two hollow semicircular *electrodes*, called dees, by means of a suitable *magnetic field* applied perpendicularly to the plane of the dees. At each half-revolution the particles receive and energy increase of some tens of thousands of electronvolts from an oscillating *voltage* applied between the dees. See also *synchrocyclotron*.

cylinder A solid figure traced out by a rectangle rotating round one side as *axis*. For a cylinder having vertical height h and radius of base r, the volume is $\pi r^2 h$ and the total surface area $2\pi r(h + r)$.

cysteine A crystalline *amino acid*, present in most *proteins*. See Appendix, Table 5.

cystine An *insoluble* crystalline *amino acid*, m.p. 247–249°C., which forms *cysteine* on *reduction*. See Appendix, Table 5.

cytochemistry The *chemistry* of living *cells*.

cytochrome A *respiratory pigment* widely distributed in *aerobic organisms*. It consists of *proteins* with an iron *prosthetic group* similar to that of *haemoglobin*. The *oxidation* of cytochrome by molecular oxygen, and its subsequent *reduction* in the cell, is the principal route by which atmospheric oxygen enters into cellular *metabolism*.

cytokinins *Plant hormones* that promote *cell* division in plants. They have potential uses in prolonging the freshness of vegetables and cut flowers.

cytology The study of the structure and function of living *cells*.

cytolysis The dissolution of *cells*, particularly by the destruction of their surface membranes.

cytoplasm The *protoplasm* of a living *cell* outside its *nucleus*.

cytosine Aminopyrimidone. $C_4H_5N_3O$. A white crystalline substance, m.p. 320–325°C. One of the *pyrimidine* bases that occur in the *nucleotides* of the *nucleic acids* and play a part in the formulation of the *genetic code*.

cytotoxic Toxic to living *cells*. Cytotoxic drugs are used to treat some cancers because they inhibit cell division.

D

daily variation of the Earth's magnetic field (see *magnetism, terrestrial*). The small variation of the *horizontal intensity, magnetic declination*, and *dip* recurring over a period of a day.

dalton An alternative name for the *atomic mass unit*.

Dalton's atomic theory See *atomic theory*.

Dalton's law of partial pressures. The total *pressure* of a mixture of two or more *gases* or *vapours* is equal to the sum of the pressures that each component would exert if it were present alone, and occupied the same *volume* as the whole mixture. It applies to *perfect gases*. Named after John Dalton (1766–1844).

damping A decrease in the *amplitude* of an oscillation or *wave motion* with time. See also *critical damping*.

Daniell cell A *primary cell* having a negative pole of amalgamated zinc, standing in a porous pot containing dilute *sulphuric acid*. This pot stands in *copper(II)* sulphate solution, which also contains the positive pole, a copper plate. On completion of the external circuit, a current flows and the following reactions take place: at the negative pole, zinc is dissolved, *zinc sulphate* being formed; at the positive pole, copper is deposited. The *EMF* is 1.1 *volts*. Named after J. F. Daniell (1790–1845).

daraf The practical unit of *elastance*; reciprocal of the *farad*.

dark ground illumination A device used in microscopy, whereby transparent or unstained objects are made to appear as bright particles on a black background.

dark matter See *missing mass*.

Darwin's theory The theory of evolution proposing that different species arise by the process of *natural selection* acting on the variations found in normal populations. See also *neo-Darwinism*. Named after Charles Darwin (1809–82).

dash-pot A mechanical *damping* device that depends upon the fact that when a body moves through a *fluid* medium, viscous *forces* are set up, which damp the motion of the body. It usually consists of a piston, attached to the part whose movements are to be damped, fitting loosely into a cylinder containing either air or oil.

dasymeter An instrument for determining the *density* of a *gas*.

database A collection of data stored in a *computer* and coded in such a way that it can be extracted in various ways, under various headings. For example, the information in this dictionary is stored in a computer and coded so that all the entries relating to chemistry, say, can be extracted.

dating The determination of the age of *mineral, fossil*, or wooden objects by measuring some parameter of the specimen that varies at a known rate. In *chemical dating* this parameter relates to the chemical composition of the specimen; in *radiometric dating* it refers to a change in the radioactivity of the specimen (see *fission-track dating; radiocarbon dating; potassium-argon dating; rubidium-strontium dating; uranium-lead dating; radioactive age; thermo-*

luminescence). Some dating techniques depend on seasonal variations in the growing period (see *dendrochronology*; *varve dating*).

dative bond Coordinate bond. See *chemical bond*.

daughter A radioactive *decay* product or the product of a chemical *dissociation*.

Davy lamp See *safety lamp*.

day The time for the *Earth* to revolve once on its axis. The solar day is the time taken for one revolution between two successive returns of the *Sun* to the *meridian*. The mean solar day of 24 hours is the mean value of the solar day for one year. The sidereal day of 23 hours 55.91 minutes is measured with respect to the fixed stars and therefore the Earth's orbital motion has also to be taken into account.

d.c. See *direct current*.

DDT Dichlorodiphenyltrichloroethane. $(C_6H_4Cl)_2.CH.CCl_3$. A white powder, m.p. 107°C., with a fruity smell. It is used as a contact *insecticide* although restrictions have now been placed on its use as it accumulates in the soil and can accumulate in cattle and other animals high in the *food chain*.

deaminase An *enzyme* that catalyses the removal of an *amino group* from a *compound*.

deamination The removal of an *amino group* from a *compound*.

Dean and Stark method A method of estimating the quantity of *water* in an *oil* or other *liquid* substance. The liquid under examination is distilled into a special *reflux condenser* so constructed that the water is prevented from running back into the distillation flask. The *volume* of water collected is measured and thus the water content of a known *mass* of initial liquid can be calculated.

de Broglie wavelength A moving particle, whatever its nature, has wave properties associated with it (see *complementarity*). For a particle of *mass m* moving with *velocity v*, the *wavelength* of the associated de Broglie wave is given by $\lambda = h/mv$, where *h* is the *Planck constant*. Named after Louis Victor de Broglie (1892–1987).

debye A unit of molecular *dipole* moment equal to 1×10^{-18} *electrostatic unit* or $3.335\ 64 \times 10^{-30}$ *coulomb metre*. Named after Peter J. W. Debye (1884–1966).

Debye–Huckel theory of electrolytic dissociation. See *electrolytic dissociation*.

deca- Prefix denoting ten times in the *metric system*. Symbol da, e.g. dam = 10 *metres*.

decane $C_{10}H_{22}$. A *liquid alkane*, b.p. 174.1°C., that occurs in several *isomeric* forms.

decanoic acid Capric acid. $C_9H_{19}COOH$. A white crystalline *organic acid* with an unpleasant odour, m.p. 31.5°C. Decanoate *esters* are used in the manufacture of flavouring substances and perfumes.

decantation The separation of a *solid* from a *liquid* by allowing the former to settle and pouring off the latter.

decay 1. The *transformation* of a radioactive substance into its decay (or daughter) products. (See *disintegration constant*; *radioactivity*.) **2.** The transformation of a particle into a more stable particle. See also *beta decay*; *half-life*.

decay constant See *disintegration constant*.

deci- Prefix denoting one tenth; 10^{-1}. Symbol d, e.g. dm = 0.1 *metre*.

decibel One tenth of a *bel*. A unit for comparing levels of *power*. Two power levels, P_1 and P_2, are said to differ by n decibels when:

$$n = 10 \log_{10} P_2/P_1$$

This unit is often used to express *sound* intensities. In this case, P_2 is the intensity of the sound under consideration and P_1 is the intensity of some reference level, often the intensity of the lowest audible note of the same *frequency*.

deciduous 1. Denoting a plant that sheds all its leaves at the end of each growing season. Compare *evergreen*. **2.** Denoting the teeth (milk teeth) that are the first set of teeth of those mammals that have two sets of teeth in their lifetime.

declination (astr.) The *angular distance* of a heavenly body from the *celestial equator*.

declination, magnetic See *magnetic declination*.

decomposer An organism, such as certain *bacteria* and *fungi*, that breaks down dead organisms or organic waste materials. They therefore return decaying organic matter to the environment as inorganic matter, thus fulfilling a useful ecological role.

decomposition (chem.) The breaking up of chemical compounds under various influences; e.g. by chemical action, by heat (*pyrolysis*), by an electric current (*electrolysis*), by biological agents (*biodegradation*), etc. See also *degradation*.

decrepitation The bursting or cracking of *crystals* of certain substances on heating, mainly due to expansion of water within the crystals.

defect A discontinuity in the pattern of *atoms*, *ions*, or *electrons* in a *crystal*. A 'point defect' consists of a *vacancy* or an *interstitial* (see also *Frenkel defect*). A 'line defect' is caused by a *dislocation*. In a *semiconductor*, 'defect conduction' is a result of *hole* conduction in the *valence band*.

deferent See *epicycle*.

deficiency diseases Diseases produced by lack of a particular *vitamin*, *amino acid*, *mineral*, or other essential food factor in the diet; e.g. scurvy, caused by the deficiency of vitamin C.

definite integral See *integration*.

definition 1. The sharpness of an *image* formed by a *lens*, *mirror*, or other optical system. **2.** The accuracy of *sound* or vision reproduction in a *radio* or *television* set.

deformation An alteration in the size or shape of a body.

deformation potential The *electric potential* that acts on a *free electron* in a *conductor* or *semiconductor* as a result of *deformation* of the *crystal lattice*.

degassing The removal of *adsorbed* gases from a solid or dissolved gases from a liquid. In high-vacuum techniques it is important to degas the walls of the low-pressure vessel as the gases adsorbed will begin to desorb as the pressure falls.

degaussing The *demagnetization* of a magnetized substance, achieved by surrounding the substance with a coil carrying an *alternating current* of ever-decreasing magnitude.

degenerate gas 1. A state of *matter* in which *electrons* and atomic *nuclei* are packed too closely together for the evolution of *nuclear energy*; it is believed to occur in *stars* of the *white dwarf* class. **2.** A gas in which the concentration of particles causes a significant departure from the classical *Maxwell–Boltzmann distribution*.

degenerate semiconductor A doped *semiconductor* (see *doping*) in which the *Fermi level* is in the *valence band* or the *conduction band* causing the semiconductor to behave as a metal.

degenerate states *Quantum states* in a system that have the same *energy*.

degradation (chem.) In general, the breakdown of *molecules* into simpler fragments; especially the stepwise decrease of the length of *polymer macromolecules*.

degree 1. A subdivision of an interval in a scale of measurement; e.g. the *Celsius* degree. **2.** A measure of *angle*. One three hundred and sixtieth of the angle traced by the complete revolution of a line OA about the point 0, until it returns to its original position. **3.** The sum of the *exponents* of the *variables* in a mathematical expression; the exponent of the *derivative* of highest *order* in a *differential equation*.

degree of latitude and longitude See *latitude*; *longitude*.

degrees of freedom (chem.) **1.** The least number of independent variables in the *phase rule* that defines the state of a system (e.g. the *temperature* and *pressure* in the case of a *gas*) that must be given definite values before this state is completely determined. **2.** The number of independent ways in which a *molecule* may possess translational, vibrational, or rotational *energies*.

dehydration The elimination or removal of *water*; usually the removal of chemically combined water. E.g. concentrated *sulphuric acid*, H_2SO_4, acts as a dehydrating agent on substances that contain hydrogen and oxygen and removes these in the proportions in which they occur in water.

dehydrogenase An *enzyme* that catalyses *oxidation* reactions by the removal of hydrogen from the *substrate*.

dekatron A gas-filled emission tube with a central *anode* usually surrounded by ten *cathodes* and associated transfer *electrodes*. Incoming pulses cause a *glow discharge* to be transferred from one cathode to the next so that the tube may be used for counting or switching.

delayed neutrons *Neutrons* resulting from *nuclear fission* that are emitted with a measurable time delay. Only a small proportion of neutrons are delayed, but the average delay period must be taken into account in the control of *nuclear reactors*. See *prompt neutrons*.

delay line A component or circuit designed to introduce a calculated delay in the transmission of a signal. See also *acoustoelectronics*; *fluidics*.

deliquescent Having the property of picking up moisture from the air to such an extent as to dissolve in it; becoming *liquid* on exposure to air.

delocalization In some chemical compounds the *valence electrons* cannot be regarded as belonging to a particular *chemical bond*, but are delocalized over several atoms. This can occur when the molecule contains a *conjugated double bond* or a *triple bond*, the delocalized electrons being those in the π bond (see *orbital*). Such molecules are more stable than they would be if these electrons were localized, which accounts for many of the properties of *aromatic* compounds.

delta connection A method of connecting the three windings of a three-*phase* electrical system. The windings are connected in series, the three-phase supply being taken from, or put into, the three junctions.

delta-iron An allotropic (see *allotropy*) from of pure iron that exists between 1400°C. and the *melting point*.

delta metal An *alloy* of copper (55%) and zinc (43%) with small amounts of iron and other *metals*.

delta ray An *electron* ejected from an *atom* by a fast moving ionizing particle.

demagnetization The process of depriving a body of its *ferromagnetic* properties. The 'demagnetization energy' is the *energy* that would be required to completely demagnetize a body. Demagnetization is achieved by disordering the *magnetic domains*, for example by subjecting the body to an *alternating current*, the magnitude of which is gradually decreased.

demodulation The process, in a *radio*, *television*, or *radar* receiver, of separating information from a *modulated carrier wave*. The equipment used is called a demodulator or a *detector*.

denature **1.** To add some poisonous substance to *ethanol* to make it unfit for human consumption, e.g. *methylated spirits*. **2.** To add another *isotope* to a *fissile* material to render it unsuitable for use in a *nuclear weapon*. **3.** To produce a structural change in a soluble protein, either chemically or by heating, so that it loses most of its solubility. It usually involves an unfolding of the *polypeptide* chain. **4.** To unwind two interwound helical chains of *deoxyribonucleic acid*. It is usually achieved by heating or by adding *alkali*.

dendrite **1.** (chem.) A many-branched *crystal*. **2.** (bio.) The branching processes of a *neurone* that carry impulses into the *cell* body and form *synapses* with the *axons* of other neurones.

dendrochronology A method of *dating* based on the growth rings of trees. It presupposes that trees grown in the same climatic conditions have a characteristic pattern of growth rings. This enables some fossil trees or wooden archaeological specimens to be dated.

dendrology The branch of *botany* concerned with trees and shrubs.

denitrifying bacteria *Bacteria* in the *soil* that, in the absence of oxygen, break down *nitrates* and *nitrites* with the evolution of free nitrogen (see *nitrogen cycle*).

denominator (math.) The number below the line in a *vulgar fraction* e.g. 4 in ¾.

densitometer An instrument for the measurement of the photographic *density* of an image produced by *light*, *X-rays*, *gamma rays*, etc., on a photographic film or plate. They consist essentially of a light source, a *photosensitive* cell, and a measuring instrument, enabling the reading with the specimen in place to be compared with the reading in its absence.

density The *mass* of unit *volume* of a substance. In *SI units* density is expressed in *kilograms* per cubic *metre*, in *c.g.s. units* in *grams* per cubic centimetre, and in *f.p.s.* units in *pounds* per cubic foot. $1 \text{ kg m}^{-3} = 10^{-3} \text{ g cm}^{-3} = 0.0624\,28 \text{ lb ft}^{-3}$. See also *relative density*; *vapour density*.

density, optical If one medium has a greater *refractive index* than another for *light* of a given *wavelength*, then it has the greater optical density for that wavelength.

density, photographic The common *logarithm* of the *opacity* of a part of a *negative* or transparency.

dentine The bony substance of which teeth and tusks are made. It is similar to bone in composition but has tiny channels through which pass the nerve fibres, blood vessels, and processes of the dentine-making cells.

deoxyribonucleic (desoxyribosenucleic) acid DNA. Long thread-like *molecules* found in *chromosomes* and some *viruses*, consisting of two interwound helical chains of *polynucleotides*. The *sugar* of all the *nucleotides* is 2-deoxy-D-*ribose*, but each nucleotide is characterized by one of the four following nitrogenous bases: *adenine*, *cytosine*, *guanine*, and *thymine*. The structure of a DNA molecule has been likened to a twisted rope-ladder, the sides of which consist of sugar–phosphate chains, the rungs of linked nitrogenous bases. The rungs consist of complementary base pairs linked by *hydrogen bonds*. DNA molecules are important because they are capable of self-replication (the sequence of bases in one of the two strands determines the sequence of complementary bases in a newly synthesized adjacent strand) and they determine the *proteins* made by the cell (see *genetic code*). Hence DNA determines all the inherited characteristics of the organism. See also *complementary DNA*.

depilatory A substance used for removing hair.

depleted material A material that contains less of a particular *isotope* than it normally possesses, especially a *nuclear fuel* that contains less *fissile* isotopes than natural uranium, e.g. the residue from an *isotope separation* plant or a *nuclear reactor*.

depletion layer Depletion region. A region of a *semiconductor* that is depleted of mobile charge *carriers*. It forms at an interface between two regions of different conductivity.

depolarization The prevention of electrical *polarization* in a *cell*. In the *Leclanché cell* polarization is reduced by surrounding the positive carbon pole with *manganese(IV) oxide*, MnO_2. This oxidizes the hydrogen liberated at the pole, the chief cause of polarization.

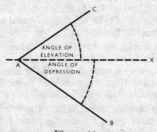

Figure 11.

depression, angle of If *B* is a point below the level of another point *A*, the angle of depression of *B* from *A* is the *angle* that *AB* makes with the horizontal *plane AX* through *A*. See Fig. 11.

depression of freezing point Lowering of the *freezing point* of a liquid when a *solid* is dissolved in it. With certain exceptions, the depression is proportional to the number of *molecules* or *ions* present (see *colligative property*), and the depression produced by the same *molar concentration* of any substance is a constant for a given *solvent*. This gives rise to the *cryoscopic method* for the determination of *relative molecular masses*.

depth of field The range of distances in front of and behind an object over which a *camera*, or other optical instrument, will produce a distinct *image*. Compare *depth of focus*.

depth of focus The range of distances between the lens and film in a camera over which the image remains distinct. Compare *depth of field*.

derivative (chem.) A *compound* derived from (but not necessarily prepared from) some other compound, usually retaining the general structure of the parent compound; e.g. *nitrobenzene*, $C_6H_5NO_2$, is a derivative of *benzene*, C_6H_6, one *hydrogen atom* in the *molecule* of the latter being replaced by a *nitro* group.

derivative (math.) Derived function. The result of *differentiation* of a mathematical function. See Appendix, Table 9.

derived unit See *base unit*.

desalination The process of removing *salt* from *sea-water* to make it suitable for agricultural purposes or for drinking. Various methods are possible, but to make the process commercially viable on a large scale, the waste heat from a *nuclear power* station is used to provide the *energy* for *distillation*, *freezing*, *electrodialysis*, *ion exchange*, or *reverse osmosis*. In some countries *solar energy* can also be used.

desiccation Drying; removal of moisture.

desiccator An apparatus used in laboratories for drying substances and for preventing *hygroscopic* substances from picking up moisture. It consists of a glass vessel, with a close-fitting ground lid, that contains some hygroscopic substance, e.g. *phosphorus(V) oxide*, P_2O_5. There is usually a tap in the lid to enable the vessel to be evacuated.

desorption The removal of *molecules*, *ions*, etc., from the surface of a solid so that they become gaseous; the reverse of *adsorption*.

destructive distillation Carbonization. Heating a complex substance to produce chemical changes in it, and distilling off the *volatile* substances so formed. E.g. the destructive distillation of *coal* was formerly used to produce *coal-gas* and many other valuable products.

detector 1. Any device that detects the passage of *radiation*, especially *elementary particles* and *electromagnetic radiation*. See *bubble chamber*; *cloud chamber*; *scintillation counter*; *wire chamber*. **2.** That part of a *radio* receiver in which the information is separated from the *modulated carrier wave*. It is now more usually called a demodulator. See *demodulation*; *crystal detector*.

detergents Washing or cleaning agents, usually consisting of *surfactants* or mixtures of these with other agents for specific purposes. Detergents cause such nonpolar substances as *grease* and *oil* to dissolve in *water*. In *soaps*, an *ion* of a long-chain fatty acid, such as the *stearate* (octadecanoate) ion, $CH_3(CH_2)_{16}COO^-$, has two parts: the nonpolar hydrocarbon chain, which attaches to the oil, and the $-COO^-$ group, which is attracted to the water. Synthetic detergents, such as sodium dodecylbenzene sulphonate, act in a similar fashion and are made from petrochemicals.

determinants An algebraic method of solving *simultaneous equations* in which an expression is written out in a square array. Thus, the determinant of $a_1b_2 - a_2b_1$ is written:

$$\begin{vmatrix} a_1b_1 \\ a_2b_2 \end{vmatrix}$$

detinning The recovery of metallic tin from scrap tin-plate by the action of chlorine, which combines with the tin to form *volatile tin(IV) chloride*, $SnCl_4$.

detonating gas A *mixture* of hydrogen and oxygen in a *volume* ratio of 2:1; i.e. in the volume ratio required to form *water*. It is extremely explosive when ignited.

detonation Extremely rapid *combustion* that takes place within a high speed *shock wave*. Also used loosely to describe the combustion reactions that occur during *knocking* or 'pinking' in an *internal-combustion engine*.

deuterated compound A compound containing some *deuterium* in place of hydrogen.

deuterium Heavy hydrogen. D. ^2_1H. The *isotope* of hydrogen with *mass number* 2, and *atomic mass* 2.0144. The *abundance* of deuterium in natural hydrogen is 0.0156%. It occurs in *heavy water* as the oxide HDO (about one molecule in 6000) and as D_2O (about one molecule in 36×10^6).

deuterium oxide D_2O. See *heavy water*.

deuteron The *nucleus* of the *deuterium atom*.

Devarda's alloy An *alloy* of 50% copper, 45% aluminium, and 5% zinc.

developing, photographic The action of certain chemicals, usually organic *reducing agents*, on an exposed photographic plate or film in order to bring out the latent image. The developer reduces those areas of the silver *salts* that had been exposed to *light* to metallic silver. This remains as a black deposit. See *photography*.

deviation (statistics) The difference between one of a *set* of values and the *mean* of the set. The 'mean deviation' is the mean of all the individual deviations of the set. See also *standard deviation*.

deviation, angle of The difference between the angle of *incidence* and the angle of *refraction* when a ray of *light* passes from one medium to another. See Fig. 13.

devitrification The crystallization of *glass*, which is normally an *amorphous* mixture in a *metastable* state; when crystallization takes place, the glass loses its characteristic state of clear transparency.

dew Liquid *water* produced by *condensation* of *water vapour* in the air when the *temperature* falls sufficiently for the vapour to reach saturation. See also *dew point*.

Dewar flask A glass vessel used for keeping *liquids* at *temperatures* differing from that of the surrounding air. This is done by reducing to a minimum the transfer of *heat* between the liquid and the air. It consists of a double-walled flask with the space between the two walls exhausted to a very high *vacuum*, to minimize transfer of heat by *convection* and *conduction*. The inner surfaces of the walls are silvered to reduce transfer of heat by *radiation*; areas of contact between the two walls are kept at a minimum to keep down conduction of heat. See Fig. 12. Named after James Dewar (1842–1913).

dew point The *temperature* at which the *water vapour* present in the air saturates the air and begins to condense, i.e. *dew* begins to form.

dew-point hygrometer A device used to measure the relative *humidity* of the atmosphere. The temperature of a polished surface is reduced until water from the atmosphere is seen to condense on it. The temperature at which this *dew point* occurs enables the relative humidity to be calculated.

dextrin British gum, starch gum. A *mixture* of gummy *polysaccharide carbohydrates* obtained by the partial *hydrolysis* of *starch*.

MODERN THERMOS FLASK

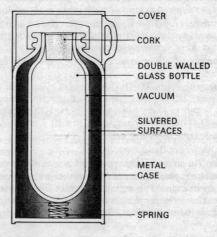

Figure 12.

dextrorotatory Rotating or deviating the plane of vibration of polarized light to the right (observer looking into the oncoming light). See *laevorotatory*; *optical activity*.

dextrose See *glucose*.

diacetyl See *butanedione*.

diagonal A line joining the intersections of two pairs of sides of a *rectilinear* figure.

dialysis The separation of *colloids* in *solution* from other dissolved substances by selective *diffusion* through a *semipermeable membrane*. Such a membrane is slightly permeable to the *molecules* of the dissolved substances, but not to the larger molecules or groups of molecules in the colloidal state. Dialysis is the process that takes place in the kidneys to remove nitrogenous waste material from the body. If the kidneys are not functioning properly an artificial kidney, or renal *dialyzer*, can be used to take over their function.

dialyzed iron A *colloidal solution* of iron(III) hydroxide, $Fe(OH)_3$. A deep red liquid, used in medicine.

dialyzer 1. An arrangement for effecting *dialysis*. The *solution* to be dialyzed is placed in a vessel in which it is separated from *water* by a *semipermeable membrane*; this is not permeable to the substance in the *colloidal state*, which will eventually remain as a pure solution on its side of the membrane. **2.** An artificial kidney that replaces the function of diseased kidneys, by separating the body's nitrogenous waste by dialysis.

diamagnetism The property of a substance that has a small negative *magnetic susceptibility*. This type of magnetism is due to a change in the orbital motion of

114

the *electrons* in the *atoms* of the substance consequent on the application of an external *magnetic field*. The phenomenon occurs in all substances, although the resulting diamagnetism is often masked by the much greater effects due to *paramagnetism* or *ferromagnetism*.

diameter See *circle*.

1,6-diaminohexane Hexamethylenediamine. $H_2N(CH_2)_6NH_2$. A *soluble* organic substance, m.p. 41.2°C., b.p. 204°C., used in the manufacture of *nylon*.

diamond A natural crystalline allotropic form (see *allotropy*) of carbon. It is colourless when pure, but is sometimes coloured by traces of impurities; it has a very high *refractive index* and *dispersive power*. Diamond is one of the hardest substances known (see *Mohs scale of hardness*) and is transparent to *X-rays* (imitations are not). It is used for cutting tools and drills, and as a gem. Industrial diamonds can be produced synthetically.

diaphragm 1. An opaque disc used in an optical system to control the amount of light passing through it. It has a circular *aperture* at its centre through which the light can pass. The amount of light permitted to enter can be varied in one of two ways: by replacing the diaphragm with another with an aperture of different diameter or by making use of an *iris* diaphragm. Diaphragms are also used to reduce *aberration* by restricting the light path to the centre of a *lens*. **2.** The muscular membrane in mammals that separates the abdomen from the thorax.

diastase An *enzyme* contained in *malt*, which converts *starch* into *maltose* during *brewing*. See also *amylase*.

diathermancy The property of being able to transmit *heat radiation*; it is similar to transparency with respect to *light*.

diatomaceous earth See *kieselguhr*.

diatomic (chem.) Consisting of two *atoms* in a *molecule*; e.g. molecular hydrogen, H_2.

diazo compounds Organic *compounds* containing two adjacent nitrogen atoms, which may form an *azo group*, but only one is attached to a carbon atom; e.g. benzenediazonium chloride $C_6H_5N^+\equiv NCl^-$, *diazomethane* $CH_2=N\rightleftarrows N$ [where $\rightarrow$ denotes a *coordinate* (dative) *bond*], and benzenediazohydroxide, $C_6H_5N{:}NOH$. *Aromatic* diazo compounds are of great importance; by *azo coupling* they give azo compounds used as dyes, drugs, etc. They are prepared from aromatic amines containing $-NH_2$ groups, the simplest of which is *phenyl-amine* (aniline). A salt of the amine is treated with *nitrous acid*, which converts $-NH_2$ into the *diazonium* group, $-N^+\equiv N$, a process known as diazotization. The resulting *diazonium* salt can be used for azo coupling.

diazomethane CH_2N_2. A highly poisonous and explosive yellow gas. It is used as a methylating agent and prepared for this purpose as required, usually in solution.

diazonium compounds *Organic compounds* of the general formula $RN_2^+X^-$, where R is an aryl radical, RN_2^+ is a *cation*, and X^- is an *anion*. E.g. benzenediazonium chloride, $C_6H_5N_2^+Cl^-$, is a typical diazonium salt. Diazo-nium salts are prepared by diazotization (see *diazo compounds*) of *amines*, an important stage in the production of *azo dyes*.

diazotization See *diazo compounds*.

dibasic acid. An *acid* containing two *atoms* of *acidic hydrogen* in a *molecule*; an acid giving rise to two series of *salts*, normal and acid salts; e.g. *sulphuric acid*, H_2SO_4, which gives rise to normal *sulphates* and hydrogensulphates.

di(benzenecarbonyl) peroxide Benzoyl peroxide. $(C_6H_5CO)_2O_2$. An insoluble crystalline explosive solid, m.p. $106-8°C$., used in bleaching flour, fats, oils, etc., and as a catalyst.

diborane See *boranes*.

dibromoethane Ethylene dibromide. $C_2H_4Br_2$. A *volatile liquid* existing in two *isomeric* forms. The common isomer 1,2-dibromoethane, m.p. $10°C$., b.p. $121°C$., is used in conjunction with anti-knock compounds in *petrol* and as a *solvent*.

dibutyl oxalate $(C_4H_9OOC)_2$. A colourless *liquid*, b.p. $243.4°C$., used as a *solvent* and in organic synthesis.

dicarbide See *carbide*.

dicarboxylic acid A *carboxylic acid* containing two *carboxyl* groups in its molecule, e.g. *tartaric acid*, $COOH(CHOH)_2COOH$. In systematic nomenclature such acids end in -dioic, e.g. tartaric acid is 2,3-dihydroxybutanedioic acid.

dichlorodifluoromethane CCl_2F_2. A colourless *gas*, b.p. $-29°C$., used as a *refrigerant* and as a *propellant* for *aerosols*. See *chlorofluorocarbon*.

dichloroethane Ethylene dichloride, Dutch liquid. $C_2H_4Cl_2$. An oily toxic *liquid* existing in two *isomeric* forms. The common isomer 1,2-dichloroethane, m.p. $-35°C$., b.p. $84°C$., is used as a *solvent* and in the manufacture of *polyvinyl chloride*.

dichloromethane Methylene chloride. CH_2Cl_2. A colourless *volatile liquid*, b.p. $40.1°C$., used as a *solvent*, a *refrigerant*, and an *anaesthetic*.

dichroism The property of some *crystals*, such as *tourmaline*, that makes them appear different colours if light falls on them from different directions. It is caused by a difference in the extent to which the *ordinary ray* and the extraordinary ray are absorbed.

dichromate A *salt* of the hypothetical dichromic(VI) acid, $H_2Cr_2O_7$, e.g. *potassium dichromate(VI)*, $K_2Cr_2O_7$.

dichromate cell Bichromate cell. A *primary cell* having a positive pole of carbon and a negative pole of zinc in a *liquid* consisting of a *solution* of *sulphuric acid*, H_2SO_4, and *potassium dichromate*, $K_2Cr_2O_7$, the latter acting as a depolarizing agent (see *depolarization*) by its oxidizing action. The *EMF* is 2.03 *volts*.

dichromatism A form of *colour blindness* in which only two *colours* of the *spectrum* can be distinguished.

dicyanodiamide Cyanoguanidine. $H_2NC(NH)NHCN$. A white crystalline substance, m.p. $208°C$., used in the manufacture of *melamine* and of *barbiturates* and other drugs.

dielectric A nonconductor of electric current; insulator. A substance in which an *electric field* gives rise to no net flow of *electric charge* but only to a displacement of charge, i.e. the electrons do not flow as an *electric current* but are displaced with respect to the nuclei of the atoms, causing the atoms to act as a *dipole* with an electric moment in the direction of the field (see *electric polarization*).

dielectric constant See *permittivity*.

dielectric heating A form of heating in which electrically insulating material is heated by being subjected to an alternating *electric field*. It results from *energy* being lost by the field to *electrons* within the *atoms* and *molecules* of the material. In industrial dielectric heating the material to be heated is placed between the plates of a *capacitor* connected to a *high frequency* power source.

dielectric strength The maximum *voltage* that can be applied to a *dielectric* material without causing it to break down; usually expressed in *volts* per mm. See table at *permittivity* for the dielectric strengths of some common dielectric materials.

dielectrophoresis The motion of electrically polarized (see *electric polarization*) particles in a nonuniform *electric field*.

Diels–Alder reaction A method of preparing a *ring compound* from a conjugated *diene* and a compound containing a single *double bond* (e.g. *butenedioic acid*) called a dienophile. Named after Otto Diels (1876–1954) and Kurt Alder (1902–58).

diene An *unsaturated hydrocarbon* containing two *double bonds*, e.g. *buta-1,3-diene*. If the two double bonds are separated by a single bond, the compound is called a 'conjugated diene'.

dienophile See *Diels–Alder reaction*.

Diesel engine A type of *internal-combustion engine* that burns heavy oil. A mixture of air and oil is compressed and thereby heated to the *ignition temperature* of the oil (about 540°C.). Named after Rudolf Diesel (1858–1913). Because the *compression ratio* (12–18:1) is much higher than in a petrol engine, Diesel engines have to be much heavier than petrol engines. But because they work by compression ignition, they are considerably simpler as no electrical spark is needed.

Diesel oil See *gas oil*; *petroleum*.

1,1-diethoxyethane See *acetals*.

diethylamine $(C_2H_5)_2NH$. A colourless liquid with a smell resembling ammonia, b.p. 55°C., used in pharmaceuticals and in the rubber industry.

diethyl ether See *ethoxyethane*.

differential calculus A branch of mathematics that deals with continuously varying quantities. It is based upon the *differential coefficient* of one quantity with respect to another of which it is a *function*. Used for solving problems involving the rates at which processes occur and for obtaining maximum and minimum values for continuously varying quantities.

differential coefficient Derived function, derivative. See *differentiation* and Appendix, Table 9.

differential equation An equation that involves *differential coefficients*. An ordinary differential equation is one in which only one independent variable is involved. The *order* of a differential equation is the same as that of the *derivative* of the highest order appearing in it; the *degree* is given by the largest *exponent*.

differentiation (bio.) 1. The development of *cells* so that they are capable of performing specialized functions in the organs and *tissues* of the *organisms* to which they belong. 2. In microscopic specimens, the removal of the excess stain from certain parts to show up the structure of the whole.

differentiation (math.) The operation, used in the *calculus*, of obtaining the *differential coefficient*; if $y = x^n$, the differential coefficient, $dy/dx = nx^{n-1}$. See Appendix, Table 9.

diffraction When a beam of *light* passes through an *aperture* or past the edge of an opaque obstacle and is allowed to fall upon a screen, patterns of light and dark bands (with *monochromatic light*) or coloured bands (with *white light*) are observed near the edges of the beam, and extend into the geometrical shadow. This phenomenon, which is a particular case of *interference*, is due to the wave nature of light, and is known as diffraction. The phenomenon is common to all *wave motions*. See also *electron diffraction*; *Fraunhofer diffraction*; *Fresnel diffraction*.

diffraction grating A device used to disperse a beam of *light*, *X-rays*, or other *electromagnetic radiation* into its constituent *wavelengths*, i.e. for producing its *spectrum*. It may consist of any device that acts upon an incident *wave front* in a manner similar to that of a regular array of parallel slits where the slit width is of the same order as the wavelength of the incident radiation. Such gratings may be prepared by ruling equidistant parallel lines on to a *glass* (transmission grating) or *metal* surface (reflecting grating). The grating may be plane or *concave*, the latter having self-focusing properties.

diffusion of gases The free and random movement of the *molecules* of all *gases*, which tends to make them distribute themselves equally within the limits of the vessel enclosing the gas; thus all gases diffuse within the limits of any enclosing walls, and are all perfectly miscible with one another. The rates of diffusion of gases through porous bodies are inversely proportional to the square roots of their *densities*. (See *Graham's Law*.)

diffusion of light The *scattering* or alteration of direction of *light* rays produced by transmission through frosted glass, fog, etc., or by irregular reflections at matt surfaces, such as blotting paper.

diffusion of particles The passage of *elementary particles* through *matter* in such a way that the *probability* of *scattering* is large compared to that of *capture*.

diffusion of solutions *Molecules* or *ions* of a dissolved substance move freely through the *solvent*, the *solution* becoming uniform in *concentration*; the phenomenon is similar to *diffusion of gases*.

diffusion plant A plant for separating *isotopes*, based on their different rates of *diffusion* in the gaseous state through a membrane.

diffusion pump See *condensation pump*.

digestion 1. The breakdown of food by an organism into a form that enables it to be absorbed by the body. The process normally requires the appropriate *enzymes* and in mammals takes place in the alimentary canal. **2.** The treatment of a material with heat, solvents, acid, etc., to cause it to decompose.

digit (astr.) One twelfth of the diameter of the *Sun* or *Moon*; used to denote the extent of an *eclipse*.

digit (math.) A single figure or numeral; e.g. 325 is a number with 3 digits.

digital computer See *computer*.

digital display A method of indicating the reading of a measuring instrument (e.g. a *voltmeter*), clock, etc., in which numbers appear on a screen, as opposed to a pointer moving round a scale. It is often based on a *digitron* or a *light-emitting diode*. See also *liquid-crystal display*.

digitalis A *mixture* of *glucosides* (*e.g. digitonin, digitoxin*) derived from vegetable sources, especially the leaves of foxgloves (*Digitalis purpurea*); used in the treatment of certain heart conditions.

digital recording A technique for recording or transmitting sound in which the pressure of the sound wave is sampled some 30 000 times per second and the resulting digits are recorded or transmitted. The player or receiver converts the digits back into the analogue form. This method is used for very high fidelity recordings as it minimizes distortion and interference.

digitoxin $C_{41}H_{64}O_{13}$. A white crystalline *glucoside*, m.p. 252–3°C., obtained from *digitalis* and used as a heart stimulant.

digitron A gas-discharge tube (see *discharge in gases*) used to give a *digital display* of a numerical value. It usually has ten *cathodes* shaped to form the digits 0–9. A *pulse* to the appropriate cathode produces an illuminated digit. Digitrons have now been replaced by *light-emitting diodes* and *liquid-crystal displays*.

dihedral Formed by two intersecting *planes*.

dihydric (chem.) Containing two *hydroxyl* groups in a molecule; e.g. a *diol*.

1,2-dihydroxybenzene Catechol, pyrocatechol. $C_6H_4(OH)_2$. A solid *dihydric phenol*, m.p. 105°C., used in *photography*. See also *catecholamines*.

2,3-dihydroxybutanedioic acid See *tartaric acid*.

dilatancy The tendency of some *non-Newtonian fluids*, such as *colloidal* solutions, to solidify or become more rigid under pressure. Compare *thixotropy*.

dilation 1. Dilatation (phys.). A change in *volume*. **2.** See *time dilation*.

dilatometer An apparatus used for measuring *volume* changes of substances. It generally consists of a bulb with a graduated capillary stem.

dilute Containing a large amount of *solvent*, generally *water*.

dilution 1. The further addition of *water* or other *solvent* to a *solution*. **2.** The *reciprocal* of *concentration*; the *volume* of *solvent* in which unit quantity of *solute* is dissolved.

dimagnesium trisilicate See *magnesium trisilicate*.

dimensional analysis A method of checking an expression, equation, or solution to a problem by analysing the *dimensions* of the units in which it is expressed. For example, if an expression is derived to evaluate a physical quantity, its validity can be checked by analysing its dimensions to ensure that they are those of the physical quantity to be evaluated.

dimensions of units The dimensions of a physical quantity are the *powers* to which the *fundamental units* (length l, mass m, time t, etc.) expressing that quantity are raised. E.g. *volume*, l^3, is of dimensions three in length; *speed*, i.e. length per unit time, l/t, is of dimensions one in length and -1 in time.

dimer A substance composed of *molecules* each of which comprises two molecules of a *monomer*.

dimethoxymethane Methylal. $(CH_3O)_2CH_2$. A colourless flammable *liquid*, b.p. 45.5°C., used as a *solvent* and in perfumes.

dimethylbenzene See *xylene*.

dimethylformamide DMF. $HCON(CH_3)_2$. A colourless *liquid*, b.p. 153°C., widely used as an organic *solvent*.

dimorphism The existence of a substance in two different crystalline forms.

dimorphous Existing in two different cryalline forms.

di-neutron An unstable system comprising two *neutrons*.

dinitrobenzene $C_6H_4(NO_2)_2$. A colourless *insoluble* crystalline substance that exists in three *isomeric* forms, the most important of which is 1,2-dinitroben-zene, m.p. 90°C., which is used in the manufacture of *dyes*.

dinitrogen oxide See *nitrogen oxides*.

dinitrogen tetroxide See *nitrogen oxides*.

diode An electronic device that allows an *electric current* to flow in one direction only. It is now invariably a *p-n semiconductor junction*, but formerly a *thermionic valve* containing two *electrodes* (*anode* and *cathode*) was used.

-dioic See *dicarboxylic acid*.

diols Glycols. *Dihydric alcohols* derived from *aliphatic hydrocarbons* by the substitution of *hydroxyl groups* for two of the hydrogen *atoms* in the *molecule*. General *formula* $C_nH_{2n}(OH)_2$. See also *ethanediol*.

dioptre A unit of power of a *lens*; the power of a lens in dioptres is the *reciprocal* of its *focal length* in *metres*. The power of a *converging lens* is usually taken to be positive, that of a *diverging lens* negative.

dioxan(e) $(C_2H_4)_2O_2$. A colourless flammable *liquid* cyclic *ether*, b.p. 101°C., used as a *solvent* and a dehydrating agent.

dioxin TCDD. 2,3,7,8-tetrachlorodibenzo-*p*-dioxin. An extremely toxic *hydrocarbon* belonging to a class of similar compounds, called dioxins. It is a by-product occurring in the manufacture of some *herbicides* and *bactericides* and is produced on the incineration of some chemicals. Large amounts were released at Seveso in Italy in 1976 as a result of an industrial accident. It causes *chloracne* in humans and cancers in some animals.

dioxonitric(III) acid See *nitrous acid*.

dioxygenyl The positive ion $O_2{}^+$, which forms compounds of the type O_2PtF_6.

dip, magnetic See *magnetic dip*.

dip circle Inclinometer. An instrument for measuring the *angle* of *magnetic dip*. It consists of a magnetized needle mounted to rotate in a vertical plane, the angle being measured on a circular scale, marked in degrees.

dipeptide A *peptide* consisting of two *amino acids*.

diphenylamine $(C_6H_5)_2NH$. A colourless crystalline substance, m.p. 52.8°C., used in the manufacture of *dyes* and in analytical chemistry as a detector of *oxidizing agents*.

diphenylmethanone Benzophenone. Diphenyl ketone. $C_6H_5COC_6H_5$. A crystalline *insoluble* solid, m.p. 48.1°C., used in organic synthesis.

diploid cell A *cell* in which the *nucleus* contains two sets of *chromosomes* that can be arranged in pairs, each set inherited from one parent, when the *gametes* unite to form a *zygote*. Nearly all animal cells are diploid, except *gametes*. See *haploid*.

dipole 1. Two equal point *electric charges* (electric dipole) or *magnetic poles* (magnetic dipole) of opposite sign, separated by a small distance. The dipole moment is the product of either charge (or pole) and the distance between the two. It may also be expressed as the *couple* that would be required to maintain the dipole at right angles to a *field* (electric or magnetic) of unit intensity. *Molecules* in which the centres of positive and negative *charge* are separated constitute dipoles, the dipole moments of which are measured in *coulomb metres*. Dipole moments can often provide evidence as to the shape of molecules, e.g. *water* has a dipole moment of 6.1×10^{-30} C m, which indicates that it is triangular in shape with an angle of 105° between the two O-H bonds. **2.** A *radio aerial* consisting of two rods.

dipole moment See *dipole*.

dippel's oil See *bone oil*.

di-proton An unstable system comprising two *protons*.

direct current d.c. An *electric current* flowing always in the same direction. Compare *alternating current*.

direct dyes Cotton dyes, substantive dyes. A group of *dyes* that dye cotton, viscose *rayon*, and other *cellulose* fibres direct, without the use of *mordants*. Generally used with 'assistants' such as common *salt* or *sodium sulphate*, which assist absorption by the fibre. They are usually *salts* of *sulphonic acid*.

direct motion (astr.) **1.** The motion of a *planet* or other celestial body round the *Sun* in the same direction as the *Earth*. All the planets have direct motion, but some *comets* and *satellites* do not, and they are said to have 'retrograde motion'. **2.** Motion across the sky from west to east.

directrix A fixed line used to describe and define a curve, by relating the distance of a point on the curve to this line and to the *focus* of the curve. See *parabola*; *hyperbola*.

direct vision spectroscope A *spectroscope* designed for compactness and portability. In this instrument, the middle portion of the *spectrum* (the yellow) remains undeviated. The eye thus looks in the direction of the source when observing the spectrum.

disaccharides. A group of *sugars* the *molecules* of which are derived by the *condensation* of two *monosaccharide* molecules with the elimination of a molecule of *water*. On *hydrolysis* disaccharides yield the corresponding monosaccharides. E.g. *cane sugar*, sucrose, $C_{12}H_{22}O_{11}$, is a disaccharide which, on hydrolysis with dilute *acids*, gives a *mixture* of *glucose* and *fructose*, both monosaccharides having the formula $C_6H_{12}O_6$. (See *inversion of cane-sugar*.) Other important disaccharides are *lactose* and *maltose*.

discharge, electrical 1. The release of the *electric charge* stored in a *capacitor* through an external circuit. **2.** The use of the *chemical energy* stored in an electric *cell* to do work electrically.

discharge in gases The passage of *electricity* through a tube (usually called a gas-discharge tube) containing a *gas* at low pressure. *Electrons* and *ions* present in the tube are accelerated towards their respective *electrodes* by the applied *potential difference*, the net transfer of *electric charge* constituting the current. The electrons are accelerated sufficiently to produce ions by collision with the gas *molecules*. Recombination of oppositely charged ions gives rise to luminous glows at certain parts of the tube. The study of this phenomenon led to many important results, including the discovery of the electron and of *isotopes*.

discriminator An *electronic* circuit that converts *frequency* or *phase modulation* into *amplitude modulation*.

disinfectant A substance, such as *cresol* or *phenol*, that is capable of destroying disease-producing microorganisms. Because they can also damage human tissues these are used to clean lavatories, drains, hospitals, etc., and are only applied to humans in very dilute solutions. Compare *antiseptic*.

disintegration (phys.) Any process in which the *nucleus* of an *atom* emits one or more particles or *photons*, either due to spontaneous *radioactivity* or as the result of a collision.

disintegration constant Decay constant. Transformation constant. The *probability* of the *decay* of an atomic *nucleus* per unit time that characterizes a *radioactive nuclide*. It determines the exponential decrease with time, t, of the *activity*, A, given by:

$$A = A_0 e^{-\lambda t}$$

where A_0 is the activity when $t = 0$, and λ is the disintegration constant. The reciprocal of the disintegration constant is called the 'mean life'. See also *half-life*.

dislocation A line *defect* in a *crystal*, the result of a slip along a surface of one or more *lattice* constants.

disodium hydrogenphosphate(V) See *sodium phosphates*.

disodium tetraborate Sodium tetraborate, borax. $Na_2B_4O_7.10H_2O$. A white soluble crystalline *salt*, occurring mainly in *tincal*. On heating it loses *water of crystallization* and melts to a clear glass-like solid (see *borax bead test*). It is used as an *antiseptic* and *flux* and in making glass and ceramics.

d-isomer See *optical activity*.

disordering The displacement of *atoms* from their position in a *crystal lattice* (e.g. as a result of the effect of *ionizing radiation*) to positions that are not part of the lattice.

disperse dyes Dyes of all chemical types that are applied in the form of fine suspensions in water to man-made fibres, such as *cellulose acetate*, *nylon*, and *polyester* fibres. They are insoluble in water, but are usually soluble in organic solvents, such as *esters*.

disperse phase The dissolved or suspended substance in a *colloidal solution* or *suspension*.

dispersion (chem.) A *disperse phase* suspended in a *dispersion medium*; a system of particles dispersed and suspended in a *solid*, *liquid*, or *gas*.

dispersion medium A medium in which a substance in the *colloidal state* is dispersed; the *solvent* in a *colloidal solution*.

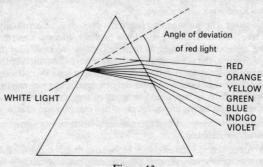

Figure 13.

dispersion of light The splitting of *light* of mixed *wavelengths* into a *spectrum*. A beam of ordinary *white light*, e.g. sunlight, on passing through an optical *prism* or a *diffraction grating*, is divided up or dispersed into light of the different wavelengths of which it is composed; if the beam that emerges after dispersion is allowed to fall upon a screen, a coloured band or spectrum is observed. Dispersion by a prism is due to the fact that lightwaves of different wavelengths are refracted (see *refraction*) or bent through different angles (see *deviation, angle of*) on passing through the prism, and are thus separated. See Fig. 13, in which the dispersion is greatly exaggerated for clarity.

dispersive power of a medium. A measure of the *dispersion of light* produced by a *prism* or a particular medium with respect to light of two specified *wavelengths* ('1' and '2'); given by the ratio

$$(n_1 - n_2)/(n - 1),$$

where n_1 is the *refractive index* of the medium for wavelength 1, n_2 that for wavelength 2, and n is the average of n_1 and n_2. When considering the dispersive power of media for ordinary *white light*, the dispersive power is often defined as

$$(n_b - n_r)/(n_y - 1)$$

where n_b, n_r, and n_y are the refractive indices for blue, red, and yellow light respectively.

displacement 1. *s*. The *vector* of distance; a specified distance in a specified direction. The *SI unit* is the *metre*. **2.** See *electric displacement*.

disproportionation A *chemical reaction* in which a molecule is both oxidized and reduced at the same time. For example, copper(I) salts that are soluble in water form equal quantities of copper(0) and copper(II).

dissociation (chem.) A temporary, reversible *decomposition* of the *molecules* of a *compound*, which occurs under some particular conditions. In *electrolytic dissociation*, the molecules are split into *ions*. In *thermal dissociation*, the effect of raising the *temperature* is to decompose a definite fraction of the molecules; e.g. *ammonium chloride*, NH_4Cl, dissociates into *ammonia*, NH_3, and *hydrogen chloride*, HCl, on heating. The products recombine on cooling, and the degree of dissociation depends on the *temperature*. The ratio of the product of the *active masses* of the molecules resulting from the dissociation, to the active mass of the undissociated molecules, when *chemical equilibrium* has been reached under a particular set of physical conditions, is called the 'dissociation constant'. See also *acidity constant*; *basicity constant*: *equilibrium constant*.

dissociation constant See *dissociation*.

distance ratio See *velocity ratio of a machine*.

distillate The *liquid* obtained by the *condensation* of the *vapour* in *distillation*.

distillation 1. The process of converting a *liquid* into *vapour*, condensing the vapour, and collecting liquid or *distillate*. It is used for separating *mixtures* of liquids of different *boiling points* or for separating a pure liquid from a non-*volatile* constituent. (See *fractional distillation*). It is also used in the separation of *isotopes*. See *isotopes, separation of*. **2.** See *destructive distillation*.

distilled water *Water* that has been purified by *distillation* of the substances dissolved in it. See also *conductivity water*.

distributive law The law of algebra stating that the order in which operations are carried out is not important. For example $a(b + c) = ab + ac$. Compare *associative law*; *commutative law*.

disulphuric(VI) acid See *sulphuric acids*.

dithionate A salt of the hypothetical dithionic acid, $H_2S_2O_6$. Dithionates contain the *ion* $S_2O_6{}^{2-}$.

diurnal Daily; performed or completed once every 24 hours.

divalent Bivalent. Having a *valence* of two.

divergent series A *series* in which the sum tends to plus or minus *infinity* or to oscillate. Compare *convergent series*.

diverging lens A *lens* that causes a parallel *beam* of *light* passing through it to diverge or spread out; *concave* lens. See Fig. 25 under *lens*.

Divers' liquid A solution of *ammonium nitrate* in liquid *ammonia*. It is used as a *solvent* for some *metals* and their *oxides* and *hydroxides*.

divinyl ether Vinyl ether. $H_2C{:}CHOCH{:}CH_2$. A colourless inflammable *liquid*, b.p. 39°C., used as an *anaesthetic*.

division An arithmetic operation in which a dividend is divided by a divisor to give a quotient and a remainder.

dl-form See *racemic mixture*.

D-lines Two lines in the visible *spectrum* of sodium which, because they are easy to recognize, are used as a standard in *spectroscopy*. They have the *wavelengths* 589.0 and 589.6 nanometres.

DMF See *dimethylformamide*.

DNA See *deoxyribonucleic acid*.

docosanoic acid Behenic acid. $CH_3(CH_2)_{20}COOH$. A crystalline *saturated fatty acid*, m.p. 80°C., used in the manufacture of cosmetics and *waxes*, and as a *plasticizer*.

dodecahedron A *polyhedron* with twelve faces.

dodecanoic acid Lauric acid. $CH_3(CH_2)_{10}COOH$. A white crystalline *insoluble* substance, m.p. 44°C., used in the manufacture of *soaps*, *detergents*, and cosmetics.

dodecanol Lauryl alcohol. $CH_3(CH_2)_{11}OH$. A white crystalline *insoluble* substance, the commercial form of which consists of a mixture of *isomers* with m.p. in the range 20–30°C. It is used in the manufacture of *detergents*.

dolomite Pearl spar. A natural double *carbonate* of magnesium and calcium, $MgCO_3.CaCO_3$. A whitish solid that occurs naturally in vast amounts, comprising whole mountain ranges.

domain See *magnetic domain*.

dominant Denoting the *allele* that functions when two different alleles of a *gene* are present in a *cell*. Compare *recessive*.

donor 1. An imperfection in a *semiconductor* that causes *electron* conduction. **2.** An *ion* or *molecule* that provides a pair of *electrons* to form a *coordinate bond* (see *chemical bond*).

dopamine A *catecholamine* that is the precursor of *adrenaline* and *noradrenaline*. It also functions as a *neurotransmitter* in the brain. It is formed from dopa, dihydroxyphenylalanine.

doping The addition of a small quantity of impurity to a *semiconductor* to achieve a particular characteristic.

Doppler broadening The broadening of spectral emission or absorption lines (see *spectrum*) due to random motion of the emitting or absorbing *molecules*, *atoms*, or *nuclei*. See *Doppler effect*.

Doppler effect Doppler shift. Doppler's principle. The apparent change in the *frequency* of *sound* or *electromagnetic radiation* due to relative motion between the source and the observer. The *pitch* (*frequency*) of the sound emitted by a moving object (e.g. the whistle of a moving train) appears to a stationary observer to increase as the object approaches him and to decrease as it recedes from him. The *light* emitted by a moving object appears more red (red light being of lower frequency than the other colours) when it is receding from the observer (or the observer receding from it). Thus the fact that the light emitted by the *stars* of distant *galaxies* suffers a *red shift*, when observed from the Earth, is taken to mean that these distant galaxies are receding from our *Galaxy*. This is the principal evidence for the widely accepted hypothesis concerning the *expansion of the Universe*. The Doppler effect is also used in *radar*, to distinguish between stationary and moving targets and to provide information concerning their *velocity*, by measuring the frequency shift between the emitted and the reflected radiation. Named after C. J. Doppler (1803–53).

dose (phys.) The 'absorbed dose' is the *energy* imparted by *ionizing radiation* to unit *mass* of irradiated *matter*. It is measured in *grays* or *rads* (0.01 *joule* per *kilogram*). 1 gray = 100 rad. The 'maximum permissible dose (or level)' is the recommended upper limit for the absorbed dose that a person should receive during a specified period. For 'dose equivalent' see *sievert*. See also *linear energy transfer*.

dosemeter Dosimeter. A device for measuring a *dose* of *ionizing radiation*. Several methods are used (see *dosimetry*).

dosimetry The measurement of *doses* of radiation, usually for medical or safety purposes. The most common method of measurement is the *ionization chamber*. Other methods include the blackening of photographic film (film dosimetry) as in the *film badge*, the change in transmission density of Perspex* (see *polymethyl methacrylate*) after irradiation (Perspex dosimetry), and the *thermoluminescence* of lithium fluoride after irradiation (lithium fluoride dosimetry).

double bond (chem.) Two covalent bonds (see *chemical bonds*) linking two *atoms* in a chemical *compound*; they are characteristic of an *unsaturated compound*.

double decomposition (chem.) Metathesis. A *chemical reaction* between two *compounds* in which each of the original compounds is decomposed and two new compounds are formed. E.g. the action of *sodium chloride* on *silver nitrate* according to the equation

$$NaCl + AgNO_3 = AgCl + NaNO_3$$

double refraction. Birefringence. The formation of two refracted rays of *light* (see *refraction*) from a single incident ray; a property of certain *crystals*, notably calcite. The *ordinary ray* follows the normal laws of refraction and is polarized at right angles to the light in the extraordinary ray, which follows different laws.

double salt A compound of two *salts* formed by *crystallization* from a *solution* containing both of them. When redissolved the double salt *ionizes* as two salts. For example, *potassium sulphate* and *aluminium sulphate* in solution together will crystallize as the double salt $K_2SO_4.Al_2(SO_4)_3.24H_2O$.

double star Two *stars* held very close to each other as a result of their mutual gravitational attraction, which move through *space* together giving the appearance, to the naked eye, of being one star.

doublet A pair of associated lines in a *spectrum* characteristic of the *alkali metals*, e.g. the sodium *D-lines*.

Down's process An electrolytic process for making sodium and chlorine from molten *sodium chloride*. The salt is melted electrically and kept molten by the current passing through the brick-lined steel cell. Chlorine is evolved at the *anode* and molten sodium floats to the top of the salt at the *cathode*. Up to 60% of *calcium chloride* is added to the salt to lower its *melting point*.

drachm, fluid A former British unit of *volume* in the *apothecaries' fluid measures*; 60 *minims*; 3.55 cm³.

drain The electrode in a *field-effect transistor* through which *charge carriers* leave the inter-electrode region.

drift chamber See *wire chamber*.

drug 1. Any chemical substance used in medicine to cure or prevent disease. **2.** A habit-forming *narcotic*; any substance that causes physiological or psychological dependence.

dry cell Dry battery. A type of small *Leclanché cell* containing no free *liquid*. The *electrolyte* of *ammonium chloride* is in the form of a paste, and the negative zinc pole forms the container of the cell within a plastic outer cover. It is used for torch batteries, *radio* batteries, etc.

dry ice Solid *carbon dioxide*, CO_2, used as a *refrigerant*. It is called 'dry' because it *sublimes* at $-78°C$., without forming a liquid.

drying oil An animal or vegetable *oil* that will harden to a tough film when a thin layer is exposed to the air. The hardening is due to *oxidation* or *polymerization* of the *unsaturated fatty acids* of which these oils partially consist. They are used in *paints* and varnishes (e.g. *linseed oil*, dehydrated *castor oil*, and certain fish oils).

D-series See *optical activity*.

ductility A property, especially of *metals*, of being capable of being drawn out into a wire.

ductless gland See *endocrine gland*.

Dulong and Petit's law For a *solid element*, the product of the *relative atomic mass* and the *specific heat capacity*, i.e. the *molar heat capacity*, is a constant, approximately equal to 25 joules per *mole* per *kelvin*. This is an approximate law. Named after P. L. Dulong (1785–1838) and A. T. Petit (1791–1820).

Dumas' method 1. A method of determining the *relative molecular mass* of a volatile liquid. A thin glass vessel of known volume is weighed full of air and then full of the vapour of the liquid. Using the known density of air the relative molecular mass of the liquid can be calculated. **2.** A method of determining the total amount of nitrogen in an organic substance. A known mass of the substance is heated with copper(II) oxide to convert the nitrogen present into its oxides, which are then reduced to nitrogen by passing the gas over hot copper. The volume of nitrogen is measured, enabling the mass of nitrogen per mass of substance to be calculated.

duplet A pair of *electrons* shared between two *atoms* forming a single covalent bond. See *chemical bond*.

Duralumin* A light hard aluminium *alloy* containing about 4% copper, and small amounts of magnesium, manganese, and silicon.

dust core A core for magnetic devices made of powdered *metal* (often molybdenum) held together with a suitable binder. They are particularly suitable for *high frequency* equipment.

Dutch liquid See *dichloroethane*.

Dutch metal An *alloy* of copper and zinc; a variety of *brass*.

dwarf star A *star* of small *volume*, high *density*, and usually low *luminosity*. See also *white dwarf star*.

dyad (chem.) An *element* having a *valence* of two.

dyes Coloured substances that can be fixed firmly to a material to be dyed, so as to be more or less 'fast' to *water*, *light*, and *soap*. See *acid dyes*; *azo dyes*; *direct dyes*; *mordants*; *reactive dyes*; *vat dyes*.

dynamic equilibrium. If two opposing processes are going on at the same rate in a system, thus keeping the system unchanged, the system is said to be in dynamic equilibrium. E.g. a *liquid* in equilibrium with its *saturated vapour*; the rate of *evaporation* from the liquid surface is equal to the rate of *condensation* of the *vapour*.

dynamics. A branch of *mechanics*; the mathematical and physical study of the behaviour of bodies under the action of *forces* that produce changes of motion in them.

dynamite An *explosive* originally consisting of *nitroglycerin* absorbed in *kieselguhr*. Modern blasting dynamites contain nitrates sensitized with nitroglycerin absorbed on wood pulp.

dynamo A device for converting *mechanical energy* into *electrical energy*, depending on the fact that if an electrical *conductor* moves across a *magnetic field*, an *electric current* flows in the conductor. (See *induction*). The simplest form of dynamo consists of a powerful *electromagnet*, called the *field magnet*, between the *poles* of which a suitable conductor, usually in the form of a coil or coils, called the *armature*, is rotated. The mechanical energy of the rotation is thus converted into electrical energy in the form of a current in the armature. The word 'dynamo' tends to be used for small d.c. *generators*. A.c. generators are usually called *alternators*.

dynamometer Any instrument designed for the measurement of a *force* or the *power* of an engine.

dynatron oscillator An oscillator, using a *tetrode* (screen grid valve) in such a way that the *anode* current increases as the anode *voltage* is reduced.

dyne *c.g.s. system unit* of *force*; the force that, acting upon a *mass* of 1 g, will impart to it an *acceleration* of 1 cm per second per second. 1 dyne = 10^{-5} *newton* = 7.233×10^{-5} *poundal*.

dysprosium Dy. Element. R.a.m. 162.50. At. No. 66. r.d. 8.551, m.p. 1412°C., b.p. 2567°C. A soft silvery metal with seven natural isotopes. See *lanthanides*.

dystectic mixture. A *mixture* that has a constant maximum *melting point*.

E

e See *exponential*.

Earth A *planet* having its *orbit* between those of *Venus* and *Mars*. It is a sphere, slightly flattened towards the poles (i.e. approximating to an oblate *spheroid* in shape). Equatorial radius 6378.388 kilometres (3963.34 miles); polar radius 6356.912 kilometres (3949.99 miles). Mean *density* 5.52×10^3 kg m^{-3}; mass 5.976×10^{24} kg. The Earth consists of a gaseous *atmosphere* (see also *upper atmosphere*), a liquid *hydrosphere*, and a solid lithosphere. The lithosphere, in turn, consists of three separate layers: the crust (also sometimes called the lithosphere), the mantle, and the core. The crust consists of the Earth's outer layer of *soil* lying on a mass of rock. It is about 30 km (19 miles) thick on land and about 10 km (6 miles) thick under the seas. Its composition by *mass* is approximately: oxygen 47%, silicon 28%, aluminium 8%, iron 4.5%, calcium 3.5%, sodium and potassium 2.5% each, magnesium 2.2%, titanium 0.5%, hydrogen and carbon 0.2% each, phosphorus and sulphur 0.1% each. The mantle extends some 2900 km (1800 miles) below the crust and is believed to consist of silicate rocks. The core, part of which is probably liquid, is believed to have a *relative density* of 13 and a *temperature* in excess of 6000 K.

earthing a conductor. Making an electrical connection between the *conductor* and the Earth; the Earth is assumed to have zero *potential*.

Earth sciences The sciences concerned with the study of the *Earth*, including *geology* (*geophysics* and *geochemistry*), *geography*, *meteorology*, and *oceanography*.

earthshine A faint illumination of the dark side of the *Moon* during a crescent *phase*, due to sunlight reflected from the *Earth's* surface.

Earth's magnetism See *magnetism, terrestrial*.

east–west asymmetry of cosmic rays The observed intensity of *cosmic ray* particles coming from the West is greater than that coming from the East at any given *latitude*. This asymmetry is due to the deflection of the primary charged cosmic ray particles by the *magnetic field* of the Earth, and indicates a preponderance of positively charged particles in the incoming radiation.

ebonite Vulcanite. A hard black insulating material made by vulcanizing *rubber* with high proportions of sulphur. It contains about 30% combined sulphur.

ebullioscopic method See *elevation of boiling point*.

ebullition See *boiling*.

eccentricity 1. (math.) A *constant* used to describe a *conic*, equal to the distance from a point on the curve to the *focus* divided by the distance from that point to the *directrix*. **2.** (astr.) A measure of the extent to which an *ellipse* is elongated, equal to the distance between the foci divided by the length of the *major axis*. This value is used to express the eccentricity of a *planet's orbit* round the *Sun*: e.g. the eccentricity of the *Earth's* orbit is 0.0167.

ECG See *electrocardiograph*.

echelon (phys.) The type of grating that replaces the ordinary *diffraction grating* in *spectroscopy* when very high resolution is required. It consists essentially of a pile of glass plates of exactly equal thickness arranged in stepwise formation with a constant offset. The echelon can be used either as a transmission or when the steps are metallized as a reflection grating.

echo The effect produced when *sound* or other *radiation* is reflected or thrown back on meeting a solid obstacle or a reflecting medium.

echolocation The location of an object by determining the direction of an *echo* reflected from it, or the time taken for the echo to return. E.g. *radar*, *echo sounding*, etc. It is made use of by such animals as bats and oilbirds.

echo sounding A method of estimating the depth of the sea beneath a ship by measuring the time taken for a *sound* pulse to reach the sea bed and for its *echo* to return. See also *sonar*.

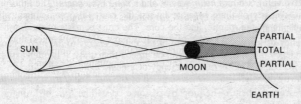

Figure 14.

eclipse The passage of a non-luminous body into the *shadow* of another. An 'eclipse of the *Moon*', or lunar eclipse, occurs when the *Sun*, the *Earth*, and the Moon are in line so that the shadow of the Earth falls upon the Moon. An 'eclipse of the Sun', or solar eclipse, is said to occur when the shadow of the Moon falls on the Earth. See Fig. 14, which also illustrates the areas of partial and total eclipse.

ecliptic The *Sun's* apparent path in the sky relative to the *stars*; the *great circle* described by the Sun on the *celestial sphere* in the course of a year.

ecology The study of the relation of all living organisms to their *environment* and to each other.

ecosphere 1. The part of the Earth's *atmosphere* in which life can exist: also called the 'biosphere'. **2.** The part of the atmosphere surrounding any *planet* on which life could exist. **3.** The part of *space* surrounding any *star* in which life could be possible.

ecosystem A biological community and its environment. Matter and *energy* pass round an ecosystem in cycles or loops. See *carbon cycle*; *nitrogen cycle*.

ectoplasm The outer layer of the *cytoplasm* of some living cells (e.g. amoeboid protozoa and leucocytes). Usually a semi-solid *gel* containing relatively few *organelles*.

eddy current heating See *induction heating*.

eddy currents Foucault currents. Induced (see *induction*) *electric currents* set up in the iron cores of *electromagnets* and other electrical apparatus. These currents cause considerable waste of *energy* in the cores of *armatures* of *dynamos* and in *transformers*. As the current will be large in a solid block of metal with low resistance, cores are made up of insulated metal sheets, the resistance between laminations serving to reduce the current and the consequent eddy-current loss.

Edison accumulator* A nickel-iron *accumulator*. Named after T. A. Edison (1847–1931).

EDTA Ethylenediaminetetraacetic acid $(NCH_2CH_2N)_2(HCOOCH_2)_4$. An important chelating agent (see *chelation*), used generally in the form of the tetrasodium salt for *complexometric analysis*.

EEG See *electroencephalograph*.

effective resistance The *resistance* of a *conductor* to *alternating currents*; in addition to the *direct current* resistance it includes the effect of any losses caused by the current (e.g. *eddy currents*). It is measured by the ratio of the total loss to the square of the *root mean square* of the current.

effective value See *root mean square of an alternating quantity*.

effervescence The escape of small *gas* bubbles from a *liquid*, usually as the result of chemical action.

efficiency of a machine The ratio of the output *energy* to the input energy. The efficiency of a machine can never be greater than unity. It is often expressed as a percentage. The thermal efficiency of a *heat engine* is the ratio of the *work* done by the engine to the *heat* supplied by the *fuel*. See *Carnot's principle*.

efflorescence (chem.) The property of some crystalline *salts* of losing a part of their *water of crystallization*, and becoming powdery on the surface. E.g. crystals of *sodium carbonate*, $Na_2CO_3.10H_2O$.

effusion of gases The passage of *gases* through small apertures under pressure. The relative rates of effusion of different gases under the same conditions are inversely proportional to the square roots of their *densities*.

EHF See *extremely high frequency*.

eichosanoic acid See *arachidic acid*.

eigenfunction An allowed *wave function* enabling a meaningful solution to be obtained from *Schrödinger's wave equation*. For each eigenfunction there is a fixed energy value (eigenvalue) for the system.

einsteinium Es. *Transuranic element*, At. No. 99. The most stable *isotope*, einsteinium-252, has a *half-life* of 270 days.

Einstein's equation 1. $E = mc^2$, where E is the energy equivalent of a mass m, and c is the *speed of light*. A direct consequence of Einstein's special theory of *relativity*, this equation refers to all types of energy. **2.** See *photoelectric effect*.

Einstein shift A slight displacement towards the red of the lines of the solar *spectrum* due to the *Sun's gravitational field* (see *red shift*). It was predicted by Einstein's general theory of *relativity* and subsequently verified experimentally. Named after Albert Einstein (1879–1955).

elastance The *reciprocal* of *capacitance*; it is measured in reciprocal *farads*, which are sometimes called *darafs*.

elastic collision A collision between bodies under ideal conditions, such that their total *kinetic energy* before collision equals their total kinetic energy after collision. In *nuclear physics*, an elastic collision is one in which an incoming particle is scattered without causing the *excitation* or breaking up of the struck *nucleus*.

elastic cross-section See *cross-section*.

elasticity The property of a body or material of resuming its original form and dimensions when the *forces* acting upon it are removed. If the forces are sufficiently large for the deformation to cause a break in the molecular structure of the body or material, it loses its elasticity and the *elastic limit* is said to have been reached; thereafter the body ceases to be elastic, becoming instead plastic (see *yield point*). *Hooke's law* applies only within the limit of proportionality, which occurs before the elastic limit is reached. Between these two points the *strain* is no longer proportional to the *stress* but there will be no permanent strain after the stress has been removed. See also *elastic modulus*.

elastic limit The limit of *stress* within which the *strain* in a material completely disappears when the stress is removed.

elastic modulus Modulus of elasticity. The ratio of *stress* to *strain* in a given material. The strain may refer to a change in length (see *Young's modulus*), a twist or *shear* (see *shear stress*), or a change in volume (see *bulk modulus*); the stress divided by the strain being in each case expressed in *newtons* per square *metre*.

elastic scattering See *scattering*.

elastin Elastic fibrous *protein* found in the connective *tissues* of vertebrates.

elastomer A material that after being stretched will return to approximately its original length. Elastomers include *natural rubber*, *synthetic rubbers*, and rubber-like *plastics*.

electret A *dielectric* possessing a permanent electric *moment*.

electrical capacity See *capacitance*.

electrical condenser See *capacitor*.

electrical energy The *energy* associated with an *electric charge* in an *electric field*. If a charge Q moves through a *potential difference* V, the energy transformed is QV. It is measured in *watt seconds* (*joules*) or *kilowatt-hours*. One kilowatt-hour equals 3.6×10^6 joules or $8.598\ 45 \times 10^5$ *calories*.

electrical image A set of point charges on one side of a conducting surface that would produce the same *electric field* on the other side of the surface (in its absence) as the actual electrification of that surface.

electrical induction See *induction*.

electrical line of force A line in an *electric field* whose direction is everywhere that of the field.

electric arc See *arc*.

electric-arc furnace A steel-making furnace in which an electric arc provides the source of heat. In direct-arc furnaces, the arc is formed between an electrode and the metal being heated. The Héroult furnace is an example of this type. Three graphite or amorphous-carbon electrodes are used and arcs form between each electrode and the metal charge. In the indirect-arc furnaces, heat is produced by a discharge between two electrodes and is radiated onto the charge. The Stassano furnace is an example of this type.

electric bell See *bell, electric*.

electric cell See *cell*.

electric charge A property of some *elementary particles* that gives rise to well-documented interactions between them. Science is unable to offer any explanation regarding the nature of an electric charge, but it is able to describe in some detail the properties of *matter* that is so charged. The *electron* is said to be negatively charged with electricity, and the *proton* is said to be positively charged to an equal but opposite extent. These entities represent the basic units of electrically charged matter. Therefore, matter containing an equal number of protons and electrons is electrically neutral, but matter containing an excess of electrons possesses an overall negative charge; similarly matter that has a deficiency of electrons (i.e. an excess of protons) possesses an overall positive charge. These positive and negative conventions are purely arbitrary, but much of science is based upon them. A *force* of repulsion acts between like charges and a force of attraction acts between unlike charges: the region in which these forces act is called an *electric field*. Electric charges are also acted upon by forces when they move in a *magnetic field* that possesses a component at right angles to their direction of motion. The size of an electric charge is measured in the derived *SI unit*, the *coulomb*; the negative charge on the electron is $1.602\ 177 \times 10^{-19}$ coulomb. Symbol Q.

electric constant Permittivity of free space. ε_0. The fundamental constant that has the value $8.854\ 187\ 817 \times 10^{-12}$ farad per metre. It arises as the constant of proportionality in *Coulomb's law*, its value depending on the choice of units. See also *electric field*; *permittivity*.

electric current A flow of *electric charge* through a *conductor*. The *charge carriers* may be *electrons*, *ions*, or *holes*. The magnitude of a current through a particular cross-section is equal to the rate of flow of charge. The *SI unit* of current is the *ampere*. Symbol I.

electric current, heating effect of Joule heating. When an *electric current* flows through a *conductor* of finite *resistance*, *work* is done on the conductor. The quantity of heat produced is proportional to the resistance of the conductor, and is equal to VI or I^2R watts (*joules* per *second*), V being the *potential difference* in *volts, I* the current in *amperes*, and R the resistance in *ohms*. See also *Joule's laws (2)*.

electric displacement Electric flux density. D. The *electric charge* per unit area displaced across a layer of conductor placed in an *electric field*. Consider a uniform *electric field* of strength E in free *space*; i.e. the electric *flux* through unit area perpendicular to the field is E. When a *dielectric* medium is introduced into the field the electric flux at any point in the medium becomes modified owing to the interaction between E and the *atoms* of the dielectric, and assumes a new value D. This is the electric displacement.

electric field The region surrounding an *electric charge*, in which a *force* is exerted on a charged particle; the electric field strength (or electric intensity) is completely defined in magnitude and direction at any point by the force upon unit positive charge situated at that point. The electric field strength E, or *force* exerted upon a unit charge at a distance r from a charge Q, is given by:

$$E = Q/4\pi r^2 \varepsilon,$$

where ε is the *permittivity* of the medium between them. For free space (i.e. a vacuum) ε becomes ε_0, the *electric constant*, and has the value $8.854\ 187\ 817 \times 10^{-12}$ F m^{-1}. The strength of the field can also be described by the *electric displacement D*. In a vacuum $D/E = \varepsilon_0$.

electric flux The quantity of electricity displaced across unit area of a dielectric. It is the scalar product of the *electric displacement* and the area. Symbol Ψ.

electric flux density See *electric displacement*.

electricity A general term used for all phenomena caused by *electric charge* whether static or in motion.

electricity, frictional Triboelectricity. A separation of *electric charge* that results from the rubbing together of different materials; e.g. on rubbing *celluloid* with rabbit's fur, the fur is found to possess a positive charge, and the celluloid receives an equal negative charge. The rubbing motion strips some of the *electrons* from the *atoms* or *molecules* of the fur, which collect on the surface of the celluloid.

electricity, static Phenomena associated with *electric charges* at rest, due purely to the *electrostatic field* produced by the charge, whereas in the case of current electricity other effects, in particular a *magnetic field*, are added.

electric light Illumination produced by the use of *electric charges*; it may be produced by virtue of the heating effect of an *electric current* on a wire or *filament* (see *electric-light bulb*), by an electric arc (see *arc lamp*), or by the passage of charged particles through a *vapour*, as in the *mercury vapour lamp*, noble gas, or *fluorescent lamps*.

electric-light bulb A glass bulb, often filled with nitrogen or some other chemically inert *gas*, containing a wire or *filament*, usually made of tungsten. The passage of an *electric current* through the filament heats it to a white heat.

electric motor A device for converting *electrical energy* into *mechanical energy*, depending on the fact that when an *electric current* flows through a *conductor* placed in a *magnetic field* possessing a component at right angles to the conductor, a mechanical *force* acts upon the conductor. In its simplest form, it consists of a coil or *armature* through which the current flows, placed between the poles of a powerful *electromagnet*, the *field magnet*; the mechanical force upon the conductor causes the armature to rotate. See *induction motor*; *synchronous motor*; *universal motor*.

electric organ (bio.) The muscles of certain fish that have been modified to generate an *electric field* in the surrounding water; sensing changes in that field allows the fish to navigate or find prey. In some fish the electric organs can generate high voltage pulses to stun prey or potential predators.

electric polarization *P.* When an *electric field* is applied to an electrically neutral *atom*, a displacement of the *electrons* with respect to the positive *nucleus* occurs. This gives rise to a small electric *dipole* possessing an electric *moment* in the direction of the field. This effect occurs when a *dielectric* is placed in an electric field, the electric field acting upon each individual atom of the dielectric. The electric polarization is given by:
$$P = D - E\varepsilon_0,$$
where D is the *electric displacement*, E is the *electric field* strength, and ε_0 is the *electric constant*.

electric potential *V* The energy needed to move unit *electric charge* from infinity to the point in an *electric field* at which the potential is specified. The *unit* of electric potential is the *volt*. See also *potential difference*.

electric power The rate of doing work, measured in *watts*. A power of 1 watt does 1 *joule* of work per second. The power in watts in a d.c. circuit is given by the product of the *potential difference* in *volts* and the *current* in *amperes*. In an a.c. circuit the *power factor* must be taken into account.

electric spark A transitory discharge of *electricity* produced by a high *potential difference*, accompanied by *light* and *sound*, through a *dielectric* or *insulator*.

electric susceptibility X_e. The ratio of the *electric polarization (P)* produced in a substance to the product of the *electric field* strength (E) to which it is subjected and the *electric constant* (ε_0), i.e.

$$X_e = P/E\varepsilon_0.$$

The susceptibility is related to the relative *permittivity* (ε_r) by

$$X_e = \varepsilon_r - 1.$$

electrocardiograph Electrocardiogram. ECG. An instrument for recording the *current* and *voltage* waveforms associated with the contraction of the heart muscle.

electrochemical equivalent z. The *mass* of the *ions* of an element liberated or deposited from a solution by 1 *coulomb* of *electric charge* in *electrolysis*.

electrochemical series See *electromotive series*.

electrochemistry The study of the processes involved in *electrolysis* and electric *cells*.

electrode 1. A *conductor* by which an *electric current* enters or leaves an *electrolyte* in *electrolysis*, an electric *arc*, or a vacuum tube (see *discharge in gases* and *thermionic valve*): the positive electrode is the *anode*, the negative electrode the *cathode*. **2.** In a *semiconductor* device, an element that emits or collects *electrons* or *holes*, or controls their movement by an *electric field*.

electrodeposition The process of depositing by *electrolysis*, especially the deposition of one *metal* on another as in *electroplating* and *electroforming*.

electrode potential The *electric potential* developed on an electrode that is in equilibrium with a solution of its *ions* (see also *half cell*). Electrode potentials cannot be measured absolutely and are usually specified by comparison with a *hydrogen electrode*, which is assumed to have an electrode potential of zero. In practice a number of more convenient electrodes are used, of known standard electrode potential. These are calibrated against the standard hydrogen electrode. See *calomel electrode*.

electrodialysis A method of *desalination*. Water containing salt is fed into a *cell* containing two *electrodes* separated by an array of *semipermeable membranes*, which are alternately semipermeable to positive and negative *ions*. The ions collect between alternate pairs of membranes enabling a stream of pure water to be extracted from the cell.

electrodynamics The study of the relationship between electric and magnetic *forces* and their mechanical causes and effects.

electrodynamometer An instrument for measuring *current*, *voltage*, or *power*, in both *direct current* and *alternating current* circuits. It depends upon the interaction of the *magnetic fields* of fixed and movable coils.

electroencephalograph Electroencephalogram. EEG. An instrument used for recording the rhythmical *electric currents* generated by the brain. The pattern obtained can be correlated with certain human physiological states (e.g. sleep) and pathological states (e.g. epilepsy).

electroforming The production, or reproduction, of *metal* articles by the deposition of a metal upon a removable *electrode* during *electrolysis*.

electrokinetic potential Zeta-potential, ζ-potential. The *potential difference* across the *interface* between a moving liquid and the fixed liquid layer attached to the solid surface over which the liquid moves.

electrokinetics The study of *electric charges* in motion and their behaviour in *electric* and *magnetic fields*, as opposed to *electrostatics*.

electroluminescence *Fluorescence* resulting from bombardment of a substance with *electrons*.

electrolysis The chemical *decomposition* of certain substances (*electrolytes*) by an *electric current* passed through the substance in a dissolved or molten state. Such substances are ionized into electrically charged *ions*, and when an electric current is passed through them by means of conducting *electrodes*, the ions move towards the oppositely charged electrodes, there give up their electric charges, become uncharged *atoms* or groups, and are either liberated or deposited at the electrode, or react chemically with the electrode, the *solvent*, or each other, according to their chemical nature. See *electrolytic dissociation*.

electrolysis, Faraday's laws of 1. The chemical action of an *electric current* is proportional to the quantity of *electric charge* that passes. **2.** The masses of substances liberated or deposited by the same quantity of electric charge are proportional to their *chemical equivalents*. See *electrochemical equivalent*. In a more modern form, the mass m liberated or deposited by a charge Q is given by:
$$m = QM/Fz,$$
where M is the relative ionic mass, z is the ionic charge, and F is the *Faraday constant*. Named after Michael Faraday (1791–1867).

electrolyte A *compound* that, in *solution* or in the molten state, conducts an *electric current* and is simultaneously decomposed by it. The current is carried not by *electrons* as in *metals*, but by *ions* (see *electrolysis*). Electrolytes may be *acids*, *bases*, or *salts*.

electrolytic capacitor (condenser) A fixed electrical *capacitor* in which one *electrode* is a *metal* (usually aluminium) foil coated with a thin layer of the metal *oxide*, and the other electrode is a non-corrosive *salt solution* or paste. The metal foil is maintained positive to prevent the removal of the oxide film by the hydrogen liberated. Tantalum sheets are also used as electrodes, immersed in an *electrolyte* of *sulphuric acid*. The advantage of electrolytic capacitors is that they provide a high *capacitance* in a limited space.

electrolytic cell See *cell*.

electrolytic dissociation The *dissociation* of the *molecules* of *electrolytes* into charged *ions*. The degree of dissociation determines the electrical *conductivity* of the solution and also other properties, which can be related theoretically to the total number of molecules and ions of the electrolyte formed in the solution. Many compounds, e.g. *ethanoic acid*, have low degrees of dissociation and are called weak electrolytes. Others have high degrees of dissociation (strong electrolytes). In the Debye–Hückel theory (named after Peter Debye, 1884–1966 and Erich Hückel, 1896–) it is assumed that electrolytes are completely dissociated and that there is electrostatic attraction between the ions. The theory gives a good approximation for nonideal behaviour in *dilute* solutions but not for *concentrated* electrolytes.

electrolytic gas Detonating gas. A mixture of hydrogen and oxygen, in a ratio of 2 to 1 by *volume*, formed by the *electrolysis* of *water*.

electrolytic rectifier. A *rectifier* consisting of two *electrodes* immersed in an *electrolyte*, which is used to convert an *alternating current* into a *direct current*. It depends on the properties of certain *metals* and *solutions* to allow current to flow in one direction only.

electrolytic refining The purification of copper and some other metals by *electrolysis*. An impure copper *anode*, a pure copper *cathode*, and an *electrolyte* of *copper(II) sulphate* form a cell. When current is passed, pure copper is transferred from the anode to the cathode, while impurity metals, the anode sludge, are deposited at the bottom of the cell (for further recovery, especially if they contain gold and silver).

electromagnet A temporary *magnet* formed by winding a coil of wire (see *field coil*) round a piece of soft iron; when an *electric current* flows through the wire, the iron becomes a magnet. Electromagnets are widely used in switches, metal-lifting cranes, etc.

electromagnetic induction See *induction, electromagnetic*.

electromagnetic interaction The *fundamental interaction* that occurs between electrically charged *elementary particles*. It can be explained by the exchange of virtual *photons* (see *virtual state*) between the interacting particles. See also *electroweak theory*.

electromagnetic moment See *magnetic moment*.

electromagnetic pump A device used for pumping *liquid metals*. A current is passed through the liquid metal, which is contained in a flattened pipe placed between the poles of an *electromagnet*. The liquid metal is thus subjected to a *force*, which acts along the axis of the pipe. Electromagnetic pumps are used in fast *nuclear reactors* to move the *coolant* of liquid sodium around the system.

electromagnetic radiation *Radiation* consisting of waves of *energy* associated with *electric* and *magnetic fields*, resulting from the acceleration of an *electric charge*. These electric and magnetic fields, which require no supporting medium and can be propagated through *space*, are at right angles to each other and to the direction of propagation. Electromagnetic waves travel through space with a uniform *speed* of $2.997\,924\,58 \times 10^8$ *metres* per *second*, or 186 282 miles per second. The nature of electromagnetic radiations depends upon their *frequency* (see *electromagnetic spectrum*). Electromagnetic radiation is emitted by *matter* in discontinuous units called *photons*.

electromagnetic spectrum The range of *frequencies* over which *electromagnetic radiations* are propagated. The lowest frequencies are *radio* waves, increases of frequency produce *infrared radiation*, *light*, *ultraviolet radiation*, *X-rays*, and *gamma-rays*. See Appendix, Table 10.

electromagnetic units E.M.U. A system of electrical *units*, within the *c.g.s. system*, based on the unit *magnetic pole*, which repels a similar pole, placed 1 cm away, with a *force* of 1 *dyne*. The E.M.U. of *electric current* is that current that, flowing in an arc of a *circle* of unit length and radius (i.e. 1 cm), exerts a force of 1 dyne on a unit magnetic pole placed at the centre. The E.M.U. of *resistance* is that resistance in which *energy* is dissipated at the rate of 1 *erg* per second by the flow of 1 E.M.U. of current. The E.M.U. of *electromotive force* or *potential* is the potential that, applied across the ends of a *conductor* of 1 E.M.U. resistance, causes 1 E.M.U. of current to flow. Electromagnetic units have the prefix *ab-* attached to the corresponding practical units (e.g. abampere, abvolt). These units have now been replaced by *SI units*. Compare *electrostatic units*.

electromagnetic waves See *electromagnetic radiation; electromagnetic spectrum*.

electrometallurgy The study of electrical processes used in separating a *metal* from its *ore*; or refining, plating, or shaping metals by electrical means. See *electrolytic refining; electroplating; electroforming*.

electrometer An instrument for measuring *voltage* differences, which draws no current from the source. They are essential for measuring *electrostatic* voltage differences. They were formerly based on *electroscopes* but now use *solid-state* devices.

electromotive force EMF. The source of *electrical energy* required to produce an *electric current* in a circuit. It is defined as the rate at which electrical energy is drawn from the source and dissipated in a circuit when unit current is flowing in the circuit. The *SI unit* is the *volt*. See *potential difference*.

electromotive series Electrochemical series. A list of metals arranged in order of the magnitudes of their *electrode potentials*. Metals with high negative electrode potentials stand at the head of the electromotive series. The list represents the order in which the metals replace one another from their salts, a metal higher in the series replacing one lower down; similarly, metals placed above hydrogen will liberate it from *acids*. Metals having a greater tendency than hydrogen to lose electrons to their solutions are said to be electropositive. Elements that gain electrons are electronegative. The chief metals in order are potassium, calcium, sodium, magnesium, aluminium, manganese, zinc, cadmium, iron, cobalt, nickel, tin, lead, hydrogen, copper, mercury, silver, platinum, gold.

electron An *elementary particle* classed as a *lepton* having a *rest mass* of $9.109\ 389\ 7 \times 10^{-31}$ *kilogram*, approximately 1/1836 that of a hydrogen *atom*, and bearing a negative *electric charge* of $1.602\ 177\ 33 \times 10^{-19}$ *coulomb*, called the electronic charge. The radius (classical, calculated from $e^2/4\pi\varepsilon_0 m_e C^2$) of the electron is $2.817\ 940\ 92 \times 10^{-15}$ *metre*. The electron is a constituent of all atoms (see *atom*, *structure of*). The positively charged *antiparticle* of the electron is the *positron*, and the word 'electron' is sometimes used to include both negative electrons (negatrons or negatons) and positive electrons (positrons or positons). A *free electron* is one that has been detached from its atomic *orbit*.

electron affinity 1. The tendency of an *atom* or *molecule* to accept an *electron* and form a negative *ion*. **2.** The *energy* (often given in electronvolts) liberated when one *mole* of an *element* in the form of gaseous atoms is converted into negative ions. The *halogens* have high electron affinities.

electron capture 1. The formation of a negative *ion* when a *free electron* is captured by an *atom* or *molecule* (also referred to as 'electron attachment'). **2.** A *radioactive transformation* as a result of which a *nucleus* captures an inner *orbital electron* and the resulting excited ion then emits an *X-ray photon* or an Auger electron (see *Auger effect*) as the vacancy in the inner orbit is filled by an outer electron.

electron carrier system A complex of *enzymes* and *coenzymes* in a living cell that accepts electrons at a high energy level and passes them in an orderly sequence of *redox reactions* to an acceptor molecule at a lower energy level. These reactions generate *chemical energy* in the form of *adenosine triphosphate* (ATP). Important examples are the *respiratory chain* in *mitochondria* and the similar system in *cyclic* and *non-cyclic photophosphorylation* in *chloroplasts*.

electron density The density of *electronic charge* at a given point in a *molecule*; alternatively defined as the probability of finding an *electron* at the particular point.

electron diffraction A *diffraction* effect resulting from the passage of *electrons* through *matter*, analogous to the *diffraction* of visible *light* or *X-rays*. The phenomenon of electron diffraction is the principal evidence for the existence of waves associated with electrons (see *de Broglie wavelength*). The diffraction of electrons when passed through *crystals* or thin *metal* foils is used as a method of investigating *bond length* and solid surfaces.

electronegative elements and groups Elements and groups that take up *electrons*, thus acquiring a negative *electric charge*, when united with other radicals by electrovalent bonds (see *chemical bond*). The *halogens*, oxygen, sulphur, and other *non-metallic elements* are generally electronegative. Electronegativity is usually calculated on the Pauling scale, based on bond dissociation energies in which fluorine, the most electronegative element is assigned a value of 4. On this scale other values are: $Si = 1.8$, $P = 2.1$, $C = 2.5$, $Br = 2.8$, $N = 3.0$, $O = 3.5$. See also *electromotive series*.

electron exchanger See *redox exchanger*.

electron gun The source of *electrons* in a *cathode-ray tube* or *electron microscope*. It consists of a *cathode* emitter of electrons (see *thermionic emission*), a control grid that regulates the intensity of the beam, one or more annular *anodes* through which the beam of electrons can pass, and one or more focusing and control *electrodes*.

electronic charge See *electron*.

electronics An applied physical science concerned with the development of electrical circuits using *semiconductors*, *thermionic valves*, and other devices in which the motion of *electrons* is controlled for purposes of communications, control, or computing.

electron lens A system of *electric* or *magnetic fields* used to focus a beam of *electrons* in a manner analogous to an optical *lens*. They are used in *electron microscopes*, *cathode-ray tubes*, etc.

electron micrograph A photograph of an object obtained with an *electron microscope*.

electron microscope An instrument similar in purpose to the ordinary *light microscope*, but with a much greater *resolving power*. Instead of a beam of light to illuminate the object, a parallel beam of *electrons* from an *electron gun* is used. In the transmission electron microscope, the object, which must be in the form of a very thin film (<50 nm) of the material, allows the electron beam to pass through it; but, owing to differential *scattering* in the film, an image of the object is carried forward in the electron beam. The latter then passes through a magnetic or *electrostatic* focusing system (see *electron lens*) which is equivalent to the optical lens system in an ordinary microscope, i.e. it produces a much magnified image. This is received on a fluorescent screen and recorded by a camera. Magnifications up to $\times 10^6$ can be achieved. In the scanning electron microscope (SEM) a thick sample can be used and the sample is scanned by the electron beam. Secondary electrons emitted from the surface of the sample are focused into a screen. The magnification is less with this type of instrument, but a three-dimensional image is formed. See also *field-emission microscope*; *field-ionization microscope*.

electron multiplier See *photomultiplier*.

electron optics The study of the methods used to focus beams of *electrons* by means of *electron lenses*, etc. The design of *electron microscopes* and similar devices relies upon electron optics.

electron-probe microanalysis A method of analysing a very small quantity of a substance by directing a finely focused *electron* beam on to it so that an *X-ray* emission is produced characteristic of the *elements* present in the sample. The diameter of the beam is usually about 1 μm and quantities as small as 10^{-13} g can be detected by this means. The method may be used quantitatively for elements whose *atomic numbers* exceed 11.

electron-spin resonance ESR. A phenomenon exhibited by paramagnetic substances (see *paramagnetism*) due to their unpaired *electrons*. The *spin* of an unpaired electron is associated with a *magnetic moment* that may align itself in one of two ways with respect to an applied *magnetic field*, each possible alignment corresponding to a different *energy level*. By applying magnetic field and irradiating the specimen with suitable *electromagnetic radiation*, transitions between these two energy levels can be made, falling in the *microwave* region of the *electromagnetic spectrum*, thus producing the phenomenon known as electron-spin resonance. If, however, the paramagnetic *molecule* includes magnetic nuclei, these transitions will interact with the nuclear spin (see *nuclear magnetic resonance*) producing a series of lines rather than a single resonance. Electron-spin resonance spectroscopy consists of analysing this *hyperfine structure* so that the electron can be located within the molecule, thus providing information about the molecule's structure. ESR is also used for studying *free radicals*.

electron transport chain Respiratory chain. An *electron carrier system* in *mitochondria* important in cellular *respiration*. *Electrons* from hydrogen atoms from the breakdown of *sugars* in the *citric-acid cycle* are passed along the electron transport chain to molecular oxygen, with the formation of water, and of ATP from ADP (see *adenosine triphosphate*). The process is called oxidative phosphorylation.

electronvolt eV. A unit of *energy* widely used in *nuclear physics*. The increase in energy or the *work* done on an *electron* when passing through a *potential* rise of 1 *volt*. 1 electronvolt = $1.602 \ 10 \times 10^{-19}$ *joule*. 1 MeV = 10^6 electronvolts; 1 GeV = 10^9 electronvolts.

electrophilic reagents Cationoid reagents. *Reagents* that react at centres of high *electron density*. They are essentially electron acceptors (e.g. *halogens*) that gain or share electrons from an outside *atom* or *ion*. Compare *nucleophilic reagents*.

electrophoresis Cataphoresis. The migration of the electrically charged *solute* particles present in a *colloidal solution* towards the oppositely charged *electrode*, when two electrodes are placed in the solution and connected externally to a source of *EMF*. The technique can be used for analysis of mixtures of *proteins* as the rate at which components of a mixture migrate depends on the charge, size, and shape of the particles.

electrophorus A laboratory demonstration apparatus for showing electrostatic charging by *induction*.

electroplating Depositing a layer of *metal* by *electrolysis*, the object to be plated forming the *cathode* in an electrolytic tank or bath containing a *solution* of a *salt* of the metal that is to be deposited, a rod of which also forms the anode. This form of *electrodeposition* is used for providing a corrosion-resistant cover on a ferrous metal and for silver- and gold-plating for decorative purposes.

electropositive elements and groups Elements and groups that give up *electrons*, thus acquiring a positive *electric charge*, when united with other radicals by electrovalent bonds (see *valence*). The *metals* and *acidic hydrogen* are generally electropositive. See also *electromotive series*.

electroscope An instrument for detecting the presence of an *electric charge*. The gold-leaf electroscope consists of two rectangular leaves of gold foil attached to a conducting rod of *metal* held by an insulating plug; when the rod and leaves acquire an electric charge, the leaves diverge owing to the mutual repulsion of charges of like sign.

electrostatic field A region in which a stationary electrically charged particle would be subjected to a *force* of attraction or repulsion as a result of the presence of another stationary *electric charge*. See *electric field*.

electrostatic generator A machine designed for the continuous separation of *electric charge*. Examples include the *Wimshurst machine* and the *Van de Graaff generator*.

electrostatic precipitation A widely used method of controlling the pollution of air (or other *gases*). The gas, containing *solid* or *liquid* particles suspended in it, is subjected to a uni-directional *electrostatic field*, so that the particles are attracted to, and deposited upon, the positive *electrode*. See *Cottrell precipitator*.

electrostatics The study of static *electric charges* and the *forces* and *fields* associated with them. Compare *electrodynamics*.

electrostatic units ESU. A system of electrical *units* based upon the electrostatic unit of *electric charge*. The electrostatic unit of charge is that quantity of electricity that will repel an equal quantity, 1 cm distant from it in a *vacuum*, with a *force* of 1 *dyne*. Electrostatic units have the prefix *stat-* attached to the corresponding practical units (e.g. statcoulomb). These units have now been replaced by *SI units*. Compare *electromagnetic units*.

electrostriction The change in the dimensions of a *dielectric* when placed in an electric *field*. An example is the contraction of a *solvent* due to the *electrostatic field* of a dissolved *electrolyte*.

electrotyping The production of copies of plates of type, etc., by the electrolytic deposition of a layer of *metal* on a previously prepared mould. This is a cast of the object to be copied, made of *plastic* material and coated with a layer of *graphite*, which acts as a *conductor* of electricity. It is then suspended to act as a *cathode* in an electrolytic bath (see *electroplating*) containing a *solution* of a *salt* of the metal required, usually copper. The passage of an *electric current* will deposit a layer of any required thickness of metal upon the cathode, the layer being a replica of the original type.

electrovalent bond See *chemical bond*.

electrovalent crystal See *ionic crystal*.

electroweak theory Glashow–Weinberg–Salam theory. Quantum flavourdynamics. QFD. A *gauge field theory* that combines the *electromagnetic interaction* and the *weak interaction*. It is thus a first step towards a *unified field theory*. It has been confirmed by many experiments in particle physics, culminating in the discovery of the *Z particle*, the electrically neutral carrier of weak interactions. In this theory, electroweak interactions arise from the exchange of *photons*, the massive *W particle*, and neutral Z particle between *quarks* and *leptons*. See also *Higgs particle*.

electrum A natural *alloy* of gold (55%–85%) and silver.

element (chem.) A substance consisting entirely of *atoms* of the same *atomic* *number*. The elements are listed in the Appendix, Table 3. See also *periodic table*.

elementary particles Fundamental particles. The basic units of which all matter is composed. By 1932, with the discovery of the *neutron*, it was accepted that *atoms* consisted of a central *nucleus* composed of *protons* and neutrons, surrounded by one or more *electrons*. At that time these three particles were assumed to be the elementary particles of which the Universe is composed. However, the stability of the nucleus could not be explained on existing theories, as no interaction between the charged proton and the uncharged neutron was known. The elucidation of the four *fundamental interactions* led to the discovery of a number of new particles, some of which turned out to be more elementary than others. In the current method of classification there are two classes of particles: *leptons*, which interact by means of the *electromagnetic interaction* or the *weak interaction*, and *hadrons*, which also interact by the *strong interaction*. Leptons, which include the *neutrino*, *muon*, and *tauon*, as well as the electron, have no internal structure and are therefore truly elementary. Hadrons, including protons, neutrons, pions, etc., do have an internal structure and are therefore not really elementary.

The structure of hadrons is believed to consist of elementary particles called *quarks*, a concept introduced into physics in the early 1960s by Murray Gell-Mann and named after a sentence in James Joyce's *Finnigans Wake* ("Three quarks for Muster Marks."). In this theory, hadrons are either *baryons* (which decay into protons) or *mesons* (which decay into *photons* or leptons). Baryons consist of three quarks and mesons consist of two quarks, a quark and its corresponding antiquark.

Quarks, themselves, have unique properties: unlike all previously known particles they have charges that are a fraction of the charge on an electron, i.e. $+\frac{2}{3}$ or $-\frac{1}{3}$ of the electronic charge. They also occur in six *flavours* (the word used is not intended to imply a connection with taste): up (u, $+\frac{2}{3}$ charge), down (d, $-\frac{1}{3}$), charmed (c, $+\frac{2}{3}$), strange (s, $-\frac{1}{3}$), top (t, $+\frac{2}{3}$), and bottom (b, $-\frac{1}{3}$). The *nucleons* (protons and neutrons) consist of the following combinations:

$$p = uud \ (charge \ \tfrac{2}{3} + \tfrac{2}{3} - \tfrac{1}{3} = +1)$$
$$n = udd \ (charge \ \tfrac{2}{3} - \tfrac{1}{3} - \tfrac{1}{3} = 0)$$

For each flavour there is a corresponding antiquark with opposite charge, e.g. ū, charge $-\frac{2}{3}$, d̄, charge $+\frac{1}{3}$, etc.

In addition to flavour it has proved necessary to add to this model the concept of *colour charge*. Thus, each flavour occurs in three primary colours: red, green, and blue. The corresponding antiquarks have the complementary anticolours: cyan, magenta, and yellow. There are thus 36 quark particles, 18 quarks and 18 antiquarks. These, together with the leptons, are the only truly elementary particles. According to this aspect of the theory, known as *quantum chromodynamics*, hadrons are composed of quarks that produce the colour white, i.e. the baryons comprise three primary colours and the mesons consist of a primary colour and its complementary anticolour.

The strong interaction between quarks, according to this theory, is mediated by the exchange of particles with zero *rest mass* called *gluons*, in much the same way as the electromagnetic interaction can be visualized as being mediated by the exchange of particles with zero rest mass called photons. Gluons, of which there are eight types, are uncharged but they do have a colour charge, each gluon carrying a colour and an anticolour. In an interaction a quark can change colour, but all changes of colour are accompanied by the emission of a gluon, which is absorbed by another quark, whose colour in turn makes a compensatory change. Throughout any system of interactions all hadrons remain white, even though colour charges move around between them.

Neither quarks nor gluons have been detected individually, but there is widespread experimental evidence for their existence. In experiments in which quarks and gluons are created, they always immediately form clusters of hadrons so that the overall colour is white. It appears that the strong interaction does not allow coloured objects to exist alone.

elements (astr.) The numerical values required to define the elliptical *orbit* of a *planet* or *satellite*: they include the semi-*major axis* of the *ellipse* and its *eccentricity*. The *plane* of the orbit is defined by the angle its plane makes with the plane of the *ecliptic*. The position of the planet or satellite in its orbit is defined by its eccentric *anomaly*.

elements, magnetic See *magnetic elements*.

elevation, angle of If C is a point above the level of another point A, the angle of elevation of C from A is the *angle* that AC makes with the horizontal plane AX through A. See Fig. 11 under *depression, angle of*.

elevation of boiling point The rise in the *boiling point* of a *solution* produced by a non-volatile substance dissolved in a *solvent*. For a *dilute* solution the elevation is proportional to the number of *molecules* or *ions* present (see *colligative properties*), and the elevation produced by the same *molecular concentration* (or ionic concentration in the case of an *electrolyte*) is a constant for a particular solvent. This forms the principle of the boiling-point method (ebullioscopic method) for the determination of *relative molecular masses*.

eleven-year period A periodic change in the occurrence of *sunspots*, the cycle being complete in approximately eleven years; associated with this is a cyclic variation in the magnitude of the *daily variation*.

Elinvar* A variety of *steel* containing 36% nickel and 12% chromium. The *elasticity* is almost unaffected by changes of *temperature*; it is used for hairsprings of watches.

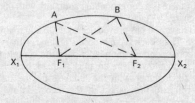

Figure 15.

ellipse A closed *plane* figure formed by cutting a right circular *cone* by a *plane* obliquely through its *axis* (see *conic sections*). The sum of the distances from any point on the *perimeter* of an ellipse to its two *foci* is constant. In Fig. 15, X_1X_2 is the *major axis*, F_1 and F_2 are the foci, and A and B are any points on the perimeter such that

$$AF_1 + AF_2 = BF_1 + BF_2$$

For an ellipse centred at the origin, in *Cartesian coordinates*, the equation is
$$x^2/a^2 + y^2/b^2 = 1$$
with the foci at $(\pm ea, 0)$, where e is the *eccentricity*.

ellipsoid A solid figure traced out by an *ellipse* rotating about one of its *axes*. A prolate ellipsoid is produced if it is rotated about the major axis and an oblate ellipsoid about the minor axis.

elliptically polarized light *Light* that can be resolved into two vibrations of equal frequency lying in *planes* at right angles to each other. The electric *vector* at any point in the path of the wave describes an *ellipse* about the direction of propagation of the light. The form of this ellipse is determined by the *amplitudes* of these two vibrations and by the difference of *phase* between them. (See also *polarization of light*.)

elution The removal of an *adsorbate* from an adsorbent by dissolving it in a *liquid* (the eluent). The resulting solution is called the eluate.

elutriation The washing, separation, or sizing of fine particles of different mass by suspending them in a current of air or *water*.

emanation Radium emanation. The former name for the *gas* formed by the *radioactive disintegration* of certain substances, consisting principally of *radon* (Rn-222), *thoron* (Rn-220), and *actinon* (Rn-219).

embryo An animal in its earliest stages of development, i.e. from the first division of the fertilized ovum until birth or hatching. A human embryo is known as a *foetus* after it is eight weeks old.

embryology The branch of *biology* concerned with the study of the growth and development of embryos.

emerald See *beryl*.

emery A mixture of *corundum* and iron oxide, usually *magnetite*, Fe_3O_4. It is used as an *abrasive*.

emetine $C_{29}H_{40}O_4N_2$. An *alkaloid* obtained from the roots of Brazilian ipecacuanha, m.p. $68°C$. It is used as an emetic and as a remedy for amoebic dysentery.

EMF See *electromotive force*.

emission of radiation The net rate at which a body emits *heat radiation* to its surroundings depends on the *temperature* of the body, the temperature of its surroundings, and the nature of the surface of the body. Dull black surfaces have the greatest *exitance* while brightly polished reflecting surfaces have least. See *Stefan's law*.

emission spectrum The *spectrum* observed when electromagnetic radiation coming directly from a source is examined with a *spectroscope*. The source must be heated or bombarded with particles in order to excite the atoms and molecules of which it consists. The emission occurs when these excited atoms or molecules decay to a lower energy state. See also *absorption spectrum*.

emissivity The ratio of the *power* per unit area of a body to the total power per unit area emitted by a perfect black body at the same *temperature* (see *black body radiation*). The emissivity is a pure numeric, equal to the *absorptance*. Symbol ε.

emittance See *exitance*.

emitter One of the three *electrodes* in a *transistor*.

empirical Based upon the results of experiment and observation only.

empirical formula The simplest type of chemical *formula*, giving only the proportion of each *element* present, but no indication of the *relative molecular mass* or the molecular structure, e.g. $(C_4H_3O_2N)_n$.

emulsifier Emulsifying agent. A substance, small quantities of which help to form or stabilize an *emulsion*. *Detergents* are frequently used.

emulsion A *colloidal solution* in which the *disperse phase* consists of minute droplets of *liquid*.

emulsion, photographic The light-sensitive coating on a *film* or plate (see *photography*). A 'nuclear emulsion' is a photographic emulsion specially prepared to record the tracks of *elementary particles* and nuclear fragments that pass through it.

emulsoid sol See *colloidal solutions*.

enamel 1. A class of substances (vitreous enamels) having similar composition to *glass* with the addition of *tin(IV) oxide*, SnO_2, or other *infusible* substances to render the enamel *opaque*. **2.** A finely ground oil *paint* containing a *resin*. **3.** The external layer of teeth consisting mainly of calcium phosphate carbonate *salts*.

enantiomorphism The occurrence of substances in two crystalline forms, one being a mirror image of the other. See also *optical isomerism*.

enantiotropism The occurrence of substances in two different physical forms, one being *stable* below a certain *temperature* (the *transition* temperature), the other above it. E.g. sulphur exists as alpha-sulphur at all temperatures below 96°C.; above this, the stable form is beta-sulphur.

endangered species A plant or animal that is in danger of becoming extinct, many as a result of human activities. In 1948, the International Union for the Conservation of Nature and Natural Resources was founded to protect endangered species. Some endangered animal species are the orang-utan, tiger, European otter, Indian elephant, blue whale, Californian condor, wild yak, and peregrine falcon. See also *conservation*.

endocrine gland Ductless gland. A gland in an animal that produces *hormones* and secretes them directly into the bloodstream. They include pituitary, adrenal, ovary, testis, thyroid, and parathyroid glands. Compare *exocrine gland*.

endocrinology The study of the *endocrine glands* and the *hormones* they produce.

endoenzyme An *enzyme* that remains within a living *cell* and does not diffuse through the cell membrane into the surrounding medium.

endoergic process An *endothermic process* (often applied in the context of a *nuclear reaction*). Compare *exoergic process*.

endoplasm The central part of the *cytoplasm* of living *cells*, usually distinct from the *ectoplasm* in that it is of greater fluidity and contains more *organelles*.

endoplasmic reticulum A network of *membrane*-bounded flattened sacs within the *cytoplasm* of eukaryotic cells (see *eukaryote*). It has a variety of functions in different cells. The membranes may be covered with many *ribosomes* (rough endoplasmic reticulum); these are the sites of *synthesis* of *proteins*, especially those destined to be *secreted* by the cell. Other endoplasmic reticulum lacks ribosomes (smooth endoplasmic reticulum) and may be the site of *triglyceride* transport (e.g. in intestine) of *lipid* synthesis (e.g. in liver), of *sterol* synthesis (e.g. in liver) or of calcium storage and release (e.g. in muscle).

endoskeleton A supporting structure that lies inside an *organism*, e.g. the bony skeleton of vertebrates. It supports and protects the organism, also providing a system of levers upon which muscles can act to produce movement. Compare *exoskeleton*.

endosmosis The inward flow of *water* into a cell containing an *aqueous solution*, through a *semipermeable membrane* or a selectively permeable membrane, due to *osmosis*.

endothermic process A process accompanied by the absorption of *heat*. Compare *exothermic process*.

end point Equivalence point. The point in a *titration*, usually indicated by a change of colour of an *indicator*, at which a particular *reaction* is completed.

energy E. The capacity for doing *work*. The various forms of energy, interconvertible by suitable means, include *potential*, *kinetic*, *electrical*, *heat*, *chemical*, *nuclear*, and *radiant energy*. Interconversion between these forms of energy can only occur in the presence of *matter*. Energy can only exist in the absence of matter in the form of radiant energy. The derived *SI unit* of energy is the *joule*. See also *internal energy*; *kinetic energy*; *potential energy*.

energy bands *Orbital electrons* in *atoms* are associated with specific amounts of *energy*, the change from one *energy level* to another taking place in *quantized* steps. In a crystalline *solid* the energies of all the electrons and atoms fall into several 'allowed' energy bands between which lie 'forbidden' bands. These bands may be depicted on an 'energy level diagram'. The range of energies corresponding to states in which the electrons can be made to flow, by an applied *electric field*, is called the *conduction band*. The range of energies corresponding to states that can be occupied by *valence electrons*, binding the *crystal* together, is called the *valence band*. The valence band in an ideal crystal is completely occupied at the *absolute zero* of temperature, but in real crystals above absolute zero some electrons are missing from the valence band, and it is these electrons that give rise to *holes*. In *insulators*, the conduction band is separated from the valence band by a wide forbidden band. Such substances do not conduct electricity because the electrons do not have the energy required to cross the forbidden band. In an intrinsic *semiconductor* the forbidden band is sufficiently narrow for electrons at the top of the valence band to jump to the conduction band as a result of thermal agitation at normal temperatures. In a good conductor, such as a metal, there is no forbidden band, the conduction band overlapping the valence band.

energy flux The rate of flow of energy per unit area. See *flux*.

energy forest See *biomass energy*.

energy levels An *atom*, *molecule*, or *nucleus* can exist only in certain definite states (*quantum states*) each having definitive energies. Thus, for each different atom or nucleus, there exists a series of energy levels corresponding to these permissible states. The lowest stable energy level is referred to as the *ground state*; systems at higher energy levels than the ground state are said to be excited (see *excitation*). In molecules, there are also quantized energy levels for vibrational and rotational energy.

energy-rich bonds A term used in *biochemistry* to distinguish between chemical *bonds* that when broken yield a large amount of *free energy* and those that give only a small yield of free energy ('energy-poor bonds'). The energy referred to in this context is the free energy liberated on *hydrolysis*. Energy-rich bonds usually involve *phosphate* groups and in this respect *adenosine triphosphate* (ATP) is of particular significance.

energy value of a food A measure of the energy made available by the complete combustion of a stated *mass* of the food; it is often given in joules per kilogram or large *Calories* per pound. It takes no account of the value of the food from any other point of view, or sometimes even of the suitability of the food for human consumption.

engine A device for converting one form of *energy* into another, especially for converting other forms of energy into *mechanical* (i.e. *kinetic*) *energy*. See *internal-combustion engine*; *ion engine*; *steam engine*.

enols *Organic compounds* containing the group –CH=C(OH)–. See *keto-enol tautomerism*.

enrich To increase the *abundance* of a particular *isotope* in a mixture of isotopes, especially to increase the abundance of the *fissile* isotope of a *nuclear fuel*. It usually refers to increasing the proportion of U-235 in natural uranium.

enthalpy H. A *thermodynamic* property of a substance given by $H = U + pV$, where U is the internal *energy*, p the *pressure*, and V the *volume*. In a *chemical reaction* carried out at constant pressure, the enthalpy change, ΔH, of the reaction (taken as negative for an *exothermic process*) is $\Delta U + p\Delta V$.

entrainment The transport of particles (e.g. fine droplets) in a moving stream of a *fluid* (e.g. the *vapour* of a boiling *liquid*).

entropy S. A quantity introduced in the first place to facilitate the calculations, and to give clear expression to the results of *thermodynamics*. When a system experiences a reversible change, its entropy (S) changes by an amount (ΔS) equal to the energy transferred to the system (ΔQ) by heat divided by the *thermodynamic temperature* (T) at which this happens, i.e. $\Delta S = \Delta Q/T$. Entropy changes for actual irreversible processes are calculated by postulating equivalent theoretical reversible changes. The entropy of a system is a measure of its degree of disorder on the molecular scale, i.e. it is a measure of the system's inability to do *work*. The total entropy of any isolated system can never decrease in any change; it must either increase (irreversible process) or remain constant (reversible process). The total entropy of the *Universe* therefore is increasing, tending towards a maximum, corresponding to complete disorder of the particles in it (assuming that it may be regarded as an isolated system). See *heat death of the Universe*.

environment The external surroundings, chemical, physical, and biological, in which an organism exists, and which can influence its behaviour and development.

enyne A *hydrocarbon* with a double (-ene) and a triple (-yne) *bond* between *carbon atoms* in its molecule.

enzyme A large group of *proteins* produced by living *cells*, which act as *catalysts* in the *chemical reactions* upon which life depends. Certain parts of the enzyme *molecule* (called the 'active centres') combine with the *substrate* molecule in such a way that the substrate undergoes chemical changes very much more rapidly than it would in the absence of the enzyme, while the enzyme itself remains unchanged. As enzymes are not consumed in these reactions they are effective in only minute quantities. Nearly all enzymes are highly specific in their action and therefore enormous numbers of them are found in nature. Many enzymes require the assistance of certain accessory substances (e.g. *coenzymes*) for their proper functioning and some require precisely defined conditions of *temperature* and *pH* for their optimum performance. The enzyme production of a cell is controlled by its *genes*.

Enzymes are usually named by adding the suffix -ase to a word indicating the nature of the substrate (e.g. *amylase*) or the type of reaction involved (e.g. *dehydrogenase*). A few enzymes retain old names that relate to neither of these rules (e.g. *pepsin, trypsin*).

enzymolysis The *catabolism* of a substance catalyzed by an *enzyme*.

eosin Tetrabromofluorescein. $C_{20}H_8Br_4O_5$. A red crystalline *insoluble* substance obtained from *fluorescein*, m.p. $295-6°C$., used as a red *dye*.

epact 1. The difference in days between the length of a solar *year* and a lunar year. **2.** The *Moon's* age in days at the start of the calendar year.

ephedrine $C_6H_5CHOHCH(CH_3)NHCH_3$. A white crystalline *optically active alkaloid*, m.p. $40°C$., used in medicine to treat asthma, colds, etc.

ephemeris A table that gives the predicted positions and the movements of a celestial body such as a *planet* or *comet*. Also an annual publication containing astronomical data.

ephemeris time Time measured on the basis of the orbital movements of the *planets* and the *Moon*. Until 1964 the ephemeris *second* provided the fundamental unit of time.

epicentre The point on the surface of the Earth that lies directly above the focus of an earthquake.

epichlorohydrin 1-Chloro-2,3-epoxypropane. $\overline{OCH_2CHCH_2Cl}$. A colourless liquid, b.p. $116°C$. A highly reactive *epoxide* used in the manufacture of *epoxy resins* and in many other reactions of organic synthesis.

epicycle 1. (math.) A *circle* whose centre rolls round the circumference of a larger circle without slipping. **2.** (astr.) Ptolemy (A.D. $127-151$) based his *astronomy* on the theory that the *planets* moved in epicycles round a larger circle, called the deferent, at the centre of which lay the *Earth*.

epicyclic gears A system of gears in which one or more wheels move around the outside, or the inside, of another wheel whose *axis* is fixed.

epidiascope An optical projector for throwing an enlarged image of either an *opaque* object or a transparency upon a screen. Used for illustrating lectures.

epimerism A type of *optical isomerism* occurring in *carbohydrates* and some other types of compound. It is due to the formation of isomers (epimers) that differ in their molecular arrangements about an asymmetric atom in a molecule containing two or more asymmetric atoms.

epinephrine See *adrenaline*.

epitaxy The growth of one crystalline substance on another so that both have the same crystal structure. Epitaxial layers are used in the manufacture of semiconductor devices.

epithermal neutrons *Neutrons* that have *energies* in excess of the energy associated with thermal agitation. Neutrons that have speeds and energies intermediate between *fast* and *thermal neutrons* (i.e. between about 0.1 and 100 *electronvolts*).

epoxide A cyclic *ether* in which an oxygen *atom* is bound to two carbon atoms, forming a three-membered ring.

epoxy- Prefix denoting an oxygen atom whose free *valences* are attached to different atoms, which are otherwise connected; usually applied to cases in

147

which oxygen forms a three-membered ring with two carbon atoms (an *epoxide*).

epoxyethane Ethylene oxide, oxirane. $\overline{CH_2CH_2O}$. A colourless flammable *epoxide* gas (liquid below 10.7°C.), made by the *oxidation* of *ethene* in the presence of a *catalyst*. It is used to make *ethanediol*, *ethanolamines*, etc.

epoxy resins *Thermosetting resins* made by the reaction of the *epoxide epichlorohydrin* with *polyhydric* compounds, such as bisphenol A (4,4'-isopropylidenediphenol), in the presence of a *catalyst*. They are hardened by mixing with polyamines, with which they form cross-linkages. They are used in the manufacture of electrical components, as structural materials, in surface coatings, and as *adhesives*.

Epsom salts Epsomite. See *magnesium sulphate*. $MgSO_4.7H_2O$.

equation, chemical A representation of a *chemical reaction*, using the *symbols* of the *elements* to represent the actual *atoms* and *molecules* taking part in the reaction; the re-arrangement of the various atoms of the substances taking part is thus shown. E.g. the chemical equation

$$H_2 + Cl_2 = 2HCl$$

represents the reaction between hydrogen and chlorine to form *hydrogen chloride*, and states that a hydrogen molecule, consisting of two atoms of hydrogen (H_2), reacts with a similarly constituted chlorine molecule, to give two molecules of hydrogen chloride, each consisting of one hydrogen and one chlorine atom (2HCl). From a knowledge of the equation for any chemical reaction, and of the *relative atomic masses* of all the elements taking part, it is thus possible to calculate the proportions by *mass* in which the substances react, since the whole bulk of the reaction consists merely of the repetition, a vast number of times, of the process depicted by the equation.

equation of motion See *motion, equations of*.

equation of state Any equation connecting the *pressure p*, *volume V*, and *thermodynamic temperature T* of a substance. Some equations of state attempt to cover more than one *phase* of the substance, e.g. *Van der Waals' equation of state*, and are approximate. Others are intended to be applied to one particular phase of the substance, e.g. the gaseous phase, and then only within certain limits of *p*, *V*, and *T* (see *gas equation*). With these limitations, these latter equations can represent the actual behaviour of the substance with greater accuracy.

equation of time The difference between mean solar time, as given by a clock, and apparent solar time, i.e. sundial time. The time of rotation of the *Earth* upon its axis is not exactly equal to the time from noon to noon, the difference being caused by the motion of the Earth relative to the *Sun* to complete a circuit in one *year*, and also by the inclination of the *ecliptic* to the *Equator*.

equator, terrestrial The *great circle* of the Earth, lying in a *plane* perpendicular to the *axis* of the Earth, that is equidistant from the two Poles. See also *magnetic equator* and *celestial equator*.

equilateral figure A figure having all its sides equal in length. E.g. an equilateral *triangle*.

equilibrium A state of balance between opposing *forces* or effects.

equilibrium, chemical See *chemical equilibrium*.

equilibrium constant In any *chemical reaction* there is always a state of *chemical equilibrium*, at a given *temperature* and *pressure*, between the *concentration* of the reactants and the concentration of the products. The position of this equilibrium, under specific conditions, is expressed by the equilibrium constant, K, such that in the reaction:

$$aA + bB \rightleftharpoons cC + dD$$

K is given by:

$$(C_C)^c.(C_D)^d/(C_A)^a.(C_B)^b,$$

where C_A is the concentration of the substance A, 'a' *molecules* of which take part in the reaction. In gas-phase reactions, *partial pressures* can be used instead of concentrations.

equimolecular mixture A *mixture* containing substances in equal molecular proportions; i.e. in the ratio of their *relative molecular masses*. E.g. *invert sugar*, formed by the *hydrolysis* of *cane-sugar*. Each *molecule* of the cane-sugar is split into a molecule of *glucose* and a molecule of *fructose*, thus forming an equimolecular mixture of the two latter.

equinox The moment (or the point) at which the *Sun* apparently crosses the *celestial equator*; the point of intersection of the *ecliptic* and the celestial equator.

equipartition of energy In any physical system in thermal equilibrium the average *energy* per *degree of freedom* is the same, and equals $kT/2$, where $k = $ *Boltzmann constant* and $T = $ the *thermodynamic temperature* of the system. This provides a means of calculating the total thermal energy of a system. Thus, in 1 *mole* of a monatomic *gas*, each *atom* possesses three degrees of freedom (due to its *translatory motion*), and the total number of atoms is L (*Avogadro constant*). Hence the total energy per mole of the gas is $3LkT/2$ or $3RT/2$, since $k = R/L$, where R is the *gas constant*. This theory was proposed by Boltzmann but was demolished by the *quantum theory*. It provides a good approximation in some circumstances.

equipotential lines and surfaces Lines and surfaces having the same *electric potential*.

equivalence point The *end point* in a *titration*.

equivalent, electrochemical See *electrochemical equivalent*.

equivalent weight See *chemical equivalents*.

equivocation A term used in *information theory* to indicate the rate of loss of information (per second or per symbol) at the receiving end of a *channel* of information due to *noise*.

erbium Er. Element. R.a.m. 167.26. At. No. 68. A silvery metal r.d. 9.066, m.p. 1529°C., b.p. 2868°C., used in some *alloys* and as a *neutron* absorber. See *lanthanides*.

erecting prism A right-angled optical *prism* used in *prismatic optical instruments* to render an inverted image upright.

erg A unit of *work* or *energy* in the *c.g.s. system* of units; the work done by a *force* of 1 *dyne* acting through a distance of 1 cm. 1 erg = 10^{-7} *joule*.

ergocalciferol Vitamin D, calciferol. See *vitamins*.

ergometrine Ergonovine. $C_{19}H_{23}N_3O_2$. A colourless crystalline *alkaloid*, obtained from ergot, and used in medicine to prevent haemorrhage.

ergonomics The engineering aspects of the study of the relation between human workers and their working environment.

ergosterol $C_{28}H_{43}OH$. A white crystalline *sterol*, m.p. 163°C., that occurs in small amounts in the *fats* of animals; it is converted into *vitamin* D_2 (*calciferol*) by the action of *ultraviolet radiation*.

ergotamine $C_{33}H_{35}N_5O_5$. A crystalline *insoluble polypeptide*, m.p. 212.4°C., obtained from ergot, and used in the form of its *tartrate* in medicine as a uterine stimulant.

ergotoxine $C_{35}H_{41}N_5O_6$. A white crystalline *insoluble alkaloid*, obtained from ergot, and used in medicine as a uterine stimulant.

Erinoid* A *thermoplastic* material prepared from *casein* and *formaldehyde*.

Erlenmeyer flask A flat-bottomed conical laboratory flask with a narrow neck. Named after E. Erlenmeyer (1825–1909).

erythritol 1,2,3,4-Butanetetrol. $(CH_2OHCHOH)_2$. An *optically active* white crystalline *polyhydric* alcohol, m.p. 121.5°C., used as an *intermediate* in organic synthesis. The tetranitrate *ester* is used in medicine for treatment of heart disease and high blood pressure.

erythrocytes Red blood cells. The cells of the *blood* that contain *haemoglobin* and whose function it is to transport oxygen through the body. Erythrocytes, which are produced in red bone marrow, have no means of propulsion, and in mammals the cells have no *nuclei*. Human blood contains approximately five million erythrocytes per cubic millimetre.

Erythromycin* $C_{37}H_{67}NO_{13}$. An *antibiotic* produced by the *Streptomyces* bacteria, used to combat a variety of bacterial infections.

Esaki diode See *tunnel diode*.

escape velocity The *velocity* that a projectile or *space probe* would need to attain in order to escape from a particular *gravitational field*. The escape velocity from the surface of a *planet* (or moon) depends on the planet's (or moon's) *mass* and diameter. The escape velocity from the *Earth's* surface is about 11 200 metres/s (25 000 m.p.h.) and from the *Moon's* surface about 2370 metres/s (5300 m.p.h.).

ESR See *electron-spin resonance*.

essential amino acid An *amino acid* that must be present in the diet of an organism, i.e. one that it is unable to synthesize.

essential element Any *element* required by an *organism*. All organisms require the elements carbon, hydrogen, oxygen, and nitrogen, which form the basis of organic compounds. Calcium, phosphorus, sodium, potassium chlorine, magnesium, and sulphur are also required in relatively large amounts (more than 1%). Elements required in smaller quantities are known as *trace elements*.

essential oils Natural *oils* obtained from plants, mostly *benzene derivatives* or *terpenes*. They are used for their flavour or odour.

esterases *Enzymes* that control *hydrolysis* of *esters*.

ester gums Rosin esters. Products made by *esterification* of *organic acids* in *rosin* with *polyhydric alcohols*, especially *glycerol*. They are used in varnishes.

esterification The formation of an *ester* by the *chemical reaction* of an *acid* with an *alcohol*; e.g. the action of *ethanol* on *ethanoic acid* to form *ethyl ethanoate* and *water*. $C_2H_5OH + CH_3COOH \rightleftharpoons CH_3COOC_2H_5 + H_2O$.

esters *Organic compounds* corresponding to *inorganic salts*, derived by replacing hydrogen of a *carboxylic acid* by an organic group, giving them the general formula RCOOR′. E.g. *ethyl ethanoate*, $CH_3COOC_2H_5$, is the ethyl ester of *ethanoic acid*, CH_3COOH. Many esters are pleasant-smelling *liquids* used for flavouring essences. Many vegetable and animal *fats and oils* also belong to this class.

etalon An *interferometer* used for studying fine *spectrum* lines. It depends upon the interference effects produced by multiple reflection between fixed, parallel, half-silvered *glass* or *quartz* plates.

ethanal Acetaldehyde. CH_3CHO. A colourless *liquid* with a pungent fruity smell, b.p. 21°C. Formed by the *oxidation* of *ethanol*; further oxidation gives *ethanoic acid*. It is made from *ethene* by the *Wacker process*. It is used as an intermediate in the manufacture of many *organic compounds*. If dilute acid is added to the compound it polymerizes to the *trimer*, ethanal trimer (see *paraldehyde*).

ethanamide Acetamide. CH_3CONH_2. A colourless crystalline substance, m.p. 82°C., odourless when pure. It is used industrially as a *solvent*, etc.

ethane C_2H_6. The second member of the *alkane* series. A colourless odourless *gas*. B.p. −89°C. It is used chiefly in organic synthesis.

ethanedioic acid See *oxalic acid*.

ethanediol Ethylene glycol, glycol. $(CH_2OH)_2$. A colourless *viscous liquid* with a sweet taste, b.p. 197°C. It is used as an *antifreeze*, in the manufacture of *polyesters* and *plasticizers*, and as a *solvent*.

ethanethiol Ethyl mercaptan. C_2H_5SH. A colourless flammable *liquid*, b.p. 37°C., that is responsible for the odour of garlic and is used in the manufacture of *rubber accelerators*.

ethanoate A *salt* or *ester* of *ethanoic acid*.

ethanoic acid Acetic acid. CH_3COOH. The *acid* contained in *vinegar* (3 to 6%). A colourless corrosive *liquid* with a pungent smell; m.p. 16.6°C., b.p. 118.1°C. It solidifies at low temperatures to *glacial ethanoic acid*. It is made by the *oxidation* of *ethanol* or *butane* and is used in the manufacture of *cellulose ethanoate* and in other industries.

ethanoic anhydride Acetic anhydride. $(CH_3CO)_2O$. A colourless pungent *liquid*, the *anhydride* of *ethanoic acid*, b.p. 140°C., used in the manufacture of *plastics*.

ethanol Ethyl alcohol. C_2H_5OH. A colourless flammable *liquid*, b.p. 78.3°C. It is the substance produced in *fermentation* of *sugars* and is the active constituent of alcholic drinks; it is used as a *fuel* and as a solvent. Industrially it is made from *ethene*. See *proof spirit*; *absolute alcohol*.

ethanolamines Organic *amines* derived from *ethanol*: monoethanolamine, a colourless viscous liquid, $NH_2CH_2CH_2OH$, m.p. 10.3°C., b.p. 172°C.; diethanolamine, $NH(CH_2CH_2OH)_2$, a viscous liquid or *deliquescent* white solid, m.p. 28°C.; triethanolamine, $N(CH_2CH_2OH)_3$, a highly *hygroscopic* viscous colourless liquid, m.p. 21°C. They are manufactured by the action of *ammonia* on *epoxyethane* and are used for the absorption of *acid* gases, and as *intermediates* in the production of *surfactants*.

ethanoyl group Acetyl group. The organic group CH_3CO–.

ethene Ethylene. $H_2C:CH_2$. The first member of the *alkene* series. A colourless *flammable gas* with a sweetish smell, b.p. −103.7°C., it is made from petroleum and used in the manufacture of *ethanol* and many other organic chemicals. It *polymerizes* to *polythene*.

ethene-propene rubber EPR. A fully *saturated*, *stereoregular*, *synthetic rubber* prepared by the solution *polymerization* of approximately equal proportions of *ethene* and *propene*. It cannot be cured by sulphur vulcanization but satisfactory vulcanization can be achieved by using peroxide curing systems.

ethenoid plastics A class of *thermoplastic resins* made from substances containing a *double bond*, e.g. *acrylic*, *styrene*, and *vinyl resins*.

ethenyl ethanoate Vinyl acetate. $CH_2:CHOOCCH_3$. A colourless *insoluble liquid*, b.p. $71 - 72°C.$, that polymerizes to form *polyvinyl acetate* (ethanoate).

ether (aether) The hypothetical medium that was supposed to fill all *space*: it was postulated as a medium to support the propagation of *electromagnetic radiations*. Once the subject of controversy, it is now regarded as an unnecessary assumption.

ethers A group of *organic compounds* with the general formula R-O-R′ formed by the *condensation* of two *alcohol molecules*. The compound commonly called 'ether' is ethoxyethane (diethyl ether), $C_2H_5.O.C_2H_5$, b.p. $34.6°C.$, made by dehydrating *ethanol* by means of concentrated *sulphuric acid*. Ethoxyethane is used as an *anaesthetic* and as a *solvent*.

ethoxy The *univalent* group C_2H_5O-.

ethoxyethane See *ethers*.

ethyl acetate See *ethyl ethanoate*.

ethyl alcohol See *ethanol*.

ethyl butyrate Butyric ether. $C_3H_7COOC_2H_5$. A *volatile liquid*, b.p. $120°C.$, used in flavouring and in perfumes.

ethyl carbamate Urethan(e). $NH_2COOC_2H_5$. A white crystalline solid, m.p. $48°C.$ Used in the molten state as a *solvent*; it is also used as an *intermediate* in the manufacture of *resins* and in medicine.

ethylene See *ethene*.

ethylenediaminetetraacetic acid See *EDTA*.

ethylene dibromide See *dibromoethane*.

ethylene dichloride See *dichloroethane*.

ethylene oxide See *epoxyethane*.

ethyl ethanoate Ethyl acetate. $CH_3COOC_2H_5$. A colourless liquid with a pleasant fruity smell, b.p. $77°C$. It is used as a *solvent* and in flavourings and perfume.

ethyl fluid A *solution* of *tetraethyllead*, $Pb(C_2H_5)_4$, and *dibromoethane*, $C_2H_4Br_2$, used as an anti-*knock* compound in motor *fuel*.

ethyl group The *univalent alkyl* group $-C_2H_5$.

ethyl nitrite Nitrous ether. $C_2H_5NO_2$. A *volatile liquid* with a sweet smell, b.p. $17°C.$, used in medicine.

ethyne Acetylene. A colourless poisonous flammable *gas*; the first member of the *alkyne* series. It is made by the action of water on *calcium dicarbide*, CaC_2, or by the action of an electric *arc* on other *hydrocarbons*. It is used as a starting material for many *organic compounds*, and for *welding* on account of the high flame temperature (about $3300°C.$) it produces when burnt in oxygen (see *oxy-acetylene burner*).

euchlorine A gaseous *mixture* of chlorine, Cl_2, and explosive chlorine peroxide, ClO_2.

eudiometer A glass tube for measuring volume changes in *chemical reactions* between *gases*.

eugenol $C_{10}H_{12}O_2$. A colourless oily *liquid*, m.p. 9.2°C., b.p. 255°C., extracted from oil of cloves. It is used in perfumes and as an *antiseptic*.

eukaryote Eucaryote. An *organism* whose *cells* contain a *nucleus* and other *membrane*-bounded *organelles*. Most multicellular organisms are eukaryotic.

europium Eu. Element. R.a.m. 151.96. At. No. 63. A silvery metal, r.d. 5.24, m.p. 830°C., b.p. 1430°C., two *isotopes* of which occur in tiny quantities in nature. Its oxide is used in *phosphors* for television screens. See *lanthanides*.

eutectic mixture A *solid solution* of two or more substances, having the lowest *freezing point* of all the possible mixtures of the components. This is taken advantage of in *alloys* of low *melting point*, which are generally eutectic mixtures.

eutectic point Two or more substances capable of forming *solid solutions* with each other have the property of lowering each other's *freezing point*; the minimum freezing point attainable, corresponding to the *eutectic mixture*, is the eutectic point.

evaporation The conversion of a *liquid* into *vapour*, without necessarily reaching the *boiling point*; it is used in concentrating *solutions* by evaporating off the *solvent*. As it is the fastest moving *molecules* that escape from the surface of a liquid during evaporation, the average *kinetic energy* of the remaining molecules is reduced, and therefore evaporation causes cooling.

evaporometer See *atmometer*.

even-even nucleus A *nucleus* that contains both an even number of *protons* and an even number of *neutrons*.

even-odd nucleus A *nucleus* that contains an even number of *protons* but an odd number of *neutrons*.

event horizon See *black hole*.

evolute A curve that is formed from the *locus* of the *centres of curvature* of another curve (the *involute*). The end of a stretched string unwound from the evolute traces the involute.

evolution (bio.) The process of gradual change in the characteristics of *organisms* over many generations. See *Darwin's theory*; *Lamarkism*; *natural selection*; *neo-Darwinism*.

exa- Prefix denoting one million million million times; 10^{18}. Symbol E, e.g. Em $= 10^{18}$ *metres*.

excess (chem.) A greater quantity of one substance or *reagent* than is necessary to react with a given quantity of another.

excess electron An *electron* in a *semiconductor* donated by an impurity, which is not required in the bonding system of the *crystal* and which is therefore available for *conduction* ('excess conduction').

exchange force 1. The type of *force* that results as a consequence of a continuous exchange of particles between two objects. In chemistry, a covalent bond (see *chemical bond*) can be regarded as resulting from an exchange of electrons, e.g. in molecular hydrogen the two protons can be regarded as continually exchanging their shared electron. In particle physics, the four *fundamental forces* can be seen as the result of the exchange of *gauge bosons*. **2.** A force occurring in ferromagnetic materials. See *ferromagnetism*.

exchanges, Prevost's theory of Bodies at all *temperatures* are constantly radiating *energy* to each other, those at constant temperature receiving in a given time as much energy as they emit.

excimer An excited *dimer*, formed by the association of excited and unexcited *molecules* (see *excitation*), which in the *ground state* would remain dissociated. Excimer *fluorescence* occurs in many *polycyclic hydrocarbons*.

excitation The addition of *energy* to a *nucleus*, an *atom*, or a *molecule*, transferring it from its *ground state* to a higher *energy level*. The 'excitation energy' is the difference in energy between the ground state and the excited state.

exciton A non-conduction, non-localized, excited *electron* state in a *semiconductor*. It may be regarded as a bound *electron-hole* pair, or alternatively as an atomic *excitation* passed from *atom* to atom.

exclusion principle See *Pauli exclusion principle*.

excretion The removal of toxic waste products from a *cell* or *organism*.

exitance Emittance (formerly). *M*. **1.** The radiant exitance (M_e) is the *radiant flux* emitted per unit area of a surface, measured in W m^{-2}. **2.** The luminous exitance (M_v) is the *luminous flux* emitted per unit area of a surface, measured in lm m^{-2}.

exocrine glands Glands that discharge their *secretions* into ducts, such as tear and salivary glands. Compare *endocrine glands*.

exoenzyme An *enzyme* that functions outside the *cell* that produces it, e.g. *pepsin*.

exoergic process An *exothermic process* (often applied in the context of a *nuclear reaction*). Compare *endoergic process*.

exoskeleton A supporting structure on the outside of an *organism* e.g. the cuticle of insects and other arthropods. It supports and protects the organism, also providing points of attachment for muscles. Compare *endoskeleton*.

exosmosis Outward osmotic flow. See *osmosis*.

exosphere The outermost layer of the *Earth's atmosphere*, in which the *density* is such than an air *molecule* moving directly outwards has a 50% chance of escaping rather than colliding with another molecule. The exosphere lies beyond the *ionosphere* and starts some 400 kilometres above the Earth's surface. See Fig. 44, under *upper atmosphere*.

exothermic process A process in which *energy* in the form of *heat* is released. Compare *endothermic process*.

expansion of gases A *perfect gas* expands by 1/273 of its volume at 0°C. for each degree rise in *temperature*, the *pressure* being constant (*Charles' Law*). Real gases obey this law only approximately at ordinary pressures, but the approximation becomes more and more valid as the pressure is reduced, i.e. as the gas tends towards a *perfect gas*.

expansion of liquids The directly observed expansion is the *apparent expansivity*, since the vessel containing the *liquid* also expands. The absolute expansivity is the sum of the apparent expansivity, and the volume expansivity of the containing vessel.

expansion of the Universe The widely accepted theory that the Universe is expanding, i.e. that clusters of *galaxies* are receding from each other as a result of the big bang with which the Universe originated (see *big-bang theory*). It is based upon the evidence of the *red shift* (see also *Doppler effect*) and the theory of *relativity*. See *Hubble constant*.

expansivity Coefficient of thermal expansion. **1.** Linear expansivity. The increase in length per unit length, caused by a rise in *temperature* of 1°C. **2.** Superficial expansivity. The increase in area per unit area caused by a rise in temperature of 1°C. **3.** Volume expansivity. The increase in *volume* per unit volume caused by a rise in temperature of 1°C. For *isotropic* media, the area and volume expansivities are approximately double and treble the linear expansivity, respectively, for the same substance. See also *expansion of liquids*.

explicit function (math.) A variable quantity, x, is said to be an explicit function of y, when x is directly expressed in terms of y.

explosion A violent and rapid increase of *pressure* in a confined space. It may be caused by an external source of *energy* (e.g. *heat*) or by an internal *exothermic chemical reaction* in which relatively large *volumes* of *gases* are produced. Explosions may also occur as the result of the release of internal energy during an uncontrolled *nuclear reaction* (either *fission* or *fusion* or both).

explosives Substances that undergo a rapid chemical change, with production of *gas*, on being heated or struck. The *volume* of gas produced being very great relative to the bulk of the *solid* explosive, great *pressures* are set up when the action takes place in a confined space. Examples include *gunpowder*, *TNT*, and nitroglycerin.

exponent (math.) The number indicating the *power* of a quantity. Thus the exponent of x in x^4 is 4.

exponential e. The mathematical *series*,
$$e^x = 1 + x + x^2/2! + x^3/3! + \ldots x^n/n!$$
is called an exponential series. When $x = 1$,
$$e = 1 + 1 + \frac{1}{2} + 1/6 + 1/24 + \ldots = 2.71828 \text{ (approx)}.$$
The function of x, defined by $y = e^x$ is called an exponential function and e^x is the exponential of x. The constant e is the base of natural or Napierian *logarithms*.

exposure meter (phot.) A *photocell* operating a suitable indicating meter, used in *photography* to assess the amount of *light* available, so that the correct shutter speed and *aperture* may be chosen for a given 'speed' of *film*.

expression (math.) A representation of a value, or relationship, in symbols.

extender 1. An *inorganic* powder added to *paints* to improve such properties as *film* formation, and to avoid settlement on storage. **2.** A substance added to *glue* or synthetic *rubber* that reduces their cost or to some extent modifies their properties (e.g. *viscosity*).

extensometer An instrument for measuring the extension produced in a body under an applied *stress*.

extinction coefficient A measure of the amount of *light* absorbed by a substance in *solution*. If light of intensity I_0 is passed through a distance d of a solution containing a *molar concentration* c of the dissolved substance, so that its intensity is reduced to I_T, then the extinction coefficient is given by:
$$[\log_{10}(I_0/I_T)]/cd$$

extraction 1. The process of separating a desired constituent from a *mixture*, by means of selective *solubility* in an appropriate *solvent*. **2.** Any process by which a pure *metal* is obtained from its *ore*.

extraordinary ray See *ordinary ray*.

extrapolation The process of filling in values or terms of a series on either side of the known values, thus extending the range of values.

extremely high frequencies EHF. *Radio frequencies* in the range 30 000 to 300 000 *megahertz*.

extrinsic semiconductor A *semiconductor* in which the *carrier* density results mainly from the presence of impurities or other imperfections, as opposed to an intrinsic semiconductor in which the electrical properties are characteristic of the *ideal crystal*.

eyepiece In optical instruments, the *lens* or system of lenses nearest the observer's eye; generally used to view the image formed by the *objective*.

F

face-centred See *body-centred*.

factor (math.) A number or quantity is exactly divisible by its factors; thus the factors of 12 (i.e. the *integral* or whole-number factors) are 1, 2, 3, 4, 6, 12.

factor, prime The prime factors of a quantity are the *prime numbers* that, when multiplied together, give the quantity. Thus, the prime factors of 165 are 3, 5, and 11.

factorial The *product* of a number and all the consecutive positive whole numbers below it down to 1. Thus, factorial 5, written 5! is
$$5 \times 4 \times 3 \times 2 \times 1 = 120.$$

faculae Large bright areas of the *photosphere* of the Sun, whose *temperatures* are higher than the average of the Sun's surface.

Fahrenheit scale The *temperature* scale in which the *melting point* of *ice* is taken as 32°F. and the *boiling point* of *water* under standard atmospheric pressure (760 mm) as 212°F. The scale is no longer in scientific use. To convert to the *Celsius temperature* the formula is:
$$9C = 5(F - 32).$$
Named after G. D. Fahrenheit (1686–1736).

Fajans' rules See *rules of Fajans*.

fall-out *Radioactive* substances deposited upon the surface of the Earth from the *atmosphere*. Three types of fall-out, subsequent to the *explosion* of a *nuclear weapon*, are recognized. 'Local fall-out' as a result of which large particles from the fire ball are deposited within a range of approximately 100 miles during the first few hours after the explosion. 'Tropospheric fall-out', during which fine particles are deposited around the globe, in the approximate *latitude* of the explosion, within a week or so. 'Stratospheric fall-out' consisting of the ultimate worldwide deposition, over a period of years, of the particles that were carried by the explosion into the *stratosphere*.

family (bio.) A group (*taxon*) of similar *genera*. See *classification*.

farad The derived *SI unit* of *capacitance* defined as the capacitance of a *capacitor* between the plates of which there appears a *potential difference* of 1 *volt* when it is charged with 1 *coulomb*. Equivalent to 10^9 *electromagnetic units* and 8.99×10^{11} *electrostatic units*. The practical unit is the microfarad, which is 10^{-6} farad. Symbol F. Named after M. Faraday (1791–1867).

Faraday cage A cage of metal wire or rods, which is earthed, and used to screen electric equipment in order to protect it from extraneous *electric fields*.

Faraday constant *F*. The *electric charge* carried by unit *amount of substance* (one *mole*) of *electrons*, i.e. the product of the *Avogadro constant* and the charge on an electron. It has the value $9.648\,453\,1 \times 10^4$ *coulombs* per mole.

Faraday effect Faraday rotation. The rotation of the plane of vibration (see *polarization of light*) of polarized light on traversing an *isotropic* transparent medium placed in a *magnetic field* possessing a component in the direction of the light ray. Although originally restricted to light, the Faraday effect is now known to apply to other *electromagnetic radiations*. Thus, the plane of polarization of a *radar pulse* travelling through the *ionosphere* is rotated by the combined effects of the *ionization* and the Earth's magnetic field (see *magnetism, terrestrial*). By reflecting radar pulses from the *Moon*, or other Earth *satellites*, and measuring the total rotation, the extent of the ionization in the ionosphere can be calculated.

Faraday's laws 1. See *electrolysis, Faraday's law of.* **2.** See *induction, electromagnetic.*

fast fission See *fast neutrons.*

fast neutrons *Neutrons* resulting from *nuclear fission* that have lost little of their *energy* by collision and therefore travel at high speeds. It is usual to describe neutrons with energies in excess of 0.1 *MeV* as 'fast'. However, fission induced by fast neutrons is often described as 'fast fission' and in this context the neutrons are so described if they have energies in excess of the fission threshold of uranium-238, i.e. above 1.5 MeV.

fast reactor A *nuclear reactor* in which little or no *moderator* is used and in which, therefore, the *nuclear fissions* are caused by *fast neutrons.*

fathom A unit of marine depth equal to 6 feet (1.83 metres).

fathometer A depth-sounding instrument. The depth of water is measured by noting the time the *echo* of a *sound* takes to return from the sea bed.

fatigue of metals The deterioration of *metals* owing to repeated *stresses* above a certain critical value; it is accompanied by changes in the crystalline structure of the metal. The fatigue limit is the stress that will cause a metal to fail after a certain number (usually 10^7) of stress cycles.

fats and oils Simple *lipids* consisting of mixtures of various *glycerides* of *fatty acids*, which occur in plants and animals and serve as storage materials. The distinction fats and oils (as distinct from *mineral oils*, which are *hydrocarbons*) is one of *melting point*; the term oil is usually applied to glycerides *liquid* at 20°C., the others being fats. The fatty acids occurring in plants and fish are predominantly *unsaturated*, while those in mammalian fats tend to be *saturated*. The former have lower *melting points* and are often softer at room temperatures. See also *essential oils.*

fatty acids *Monobasic organic acids* having the general formula R.COOH, where R is hydrogen or a group of carbon and hydrogen atoms. The *saturated* fatty acids have the general formula $C_nH_{2n+1}COOH$. *Unsaturated* fatty acids have a single *double bond*. Those with two or more double bonds are polyunsaturated fatty acids. Many fatty acids occur in living things, usually in the form of *glycerides* in *fats and oils.*

febrifuge See *antipyretic.*

feedback In general, the coupling of the output of a process to the input. In 'negative feedback' a rise in the output *energy* is arranged to cause a decrease in the input energy (e.g. a *governor*). In 'positive feedback' a rise in the output energy is caused to reinforce the input energy. In particular, these terms are applied to electronic *amplifiers*, in which a portion of the output energy is used to reduce or increase the amplification, by reacting on an earlier stage according to the relative *phase* of the return.

Fehling's solution A solution of *copper(II) sulphate*, $CuSO_4$, *sodium hydroxide*, NaOH, and potassium sodium tartrate (*Rochelle salt*). It is used for the detection and estimation of certain *sugars*, *aldehydes*, and other *reducing agents*, which act upon the solution with the formation of a red precipitate of *copper(I) oxide*, Cu_2O. Named after Herman Fehling (1812–85).

feldspar A large group of rock-forming *minerals* consisting chiefly of *aluminosilicates* of potassium and sodium. They are constituents of *granite* and other primary rocks.

femto- Prefix denoting one thousand million millionth; 10^{-15}. Symbol f, e.g. fW $= 10^{-15}$ *watt*.

Fermat's principle of least time The path taken by a ray of *light* or other *wave motion* in traversing the distance between any two points is such that the time taken is a minimum. Named after Pierre de Fermat (1601–65).

fermentation A chemical change brought about in organic substances by living *organisms* (*yeast*, *bacteria*, etc.) by *enzyme* action. It is usually applied to the alcoholic fermentation produced by the action of *zymase* on certain *sugars*, giving *ethanol* and *carbon dioxide* according to the equation:
$$C_6H_{12}O_6 = 2C_2H_5OH + 2CO_2$$

fermi A unit of length, used in *nuclear physics*, equal to 10^{-13} cm. Named after Enrico Fermi (1901–54).

Fermi-Dirac statistics The branch of *quantum statistics* used with systems of identical particles having the property that their *wave function* changes sign if any two particles are interchanged. See *fermions*. Named after Enrico Fermi (1901–54) and P. A. M. Dirac (1902–84).

Fermi level The *energy level* in a solid at which the *probability* of finding an *electron* is 1/2. At *absolute zero* all the electrons would occupy levels below the Fermi level. But at real temperatures, in *conductors* the Fermi level lies in the conduction band (see *energy bands*), in *insulators* and in *semiconductors* it lies in the forbidden band between the conduction band and the lower occupied band.

fermions *Elementary* particles that conform to *Fermi-Dirac statistics*. The numbers of elementary fermions are conserved throughout all nuclear interactions, but they are divided into two groups, *baryons* and *leptons*, which are distinguished from each other in that members of one group cannot transform into members of the other group. All fermions have *spin* 1/2. Compare *bosons*. See Appendix, Table 6.

fermium Fm. *Transuranic element*. At. No. 100. The most stable *isotope*, fermium-257, has a *half-life* of only 10 days.

ferrate A *salt* of the hypothetical ferric acid, H_2FeO_4.

ferric Containing iron in its +3 *oxidation* state, e.g. ferric oxide is iron(III) oxide, Fe_2O_3. Ferric *salts* are usually yellow or brown in colour.

ferric alum Iron alum, iron(III) potassium sulphate. $Fe_2(SO_4)_3.K_2SO_4.24H_2O$. A violet *soluble* crystalline *double salt* used in chemical *analysis*.

ferric chloride. See *iron(III) chloride*.

ferric oxide. See *iron(III) oxide*.

ferric sulphate See *iron(III) sulphate*.

ferricyanide A *salt* of the unstable ferricyanic acid, $H_3Fe(CN)_6$, i.e. one containing the hexacyanoferrate(III) *ion*.

ferrimagnetism The type of *magnetism* occurring in materials in which the *magnetic moments* of adjacent *atoms* are anti-parallel, but of unequal strength, or in which the number of magnetic moments orientated in one direction outnumber those in the reverse direction. Ferrimagnetic materials therefore have a resultant magnetization similar to that of *ferromagnetism*. Typical ferrimagnetic materials are the *ferrites*.

ferrite 1. Any of several types of iron *ore*. **2.** Pure *alpha-iron*, or *solid solutions* of which *alpha-iron* is the *solvent*. **3.** Any of a group of *ceramic* materials that exhibit the property of *ferrimagnetism*. They consist of iron oxide to which small quantities of *transition metal* oxides (e.g. cobalt and nickel oxides) have been added. The *spinel* ferrites have the formula $MO.Fe_2O_3$ where M is a *divalent* transition metal ion. More complex barium-containing ferrites have also been manufactured. By suitable combinations of metallic oxides, ferrites can be made that exhibit *ferromagnetism*, but as they are electrical *insulators* and therefore do not suffer from the effects of *eddy currents*, they can be used as cores in coils and *transformers* in *electronic* equipment at *frequencies* that would be impossible with ordinary ferromagnetic materials. Ferrites are also used in the construction of memory circuits in *computers* and, an account of their light weight, in the electrical equipment of aircraft.

ferritin A *protein* found in the liver and spleen that contains iron. It acts as a reservoir of iron for the whole body.

ferro- Prefix denoting iron, especially in names of *ferroalloys*.

ferroalloy An *alloy* of iron and other elements used in making *alloy steels*. Examples include *ferroaluminium, ferrochrome, ferromanganese, ferrosilicon*, and *ferrotungsten*.

ferroaluminium An *alloy* of aluminium (up to 80%) and iron.

ferrocene $Fe(C_5H_5)_2$. An orange crystalline solid, m.p. 173°C. It consists of a *sandwich compound* in which the iron is sandwiched between the two cyclopentadiene rings. Similar complexes with other metal *ions* are known, and are called metallocenes.

ferrochrome An *alloy* of chromium with 30%–40% iron, obtained by the reduction of *chromite* with carbon in an electric furnace.

ferrocyanide A *salt* of the *unstable* ferrocyanic acid, $H_4Fe(CN)_6$, i.e. one containing the hexacyanoferrate(II) *ion*.

ferroelectrics *Dielectric* materials that have electrical properties analogous to certain magnetic properties such as *hysteresis*, e.g. *barium titanate* and *Rochelle salt*. Ferroelectric materials usually also have piezoelectric properties (see *piezoelectric effect*).

ferromagnetic substances See *ferromagnetism*.

ferromagnetism The *metals* iron, cobalt, nickel, and certain *alloys* are vastly more magnetic than any other known substance: these metals are said to be ferromagnetic. Ferromagnetism is due to unbalanced *electron spin* in the inner electron *orbits* of the *elements* concerned (see *atom, structure of*), which gives the *atom* a resultant *magnetic moment*. The ionic spacing in ferromagnetic *crystals* is such that very large *forces*, called *exchange forces*, cause the alignment of all the individual magnetic moments of large groups of atoms to give highly *magnetic domains*. In an unmagnetized piece of iron, these domains are oriented at random, their magnetic axes pointing in all directions. The application of an external field serves to line up the domain axes, giving rise to the observed magnetism. Ferromagnetic substances have very large *magnetic permeabilities*, which vary with the strength of the applied field. A given ferromagnetic substance loses its ferromagnetic properties at a certain critical temperature, the *Curie temperature* for that substance.

ferromanganese An *alloy* of manganese (70%–80%) and iron.

ferrosilicon An *alloy* of silicon (15%) and iron, used in special *steels*.

ferrotungsten An *alloy* of tungsten (up to 80%) and iron.

ferrous Containing iron in its +2 *oxidation state*, e.g. ferrous oxide is *iron(II) oxide*, FeO. Ferrous *salts* are generally pale green in colour.

ferrous chloride See *iron(II) chloride*.

ferrous oxide See *iron(II) oxide*.

ferrous sulphate See *iron(II) sulphate*.

fertile material *Nuclides* that can be transformed into *fissile material* by the absorption of *neutrons* (e.g. uranium-238 and thorium-232).

fertilization The union of two sexually dissimilar *gametes* to form a *zygote*.

fertilizers Materials put into the *soil* to provide *compounds* of *elements* essential to plant life; often to replace minerals removed from the soil by the harvesting of crops. The *essential elements* provided by fertilizers are nitrogen, phosphorus, and potassium. Nitrogen is provided in the form of *nitrates*, *ammonium salts*, *calcium cyanamide*, etc. (see *fixation of atmospheric nitrogen*); phosphorus is added in the form of *superphosphate*, *basic slag*, various *phosphates*, etc. Potassium is obtained from natural potassium salts. Products of organic *decomposition* and waste, manure, etc., contain these and other necessary elements and form valuable fertilizers.

FET See *field-effect transistor*.

fetus See *foetus*.

Feynman diagram A diagram used to aid calculations of the interactions of *elementary particles* in the theory of *quantum electrodynamics*. In the diagrams, straight lines represent the propagation of elementary particles through *space-time*, while the *virtual* particles being exchanged in the interactions are represented by wavy lines. The use of the diagrams has since been extended to cover all *fundamental interactions*. Named after Richard Feynman (1918–89).

fibre 1. Any long threadlike structure occurring in animals, e.g. a nerve fibre, or in plants, e.g. flax. **2.** Dietary fibre. The *cellulose*, *hemicellulose*, *lignin*, or *pectin* in food, which are not digested and absorbed but which pass through the digestive system. It is believed to have beneficial effects, playing an important part in the prevention of many diseases (e.g. constipation, colitis, and diverticulitis) common in civilization, where much food is highly refined and processed to remove the dietary fibre.

Fibreglass* See *glass fibre materials*.

fibre optics See *optical fibres*.

fibrin An *insoluble* substance precipitated in the *blood* of vertebrates in the form of a meshwork of fibres during the process of clotting. Fibrin is formed when *thrombin* acts upon *fibrinogen*.

fibrinogen A *soluble protein* found in the *blood* of vertebrates that causes clotting of the blood by the action of the *enzyme thrombin* as a result of which *fibrin* is formed.

fidelity A measure of the *frequency* response of a *sound*-producing system. 'High fidelity' systems are usually taken to be those that are capable of a constant frequency response from 20 hertz to 20 000 kHz, with total distortion of less than 2%. See also *digital recording*.

field The region in which an electrically charged body (see *electric field*), a magnetized body (see *magnetic field*), or a massive body (see *gravitational field*) exerts its influence. A field is thus a model for representing the way in which a *force* can exist between bodies, whether or not they are in contact. See also *quantum field theory*.

field coil A coil of wire that produces a *magnetic field* when an *electric current* passes through it. It forms a part of most *electric motors* and *generators* as well as *electromagnets*.

field-effect transistor FET. A type of *transistor* that is in wide use for a variety of purposes. The two main forms are the junction field-effect transistor (JUGFET) and the insulated-gate field-effect transistor (IGFET). The former consists of a wafer of *semiconductor* material flanked by two highly doped layers of opposing types (n^+ and p^+) the *source* travel through a channel to the *drain*, the flow being controlled by a *gate*. In the insulated-gate type, a wafer of semiconductor has an insulating layer (usually of silicon dioxide) formed on its surface between two highly doped regions of opposite polarity, which form the source and the drain. A conductor attached to the top of the insulating layer forms the gate. When a positive voltage is applied to the gate, electrons move along the surface of the p-type substrate below, producing a thin n-type surface, called an inversion layer, which forms the channel between the source and the gate. The most widely used IGFET is the metal-oxide-silicon field-effect transistor known as the MOSFET.

field emission. The emission of *electrons* from an unheated surface as a result of a strong *electric field* existing at that surface.

field-emission microscope A type of *electron microscope* for observing the surface structure of a solid. A high negative voltage (>10 kV) is applied to a metal tip placed at the centre of a spherical fluorescent screen in a vacuum. *Field emission* from the tip produces electrons, which create an enlarged image on the screen. As resolution is limited by the vibrations of the metal atoms, the tip is usually cooled with liquid helium.

field glasses See *binocular field glasses*.

field guidance A method of guiding a missile to a point within a *field* by means of the properties of that field. The field may be natural (e.g. a *gravitational field*) or artificial (e.g. an *electromagnetic* or *radio* field).

field ionization The *ionization* of atoms or molecules at the surface of an unheated solid as a result of a strong *electric field* existing at that surface. *Electrons* are transferred from the atoms or molecules to the solid, producing positive ions.

field-ionization microscope A similar type of *electron microscope* to the *field-emission microscope* except that a high positive voltage is applied to the metal tip and instead of a vacuum the tip is surrounded by a low pressure of helium gas. The image is formed on the fluorescent screen by the helium *ions* striking it. The resolution can be made sufficiently high for individual atoms to be distinguished.

field lens The *lens* in the *eyepiece* system of optical instruments farthest from the eye.

field magnet A *magnet* that provides a *magnetic field* in the *dynamo, electric motor*, or other electrical machine.

filament A thin thread. In incandescent *electric-light bulbs* and *thermionic valves*, the filament is a wire of tungsten or other *metal* of high *melting point*, which is heated by the passage of an *electric current*.

file (*computers*) A body of information that has a describable structure, allowing all, or part, of it to be retrieved from the *store* (memory) of a computer on demand.

filler A *solid* substance added to synthetic *resins, paints*, and *rubbers*, either to modify their properties or to reduce their cost.

film 1. (chem.) A thin layer of a substance formed on the surface of a *liquid* or at the interface between two immiscible liquids, usually only a few *molecules* thick. **2.** (phot.) A flexible strip (usually *cellulose ethanoate* or a *polyester*) coated with a light-sensitive *emulsion*. See *photography*.

film badge A badge containing a masked photographic *film* worn by workers in contact with *ionizing radiations* to indicate the extent of their exposure to these radiations.

filter 1. (chem.) A device for separating *solids* or suspended particles from *liquids* or *gases*. It consists of a porous material (e.g. filter-paper) through the pores of which only fluids and dissolved substances can penetrate. See also *filter press; Gooch crucible*. **2.** (phys.) A material or device inserted in the path of an *electromagnetic radiation* to alter its *frequency* distribution.

filter press An apparatus used for carrying out *filtration*; it consists of a series of frames (metal or wooden) the two sides of which are covered with filter cloth. The frames are clamped together and the *liquid* to be filtered is pumped into them so that the *solid* residue forms a cake between the cloths while the *filtrate* is drained off.

filter pump A type of *vacuum pump* used to assist *filtration*. It is similar in principle to the *condensation pump*. A jet of water entrains air molecules, thus reducing the pressure below the filter paper or filter bed. It does not reduce the pressure below the *vapour pressure* of water.

filtrate A clear *liquid* after *filtration*; a substance that has been filtered, and contains no suspended matter.

filtration The process of separating *solids* from *liquids* by passing them through a *filter*.

finder A small low-powered *telescope* fixed parallel to the axis of a large telescope (usually astronomical) so that the object to be observed may be located and set in the field of vision of the large telescope.

fineness of gold The quantity of gold in an *alloy* expressed as parts per thousand. Thus gold with a fineness of 900 is in alloy containing 90% gold. See also *carat*.

fine structure (bio.) The structure of *cells* or other parts of *organisms* when examined by *electron microscopy*.

fine structure (phys.) The structure of certain *spectrum* lines when they are examined under high resolution. Single lines may be resolved into two or more closely spaced lines. They are caused by *transitions* between *energy levels* split by the vibrational or rotational motions of a molecule or by electron *spin*. See also *hyperfine structure*.

fire A *chemical reaction* accompanied by the evolution of *heat, light*, and *flame*

163

(i.e. a glowing mass of *gas*). It is generally applied to the chemical combination with oxygen of carbon, hydrogen, and other *elements* constituting the substance being burnt. See *combustion*.

fireclay *Clay* consisting principally of *aluminium oxide*, Al_2O_3, and *silica*, SiO_2, which will only soften at high *temperatures* and which is therefore used as a *refractory* material. Fireclays often occur beneath *coal* seams.

fire-damp An explosive mixture of *methane* (CH_4) and air, formed in coal mines.

fire extinguishers Hand devices for extinguishing *fires* in their early stages. They are usually classified according to the type of fire they are intended to combat, i.e. Class A fires (*paper*, wood, furnishings, and other common solid combustibles) and Class B fires (inflammable *liquids*, e.g. *petrol*, *paraffin*, etc.). Class A fires (which do not involve electrical equipment) are best combated with *water* under pressure delivered from extinguishers in which the water is expelled by stored pressure or by *carbon dioxide* produced by the action of *sulphuric acid* on *sodium hydrogencarbonate* (the 'soda-acid' type). Also in use are dry powder extinguishers (consisting of finely ground sodium or potassium hydrogencarbonate), and the *halogenated hydrocarbon* type, e.g., bromochlorodifluoromethane (BCF) or chlorobromomethane (CB). These halogenated hydrocarbons, however, like the fires themselves, produce *toxic* products of *combustion*, but BCF and CB are less toxic than *carbon tetrachloride* (CTC), as used in the older extinguishers.

Class B fires are best extinguished by the dry powder extinguishers, halogenated hydrocarbon extinguishers, or by carbon dioxide extinguishers. Also used are air-foam extinguishers (based on slaughterhouse products high in *protein*) or chemical foam extinguishers (based on solutions of *aluminium sulphate* and sodium hydrogencarbonate, which react together on mixing, evolving carbon dioxide and producing a *foam*).

Fischer-Tropsch process A process for the manufacture of *hydrocarbon oils* from *coal*, *lignite*, or *natural gas*. The process essentially consists of the *hydrogenation* of *carbon monoxide*, CO, in the presence of *catalysts* (nickel or cobalt) at 200°C.; this results in the formation of hydrocarbons and *steam*. Named after F. Fischer (d. 1948) and H. Tropsch (d. 1935).

fissile material *Nuclides* that are capable of undergoing *nuclear fission*. Sometimes the term is restricted to nuclides that are capable of undergoing fission upon impact with a slow *neutron* (e.g. uranium-233, uranium-235, and plutonium-239). Compare *fertile material*.

fission, nuclear See *nuclear fission*.

fission products Both the stable and the unstable *nuclides* produced as the result of *nuclear fission*.

fission spectrum The *energy* distribution of the *neutrons* produced by the *nuclear fission* of a particular *fissile material*.

fission-track dating A method of *dating* minerals, glass, etc., by observing the tracks made by the *fission* of uranium atoms within them. The age of a specimen can be estimated by exposing it to *neutron* radiation and comparing the density and number of tracks created by the fission so induced with those produced by natural fission.

Fittig reaction Wurtz-Fittig synthesis. The synthesis of *alkylarene* hydrocarbons by the action of metallic sodium on a mixture of an *haloalkane* and a halogenated *benzene* derivative. For example,

$$C_6H_5Cl + CH_3Cl + 2Na \rightarrow C_6H_5CH_3 + 2NaCl$$

Named after R. Fittig (1835–1910). See also *Wurtz reaction*.

Fitzgerald-Lorentz contraction The hypothesis put forward independently by Fitzgerald (1893) and Lorentz (1895) to explain the result of the *Michelson-Morley experiment* on the supposition that a body moving with high *velocity* through the *ether* would experience a contraction in length in the direction of the motion. This contraction was later shown to be a direct consequence of the theory of *relativity*. Named after G. F. Fitzgerald (1851–1901) and H. A. Lorentz (1853–1928).

fixation of atmospheric nitrogen The formation of *compounds* of nitrogen from the free nitrogen in the air. Certain *bacteria* fix atmospheric nitrogen and build it into *amino acids*, thus contributing to the *nitrogen cycle*; the best-known nitrogen-fixing bacteria live in nodules on the roots of leguminous plants, such as peas, beans, clover. A shortage of natural nitrogen compounds (caused partly by increased cultivation of the *soil*, due to increase of human populations, and partly by the loss of nitrogen compounds from animal waste products by sewage disposal into the sea) has led to the manufacture of nitrogen compounds from atmospheric nitrogen, for use as *fertilizers*. The first practical process was the *Birkeland and Eyde process*; the *Haber* and *Serpek processes* are now the main ones used. A more recent approach has been to incorporate the *gene* for nitrogenase (the enzyme required to catalyze the fixation of nitrogen by bacteria) into crop plants by gene manipulation. These attempts are proving hopeful.

fixed air Former name for *carbon dioxide*, CO_2.

fixed alkali Former name for *potassium* or *sodium carbonate*, to distinguish them from volatile alkali, ammonium carbonate.

fixed point Any accurately reproducible equilibrium *temperature*. Examples include the *ice point*, the *steam point*, and the *sulphur point*. See *International Practical Temperature Scale*.

fixed stars True *stars*. Heavenly bodies that do not appear to alter their relative positions on the *celestial sphere* compared to the planets, formerly called 'wandering stars', which do.

fixing, photographic Rendering that portion of the sensitive *film*, plate, or paper that has not been affected by *light*, insensitive to exposure, after *developing*. It is usually carried out by the action of *sodium thiosulphate*, $Na_2S_2O_3$ (*hypo*), which reacts with the unaffected *silver bromide* to give a *soluble double salt*, silver sodium thiosulphate, which is then washed away. See *photography*.

flagellum A long motile thread projecting from a living *cell*, whose movement causes the cell to move through the surrounding fluid.

flame The glowing mass of *gas* produced during *combustion*, the light is emitted by excited ions, molecules, etc., and by incandescent carbon particles in fuel-rich flames.

flame photometry A development of the *flame test* used in qualitative analysis; photometric (see *photometer*) measurement of flame *emission* is used to determine the concentration of substances introduced into the flame.

flame test A *qualitative* test for the presence of an element by the colour it or its compounds give to a *Bunsen burner* flame. Sodium compounds colour a flame bright yellow; potassium, caesium, and rubidium give a violet colour; strontium and lithium a red colour; barium, copper, thallium, and tellurium give a green colour, except copper *halides*, which give a blue colour.

flash photography See *spark photography*.

flash photolysis See *photolysis*.

flash point The lowest *temperature* at which a substance gives off sufficient inflammable *vapour* to produce a momentary flash when a small flame is applied.

flavoproteins Yellow conjugated *proteins* in which the *prosthetic group* is either flavin mononucleotide (FMN) or flavin adenine dinucleotide (FAD). Flavoproteins are *enzymes* of the *dehydrogenase* type.

flavour See *elementary particles*.

Fleming's rules Mnemonics for relating the direction of motion, *flux*, and *EMF* in electric machines. If the forefinger, second finger, and thumb of the right hand are extended at right angles to each other, the forefinger indicates the direction of the flux, the second finger the direction of the EMF, and the thumb the direction of motion in an electric *generator*. If the left hand is used the digits indicate the conditions obtaining in an *electric motor*. Named after Sir John Ambrose Fleming (1849–1945).

flint Natural variety of impure *silica*, SiO_2. 'Flints' of automatic lighters are composed of *pyrophoric alloys* of such *metals* as cerium and iron.

flint glass A variety of *glass* containing lead silicate; it is used for optical purposes.

flip-flop Bistable circuit. An *electronic circuit* with two stable states, which can be switched from one to the other by means of a *pulse*. They form the basis of many *logic circuits* and are extensively used in digital *computers*.

flocculation The coagulation of finely divided particles into particles of greater mass.

floppy disk See *magnetic disk*.

flotation, principle of The *mass* of *liquid* displaced by a floating body is equal to the mass of the body. A particular case of *Archimedes' principle*.

flotation process The separation of a *mixture*, e.g. of *zinc blende*, ZnS, and *galena*, PbS, making use of the *surface tension* of *water*. Zinc blende is not easily wetted by water and floats, supported by the surface film of water, while galena sinks. Various agents may be added to the water to cause one of the constituents to float in the froth produced by aerating and agitating the water. See *froth flotation*.

flowers of sulphur A fine powder, consisting of very small *crystals* of *sulphur* obtained by the *condensation* of sulphur *vapour* during the *distillation* of crude sulphur.

flue gas The gaseous products of *combustion* from a boiler furnace consisting predominantly of *carbon dioxide*, *carbon monoxide*, oxygen, nitrogen, and *steam*. Analysis of the flue gases is used to check the efficiency of the furnace. See *Orsat apparatus*.

fluid A substance that takes the shape of the vessel containing it; a *liquid* or *gas*.

fluid drachm See *drachm*.

fluidics Fluidic logic. The study, design, and use of jets of fluid to carry out *amplification* and *logic* to perform tasks usually carried out by *electronics*.

Fluidic systems, which depend on the flow of *fluids* instead of *electrons*, are about 10^6 times slower than electronics, but they can operate at higher *temperatures*. They are also unaffected by *ionizing radiations* and are useful when *delay lines* are required. They have therefore found use in *nuclear reactors* and space *rockets*.

fluidity The *reciprocal* of *viscosity*. The *c.g.s. unit* is the reciprocal of the *poise* known as the *rhe*.

fluidization (chem.) A technique used in industrial chemistry, in which a mass of solid particles is brought into a state of suspension by an upward stream of gas blown through it in a *reactor*. The material in the resultant "fluidized bed," which resembles a boiling liquid, is more accessible to chemical reactions, etc., than the same solid material in the static state.

fluid measure See *apothecaries' fluid measure*.

fluid mechanics The study of the *forces* and *pressures* on *gases* and *liquids* at rest (fluid statics) and in motion (fluid dynamics).

fluon* See *fluorocarbons*.

fluorene *o*-diphenylenemethane. $C_{13}H_{10}$. A white crystalline *aromatic solid hydrocarbon*, m.p. 116°C. It is used in the manufacture of *dyes* and *resins*.

fluorescein $C_{20}H_{12}O_5$. A dark red crystalline *organic compound*, m.p. 314°C. It dissolves in *alkaline solutions* to give a *liquid* of intense green *fluorescence*. It is used as an *indicator* and in *dyes*.

fluorescence A form of *luminescence* in which certain substances (e.g. *quinine* sulphate solutions, *paraffin oil*, *fluorescein* solutions) are capable of absorbing *light* of one *wavelength* (i.e. *colour*, when in the visible region of the *spectrum*) and in its place emitting light of another wavelength or colour. Unlike *phosphorescence*, the phenomenon ceases immediately the source of light is cut off.

fluorescent tube A *light* source consisting of a glass tube the inside of which is coated with a fluorescent substance (see *fluorescence*). The tube contains mercury *vapour* and is fitted with a *cathode* and *anode* between which a stream of *electrons* can be made to flow by the application of a suitable *potential difference*. When the mercury *atoms* are struck by the electrons they are raised to an excited state (see *excitation*); when they fall back to the *ground state* they emit *ultraviolet radiation* which is converted to visible radiation by the fluorescent substance on the tube walls.

fluoridation The addition of minute quantities of *fluorides* to drinking water supplies to give protection against caries (decay) in the teeth of growing children. 1 part per million of *fluoride ion* is usually added.

fluoride A *salt* of *hydrofluoric acid*. See *fluoridation*.

fluorination The introduction of a fluorine atom into a compound by *substitution* or by an *addition reaction*. See *halogenation*.

fluorine F. Element. R.a.m. 18.998 403. At. No. 9. A pale yellowish-green *gas*, m.p. −219.62°C., b.p. −188.1°C., resembling chlorine but more *reactive*. It occurs combined as *fluorspar* and as *cryolite* and is made by the *electrolysis* of a *solution* of *potassium hydrogen difluoride* in *anhydrous hydrogen fluoride*. Fluorinated *organic compounds*, made by replacing hydrogen in organic compounds by fluorine, are of considerable industrial importance (see *fluorocarbons*). Fluorine is the most reactive element.

fluorite See *fluorspar*.

fluorocarbons A group of *synthetic organic compounds* (both *aliphatic* and *aromatic*) in which some or all of the hydrogen *atoms* have been substituted by fluorine atoms. Many of these compounds and their derivatives are nonflammable, chemically resistant, and immiscible with *water* or *oil*. Polytetrafluoroethene (Teflon* and Fluon*) is a *polymer* used as a *plastic*, while the Freons* are *monomers* used as *refrigerants* and *solvents*. See also *chlorofluorocarbon*.

fluoroscope A fluorescent screen (see *fluorescence*) for the direct visual observation of *X-ray* images; it is used diagnostically in medicine.

fluorosilicic acid Hydrofluosilicic acid. H_2SiF_6. An *acid* that is only stable in the form of its fuming *aqueous solution*. It is used as a *disinfectant* and wood preservative.

fluorspar Natural calcium fluoride, CaF_2, consisting of colourless *crystals*, often coloured by impurities. It is used as a source of fluorine and its *compounds*.

flux (chem.) A substance added to assist the fusion of metals, as in soldering or brazing, by inhibiting oxidation.

flux (phys.) The rate of flow of mass or energy per unit area normal to the direction of the flow. See also *magnetic flux*; *electric flux*; *luminous flux*. In *nuclear physics*, it is the product of the number of particles per unit volume and their average velocity.

flux density The *magnetic flux* or *luminous flux* per unit of cross-sectional area. The *SI unit* of magnetic flux density is the *tesla*.

fluxmeter An instrument for the measurement of *magnetic flux*. Essentially a moving coil *galvanometer* so designed that the coil experiences negligible restoring *torque* from its suspension system. A change in the magnetic flux through a flux coil connected to the galvanometer induces a current in the coil, thus causing a deflection of the galvanometer. Modern fluxmeters use the *Hall probe*.

FM See *frequency modulation*.

***f*-number of a lens** See *aperture*.

foam A *colloidal* suspension of a *gas* in a *liquid*. A solid foam (e.g. foam rubber) is made by making a liquid foam and allowing it to solidify.

focal length The distance from the *optical centre* or pole to the principal *focus* of a *lens* (see Fig. 25 under *lens*), or *spherical mirror* (see Fig. 27 under *mirrors*, *spherical*). The focal length of a spherical mirror is half its *radius of curvature*.

focal ratio See *aperture*.

focus 1. Focal point. (phys.) The point at which converging rays, usually of *light*, meet (real focus); or a point from which diverging rays are considered to be directed (virtual focus). The 'principal focus' of a *lens* (see Fig. 25 under *lens*) or *spherical mirror* (see Fig. 27 under *mirrors, spherical*) is the point on the principal axis through which rays of light parallel to the *principal axis* will be refracted or reflected. **2.** (math.) One of the fixed points used to define a curve, by a linear relationship with the distance from one of these fixed points to any point on the curve. See *ellipse*; *parabola*; *hyperbola*.

foetus Fetus. A mammalian *embryo*, that has developed sufficiently for the features of the adult form to be recognized. It is usually applied to human embryos from eight weeks old to birth.

fog The effect caused by the *condensation* of *water vapour* upon particles of dust, soot, etc.

folic acid Pteroylglutamic acid, PGA, folacin. $C_{19}H_{19}N_7O_6$. A yellow crystalline substance forming part of the *vitamin* B complex. It is a *coenzyme* in the *metabolism* of some *amino* acids, and is used in the treatment of anaemia. It occurs in many foods, especially vegetables, and is synthesized by bacteria in mammalian intestines.

food chain A chain of organisms through which energy and *atoms* in the form of *organic compounds* in food are transferred. It starts with green plants upon which herbivores feed. The herbivores are eaten by carnivores, which may in turn be eaten by other carnivores, and so on. As some animals feed at different levels in the chain, a food web may be required to cover all the relationships.

food preservation The prevention of chemical *decomposition* and of the development of harmful *bacteria* in foods. It is generally effected by the sterilization of the food (i.e by the destruction of bacteria in it) by heating in sealed vessels, i.e. canning; or by making the conditions unfavourable for the development of bacteria, by pickling, drying, freezing, smoking, salting, increasing the sugar concentration, and by *irradiation*.

fools' gold See *pyrites*.

foot British unit of length; one-third of a *yard*; 0.3048 *metre*.

foot-candle A unit of *illumination*. One *lumen* per square foot. It has now been replaced by the *lux*.

foot-lambert A unit of *luminance*. The luminance of a uniform diffuser emitting one *lumen* per square foot. This unit is now obsolete, except in the U.S.A.

foot-pound A former practical unit of *work*. The work done by a *force* of 1 pound weight acting through a distance of 1 foot.

foot-poundal A former unit of *work* in the *f.p.s. system*. The work done by a *force* of 1 *poundal* acting through a distance of 1 foot.

forbidden band See *energy bands*.

force *F*. An external agency capable of altering the state of rest or motion in a body; it is defined as being proportional to the rate of increase in *momentum* of the body and is a *vector* quantity, measured in *newtons* (*SI units*). The force, *F*, required to produce an *acceleration*, *a*, in a *mass*, *m*, is given by $F = ma$. If *m* is in *kilograms* and *a* in m s^{-2}, *F* will be in *newtons*.

force ratio See *mechanical advantage*.

forces, parallelogram of See *parallelogram of forces*.

forces, triangle of See *triangle of forces*.

formaldehyde See *methanal*.

formalin A 40% *solution* of *methanal* containing *methanol* as a stabilizer. It is used to preserve biological specimens and as a *disinfectant*.

formate See *methanoate*.

formic acid See *methanoic acid*.

formula (chem.) The representation of a *molecule* or smallest portion of a *compound*, using *symbols* for the *atoms* of the *elements* which go to make up the molecule. E.g. the formula of *water*, H_2O, implies that the smallest portion of water that can exist independently consists of 2 hydrogen atoms chemically bonded to 1 oxygen atom. The *structural formula* represents the way in which the atoms in a molecule are joined by *chemical bonds*. E.g. the structural formula of water is written H–O–H, indicating that 2 hydrogen atoms are both attached to the *bivalent* oxygen atom. The *empirical formula* of a compound is its simplest formula, indicating only the numerical ratio of the atoms present in a molecule, but not necessarily their actual number. Thus the empirical formula of *hydrogen peroxide* is HO while its *molecular formula* is H_2O_2.

formula (math. and phys.) A statement of facts in a symbolic form; by substitution a result applicable to particular data may be obtained. Thus the time of swing of a *pendulum* (T) is given by the formula $T = 2\pi(l/g)^{1/2}$, showing the connection between length (l) and time of swing. (g is the *acceleration of free fall*).

formula weight The *relative molecular mass* of a compound based on its *molecular formula*.

formyl The *univalent* group O:CH–, derived from *methanoic* (formic) *acid*.

Fortin barometer A mercury *barometer* that has a leather mercury reservoir the height of which can be adjusted; the height of the column is read using a *vernier* scale, which, used in conjunction with various correction tables, enables accurate measurements of atmospheric pressure to be made. Named after J. Fortin (1750–1831).

fossil The remains of an *organism* preserved in rocks in the *Earth's crust*. Usually only the hard parts (bones, shells, etc.) are so preserved, but occasionally remains of organisms having no hard parts have been recognized. Footmarks, tracks, etc., are also regarded as fossils.

fossil fuels The organic content from the remains of *organisms* embedded in the surface of the Earth, especially those with high carbon and/or hydrogen contents, that is used by man as *fuel* (e.g. *coal*, *oil*, *natural gas*). Most of the *energy* obtained from the *combustion* of fossil fuels derives from the *exothermic* conversion of carbon into *carbon dioxide* and of hydrogen into *water* (*steam*).

Foucault pendulum A *pendulum* consisting of a heavy weight attached to a long wire, which is free to swing in any direction. The slow turning of the *plane* of the pendulum's swing is a demonstration of the *Earth's* rotation. Named after its inventor, J. B. L. Foucault (1819–68).

Fourier analysis The expansion of a mathematical *function* or of an experimentally obtained curve in the form of a trigonometric series of the form:

$$f(x) = a_0 + (a_1\cos x + b_1\sin x)$$
$$+ (a_2\cos 2x + b_2\sin 2x) + \dots$$

This Fourier series is used as a method of determining the harmonic components of a complex periodic wave. Named after J. B. J. Fourier (1768–1830).

fourth dimension Ordinary *space* has three dimensions, i.e. length, breadth, and thickness, each one at right angles to both the others. Mathematically it is possible to write down equations, similar to those governing relations between points in ordinary three-dimensional space, but connecting any number of imaginary dimensions. These are sometimes said to refer to a 'hyperspace' of many dimensions. In dealing with a material particle, it is neccessary to state not only where it is, but when it is there. Thus time is somewhat analogous to a dimension of space. *Relativity* has shown in particular in what manner time may be regarded as a fourth dimension, so that all real events take place in a four dimensional *space-time continuum*.

Fowler's solution A *solution* containing potassium arsenite; formerly used in medicine.

f.p.s. system The foot-pound-second system of units. The British system of physical *units* derived from the three *fundamental units* of length, *mass*, and time, i.e. the *foot*, *pound* mass, and the *second*. It is now replaced, for scientific purposes, by *SI units*.

fractional crystallization The separation of a *mixture* of dissolved substances by making use of their different *solubilities*. The *solution* containing the mixture is evaporated until the least *soluble* component crystallizes out.

fractional distillation Fractionation. The separation of a *mixture* of several *liquids* that have different *boiling points*, by collecting separately 'fractions' boiling at different *temperatures* in a *fractionating column*. This process is used to separate the various fractions of *petroleum*.

fractionating column A long vertical column used in *fractional distillation*, containing rings, plates, or bubble caps, that is attached to a *still*. As a result of internal *reflux* a gradual separation takes place between high and low boiling 'fractions' of a liquid *mixture*. When equilibrium is reached between the rising vapour and the descending condensed liquid, the various fractions can be tapped from points on the column, the less volatile components at the lower part of the column and the more volatile components at the top.

fractionation The separation of a *mixture*, usually of chemically related or otherwise similar components, into fractions of different properties, usually by *fractional distillation* in a *fractionating column*.

frame of reference A set of reference axes for defining the position of a point or body in *space*. A frame of reference in a four-dimensional *continuum* consists of an observer, a *coordinate* system and a clock to correlate positions with times.

francium Fr. Element. At. No. 87. It has no known stable *isotope* and only one natural *radioactive* isotope, francium-223 (*half-life* 21 mins.). It belongs to the *alkali metal* group of *elements*.

Frasch process A process for extracting sulphur from deposits deep down under sand. A series of concentric pipes is sunk down to the level of the sulphur deposit, *superheated steam* is forced down to melt the sulphur, which is then forced to the surface by compressed air blown down the centre pipe.

Fraunhofer diffraction The class of *diffraction* phenomena in which both the *light* source and the receiving screen are effectively at an infinite distance from the diffracting system. Compare *Fresnel diffraction*. Named after J. von Fraunhofer (1787–1826).

Fraunhofer lines Dark lines in the continuous *spectrum* of the *Sun*, caused by the absorption of certain *wavelengths* of the *white light* from the hotter regions of the Sun by chemical *elements* present in the *chromosphere* surrounding the Sun.

free (chem.) Uncombined. Free elements occur in the Earth's crust in an uncombined state.

free electron An *electron* that is not attached to an *atom*, *molecule*, or *ion*, but is free to move under the influence of an *electric field*.

free energy A *thermodynamic* quantity representing the *energy* that would be liberated or absorbed during a *reversible process*. The Gibbs free energy (or Gibbs function, G) is defined, under conditions of constant *temperature* and *pressure*, by $G = H - TS$, where H is the heat content (*enthalpy*), T the *thermo-*

dynamic temperature, and *S* the *entropy*. Referred to chemical processes, the important quantity is not the absolute magnitude of *G*, but the change in free energy, ΔG (also called the chemical *affinity*), during a reaction, which is given by

$$\Delta G = \Delta H - T.\Delta S.$$

By convention, if a reaction gives out *heat* ΔH will be negative (as the system is losing heat to the surroundings). Therefore, if $T.\Delta S$ is not large compared to ΔH, ΔG will also be negative indicating that the reaction will proceed to *chemical equilibrium*. When equilibrium has been attained, $\Delta G = 0$, and if ΔG is positive the reaction will only occur if energy is supplied in some way to force it away from equilibrium. As the entropy, *S*, is a measure of the molecular disorder of a system, and as a change of state involves a change of molecular orderliness, the term $T.\Delta S$ is dependent upon changes of state.

The Helmholtz free energy (or Helmholtz function, *F*) is defined as $U - TS$, where *U* is the *internal energy*. Also,

$$\Delta F = \Delta U - T.\Delta S$$

and for a reversible *isothermal* process ΔF represents the maximum work available. ΔF is sometimes called the 'work function'.

free fall The fall of a body as a result of a gravitational force in a *vacuum*, i.e. a fall in a *gravitational field* through a medium that offers no resistance to the fall. The *acceleration of free fall* is the constant acceleration that would be experienced by a falling body in the Earth's gravitational field under these conditions.

free radical A group of *atoms* (see *radical*), which usually exists in combination with other atoms, but which because it has an unpaired *valence electron* may exist independently for short periods (short-lived free radicals) during the course of a *chemical reaction*, or for longer periods (free radical of long life) under special conditions. Free radicals are created by *homolytic fission* in *photolysis* or *pyrolysis* and are usually highly reactive.

free space A region in which there is neither matter nor gravitational or electromagnetic field. It has a temperature of 0 K. The *electric* and *magnetic* constants are defined for free space.

freeze drying A process of drying heat-sensitive substances, such as food or *blood plasma*, by *freezing* and then removing the frozen water by volatilization at low *pressure* and *temperature*.

freezing Change of state from *liquid* to *solid*; it takes place at a constant *temperature* (*freezing point*) for any given substance under a given *pressure*. The freezing point normally quoted is that for standard atmospheric pressure.

freezing mixtures Certain *salts* that, when dissolved in *water* or mixed with crushed *ice*, produce a considerable lowering of *temperature*. The action depends upon absorption of *heat of solution* by the dissolving salt; in the case of mixtures in contact with ice, the *melting point* of ice is lowered in the presence of a dissolved substance; *latent heat* of fusion of ice is absorbed, and the salt dissolves in the melting ice.

freezing point The *temperature* of equilibrium between *solid* and *liquid* substance at a pressure of one standard atmosphere (101 325 Pa).

freezing-point depression See *depression of freezing point*.

French chalk Powdered *talc*.

Frenkel defect A *defect* in a *crystal lattice* caused by an *atom* or *ion* being removed from its normal position in the lattice (thus causing a *vacancy*) and taking up an *interstitial* position.

Freon* Trade name for a group of *chlorofluorocarbons*.

frequency *f* The number of cycles, oscillations, or vibrations of a *wave motion* or oscillation in unit time, usually one second. In a wave motion the frequency is equal to the speed of propagation divided by the *wavelength*. The derived *SI unit* of frequency is the *hertz*.

frequency band A range of *frequencies* of *electromagnetic radiations* falling within prescribed limits. See Appendix, Table 10 for internationally agreed *radio frequency* bands.

frequency modulation FM. The type of *radio* transmission system in which the *frequency* of a *carrier wave* is modulated (see *modulation*) rather than its *amplitude* (as in *amplitude modulation*). It provides a method of transmission free from 'static' interference.

frequency of a vibrating string The fundamental *frequency*, *f*, of a stretched string of length *l*, under tension *T*, is given by
$$f = (T/\pi\rho)^{1/2}/2rl,$$
where *r* is the radius of the string and ρ its *density*.

fresnel A unit of *frequency* equal to 10^{12} *hertz*. Named after A. J. Fresnel (1788–1827).

Fresnel diffraction A class of *diffraction* phenomena in which the *light* source or the receiving screen, or both, are at a finite distance from the diffracting system. Compare *Fraunhofer diffraction*.

Fresnel lens An optical *lens* whose surface consists of a number of smaller lenses so arranged that they give a lightweight lens of large *diameter* and relatively short *focal length*. They are used in headlights, searchlights, etc. Named after A. J. Fresnel (1788–1827).

friable Easily crumbled.

friction The *force* that offers resistance to relative motion between surfaces in contact. See *friction, coefficients of*.

friction, coefficients of If F_s = the frictional resistance when a body is on the point of sliding along a specified surface, F_k = the frictional *force* when steady sliding has been attained, and R = the perpendicular force between the surfaces in contact, the static coefficient of friction $\mu_s = F_s/R$; the kinetic coefficient $\mu_k = F_k/R$. In the case of a wheel rolling on a plane surface, the frictional force $F_r = R\mu_r$, where μ_r is the coefficient of rolling friction.

Friedel-Crafts reaction Originally the *synthesis* of *aromatic hydrocarbons* by reacting *alkyl halides* with *benzene derivatives* in the presence of *anhydrous* aluminium chloride as a *catalyst*. Examples include methylation, e.g.
$$C_6H_6 + CH_3Cl \rightarrow C_6H_5CH_3 + HCl$$
and acetylation to form *ketones*, e.g.
$$C_6H_6 + CH_3COCl \rightarrow C_6H_5COCH_3 + HCl$$
It is now extended to include the addition of *alkenes* to, and the *condensation* of *alcohols* with, aromatic hydrocarbons in the presence of such catalysts as *anhydrous iron(III) chloride*, gallium chloride, boron trifluoride, and *hydrogen fluoride*. Named after Charles Friedel (1832–99) and J. M. Craft (1839–1917).

froth flotation The separation of a *mixture* of finely divided minerals by agitating them in a froth of *water* and a frothing agent, so that some float and others sink (see also *flotation process*). The process can be made selective by adjusting the nature of the froth with suitable *surface active agents*.

fructose Fruit sugar, laevulose. $C_6H_{12}O_6$. A sweet *soluble* crystalline *hexose*, m.p. $102°$–$104°C$. It occurs in sweet ripe fruits, in the nectar of flowers, and in honey.

frustum Any part of a solid figure cut off by a *plane* parallel to the base, or lying between two parallel planes.

fuel A substance used for producing *heat*, either by means of the release of its *chemical energy* by *combustion* (see *fossil fuels*) or its *nuclear energy* by *nuclear fission* or *nuclear fusion*.

fuel cell A *cell* for producing *electricity* by *oxidation* of a *fuel*. The fuel cell is similar to an *accumulator* but instead of needing recharging with electrical energy, it has to be fed with fresh fuel. The simplest fuel cell consists of supplies of gaseous oxygen and hydrogen brought together over porous catalytic *electrodes*. The two compartments containing the electrodes are separated by a third compartment containing a hot alkaline *electrolyte* (e.g. KOH). The reactions are:

$$\text{at the cathode, } 2H_2 + 4OH^- \rightarrow 4H_2O + 4e^-;$$
$$\text{at the anode, } O_2 + 2H_2O + 4e^- \rightarrow 4OH^-.$$

The electrons flowing round an external circuit constitute a current. Other cells use *hydrazine*, *ammonia*, or *methanol* to provide the hydrogen. Interest in electric cars has stimulated fuel cell development. Although hydrogen-oxygen fuel cells can provide higher energy densities than zinc-air accumulators, because they are very bulky their energy per volume is little better than that of a lead accumulator.

fuel element An element of *nuclear fuel* for use in a *nuclear reactor*, usually an *oxide* or *alloy* of uranium encased in a can.

fugacity f. A thermodynamic function defined by $d(\ln f) = d\mu/RT$, where μ is the chemical potential. It has the units of pressure and is used in place of *partial pressure* for reactions involving real gases and mixtures. In a *perfect gas* the fugacity is equal to the pressure. The ratio of the fugacity of a gas to its fugacity in a standard state (usually taken as $f = 1$) is the *activity* of the gas (in which case fugacity equals activity). The fugacity of a liquid or solid is that of the vapour with which it is in equilibrium.

fuller's earth A variety of *clay*-like materials that absorb *oil* and *grease*. They consist of *hydrated silicates* of magnesium, calcium, aluminium, and sometimes other *metals*. They are used in scouring textiles and in refining *fats and oils*. Its name derives from the former practice of 'fulling' raw wool by kneading it with fuller's earth to remove the grease and lighten its colour.

full-wave rectifier A *rectifier* that converts the negative half wave of an *alternating current* into a positive half wave, so that both halves of the swing are able to deliver a unidirectional current.

fulminate See *cyanate*.

fulminate of mercury See *mercury cyanate*.

fulminic acid See *cyanic acid*.

fumaric acid See *butenedioic acid*.

fumigation The destruction of *bacteria*, insects, and other pests by exposure to poisonous *gas* or *smoke*.

fuming nitric acid A brown fuming highly corrosive *liquid* consisting of *nitric* *acid* containing an excess of *nitrogen dioxide*. It is used as an *oxidant* in *rockets* and in organic synthesis.

fuming sulphuric acid See *sulphuric acids*.

function (math.) One quantity y is said to be a function of another quantity x, written $y = f(x)$, if a change in one produces a change in the other. Thus, in the statement

$$y = 3x^2 + 5x \text{ (i.e. } f(x) = 3x^2 + 5x),$$

y is a function of x, and a change in the value of x produces a change in the value of y.

functional group The group of *atoms* in a *compound* that are responsible for its properties. For example, –COOH is the functional group of *carboxylic acids*.

fundamental constants The physical constants that are believed to be unchanged throughout the Universe. See Appendix, Table 2.

fundamental interactions The four ways (see *gravitational, electromagnetic, strong,* and *weak interactions*) in which bodies influence each other even when they are not in contact. Science has been seeking a way of unifying these interactions in a *unified field theory*, but so far has only achieved unification of two interactions, the electromagnetic and the weak interactions, in the *electroweak theory*. See also *gauge field theory*.

fundamental note (phys.) See *quality of sound*.

fundamental units The *units* in which physical quantities are measured are not all independent; many of them are derived from a small number of fundamental or base units on which a particular system of units is based. See *SI units*.

fungi Simple organisms that contain no *chlorophyll*. They may consist of one *cell* or of many cellular filaments called *hyphae*. Important in the breakdown and recycling of plant and animal matter in the *soil*, they can cause decay of food, fabrics, and timber. Some are *parasites* causing diseases of plants and of some animals. Certain fungi are used in *brewing* and baking and for the production of *antibiotics*.

fungicide A substance capable of detroying harmful *fungi*, such as moulds and mildews.

furan Furfuran. C_4H_4O. A five-membered *heterocyclic compound* consisting four CH groups and one O atom. It is a colourless *liquid*, b.p. 32°C., used in organic synthesis and in the form of its *derivatives* in the manufacture of *furan resins*.

furan resins A group of synthetic *resins* obtained by the partial *polymerization* of *furfuryl alcohol* or by the *condensation* of furfuryl alcohol with *furfural* or *methanal*. They are used as *adhesives*, metal coatings, etc.

furfural Furfuraldehyde. C_4H_3OCHO. A *liquid organic compound*, b.p. 161.7°C.; used as a *solvent* and in synthetic *resins*. See *furfural resins*, and *furan resins*.

furfural resins *Thermosetting* resins obtained by the condensation of *furfural* and *phenol* or its homologues. They are used as *adhesives* and in the manufacture of moulding materials, varnishes, etc.

furfuryl alcohol $C_4H_3OCH_2OH$. A yellowish *liquid*, b.p. 171°C., used in the manufacture of *furan resins*.

fur in kettles An *insoluble* gritty deposit, consisting mainly of the *carbonates* of calcium, magnesium, and iron; it is formed by the *decomposition* of the *soluble hydrogencarbonates* of these metals when *hard water* is boiled.

fuse, electrical A device to prevent an unduly high *electric current* from passing

175

through a *circuit*. It consists of a piece of wire made of *metal* of low *melting point*, e.g. tin, placed in series in the circuit. An excessive current will raise the *temperature* of the fuse wire sufficiently to melt it and thus break the circuit. The fuse wire is usually encased in a glass or ceramic cartridge with metal ends.

fused (chem.) In the molten state, usually applied to *solids* of relatively high *melting point*; or, having previously been melted and allowed to solidify.

fused quartz See *quartz*.

Fusel oil A *mixture* of *butanol* and *pentanol* together with other organic substances; a *liquid* of unpleasant smell and taste. It is a *by-product* of the *distillation* of alcohol produced by *fermentation*.

fusible alloys *Alloys* of low *melting point*; generally *eutectic mixtures* of *metals* of low melting point such as bismuth, lead, tin, and cadmium. *Wood's metal* and *Lipowitz alloy* both contain all four and melt below the *boiling point* of *water*. Fusible alloys having a melting point a little above the boiling point of water are used in the construction of automatic sprinklers, *heat* from a fire melting the metal and releasing a spray of water.

fusion Melting; melting together.

fusion, cold See *nuclear fusion*.

fusion, latent heat of See *latent heat*.

fusion, nuclear See *nuclear fusion*.

fusion bomb See *nuclear weapons*.

fusion mixture A *mixture* of *anhydrous sodium* and *potassium carbonates*, Na_2CO_3 and K_2CO_3.

G

g Symbol for the value of the *acceleration of free fall*.

gadolinium Gd. Element. R.a.m. 157.25. At. No. 64. A soft silvery metal, r.d. 7.9, m.p. 1312°C., b.p. 3273°C. A strong *neutron* absorber, gadolinium is used in nuclear technology and in some *ferromagnetic* alloys. Some compounds are used in electronics. See *lanthanides*.

GAG See *glycosaminoglycan*.

gain An increase in electronic signal power; usually expressed as the ratio of the output power (for example, of an *amplifier*) to the input power in *decibels*.

galactose $CH_2OH(CHOH)_4.CHO$. A *hexose sugar*, m.p. 166°C.; it is a *stereoisomer* of *glucose* and a constituent of *lactose* and certain *polysaccharides*.

galaxies Extra-galactic nebulae. The *stars* of the *Universe* are not evenly distributed throughout *space*, but are collected by gravitational attraction (see *gravitation*) into some 10^9 giant clusters called galaxies. Each galaxy contains about 10^{11} stars. The *Sun* is one of such a number of stars in our own *Galaxy* (the *Milky Way*), which is itself a member of a *local group of galaxies*. (See also *expansion of the Universe*.) The galaxies are separated from each other by enormous distances, the nearest galaxy to the Milky Way being some 16×10^5 *light-years* away. Galaxies are either elliptical or spiral shaped (see *spiral galaxies*); a very few however appear to have no regular shape.

Galaxy, the The *Milky Way*. A cluster of some 10^{11} *stars*, one of which is the *Sun*. The Galaxy is a flat disc-shaped spiral structure, approximately 10^5 *light-years* across, with a slight bulge at the centre. The *solar system* is situated quite close to the central plane of this disc at a distance of about three fifths of its radius from the centre.

galena Natural lead(II) sulphide, PbS. A heavy crystalline *mineral* of metallic appearance; it is the principal *ore* of lead. It is used as a *semiconductor* in *crystal rectifiers*.

Galilean telescope A type of refracting *telescope* invented by Galileo Galilei (1564–1642) but no longer used in *astronomy* although its principle is still used in opera glasses. It consists of a *biconvex objective* of long *focal length* and a *biconcave eyepiece* of short focal length, producing an erect image.

gallic acid See *trihydroxybenzoic acid*.

gallium Ga. Element. R.a.m. 69.723. At. No. 31. A silvery-white *metal*, r.d. 5.9, m.p. 29.78°C., b.p. 2403°C. The only two stable *isotopes* are gallium-69 and gallium-71. *Compounds* are very rare; the metal is used in high-temperature *thermometers* and gallium arsenide is used as a *semiconductor*.

gallon A unit of volume or capacity. The British Imperial gallon is the *volume* occupied by ten pounds of *distilled water* under conditions precisely defined by the 1963 Weights and Measures Act. It is equal to 4.546 09 cubic decimetres (*litres*). The U.S. gallon is 0.832 68 British gallons.

galvanic cell See *primary cell*.

galvanized iron Sheet iron coated with a layer of zinc to prevent *corrosion*, usually made by dipping the sheet metal into the molten zinc, with small quantities of magnesium or aluminium added to prevent the formation of a brittle zinc-iron *alloy*. See also *sacrificial protection*.

galvanometer An instrument for detecting, comparing, or measuring small *electric currents*, but not usually calibrated in *amperes*; it requires calibration when an actual current measurement is needed. Galvanometers usually depend upon the magnetic effect produced by an electric current. See *ammeter*; *ballistic galvanometer*.

gamboge A yellow substance obtained from the hardened gum-resin of the tree *Garcinia hanburii*. It is used as a *pigment* and for colouring varnishes.

gamete Germ cell. A reproductive *cell*, usually *haploid* and sexually differentiated. The macrogamete (or *ovum*) unites with the microgamete (or *spermatozoon*) during *fertilization* to produce a *zygote*, which develops into a new individual.

gametocyte A *cell* that undergoes *meiosis* to form one or more *gametes*.

gamma camera A device used in *nuclear medicine* to visualize the distribution of *radioisotopes* in tissues of the human body. It consists essentially of a *scintillation crystal* with an array of *photomultipliers* mounted above it. The gamma rays from a source (often technetium-99 with a single gamma-ray energy of 140 keV) in the patient pass into the camera through a collimator creating scintillations at specific points in the crystal. These scintillations cause pulses in the photomultiplier tubes, the size of the pulses being related to the position of the scintillation. This enables an image to be built up on a *cathode-ray tube* or to be fed into a *computer* in the more sophisticated devices.

gamma globulin See *globulins*.

gamma-iron An allotropic form (see *allotropy*) of iron, which is nonmagnetic and exists between $900°C.$ and $1400°C.$ See *austenite*.

gamma rays Gamma radiation. γ-rays. *Electromagnetic radiation* of the same nature, but shorter *wavelength* than *X-rays* ($10^{-10}-10^{-13}$ metre). They are emitted by excited atomic *nuclei* during *decay* to a lower *excitation* state. Gamma ray *photons* have an energy range of 10^3 to 10^6 electronvolts.

gangue The useless stony minerals that occur with a metallic *ore*.

garnet A group of *minerals* of varying composition, mainly double *silicates* of calcium or aluminium with other *metals*. Several varieties are red in colour, and are used as gems.

gas A substance whose *physical state* (the gaseous state) is such that it always occupies the whole of the space in which it is contained. In a *perfect gas* (ideal gas), the atoms and molecules would move freely, have negligible *volume*, and have *elastic* collisions with each other but in a real gas they are subject to small inter-molecular *forces* (*Van der Waals' forces*) and have a finite volume; polyatomic gas molecules are in some cases inelastic. See also *kinetic theory of gases*.

gas carbon Retort carbon. A hard deposit consisting of fairly pure carbon that was found on the walls of the *retorts* used for the *destructive distillation* of *coal* in the manufacture of *coal-gas*.

gas chromatography Gas-liquid chromatography. A very sensitive method of analysing the components of a complex *mixture* of *volatile* substances. See *chromatography*. The apparatus consists of a long narrow tube, packed with an inert support material of uniform particle size (e.g. *diatomaceous earth*) that has been coated with a non-volatile *liquid*, called the stationary phase, the whole tube and its contents being maintained in a thermostatically controlled oven. The sample to be analysed is carried through the tube by an *inert gas* (e.g. argon) so that the progress through the tube of various components of the mixture is selectively interfered with by the stationary phase, some components passing through the tube more rapidly than others. A detector measures the electrical or thermal *conductivity*, or some other characteristic property, of the gas leaving the column, differences being recorded on a strip chart, which indicates peaks corresponding to the various components. The instrument is calibrated by analysing samples of known compositions. The technique can also be used for separating *mixtures*.

gas constant R. In the *gas equation*, $pV = RT$, the gas constant, R, equals 8.314 510 *joules* per *kelvin* per *mole*. This constant is also sometimes called the 'molar gas constant' or the 'universal gas constant'.

gas-cooled reactor A *nuclear reactor* in which the *coolant* is a *gas*. In the Mark 1, or *magnox* type, natural uranium *fuel* is used with a *graphite moderator*, the *fuel elements* being cased in *magnox*. The coolant used is *carbon dioxide* and the outlet *temperature* is about 350°C., the hot gas being used to raise *steam*. In the Mark II, or advanced gas-cooled reactor (AGR), the moderator is also graphite and the coolant is carbon dioxide, but in this type the outlet temperature is much higher (about 600°C.) and the fuel is *ceramic uranium dioxide* in a *stainless steel* casing.

gas-discharge tube See *discharge in gases*.

gaseous combination, law of See *Gay-Lussac's law*.

gas equation An *equation* connecting the *pressure* and *volume* of a quantity of *gas* with the *thermodynamic temperature*. For one *mole* of a *perfect gas*, $pV = RT$, where p = pressure, V = volume, T = thermodynamic temperature, and R = the *gas constant*. See *gas laws*.

gas laws Statements as to the *volume* changes of *gases* under the effect of alterations of *pressure* and *temperature*. *Boyle's law* states that at constant temperature the volume of a given mass of gas is inversely proportional to the pressure; i.e. pV = constant. *Charles' law* states that at constant pressure all gases expand by 1/273 of their volume at 0°C. for a rise in temperature of 1°C.; i.e. the volume of a given mass of gas at constant pressure is directly proportional to the *thermodynamic temperature*. For 1 mole of gas, the two laws may be combined in the expression $pV = RT$ (see *gas equation*), where T is the thermodynamic temperature. This gives the behaviour of a gas when both temperature and pressure are altered. The gas laws are not perfectly obeyed by ordinary gases, being strictly true only for the *perfect gas*. See *gas laws, deviations from*.

gas laws, deviations from Real *gases* do not strictly obey the *gas laws*, but follow them more and more closely as the *pressure* of the gas is reduced. Various *equations* have been derived that attempt to give a better approximation to the behaviour of actual gases. The best known of these is *Van der Waals' equation*. See also *virial equation*.

gas mantle A structure used to provide illumination by burning a gas; it is composed of the *oxides* of thorium (99%) and cerium (1%), made by impregnating a combustible fabric with a *solution* of the *nitrates* of the metals, and decomposing the nitrates by heat. They were widely used before the introduction of electric lighting and are now used in portable *butane* camping lights.

gas maser A *maser* in which *microwave radiation* interacts with *gas molecules*.

gas mask Respirator. A device for protecting the face and breathing organs against poisonous 'gases'. (These include poisonous *smokes*, etc., used in chemical warfare.) In the simplest device air is drawn through a layer of *activated carbon*, which adsorbs *vapours*, and also through a filter-pad, which retains solid particles of smokes. Such an arrangement is effective against war 'gases' and smokes, but not against gases of low *relative molecular mass*, such as *carbon monoxide*.

gas oil Diesel oil. The *oil* left after *petrol* and *kerosene* have been distilled from crude *petroleum*. It is used as a *fuel* for *diesel engines* and for carburetting *water gas*.

gasoline The name for *petrol* in the USA.

gas thermometer An apparatus for measuring *temperature* by the alteration in *pressure* produced by temperature changes in a *gas* kept at constant *volume*, or by the alteration in volume of a gas kept at constant pressure. For practical purposes, other more convenient forms of thermometer are used whenever possible. However, the gas thermometer, operated at low pressure, gives the only direct means of determining the *thermodynamic temperature*.

gas turbine An *engine* that converts the *chemical energy* of a *liquid fuel* into *mechanical energy* by internal *combustion*; the gaseous products of the fuel, burnt in compressed air, are expanded through a *turbine*. It is used as the power plant in most commercial and military aircraft, locomotives, and experimentally in motor vehicles. It is also used as an auxiliary power plant in electrical generating stations.

gate 1. A *circuit* with only one output, which has more than one input and which can be activated by various combinations of input signals. **2.** A signal that activates a circuit for a predetermined time or until another signal is received. **3.** A device for selecting portions of a *wave*, either on a time or *amplitude* basis. **4.** The electrode in a *field-effect transistor* that controls the flow of current through the channel.

gauge boson An *elementary particle* that is visualized as being exchanged (see *exchange force*) between particles that are interacting in accordance with a *gauge field theory*. For example, the photon is the gauge particle exchanged in *electromagnetic interactions* and *gluons* are exchanged in *strong interactions*.

gauge field theories Quantum field theories predicting the behaviour of all the *fundamental interactions*, except *gravitation*. Gauge field theories are based on the concept of *symmetry*, i.e. that the solution of a set of equations remains unchanged even when a characteristic of the system they describe is varied. The quantum field theory applied to the *electromagnetic interaction*, known as *quantum electrodynamics*, is a gauge field theory in which the electromagnetic force is visualized as a result of the exchange of massless *photons* between charged particles, the equations describing the motions of the charged particles remaining unchanged during local symmetry operations. The success of this theory enabled it to be extended to encompass the *electroweak theory* and to the formulation of the gauge field theory applying to *strong interactions*, known as *quantum chromodynamics* (see *elementary particles*), in which the exchanged particle is the *gluon*. In general, the particles exchanged in gauge field theories are known as *gauge bosons*.

gauss The *c.g.s. system* unit of *magnetic flux density*. If a *magnetic field* of 1 *oersted* intensity exists in a medium of unit *magnetic permeability*, e.g. air, then the induction will be 1 gauss. 1 gauss is equal to 1 *maxwell* per square centimetre or 10^{-4} *tesla*. Named after K. F. Gauss (1777–1855).

Gaussian system A c.g.s. system of electrical units in which both the *electric constant* and the *magnetic constant* are taken as 1. Although replaced by *SI units* for general use, the system still has some uses in *particle physics* and in *relativity*.

gaussmeter A *magnetometer* calibrated in *gauss*, for measuring *magnetic flux density*.

Gauss's law The total *electric flux* of a closed surface in an *electric field* is proportional to the *electric charge* within that surface. The law can be generalized for any *vector* field through a closed surface.

Gay-Lussac's law of gaseous combination When *gases* combine, they do so in simple ratio by *volume* to each other, and to the gaseous product, measured under the same conditions of *temperature* and *pressure*. It is explained by *Avogadro's law*. Named after Joseph Louis Gay-Lussac (1778–1850).

Gegenschein Counter glow. A faint elliptical patch of *light* in the sky that may be observed at night directly opposite the *Sun*, though it is rarely seen in Britain. It is caused by a reflection of sunlight by meteoric particles in *space*.

Geiger counter Geiger-Muller counter. An instrument for the detection of *ionizing radiations*, capable of registering individual particles or *photons*. It consists normally of a fine wire *anode* surrounded by a *coaxial* cylindrical *metal cathode*, mounted in a glass envelope containing *gas* at low *pressure*. A large *potential difference*, usually about 1000 *volts*, is maintained between the anode and the cathode. The *ions* produced in the counter by an incoming ionizing particle are accelerated by the applied potential difference towards their appropriate *electrodes*, causing a momentary drop in the potential between the latter. This voltage *pulse* is then passed on to various *electronic* circuits by means of which it can, if desired, be made to work a counter or cause a click in a *loudspeaker*. See Fig. 16. Named after Hans Geiger (1882–1947).

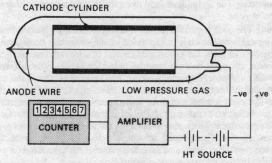

Figure 16.

Geiger-Nuttal law The approximate empirical law that the range R of an *alpha particle* emitted by a radioactive substance is given by the relation:
$$\log \lambda = B \log R + A,$$
where λ is the *disintegration constant* of the substance and A and B are constants.

Geissler tube A form of *discharge tube* for showing the luminous effects of an electric *discharge* through various rarefied *gases*. It consists of a sealed glass tube containing platinum electrodes. Named after H. Geissler (1814–79).

gel A *colloidal solution* that has set to a jelly, the *viscosity* being so great that the solution has the *elasticity* of a *solid*. The formation of a gel is attributed to a mesh-like structure of the *disperse phase* or *colloid*, with the *dispersion medium* circulating through the meshwork. *Gelatin* is an example.

gelatin Gelatine. A complex protein formed by the *hydrolysis* of *collagen* in animal cartilages and bones, by boiling with *water*. *Soluble* in hot water, the *solution* has the property of setting to a *gel*. It is used in foods (jellies, etc.), photographic emulsions, as an *adhesive*, textile size, and in a variety of other arts and industries. It is also used as a *culture medium*.

gelation 1. The process of *freezing*. See also *regelation of ice*. **2.** The formation of a *gel*.

gelignite An *explosive* consisting of a *mixture* of *nitroglycerin*, *cellulose nitrate*, *sodium* or *potassium nitrate*, and wood pulp.

gem See *vicinal*.

gene A unit, comprising part of a *chromosome*, which controls an individual inherited characteristic of an *organism* and which is capable of *mutation* as a unit. (See also *cistron*.) It is now usually regarded as the length of DNA that controls the synthesis of one *polypeptide* chain. See *operon*; *genetic code*.

general theory of relativity See *relativity, theory of*.

generation time (phys.) The average lapse of time between the creation of a *neutron* by *nuclear fission* and a subsequent fission produced by that neutron.

generator A machine for producing *electrical energy* from *mechanical energy*. See *alternator*; *dynamo*.

generatrix (math.) A point, line, or *plane* the movement of which produces a line, surface, or solid.

gene sequencing The techniques used to establish the sequence of *nucleotides* in a particular *gene*.

genetic code The code by which inherited characteristics are handed from generation to generation. The code is expressed by the molecular configuration of the *chromosomes* of *cells*. Chromosomes consist of *deoxyribonucleic acid* (DNA) and *protein*, the code-bearing material being the DNA. Four different nitrogenous bases (*adenine*, *cytosine*, *guanine*, and *thymine*) occur in the *nucleotides* of DNA, and the sequence of three of these bases constitutes a unit of the genetic code, in that each sequence of three bases codes for one of the twenty different *amino acids* that go to make up the *proteins* that control the characteristics of a cell. Chromosomes, which almost always exist in the *nuclei* of cells, transcribe (see *transcription*) their coded information to the *cytoplasm* of these cells (where the enzyme proteins and structural proteins are assembled on units called *ribosomes*) by way of messenger *ribonucleic acid*.

genetic engineering Gene manipulation. The technology involved in altering in some prescribed way the genetic constitution of an organism. Typically, 'useful' *genes*, i.e. very short sequences of DNA (see *deoxyribonucleic acid*), are isolated from one organism and inserted into the DNA of a *bacterium* or *yeast*. These *microorganisms* multiply rapidly and can be cultured easily, enabling large quantities of the gene product to be obtained. Genetic engineering has been used for the large-scale production of *antibiotics*, *enzymes*, and *hormones* (e.g. *insulin*). Genetic engineering is also used to produce genetically desirable domestic animals and crop plants by inserting the appropriate genes into the DNA of cultured animal or plant cells. Organisms into which foreign DNA has been artificially inserted are called 'transgenic organisms'.

genetic fingerprinting An identification technique based on the pattern of repetition of particular *nucleotide* sequences (marker sequences) in the DNA of an individual. DNA is extracted from the individual's *cells* (usually from blood or semen) and broken into small fragments using *enzymes*; these fragments can be compared with fragments taken from another individual (e.g. in paternity testing) or with test samples (in forensic identification).

genetic mapping The construction of a plan showing the relative positions of the individual *genes* along the length of a *chromosome*.

genetics The study of heredity in living organisms. Classical genetics, the original form of the subject, is based on the observations of pea plants by the Austrian monk Gregor Mendel (1822–84) in the 1850s. The science has expanded to include the subdisciplines of molecular genetics, microbial genetics, cytogenetics, and population genetics and has important applications in medicine, agriculture, and industry (see *genetic engineering*), as well as in all aspects of *biology*.

genome All the *genes* in a set of *chromosomes*. Parents denote their own genome to their offspring.

genotype 1. The genetic constitution of an individual *organism* or of a well defined group of organisms. **2.** A group of organisms that have the same genetic constitution. See also *phenotype*.

gentian violet A purple *dye* derived from *aniline* and used as an *indicator*, an *antiseptic*, and a dye.

genus (plural **genera**) A group (*taxon*) of similar *species*. See *classification*.

geocentric Having the *Earth* as a centre; measured from the centre of the Earth.

geochemistry The study of the chemical composition of the *Earth's* crust and the changes that take place within it.

geochronology The study of dating geological events, sediments, rocks, and biological remains. Radioactive dating techniques are used to provide actual ages, while other techniques, such as *pollen analysis*, and the correlation of *fossils* provide relative dates. See also *varve dating*.

geodesic Pertaining to the *geometry* of curved surfaces. A 'geodesic line', also called a 'geodesic', is the shortest distance between two points on a curved surface.

geodesy Surveying on a scale that involves making allowance for the curvature of the *Earth*.

geography The study of the Earth's surface and the way man has interacted with it. Physical geography is concerned with the structure of the Earth's surface (geomorphology), its climate (climatology), and its oceans (oceanology), as well as mapmaking (cartography). Human geography is concerned with such factors as economics (economic geography), politics (political geography), history (historical geography), and its major centres of population (urban geography).

GEOLOGICAL TIME SCALE

Era		Period	Time Scale millions of years
CENOZOIC	QUATERNARY	Holocene	0.01
		Pleistocene (Glacial)	1
	TERTIARY	Pliocene	10
		Miocene	25
		Oligocene	40
		Eocene	60
		Paleocene	70
MESOZOIC (SECONDARY)		Cretacious	135
		Jurassic	180
		Triassic	225
PALAEOZOIC (PRIMARY)		Permian	270
		Carboniferous	350
		Devonian	400
		Silurian	440
		Ordovician	500
		Cambrian	600
PRE-CAMBRIAN			2000

geological time scale Geological periods. A scale of time that serves as a reference for correlating various events in the history of the *Earth*; it has been built up by studying the various strata of rocks that comprise the *Earth's* crust with special reference to the *fossils* found in them. The time scale is divided into three main 'eras', based upon the general character of the life that they contain, each era being subdivided into 'periods'. The Table gives the names of these eras and periods, together with their approximate ages.

geology The scientific study of the *Earth's* crust. Physical geology includes *geochemistry, geomorphology, geophysics, petrology,* and *mineralogy,* while historical geology is concerned with *geochronology, palaeontology,* and *stratigraphy.*

geomagnetism The study of the *magnetic field* associated with the *Earth.* See *magnetism, terrestrial.*

geometrical progression A series of quantities in which each term is obtained by multiplying the preceding term by some constant factor, called the 'common ratio'. E.g. 1, 3, 9, 27, 81..., each term being three times the preceding. For a series of n terms, having common ratio r and the first term a, the sum

$$S = a(r^n - 1)/(r - 1);$$

or, if r is less than 1, a more convenient expression is
$$S = a(1 - r^n)/(1 - r).$$

geometric mean The geometric *mean* of n positive numbers, a, b, c... is $(abc...)^{1/n}$. E.g. the geometric mean of 3 and 12 is 6.

geometry The mathematical study of the properties and relations of lines, surfaces, and solids in space.

geomorphology The study of the Earth's landforms, their origin and structure.

geophysics The study of the *Earth* and its *atmosphere* by physical methods. This includes *seismology*, *meteorology*, *hydrology*, *terrestrial magnetism*, etc.

geostationary orbit See *synchronous orbit*.

geothermal energy Heat in the Earth's interior that can be tapped and made use of. Volcanoes, geysers, and hot springs are all potential sources. High-temperature porous rock can be drilled into and the heat extracted by means of a fluid, which can then be used for direct heating or for steam raising. New Zealand, USA, and Iceland have made some use of geothermal energy but at present it only supplies about 0.11 of the world's energy.

geraniol $C_{10}H_{17}OH$. A *liquid terpene alcohol*, b.p. 107°C., present either free, or as an *ester*, in many *essential oils*.

germane Germanium tetrahydride. GeH_4. A colourless gas. The germanium analogue of *methane*; the first member of a series of germanium compounds of the general formula Ge_nH_{2n+2}, corresponding to the *alkanes*.

germanium Ge. Element. R.a.m. 72.59. At. No. 32. A brittle white *metalloid*, r.d. 5.35, m.p. 937.4°C., b.p. 2830°C. It occurs in *zinc sulphide* and is obtained as a by-product in zinc smelting. *Compounds* are rare although it forms a number of compounds (see *germane*). It is used extensively in *transistors* and other electronic devices as a *semiconductor*.

German silver See *nickel silver*.

gestation The period between conception and birth for animals that bear live offspring.

getter Vacuum getter. A substance used for removing the last traces of air or other *gases* in attaining a high *vacuum*. E.g. magnesium metal is used in *thermionic valves*; after exhausting and sealing the valve a small amount of magnesium left in the valve is vaporized by *heat* and combines chemically with any remaining oxygen and nitrogen.

GeV Abbreviation for *giga electronvolt*, i.e. 10^9 electronvolts. In America this is usually written BeV where the 'B' represents the American *billion*.

ghosts (phys.) False lines appearing in a *line spectrum* due to imperfections in the ruling of the *diffraction grating* used.

giant star A *star* possessing high *luminosity*, low *density*, and a diameter 10 to 100 times that of the *Sun*. See *red giant*; *supergiant*.

gibberellins A group of diterpinoid *plant hormones* (known as growth substances) that promote the development of plant stems and fruit and can have other effects, e.g. on flowering and on seed dormancy.

gibbous The shape of the *Moon* or a *planet* when it is more than half-phase, but less than full-phase. See *phases of the Moon*.

Gibbs' function See *free energy*. Named after Josiah Willard Gibbs (1839–1903).

Gibbs' phase See *phase rule*.

giga- Prefix denoting one thousand million times; 10^9. Symbol G, e.g. $GW = 10^9$ *watts*.

gilbert The *c.g.s. unit* of *magnetomotive force* in *electromagnetic units*. It is equal to $10/4\pi$ *ampere-turns*. Named after William Gilbert (1544–1603).

gilding Covering with a thin layer of metallic gold, often by *electrolysis* (see *electroplating*).

gill A unit of capacity equal to one quarter of a *pint*.

gilsonite A pure form of *asphalt* that occurs in North America; it is used in *paints* and *varnishes*.

Giorgi system See *m.k.s. system*.

glacial acid Pure *ethanoic acid*; *solid* crystalline ethanoic acid below its *freezing point* (16.6°C.).

Glashow–Weinberg–Salam theory See *electroweak theory*. Named after S. L. Glashow (1932–), S. Weinberg (1933–), and Abdus Salam (1926–).

glass A hard brittle *amorphous mixture*, usually *transparent* or *translucent*, of the *silicates* of calcium, sodium, or other *metals*. Ordinary soda glass is made by melting together sand (*silica*), *sodium carbonate*, and *lime*. Glass for special purposes may contain lead, potassium, barium or other metals in place of the sodium, and *boric oxide* in place of the silica. See *crown glass*; *flint glass*; *optical glass*; *Pyrex**; *safety glass*.

glass-ceramics Materials that usually consist of lithium and magnesium *aluminosilicates*; they are chemically similar to glasses, but differ from glasses in consisting of very small crystals. They have high mechanical strength and very low thermal expansion, making them resistant to abrupt temperature changes. They are used for heat-resisting ovenware, for radomes, and other purposes involving exposure to drastic conditions.

glass fibre Fibreglass*. Fine *glass* fibres, usually less than a quarter of a micrometre in diameter, that are woven into a cloth and impregnated with various *resins*. Owing to their high *tensile strength* and *corrosion* resistance these materials are used in small boat building and for some motor vehicle body parts.

glass transition The change, characteristic of many *rubbers* and other *polymers*, from a plastic or rubbery to a glassy or brittle state. The temperature region of this change (the glass temperature, glass-transition temperature, or second-order transition temperature) is designated T_g.

glass wool A material consisting of very fine *glass* threads, resembling cotton wool. It is used for filtering and absorbing corrosive *liquids*.

Glauber's salt Crystalline *sodium sulphate*, $Na_2SO_4.10H_2O$, used as a laxative. Named after J. R. Glauber (1604–68).

glaze A *vitreous* covering for pottery, chemically related to *glass*.

global warming See *greenhouse effect*.

globular clusters Self-contained, approximately spherical, clusters of about one hundred thousand *stars*; some hundred of these clusters are known to be distributed about the centre of the *Milky Way*, and although they appear to be outside the *Galaxy*, they are believed to be gravitationally associated with it.

globulins Large globular *proteins* in which the *polypeptide* chains are folded to form a spherical shape; they are soluble in dilute *solutions* of mineral *salts*, such as *sodium chloride*, NaCl, *magnesium sulphate*, $MgSO_4$, etc. They occur in many animal and vegetable *tissues* and fluids; e.g. lactoglobulin in milk, serum globulin in *blood*, vegetable globulins in seeds. Serum globulins include immunoglobulins (gamma globulin), which are the *antibodies* responsible for the immune response.

glove box A metal box that provides protection to workers who have to manipulate *radioactive* materials or that enables the manipulation of substances requiring a dust-free, sterile, or *inert* atmosphere. Manipulation is carried out by means of gloves fitted to ports in the walls of the box.

glow discharge A silent electrical discharge through a *gas* at low *pressure*, usually luminous. See *discharge in gases*.

glucinum See *beryllium*.

glucocorticoid See *corticosteroid*.

gluconic acid $CH_2OH(CHOH)_4COOH$. A colourless *soluble* crystalline *optically active* substance, m.p. 125°C., obtained by the *oxidation* of *glucose*. It is used for cleaning *metals*.

glucose Dextrose, grape-sugar. $C_6H_{12}O_6$. A colourless crystalline *soluble hexose sugar*. M.p. 146°C. It occurs in honey and sweet fruits. Other sugars and *carbohydrates* are converted into glucose in the human body before being utilized to provide *energy*. Glucose is an *optically active* substance, the naturally occurring sugar invariably being *dextrorotatory* (d-glucose). Commercially prepared from *starch* and other carbohydrates by *hydrolysis*, it is used in brewing, jam-making, confectionery, etc. The *laevorotatory* form (1-glucose) is rare and does not occur in nature.

glucosides See *glycosides*.

glue A general name for *adhesives*, particularly those made by extracting hides, bones, cartilages, etc., of animals with *water*.

gluon An *elementary particle* that is the *gauge boson* by which the *strong interaction* is mediated between quarks. See *quantum chromodynamics*.

glutamate A *salt* or *ester* of *glutamic acid*.

glutamic acid A colourless crystalline *amino acid*, m.p. 206°C., used in the form of its sodium *salt* as a flavouring. See Appendix, Table 5.

glutamine A colourless *soluble amino acid*, m.p. 184–5°C. See Appendix, Table 5.

gluten A mixture of two *proteins* contained in wheat flour (8%–15%) containing some 33% of *glutamic acid* and 12% of *proline*. In coeliac disease a gluten-free diet is required.

glycan A *polymer* of *saccharides*; a *polysaccharide*. See also *glycosaminoglycan*.

glycerides *Esters* of *glycerol* with *organic acids*. Animal and vegetable *fats* are mainly composed of triglycerides of *fatty acids*, such as *stearic*, *palmitic*, and *oleic*, a *molecule* of such a triglyceride being derived by the combination of one molecule of glycerol with three fatty acid molecules.

glycerin(e) See *glycerol*.

glycerol Glycerin(e). Propane-1,2,3,-triol. $CH_2OH.CHOH.CH_2OH$. A thick syrupy sweetish *liquid triol soluble* in *water*. B.p. 290°C. It occurs combined with *fatty acids* in *fats and oils* and is obtained by the *saponification* of fats in the manufacture of *soap*. It is used in the manufacture of *explosives* (see *nitroglycerin*), *plastics*, in pharmacy, and as an *anti-freeze*.

glycerol monoethanoate Acetin. $CH_3COOC_3H_5(OH)_2$. A colourless viscous *hygroscopic liquid*, b.p. 158°C., used in the manufacture of *explosives*.

glyceryl The *trivalent* group $-CH_2(CH-)CH_2-$, derived from *glycerol*.

glycine A colourless *soluble* crystalline *amino acid*, m.p. 232°C., used in organic synthesis. See Appendix, Table 5.

glycogen Animal starch. A complex highly branched *carbohydrate polymer* formed from *glucose* in the liver and other organs of animals, serving as a *sugar* reserve. It is also a carbohydrate reserve of many *fungi*.

glycol See *ethanediol*.

glycolipids Compound *lipids* that are *glycerides* or similar molecules with two *hydrophobic* hydrocarbon chains and a *polar* head group consisting of one or more *sugar* residues; *amphipathic* and found on the outer surface of all *plasma membranes*, their role is probably in intercellular communication.

glycols See *diols*.

glycolysis The breakdown of *glucose* by a series of *enzyme*-catalysed reactions to yield *pyruvic acid* and chemical *energy* (see *adenosine triphosphate*); the process occurs in the *cytoplasm* of living *cells* and is the first stage of aerobic *respiration*. In anaerobic respiration the pyruvate is converted to *lactic acid* (e.g. in animals) or to *ethanol* (in certain *yeasts*).

glycoproteins Glucoproteins. Complex *proteins* that contain *carbohydrates*.

glycosides *Ether*-type compounds, derived from *sugars* and hydroxy compounds. If the latter component in a glycoside is a non-sugar, it is called an aglycone. Glycosides in which the sugar is *glucose* are called glucosides. Glycosides occur widely in plants. Some are heart poisons.

glycosaminoglycan GAG. A *glycan* whose repeating unit includes an amino sugar (a *monosaccharide* containing an *amino* group). Formerly called mucopolysaccharides, they include *chitin* and *molecules* important in the skin, bones, and joints of animals (including humans).

glycyl The *univalent* group NH_2CH_2CO- (from *glycine*).

glyoxal Diformyl. $(CHO)_2$. A yellow crystalline substance, m.p. 15°C., b.p. 51°C. It is used in the manufacture of *plastics*, and in textile finishing.

glyptal resins Alkyd resins. A class of synthetic *resins* obtained by the reaction of *polyhydric alcohols* with *polybasic organic acids* or their *anhydrides*; e.g. *glycerol* and *phthalic anhydride*. They are used chiefly for surface coatings.

gnotobiotics The study of germ-free life, especially in experimental conditions in which animals are subsequently inoculated with specific strains of *microorganisms*.

gold Au. Element. R.a.m. 196.9665. At. No. 79. A bright yellow soft *metal* that is extremely malleable and ductile; m.p. 1064°C., b.p. 2807°C., r.d. 19.32. Gold is not corroded by air, is unattacked by most acids, but dissolves in *aqua regia*. It occurs mainly as the free metal; most compounds are unstable and are easily reduced to gold. It is extracted from the ore by the *amalgamation process* and the *cyanide* process. *Alloys* with copper or silver, to give hardness, are used in coins, jewellery, and dentistry. Compounds are used in photography and medicine.

gold(III) chloride Auric chloride. $AuCl_3$. A red *soluble* crystalline solid, which decomposes at 254°C. It is used in *photography* and gilding *glass*.

gold leaf Gold is the most malleable of *metals*, and can be beaten into leaves

0.0001 mm thick. The leaf has the appearance of metallic gold, but transmits green *light*; i.e. appears green when held up to the light.

gold-leaf electroscope See *electroscope*.

Goldschmidt process The preparation of *metals* from their *oxides* by *alumino-thermic reduction*, i.e. by reducing the oxide by means of aluminium powder, e.g. $Cr_2O_3 + 2Al = 2Cr + Al_2O_3$. The method was discovered by Hans Goldschmidt (1861–1923).

Golgi body Golgi apparatus. A region in the *cytoplasm* of plant and animal *cells*, consisting of vesicles and folded *membranes*, that stores and transports *enzymes*, *hormones*, etc., and helps to form cell walls. Named after Camillo Golgi (1843–1926).

gonad The organ that in animals produces the *gametes*. The organs, which usually occur in pairs, are the male testes and the female ovaries. The gonads also produce the *hormones* that control secondary sexual characteristics.

goniometer An instrument for the measurement of *angles* (of *crystals*).

Gooch crucible A laboratory *filter* consisting of a shallow *porcelain* cup, the flat bottom of which is perforated with small holes over which a layer of *asbestos* fibres is placed.

governor A device for regulating the speed of an *engine* or machine, on the principle of negative *feedback*, so that its speed is kept constant under all conditions of loading. This is often achieved by controlling the *fuel* consumption, so that a rise in speed is arranged to reduce the fuel intake and a fall in speed to increase it.

gradient The degree of inclination of a slope, usually expressed as unit rise in height per number of units covered along the slope; i.e. the sine of the angle of rise (see *trigonometrical ratios*). Mathematically, the gradient is the ratio of the vertical distance to horizontal distance, i.e. the tangent of the angle. For small gradients the difference between the sine and the tangent is negligible. In *Cartesian coordinates* the gradient of the straight line given by $y = mx + c$, is m. In general, the curve $y = f(x)$ has a gradient given by dy/dx at the point (x,y).

graduation The marking that indicates the scale of an instrument, e.g. the stem of a *thermometer* is graduated in *degrees*.

Graham's Law The rate of *diffusion* of a *gas* is inversely proportional to the square root of its *density*. The principle is made use of in separating *isotopes* (see *isotopes, separation of*). Named after Thomas Graham (1805–69).

grain A British unit of *mass*. 1/7000 of a pound; 0.0648 g.

gram Gramme. One of the *fundamental units* of measurement in the *c.g.s. system* of units. The unit of *mass*, defined as 1/1000 of the mass of the International Prototype *Kilogram*, which is a platinum-iridium standard preserved in Paris. Symbol g.

gram-atom A former *c.g.s. unit* equal to the *relative atomic mass* of an *element* expressed in *grams*; e.g. 32 g of sulphur. It has now been replaced by the *mole*.

gram-equivalent See *chemical equivalents*.

gram-molecule A former *c.g.s. unit* equal to the *relative molecular mass* of a *compound* expressed in *grams*; e.g. 18 g of *water*. It has now been replaced by the *mole*.

Gram's method A method of staining and classifying *bacteria* in which *gentian violet* is used to stain a bacterial smear. If the bacteria retain the violet dye, after washing with a *solution* of iodine and *potassium iodide* in water (Gram's solution) and counterstaining with safranine, they are said to be 'gram positive'. If they do not retain the dye they are said to be 'gram negative'. Named after Hans Gram (1853–1938).

grand unified theory GUT. A unified theory that would combine the *strong*, *weak*, and *electromagnetic interactions* into one *gauge field theory*. Although the *electroweak theory* has gone some of the way to unification a single GUT has not yet been accepted. There are, however, several models, most of which assume that the three interactions are the low-energy manifestations of one interaction. The unification is seen as taking place above energies of 10^{15} GeV, which is much higher than can be achieved in current particle *accelerators*. One feature common to GUT models is that *protons* would be expected to decay (see *proton decay*), perhaps with a lifetime of 10^{35} years. There is, however, as yet no evidence to support proton decay. Other GUTs predict the existence of massive *neutrinos*.

granite Any of a class of heterogeneous igneous *rocks*, containing *quartz, feldspar*, and other *minerals*.

grape sugar See *glucose*.

graph A diagram, generally plotted between axes at right angles to each other, showing the relation of one *variable* quantity to another. E.g. the variation of rainfall with time, or the variation in the value of a mathematical *function* as different values are assigned to one of the variables in the function. See *Cartesian coordinates; polar coordinates*.

-graph A suffix applied to instruments that automatically record or write down observations; e.g. *barograph*.

graphite Blacklead, plumbago. A natural *allotropic form* of carbon, used for pencil leads, in electrical apparatus, and as a lubricant for heavy machinery. It is also used as a *moderator* in *nuclear reactors*.

graphite-moderated reactor See *nuclear reactor*.

graticule 1. A scale, or network of fine wires, in the *eye-piece* of a *telescope* or *microscope*. **2.** A network of parallel lines (longitudinal and latitudinal) on a map.

grating See *diffraction grating*.

gravimetric analysis A form of *quantitative chemical analysis*. The mass of a substance present is determined by converting it, by a suitable *chemical reaction*, into some other substance of known chemical composition, which can be readily isolated, purified, and weighed.

gravitation See *Newton's law of gravitation*.

gravitational collapse The collapse of an astronomical body as a result of the gravitational force between its components. This is thought to occur when a *star* runs out of *nuclear fuel* so that it can no longer produce energy by *nuclear fusion* to counteract the inward gravitational force. The products of such a collapse are probably *white dwarfs, neutron stars*, and *black holes*.

gravitational constant G. The fundamental constant that appears in *Newton's law of gravitation*. It has the value $6.672\ 59 \times 10^{-11}$ N m^2 kg^{-2}.

gravitational field The region in which one massive body (i.e. a body that possesses the attribute of *mass*) exerts a *force* (gravitational force) of attraction on another massive body. Einstein's general theory of *relativity* describes the way in which mass gives rise to this type of field.

gravitational interaction The *fundamental interaction* between all massive particles. It is the weakest of all known interactions, being some 10^{40} times weaker than the *electromagnetic interaction*. Compare *strong interaction* and *weak interaction*.

gravitational mass See *inertial mass*.

gravitational waves *Waves* that would be propagated through a *gravitational field* at the *speed of light*, as predicted by the general theory of *relativity*, when a *mass* is accelerated. So far there is no direct evidence to support their existence.

graviton A hypothetical particle or *quantum* of gravitational *energy* (see *gravitation*). If it exists it would be a *gauge boson* and would be expected to have zero *rest mass* and *charge*, and *spin* 2.

gravity The gravitational force (see *Newton's law of gravitation*) between the *Earth* (or other *planet* or *satellite*) and a body on its surface, or within its *gravitational field*. As gravity is proportional to the *mass* of the planet or satellite and inversely proportional to the square of the distance from its centre, the gravity on a planet or satellite in terms of the Earth's gravity is given by

$$(d_p/d_e)^2/M_p,$$

where M_p is the mass of the planet in Earth masses, and d_p and d_e are the diameters of the planet and Earth respectively. Substituting the relevant figures from Table 4 of the Appendix will show that gravity on the surface of the *Moon* is 1/6 that on the surface of the Earth.

Gravity is responsible for the *weight* of a body; the weight of a body is the gravitational force of attraction that the Earth exerts on that body. It is equal to the mass of the body multiplied by the *acceleration of free fall*. Gravity causes bodies to fall to Earth with a uniform acceleration, but the magnitude of the acceleration of free fall varies with geographical location and altitude.

gray The derived *SI unit* of absorbed *dose* of *ionizing radiation*. The energy in *joules* absorbed by one *kilogram* of irradiated material. Symbol Gy. Named after L. H. Gray (1905–65).

grease A semi-solid lubricant composed of emulsified *petroleum* oils and *soluble hydrocarbon soaps*.

great circle A circle obtained by cutting a *sphere* by a *plane* passing through the centre. E.g. regarding the Earth as a sphere, the *Equator* is a great circle, as are all the *meridians* of longitude. On the Earth's surface, an apparent straight line joining any two points is an arc of a great circle, i.e. a *geodesic*.

Greek fire A mixture of materials that caught fire when wetted; it was thought to have been used by the ancient Greeks in naval warfare. However, its invention is now ascribed to a Greek-speaking Syrian refugee (Kallinikos of Heliopolis) from the Muslim conquest in the second half of the 7th century AD. It was probably composed of sulphur, *naphtha*, and *calcium oxide* or similar materials.

greenhouse effect 1. The effect produced inside a greenhouse: solar radiation (infrared, visible, and some ultraviolet) is admitted to the greenhouse through its glass roof and is absorbed by the contents. The longer wavelength infrared radiation emitted by the contents cannot escape through the glass and the temperature of the interior rises. **2.** A similar effect that applies to the whole Earth. The short-wave solar radiation passes through the Earth's atmosphere but after being reradiated by its surface as *infrared radiation*, some of it is absorbed by gases in the *atmosphere* causing a rise in temperature (known as global warming). The so-called greenhouse gases that absorb the infrared radiation are predominantly CO_2 and CH_4. In recent decades the amount of CO_2 in the atmosphere has increased to 350 ppm (from a preindustrial level of 275 ppm) as a result of the widespread use of *fossil fuels*, which are now producing some 2.5 $\times 10^{10}$ tonnes CO_2 per year. In addition, the burning of trees as a consequence of worldwide deforestation has produced about 1×10^{10} tonnes CO_2; deforestation has also reduced the amount of CO_2 absorbed by plants from the atmosphere in *photosynthesis*. It is calculated that 55% of global warming is caused by absorption of energy by CO_2. The remaining 45% is caused mainly by methane absorption and by damage to the *ozone layer* as a result of the use of CFCs (see *chlorofluorocarbon*). Global warming is said to be causing a rise in the temperature of the Earth's oceans of 1°C per decade.

greenockite See *cadmium sulphide*.

green revolution The application, during the second half of the 20th century, of scientific methods to arable farming, which has brought about a dramatic increase in crop yields in many countries. It has included gene manipulation (see *genetic engineering*) to improve plant breeds, widespread use of *fertilizers* and *pesticides*, improvements in irrigation, and considerable advances in mechanization. These increases in crop yields have been essential to feed the world's expanding population.

green vitriol Copperas. *Iron(II) sulphate crystals*, $FeSO_4.7H_2O$.

GREGORIAN TELESCOPE

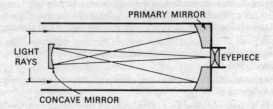

Figure 17.

Gregorian telescope A form of astronomical reflecting *telescope* similar to a *Cassegrainian telescope* in that a hole in the centre of the parabolic primary *mirror* allows *light* to pass to the *eyepiece*, but in the Gregorian telescope the secondary mirror is *concave* rather than *convex*. Although this type of telescope gives an erect image it is not now widely used as it is difficult to adjust. Named after James Gregory (1638–75). See Fig. 17.

grid 1. See *control grid*. **2.** A system of *high tension* cables (overhead or underground) by which electrical power is distributed throughout a country. **3.** A network of horizontal and vertical lines superimposed on a map to enable references to be given.

grid bias A fixed *voltage* applied between the *cathode* and the *control grid* of a *thermionic valve*, which determines its operating conditions.

Grignard reagents *Alkyl* magnesium *halides* with the structure $R_2Mg.MgX_2$, where R is an alkyl group and X is a halogen. They are prepared by the action of magnesium metal on a *haloalkane* in *ether* solution; e.g. C_2H_5MgI. They are used in organic *synthesis*. Named after François A. V. Grignard (1871–1935).

ground state The most stable *energy* state of a *nucleus*, *atom*, or *molecule*. The normal state of an atom when its circum-nuclear *electrons* move in *orbits* such that the energy of the atom is a minimum. See *atom, structure of*.

ground waves Direct waves. *Electromagnetic radiations* of *radio frequencies* that travel more or less directly from transmitting *aerial* to receiving aerial, that is without reflection from the *ionosphere*. See *sky waves*.

group 1. The set of *elements* that have similar chemical properties and constitute a vertical column in the *periodic table*. **2.** A number of covalently bonded atoms that form part of a *compound* and have characteristic properties. *Ethanol*, for example, consists of the *ethyl group* (C_2H_5-) and the *hydroxyl group* (–OH).

group speed See *wave motion*.

growth substance See *hormone*.

Grüneisen's law The ratio of the expansivity of a *metal* to its *specific heat capacity* at constant pressure is a constant at all *temperatures*.

guaiacol *o*-methoxyphenol. $CH_3OC_6H_4OH$. A yellowish crystalline substance, m.p. 28.6°C., b.p. 205°C., used in medicine as a local *anaesthetic*.

guanidine $HN:C(NH_2)_2$. A strongly *basic*, *water soluble*, crystalline *organic compound*, used in the manufacture of *plastics*, *explosives*, and rubber *accelerators*.

guanine 2-aminohypoxanthine. $C_5H_5N_5O$. A colourless *insoluble* crystalline substance; a *purine* and one of the four nitrogenous bases occurring in the *nucleotides* of *nucleic acids*, which play a part in the formulation of the *genetic code*. The crystals are used by some animals to reflect light, e.g. in the silvery reflecting scales of some fish.

guano Large deposits formed from the excrement and bodies of seabirds. They are found on islands off the coast of Peru. Very rich in nitrogen and phosphorus *compounds*, they provide a valuable *fertilizer*.

guided missile Rocket propelled missiles whose flight path can be controlled in flight by an external source of radiation (radio, radar, or laser), by a trailing wire, or by an internal homing device. They carry warheads ranging from antitank explosives to nuclear devices. Cruise missiles are driven by jet engines and have wings; they are guided by inertial systems and fly so low that they are difficult to detect by radar. See *beam riding*; *command guidance*; *field guidance*; *homing guidance*; *inertial guidance*; *terrestrial guidance*.

gum arabic Gum acacia. A *water soluble*, yellowish *gum* obtained from certain varieties of acacia. It is used in food and pharmaceutical products and as an *adhesive*.

gums A large class of substances of vegetable origin, most of which are exuded from plants.

gun-cotton See *cellulose nitrate*.

gun-metal A variety of *bronze* containing about 90% copper, 8%–10% tin, and up to 4% zinc.

gunpowder A *mixture* of *potassium nitrate*, KNO_3, powdered *charcoal*, and sulphur. When ignited, a number of *chemical reactions* take place, evolving *gases*, thus producing an *explosion* in a confined space.

GUT See *grand unified theory*.

gutta-percha A material very similar to *rubber*, obtained from the *latex* of certain Malayan trees; chemically, it consists of the *trans-form* of polyisoprene. A horny substance at ordinary temperatures; it is *thermoplastic* and at about 70°C. resembles unvulcanized rubber. It is used for golf ball covers.

gypsum Natural *hydrated calcium sulphate*, $CaSO_4.2H_2O$, that loses three quarters of its *water of crystallization* when heated to 120°C., becoming *plaster of Paris*.

gyration Motion round a fixed *axis* or centre.

gyrocompass Gyroscopic compass. A compass that does not make use of *magnetism* and is therefore not affected by magnetic storms, etc.; it consists of a universally mounted spinning wheel that has a rigidity of direction of axis and plane of rotation relative to space, the rotation being electrically maintained. See *gyroscope*.

gyromagnetic ratio γ. The ratio of the *magnetic moment* of an *atom* or *nucleus* to its *angular momentum*.

gyroscope A spinning wheel mounted in such a way that it is free to rotate about any axis; i.e. 'universally mounted'. Such a wheel has two properties upon which applications of the gyroscope depend: 1. Rigidity in space (gyroscopic inertia); the support of the wheel may be turned in any direction without altering the direction of the wheel relative to space. 2. Precession. When a gyroscope is subjected to a *couple* tending to alter the direction of its axis, the wheel will turn about an axis at right angles to the axis about which the couple was applied.

H

Haber process Haber-Bosch process. An industrial process for the preparation of *ammonia*, especially for use in *fertilizers*, from atmospheric nitrogen (see *fixation of atmospheric nitrogen*). A heated mixture of nitrogen and hydrogen is passed over an iron *catalyst* under pressure (usually 250 atmospheres); the gases combine to form ammonia gas according to the equation $N_2 + 3H_2 = 2NH_3$. It was originally devised by Fritz Haber (1868–1934) and adapted for industrial production by Carl Bosch (1874–1940) in combination with his process for producing the necessary hydrogen (see *Bosch process*). However, most of the hydrogen is now produced by *steam reforming* of *natural gas*. The nitrogen used is obtained by the liquefaction of air.

habit The external form of a *crystal*.

hadron Any *particle* that can take part in a *strong interaction*. Hadrons include the *baryons* and *mesons*. Hadrons are not truly elementary but consist of *quarks* in different arrangements (see *elementary particles*).

haem The *prosthetic group* of *haemoglobin*; it is an iron-containing *porphyrin*.

haematite Natural *iron(III) oxide*, Fe_2O_3. A valuable *ore* of iron.

haematology The study of *blood*, its constituents, and the diseases connected with it.

haemocyte A *blood cell*.

haemoglobin A red *respiratory pigment*, present in *erythrocytes*; it is a conjugated *protein*, consisting of four *polypeptide* chains arranged around four *haem* groups. The haem combines loosely and reversibly with oxygen, enabling haemoglobin to carry oxygen around the body in the bloodstream from the lungs or *gills* to the respiring *tissues*. With *carbon monoxide* it forms the irreversible compound carboxyhaemoglobin; this is why CO is so toxic to red-blooded animals.

hafnium Celtium. Hf. Element. R.a.m. 178.49. At. No. 72. A rare *metal*, r.d. 13.3, m.p. 2227°C., b.p. 4602°C. It is used in the manufacture of tungsten *alloys* for *filaments* and as a *neutron* absorber in *nuclear reactors*.

hahnium A name proposed for the element with *atomic number* 105. See *transactinides*.

hair salt Natural *aluminium sulphate*, $Al_2(SO_4)_3.18H_2O$.

half cell Half of an electrolytic cell, consisting of an electrode dipping into an *electrolyte*. The *electrode potential* of such a system is measured by comparison with a *hydrogen electrode*, which is assigned an electrode potential of zero.

half-life Half-value period. The time taken for the *activity* of a *radioactive nuclide* to *decay* to half of its original value, that is for half of the *atoms* present to disintegrate (see also *disintegration constant*). Half-lives vary from isotope to isotope, some being less than a millionth of a second and some more than a million years. Symbol $T_{1/2}$.

half-period zones The division of a *wave front* into elements of area or zones such that secondary wavelets (see *Huygens' construction*) reaching a given point ahead of the wave from adjacent zones differ in *phase* by half a period, or π. This construction is used in theoretical investigations of *Fresnel diffraction* in simple cases.

half-thickness Half-value layer. The thickness of a specified material that, when introduced into the path of a given beam of *radiation*, reduces its intensity to one half of its original value.

half-wave plate A plate (sometimes called a 'retardation plate') of double refracting material (see *double refraction*) cut parallel to the *optic axis* and of such a thickness that a *phase* difference of π or $180°$ is introduced between the *ordinary ray* and the extraordinary ray for *light* of a particular *wavelength* (usually sodium light). The half-wave plate is chiefly used to alter the plane of vibration of plane-polarized light. See *polarization of light*.

half-wave rectifier A *rectifier* using only one *semiconductor diode*, which therefore only rectifies half the *alternating current* swing, producing a pulsating current. Compare *full-wave rectifier*.

halide A *binary compound* of one of the *halogen elements* (fluorine, chlorine, bromine, or iodine); a *salt* of the *hydride* of one of these elements. Most metals form ionic halides, (e.g. *sodium chloride*, *potassium bromide*) but some form *covalent* compounds (e.g. *aluminium chloride*). Some organic compounds are also called halides, e.g. the *haloalkanes* are sometimes known as *alkylhalides*.

halite See *rock salt*.

Hall effect If an *electric current* flows in a wire placed in a strong transverse *magnetic field*, a *potential difference* is developed across the wire, at right angles to both the magnetic field and the wire. In *metals* and *semiconductors*, if E_H is the strength of the *electric field* produced, $E_H = jB/ne$, where j is the *current density*, B is the *magnetic flux density*, n is the number of charge *carriers* in unit volume, and e is the electronic charge. The constant $1/ne$ is sometimes called the 'Hall coefficient', R_H. In the quantum Hall effect, R_H is quantized and is proportional to h/e^2, where h is the *Planck constant*. Named after Edwin H. Hall (1855–1938).

Halley's comet A bright *comet* that takes about 76 years to *orbit* the *Sun*; it was last seen in 1986. It moves round the Sun in the opposite direction to the planets. Named after Edmund Halley (1656–1742) who first calculated its orbit.

Hall-Héroult cell An *electrolytic cell* for extracting aluminium from *bauxite* ($Al_2O_3.xH_2O$). After purification the oxide is mixed with *cryolite* (Na_3AlF_6) to lower its melting point; this forms the *electrolyte*, which is maintained at about 850°C. by means of the current passed through it between the *graphite* cell lining, which functions as a *cathode*, and the graphite *anodes*. Molten aluminium is tapped off from the base of the cell. Named after Charles M. Hall (1863–1914) and Paul Héroult (1863–1914), who both discovered it independently.

Hall mobility Drift mobility. The mobility of *carriers* in a *semiconductor*; numerically, it is the *velocity* of the carriers under the influence of an *electric field* of 1 *volt* per *metre*.

Hall probe A type of *fluxmeter* based on the *Hall effect*.

halo A luminous ring sometimes observed surrounding the *Sun* or the *Moon*. It is caused by the *refraction* of *light* by *ice crystals* in the *atmosphere*.

haloalkane Alkyl halide. An *alkane* in which one or more hydrogen atoms have been replaced by a *halogen*, e.g. *trichloromethane* (chloroform), $CHCl_3$.

haloform A *haloalkane* containing three *halogen* atoms, e.g. *iodoform*, CHI_3; *bromoform*, $CHBr_3$; or *chloroform*, $CHCl_3$. A haloform reaction is a reaction to produce haloforms from a *ketone*. For example, if *propanone* is treated with *bleaching powder*, the chlorinated ketone so formed reacts to form chloroform:
$$CH_3COCH_3 \rightarrow CH_3COCl_3 \rightarrow CHCl_3.$$

halogenation The introduction of *halogen atoms* into a *compound* by addition or substitution.

halogens The four *elements* fluorine, chlorine, bromine, and iodine, having closely related and graded properties and forming group 7A of the *periodic table*. (Astatine is also a member of the halogen group, but it has no stable *isotopes*.)

halon A group of compounds consisting of *alkanes* in which bromine and other *halogens* have been substituted for hydrogen atoms, e.g. bromochlorodifluoromethane, $BrClF_2C$. These unreactive compounds have been used as *fire extinguishers*, but like the *chlorofluorocarbons*, they cause depletion of the *ozone layer* and their use is deprecated.

haploid Having a set of single (unpaired) *chromosomes*; e.g. *gametes*.

Hamiltonian H. A function used to express the energy of a system in terms of its momentum and positional coordinates. It is used in *wave mechanics*.

hardening of fats The conversion of *liquid fats* (oils) consisting mainly of *triolein* into hard fats by the action of hydrogen in the presence of a nickel *catalyst*. It is used to make margarine from vegetable oils. See *hydrogenation of oils*.

hardness 1. See *hard water*. **2.** See *Brinell test*; *Mohs scale of hardness*; *sclerometer*.

hard radiation See *soft radiation*.

hardware See *software*.

hard water *Water* that does not form an immediate lather with *soap*, owing to the presence of calcium, magnesium, and iron *compounds* dissolved in the water. The addition of soap produces an *insoluble* scum consisting of *salts* of these *metals* with the *fatty acids* of the soap, until no more is left in *solution*. Removal of these salts from solution renders the water soft. Hardness is divided into two types: 1. Temporary hardness, due to *hydrogencarbonates* of the metals. These enter the water by the passage of water, containing dissolved *carbon dioxide*, over *solid carbonates* (*chalk* or *limestone* deposits, etc.). Such hardness is removed by *boiling*, the soluble hydrogencarbonates being decomposed into the insoluble carbonates (see *fur in kettles*), carbon dioxide, and water. 2. Permanent hardness, due to *sulphates* of the metals. This is destroyed by the addition of washing-soda, *sodium carbonate*, which precipitates the insoluble carbonates. All hardness may be destroyed by the use of *zeolites*. This method is used in domestic water softeners. *Sequestration* is also used, in which the calcium *ions* are reacted with sequestering agents to form complexes. This does not remove the calcium ions, but locks them up to prevent them from taking part in other reactions.

harmonic motion See *simple harmonic motion*.

harmonic oscillator See *oscillator*.

harmonic series (math.) A *series* in which the *reciprocals* of the terms are in arithmetical progression. E.g.

$$1 + \tfrac{1}{2} + \tfrac{1}{3} + \tfrac{1}{4}\ldots$$

harmonics of a wave motion Waves superimposed on a fundamental wave, having a *frequency* that is a whole multiple of the fundamental frequency. The second harmonic has a frequency twice that of the fundamental, the third harmonic three times, and so on. See also *overtone*.

hartree Atomic unit of energy. A unit of energy in the Hartree system of units equal to e^2/a_0, where e is the charge on an *electron* and a_0 is the atomic unit of length. It is equal to 4.85×10^{-18} joule or 27.2 eV.

Hartsthorn, spirits of A *solution* of *ammonia* in *water*.

Harvard classification See *spectral types*.

Hawking radiation The emission of particles by a *black hole*. The *gravitational field* of a black hole causes particle-antiparticle pairs to be formed in the vicinity of an event horizon, one member of which is pulled back into the black hole. The other, however, escapes, taking with it some of the mass of the black hole. As this Hawking radiation appears to be emitted like a *black body*, obeying Planck's *law of radiation*, the *radiation temperature* is inversely proportional to the mass of the black hole. For a body of the mass of the *Sun* $(2 \times 10^{30}$ kg), the radiation temperature is only about 10^{-7} K, making the energy loss negligible. However, a mini black hole of 10^{12} kg, as might have been created early in the Universe's history, the radiation temperature would have been some 10^{11} K, equivalent to a substantial energy loss. Named after Stephen Hawking (1942–).

health physics The branch of *physics* that deals with the effects of *ionizing radiation* on living *organisms*, with particular reference to the protection of humans from the ill-effects caused thereby. See *dose*; *nuclear waste*.

heat *Energy* that is transferred from one body or system to another as a result of a *temperature* difference. The *internal energy* of a body is also sometimes rather confusingly called its heat. Heat is measured in *joules*, like all other forms of energy, in *SI units*. In *c.g.s. units* it is measured in *calories* and in *Imperial* units it is measured in *British thermal* units. See *thermodynamics, laws of*.

heat capacity *C*. When the *temperature* of a system is increased by an amount d*T* as a consequence of the addition of a small quantity of *heat* d*Q*, the quantity d*Q*/d*T* is called the heat capacity. In practice, it is the heat in *joules* required to raise the temperature of a body or system by 1 K. See *specific heat capacity* and *molar heat capacity*.

heat death of the Universe The second law of thermodynamics (see *thermodynamics, laws of*) can be interpreted to mean that the *entropy* of a closed system tends towards a maximum and that its available *energy* tends towards a minimum. It has been held that the *Universe* constitutes a thermodynamically closed system, and if this were true it would mean that a time must finally come when the Universe 'unwinds' itself, no energy being available for use. This state is referred to as the 'heat death of the Universe'. It is by no means certain, however, that the Universe can be considered as a closed system in this sense.

heat engine Any device that makes use of *heat* to do *work*. In a *steam engine* the heat is used to raise steam, which either turns a *turbine* (steam turbine) or forces a piston to move up and down in a cylinder. In an *internal-combustion* engine, fuel is burnt inside the engine itself and the resulting heat expands a gas, which either moves a piston or turns a turbine (gas turbine). Engines work on a cycle of operations, the most *efficient* of which is the idealized *Carnot cycle*; practical engines work on less efficient cycles.

heat exchanger Any device that transfers *heat* from one *fluid* to another without allowing the fluids to come into contact with each other. The simplest type consists of a cylinder within which a coiled tube is mounted. One fluid passes through the coiled tube in one direction, while the other fluid passes through the cylinder, outside the tube, in the other direction.

heat of combustion The *heat* evolved by 1 *mole* of a substance when it is burned in oxygen.

heat of formation The *heat* given out or absorbed when 1 *mole* of a *compound* is formed from its *elements* in their normal state. The heat of formation of elements is, for the purpose of thermochemical calculations, taken as zero. See *Hess's law*.

heat of neutralization The *heat* evolved when 1 *mole* of an *acid* or *base* is exactly neutralized. For all strong acids or bases, its value is approximately 57 500 *joules* (13 700 *calories*).

heat of reaction The *heat* given out or absorbed in a *chemical reaction*, usually per *mole* of reacting substances. See *Hess's law*.

heat of solution The *heat* evolved or absorbed when 1 *mole* of a substance is dissolved in a large volume of *water*.

heat pump A machine for extracting *heat* from a *fluid* that is at a slightly higher *temperature* than its surroundings. For example, the rivers flowing through industrial towns are often slightly warmer than the ambient temperature as a result of the disposal of hot effluents in them. A heat pump transfers the heat from the low-temperature source to a high-temperature region by doing work on a working fluid. This fluid is compressed adiabatically in vapour form by means of a pump so that its temperature rises. The vapour is then passed through a radiator, where it gives off heat to its surroundings (a space or water heater, for example) and in doing so condenses to a liquid. The liquid is then expanded into an *evaporator* where it absorbs heat from the surroundings, so becoming a vapour again before re-entering the pump. This cycle is repeated continuously. Basically, the device functions as a *refrigerator* with a different set of parameters. Heat pumps are very versatile and can be used as water or space heaters or as air-conditioning units. Heat pumps can be adapted to be used for space heating in the winter or air conditioning in the summer.

heat radiation *Energy* emitted by a substance as *electromagnetic waves*. The higher the *temperature* of the substance the greater the emission of energy, most of which lies in the *infrared* part of the *electromagnetic spectrum*. See also *black-body radiation*; *Planck's law of radiation*; *Stefan's law*; *Wien displacement*.

heat shield The shielding surface or structure that protects a *spacecraft* from excessive heating on re-entering the *Earth's* atmosphere. (See *re-entry*.)

heat sink A device or system that absorbs *heat* at a constant *temperature*. It is a useful concept in *thermodynamics*.

heat transfer The transfer of *energy* from one body or place to another as a result of a difference in *temperature*. Heat is transferred by *conduction*, *convection*, or *heat radiation*.

heat transfer coefficient The rate of *heat transfer*, i.e. the flow of *heat* through unit area per unit time per unit temperature difference. It is measured in watts per square metre per kelvin. When referring to *conduction* it is called the '*thermal conductance*'.

Heaviside-Kennelly layer A region of the *ionosphere*, between 90 and 150 kilometres above the surface of the *Earth* (for more recent designation of layers see *ionosphere*) that reflects *electromagnetic radiation* of *radio frequencies*. Inter-continental radio transmission, round the curved surface of the Earth, is possible because of the reflection of *sky-waves* by the Heaviside-Kennelly layer. Named after Oliver Heaviside (1850–1925) and Arthur Kennelly (1861–1939).

heavy hydrogen See *deuterium*.

heavy metal A metallic *element* of relatively high *relative atomic mass*, e.g. platinum, gold, lead.

heavy spar See *barytes*.

heavy water Deuterium oxide. *Water* in which the hydrogen is replaced by *deuterium*, either as HDO, which is present in natural water to an extent of 1 part in 6000, or as D_2O, present 1 part in 36×10^6. Pure D_2O has a maximum relative density of 1.106 at 11.185°C., f.p. 3.82°C., and b.p. 101.42°C. Heavy water is used as a *moderator* in some *nuclear reactors*. It is separated from ordinary water by *fractional distillation* or *electrolysis*.

hectare A metric unit of area; 10 000 square *metres*, 2.471 05 *acres*.

hecto- Prefix denoting one hundred times; 10^2. Symbol h, e.g. hm = 10^2 *metres*.

Heisenberg's uncertainty principle See *uncertainty principle*.

heliocentric Having the *Sun* as a centre; measured from the centre of the Sun.

helium He. Element. R.a.m. 4.002 602. At. No. 2. A *noble gas* that occurs in certain *natural gases*, occluded in *radioactive ores* (e.g. *monazite*, *pitchblende*), and in the *atmosphere* (1 part in 200 000). It is not flammable and is very light, and therefore used for filling airships and balloons. Its b.p., –268.93°C., is the lowest of all substances. It is used as a *refrigerant* and to provide an inert atmosphere in welding, etc.

helix A curve that lies on a *cylinder* or *cone* and makes a constant angle with the line segments making up the surface of the cylinder or cone. Many large natural *molecules* (e.g. *proteins* and *nucleic acids*) are helical in shape. Compare *spiral*.

Helmholtz free energy See *free energy*. Named after Hermann von Helmholtz (1821–94).

hemicelluloses *Polysaccharides* (mainly *pentosans*) that occur in cell walls of plants associated with *cellulose* and *lignin*.

hemihydrate A *compound* that has one molecule of *water of crystallization* for every two molecules of the compound. *Plaster of Paris*, $2CaSO_4.H_2O$ or $CaSO_4.\frac{1}{2}H_2O$, is sometimes called hemihydrate plaster.

hemimorphite Natural *zinc silicate*, $2ZnO.SiO_2.H_2O$.

henry The derived *SI unit* of self- and mutual inductance (see *self-induction*; *mutual induction*). An inductance in a closed *circuit* such that a uniform rate of change of current of 1 *ampere* per second produces an induced *EMF* of 1 *volt*. Symbol H. Named after Joseph Henry (1797–1878).

Henry's law The *mass* of a *gas* dissolved by a definite *volume* of *liquid* at constant *temperature* is directly proportional to the *partial pressure* of the gas. The law holds only for sparingly *soluble* gases at low pressures. Named after William Henry (1774–1836).

heparin A complex *organic acid* related to the *polysaccharides* but containing sulphur and nitrogen, which prevents the clotting of *blood* by interfering with the formation, and action, of *thrombin*. It is a product of *mast cells* particularly in connective tissue. It is used as an anticoagulant.

hepta- Prefix meaning seven; e.g. heptahydrate, containing seven molecules of *water of crystallization*.

heptane C_7H_{16}. The seventh member of the *alkane* series. It is found in *petroleum*. B.p. 98.4°C. and r.d. 0.68.

heptavalent Septivalent. Having a *valence* of seven.

herbicides Substances that kill plants or inhibit their growth. Selective herbicides affect only particular plant types, making it possible to attack weeds growing among cultivated plants.

Héroult furnace See *electric-arc furnace*.

Herschelian telescope A form of astronomical reflecting *telescope* in which the primary mirror is *concave* and is set at an angle to the incoming light, enabling the incoming light to be reflected directly into the *eyepiece*.

hertz The derived *SI unit* of *frequency* defined as the frequency of a periodic phenomenon of which the periodic time is one *second*; equal to 1 *cycle* per second. Symbol Hz. 1 kilohertz (kHz) = 10^3 cycles per second; 1 megahertz (MHz) = 10^6 cycles per second. Named after Heinrich Hertz (1857–94).

hertzian waves *Radio* waves. *Electromagnetic radiation* covering a range of *frequency* from above 3×10^{10} *hertz*, corresponding to the shortest *radar* waves of 1 cm, to below 1.5×10^5 *hertz*, corresponding to long radio waves of 2000 metres.

Hertzsprung-Russell diagram H-R diagram. A diagram for correlating data concerning *stars*. It consists of a *graph* in which the absolute *luminosity* of a star is plotted against its spectral type (obtained by examining the *spectra* of stars and arranging them in a sequence that reflects increasing *temperature*). This graph is thus essentially a plot of total *energy* output against surface temperature. The outstanding feature of this type of diagram is that most stars are concentrated in a narrow band running across the diagram: the stars at the upper end of the band are hot, bright, and bluish-white, while those at the lower end are cooler, dimmer, and reddish in colour. This band is called the 'main sequence' and stars that fall on it are called main-sequence stars. It is mainly from H-R diagrams that the theory of *stellar evolution* has been derived. Named after Ejnar Hertzsprung (1873–1969) and Henry N. Russell (1897–1957).

Hess's law If a *chemical reaction* is carried out in stages, the algebraic sum of the *heat* evolved in the separate stages is equal to the total heat evolved when the reaction occurs directly; this is a consequence of the law of *conservation of energy* as applied to *thermochemistry*. As a result of this law, thermodynamic data can often be calculated when it cannot be measured directly. Named after G. H. Hess (1802–50).

hetero- Prefix denoting other, different.

heterocyclic compounds *Organic compounds* containing a ring structure of *atoms*

201

in the *molecule*, the ring including atoms of *elements* other than carbon. E.g. pyridine, C_5H_5N, having a *molecule* consisting of 5 carbon atoms and 1 nitrogen atom in a closed ring, with a hydrogen atom attached to each carbon atom.

heterodyne A beat effect (see *beats*) produced by superimposing two waves of different *frequency*. It is used extensively in *radio* receivers in which the received wave is combined with a wave (of slightly different frequency to the *carrier wave*) generated within the receiver. The two combining waves produce an *intermediate frequency*, which is amplified and then *demodulated*. See *super-heterodyne*.

heterogeneous 1. Not of a uniform composition; showing different properties in different portions. **2.** Consisting of more than one *phase*. Compare *homogeneous*.

heterolytic fission The breaking of a chemical bond so that charged *ions* are formed, e.g. $HCl = H^+ + Cl^-$. Compare *homolytic fission*.

heteropolar bond An electrovalent bond. See *chemical bond*.

heteropoly compound See *cluster compound*.

heuristic Denoting the method of solving mathematical problems for which no *algorithm* exists; it involves the narrowing down of the field of search for a solution by inductive reasoning from past experience of similar problems.

Heusler's alloys *Alloys* containing neither iron, nickel, nor cobalt that exhibit strong *ferromagnetism*. They are composed of copper, manganese, and aluminium. Named after Conrad Heusler.

hexa- Prefix denoting six; six times.

hexachlorocyclohexane See *benzene hexachloride*.

hexadecane Cetane. $C_{16}H_{34}$. A colourless *liquid alkane*, m.p. 18°C., b.p. 287°C. It is used as a *solvent* and in the determination of *cetane numbers*.

hexadecanoic acid See *palmitic acid*.

hexadecanol Cetyl alcohol. $C_{16}H_{33}OH$. A white crystalline *insoluble solid*, m.p. 50°C., used in cosmetics and in pharmaceutical products.

hexamine 1,3,5,7-tetraazaadamantane, urotropine. $(CH_2)_6N_4$. A white crystalline substance obtained by the *condensation* of *ammonia* with *methanal*. It is used in medicine, in the manufacture of vulcanized *rubber*, and in the manufacture of *cyclonite*.

hexane C_6H_{14}. The sixth member of the alkane series. It is found in *petroleum*. B.p. 69°C. and r.d. 0.66.

hexanedioic acid Adipic acid. $COOH(CH_2)_4COOH$. A white crystalline solid, m.p. 152°C., used in the synthesis of *nylon*.

hexanoic acid Caproic acid. $CH_3(CH_2)_4COOH$. A colourless oily liquid with an unpleasant smell, b.p. 205°C. Its *esters*, e.g. ethyl hexanoate (caproate), are used in artificial flavourings.

hexavalent Sexivalent. Having a *valence* of six.

hexogen See *cyclonite*.

hexosans *Polysaccharides* that yield *hexoses* on *hydrolysis*.

hexose A *monosaccharide* whose *molecule* contains six carbon *atoms*, e.g. the *sugars glucose*, *fructose*, and *galactose*.

hexyl group Any of the five *isomeric univalent* groups $C_6H_{13}-$.

hexylresorcinol $C_6H_{13}C_6H_3(OH)_2$. A yellow crystalline *solid*, m.p. 60°C., used as an *antiseptic* and in medicine.

HF See *high frequency*.

hidden matter See *missing mass*.

Higgs particle Higgs boson. A hypothetical *elementary particle* required by the mechanism proposed by Peter Higgs (1929–) and others to explain the appearance of massive particles of zero *spin* in the *electroweak theory*.

high fidelity See *fidelity*.

high frequency HF. *Radio frequencies* between 3000 and 30 000 kilo*hertz*. See Appendix, Table 10.

high-frequency welding Radio-frequency welding. A method of welding *thermoplastic* materials in which the *heat* required to fuse the surfaces together is generated by the application of *radio frequency electromagnetic radiation*.

high-speed steel A very hard *steel* containing 12%–22% tungsten, with chromium, vanadium, molybdenum, and small amounts of other *elements*; it is used for tools that remain hard even at red heat.

high tension High *voltage*.

histamine $C_3H_3N_2.(CH_2)_2NH_2$. A white crystalline substance, m.p. 86°C., that occurs in animal *tissues* when they are injured and as part of the allergic reaction, causing dilation of blood vessels; it also stimulates gastric secretion of *hydrochloric acid*.

histidine A crystalline *soluble amino acid*, which occurs in fish and from which *histamine* is manufactured. See Appendix, Table 5.

histochemistry The study of the chemical composition of tissues.

histogram A type of graphical representation, used in *statistics*, in which frequency distributions are illustrated by rectangles.

histology The study of the microscopic structure of the *tissues* and organs of living creatures.

hodoscope An apparatus for tracing the path of a charged particle (usually a *cosmic-ray* particle).

hole The absence of an *electron* in the *valence* structure (see *energy bands*) of a body. The process of filling these vacancies by electrons, which thereby creates new holes, gives rise to 'hole conduction'. A hole may, therefore, be regarded as a mobile vacancy. See *semiconductor*.

holmium Ho. Element. R.a.m. 164.9304. At. No. 67. A soft silvery metal, r.d. 8.797, m.p. 1470°C., b.p. 2300°C., it has one natural *isotope*, holmium-165. See *lanthanides*.

holo- Prefix denoting whole; e.g. holohedral *crystal*, a crystal having the full number of faces for perfect symmetry.

holocellulose All the *carbohydrate* components of a *cellulose* raw material.

hologram The intermediate photographic record that contains the information for reproducing a three-dimensional image by *holography*.

holography A method of reproducing three-dimensional images without *cameras* or *lenses* using photographic *film* and *coherent light*. A beam of coherent light from a *laser* is split in two by a semi-transparent mirror, so that one beam (the signal beam) can be *diffracted* by the object to be reproduced onto a photographic film or plate. The other beam (the reference beam) falls directly onto the film or plate (see Fig. 18a). The two beams form *interference* patterns on the plate thus forming the *hologram*. The fine speckled pattern on the plate contains information characteristic of the *wave fronts* themselves, rather than of the light intensities as in normal *photography*. To reproduce the image the hologram is illuminated by coherent light (usually of the same *wavelength* as the original beam). The hologram acts as a *diffraction grating* and produces two sets of diffracted waves, which form equal angles with the plate (see Fig. 18b). One set of waves forms a *real image* on a screen or photographic plate, while the other forms a three-dimensional *virtual image*.

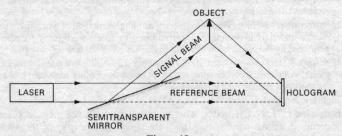

Figure 18a.

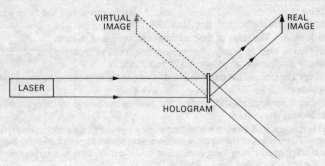

Figure 18b.

homeomorphous Having the same crystalline form but different chemical composition.

homeostasis The ability of an organism to vary certain aspects of its internal environment, such as body temperature, *pH* of its body fluids, etc., so that its physiological functions can continue at an optimum rate even when external conditions are changing.

homing guidance A method of missile or *rocket* guidance in which the missile contains equipment enabling it to detect and steer itself onto its target.

homo- Prefix denoting same; e.g. *homogeneous*.

homocyclic compounds *Organic compounds* the *molecules* of which contain a ring structure of *atoms* of the same kind (usually carbon). E.g. *benzene*, C_6H_6.

homogeneous 1. Of uniform composition throughout. **2.** Consisting of only one *phase*. Compare *heterogeneous*.

homologous chromosomes Two similar *chromosomes* in a *diploid cell* that pair together in *meiosis*; each is a homologue and has come from one parent of the organism. Though similar in appearance, they may contain different forms (*alleles*) of each *gene*.

homologous organs Parts of an *organism* that are similar in basic structure, *topological* relationship, and development (though differing in function and in superficial form) and are therefore considered to be the products of evolutionary inheritance from a common ancestral population. For example, the forelimb of mammals, though modified as grasping arm (human), running leg (dog), wing (bat), or flipper (seal) has the same arrangement of bones and most muscles as the front leg in frogs or lizards or the wing in birds; they are all therefore homologous.

homologous pair In *spectrographic analysis* an homologous pair consists of the particular spectral line (see *line spectrum*) utilized in the determination of the *concentration* of an *element* and an *internal standard line*, such that the ratio of the intensities of the *radiations* producing the lines remains unchanged with variations in the conditions of *excitation*.

homologous series A series of chemical *compounds* of uniform chemical type, showing a regular gradation in physical properties, and capable of being represented by a general *molecular formula*, the *molecule* of each member of the series differing from the preceding one by a definite constant group of *atoms*. E.g. the *alkanes*.

homologues 1. (chem.) Members of the same *homologous series*; e.g. *methane*, CH_4, and *ethane*, C_2H_6. **2.** (bio.) See *homologous chromosomes*; *homologous organs*.

homolytic fission The breaking of a chemical bond so that neutral atoms or *free radicals* are formed. Compare *heterolytic fission*.

homopolar bond A covalent bond. See *chemical bond*.

Hooke's law Within the limit of proportionality (see *elasticity*), a *strain* is proportional to the *stress* producing it. 'Ut tensio, sic vis.' Named after Robert Hooke (1635–1703).

horizontal component B_0. The horizontal component of the Earth's *magnetic field*. See *magnetism, terrestrial*.

hormone A specific *organic compound* produced in one part of the body of an *organism* and transported to other parts where it produces a response. In animals, hormones are produced by *endocrine glands*, secreted into the *blood*, and are thus carried to all parts of the body, where they regulate many functions of *metabolism*; they are quick-acting and only minute amounts can have profound effects, cascade systems of *enzymes* amplifying the effect of relatively few molecules of a hormone. Animal hormones are *proteins* (e.g. *insulin*), steroids (e.g. *cortisone*), or relatively simple compounds (e.g. *adrenaline*). Plant hormones are known as 'growth substances', e.g. *auxins*, *gibberellins*. See also *sex hormone*.

hornblende A rock-forming *mineral* consisting mainly of *silicates* of calcium, magnesium, and iron.

horn silver Cerargyrite, chlorargyrite. Natural *silver chloride*, AgCl. An important *ore* of silver.

horsepower h.p. The British unit of *power*; *work* done at the rate of 550 *foot-pounds* per second. 1 h.p. = 745.7 *watts*.

hot-wire instrument An electrical measuring instrument (*ammeter* or *voltmeter*) that depends upon the expansion, or change in *resistance*, of a wire heated by the passage of an *electric current*. It will measure both a.c. and d.c.

Hubble constant H_0. The ratio of the velocity of recession of the *galaxies* (see *expansion of the Universe*) to the distance away of the galaxies being observed. It is measured in kilometres per second per megaparsec and its value lies between 55 and 100 km s^{-1} Mpc^{-1}. The velocity of recession can be measured accurately by a galaxy's observed *redshift* but distance is more difficult to determine, hence the uncertainty in the value of H_0. $1/H_0$ is known as the Hubble time, and is a measure of the age of the Universe, assuming that its rate of expansion has been constant since the *big-bang*. As it seems likely that the rate of expansion has slowed down as a result of *gravitation*, $1/H_0$ is likely to give an upper limit; taking H_0 as 55 km s^{-1} Mpc^{-1}, gives an age of the Universe of 18×10^9 years. Named after Edwin Hubble (1889–1953).

hue The characteristic of a *colour* that is determined by its *wavelength*.

humidity of the atmosphere A measure of the *water vapour* present in the air. It may be given in terms of *relative humidity* or the *absolute humidity*. The specific humidity is the mass of water in the atmosphere per unit mass of air.

humus Dark brown *colloidal* matter present in *soil* as the result of animal and vegetable *decomposition*. It is rich in *bacteria*, *fungi*, and microarthropods and is an important source of mineral nutrients for plants.

Huygens' construction Each point of a *wave front* may be regarded as a new source of secondary wavelets. Knowing the position of the wave front at any given time, the construction enables its position to be determined at any subsequent time. Named after Christian Huygens (1629–95).

Huygens' principle of superposition The resultant displacement at any point due to the superposition of any system of waves is equal to the sum of the displacements of the individual waves at that point. This principle forms the basis of the theory of *light interference*.

hydracid See *binary acid*.

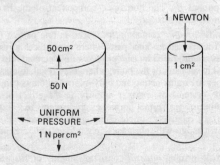

Figure 19.

hydrargyrum See *mercury*.

hydrate A *compound* containing combined *water*. It is generally applied to *salts* containing *water of crystallization*.

hydrated 1. The opposite of *anhydrous*; containing chemically combined *water*. **2.** Denoting a *salt* containing *water of crystallization*.

hydration *Solvation* in which the *solvent* is *water*.

hydraulic cement *Cement* that hardens in contact with *water*.

hydraulic press An application of *Pascal's law*; a device enabling a *force* applied by a piston over a small area to be transmitted through *water* to another piston having a large area; by this means very great forces may be obtained. See Fig. 19.

hydraulics The practical application of *hydrodynamics* and *hydrostatics* to engineering.

hydrazine $H_2N.NH_2$. A fuming strongly *basic liquid*, b.p. 113°C., a powerful *reducing agent*, it is highly reactive being used in organic synthesis and as a *rocket propellant*, either alone or mixed with the dimethyl *derivative*.

hydrazo group The *bivalent* group –HNNH–.

hydrazoic acid See *hydrogen azide*.

hydrazones *Organic compounds* containing the group $H_2NN:C:$. They are produced by reacting substituted *hydrazines* with *aldehydes* and *ketones*.

hydride A *binary compound* with hydrogen.

hydro- Prefix denoting *water*; e.g. *hydrogen*, water producer. In chemical nomenclature, it often denotes a *compound* of hydrogen; e.g. *hydrochloric acid*.

hydrobromic acid HBr. A *solution* of the pale yellow *gas*, hydrogen bromide, in *water*.

hydrobromide A *salt* formed when an *organic base* (e.g. an *alkaloid*) combines with *hydrobromic acid*. The salt so formed is usually more *soluble* than the base.

hydrocarbons *Organic compounds* that contain only carbon and hydrogen. They are classified as either *aliphatic* or *aromatic compounds* (or a combination of both). Hydrocarbons may be either *saturated* or *unsaturated compounds*.

hydrochloric acid Muriatic acid, spirits of salts. A *solution* of the colourless pungent *gas* hydrogen chloride, HCl, in *water*. The concentrated *acid* contains 35%–40% HCl by mass and is a colourless fuming corrosive *liquid*. It is manufactured by the action of *sulphuric acid*, H_2SO_4, on *sodium chloride*, or by the direct chemical combination of hydrogen and chlorine obtained by the *electrolysis* of brine. It is very widely used in chemical industry.

hydrochloride A *salt* formed when an *organic base* (e.g. an *alkaloid*) combines with *hydrochloric acid*. The salt so formed is usually more *soluble* than the base.

hydrocyanic acid Prussic acid, hydrogen cyanide. HCN. A colourless, intensely poisonous *liquid* with a smell of bitter almonds. B.p. 26.5°C. It is used in the manufacture of *acrylic resins*.

hydrodynamics The mathematical study of the *forces*, *energy*, and *pressure* of *liquids* in motion.

hydroelectric power The generation of electrical *power* by means of falling water. Using a natural waterfall, usually in conjunction with a dam across a river to provide a reservoir to enable generation to continue in a dry season, the water is fed to a water *turbine*, the output of which is used to drive an electric *generator*.

The *efficiency* of such a system is as high as 90% at full load. Once the hydroelectric power station is built, there are no fuel costs and the *energy* from this *renewable energy source* can be provided at very low cost. In pumped-storage stations energy is stored for use at peak times by using electricity generated at times of low demand to operate water pumps that raise water to a supplementary elevated reservoir, which can supply additional hydroelectric power at peak demand. Some hydroelectric stations have an output of 10^4 MW and currently hydroelectricity is producing some 6.6% of world energy needs. In the UK, however, due to lack of suitable sites it supplies less than 1% of energy requirements and there is limited scope for increasing this figure.

hydrofluoric acid 1. A *solution* of hydrogen fluoride, HF, in *water*. **2.** The compound HF itself, a colourless corrosive fuming *liquid*, b.p. 19.5°C., that attacks *glass* and is used for etching glass.

hydrogel A *colloidal gel* in which *water* is the dispersion medium.

hydrogen H. Element. R.a.m. 1.00794. At. No. 1. A colourless odourless tasteless *gas*, m.p. −259.14°C., b.p. −252.87°C., that forms *diatomic* molecules (see *orthohydrogen*; *parahydrogen*). It is the lightest substance known, is flammable, and combines with oxygen to form *water*. It occurs as water, H_2O, in *organic compounds*, and in all living things. It is manufactured by *steam reforming* of natural gas, by the *Bosch process*, and by *electrolysis*. It is used in the *oxyhydrogen burner*, as a *reducing agent*, in the manufacture of synthetic *ammonia* (see *fixation of atmospheric nitrogen*) and of synthetic *oil* (see *Fischer-Tropsch process*), and for *hydrogenation of oils*. Three *isotopes* of hydrogen are known; the two 'heavy' isotopes, *deuterium* and *tritium*, are of importance in *nuclear physics*.

hydrogen arsenide See *arsine*.

hydrogenation Subjecting to the chemical action of, or causing to combine with, hydrogen.

hydrogenation of coal The manufacture of artificial mineral *oil* from *coal* by the action of hydrogen. This depends on causing the carbon in coal to combine with hydrogen to form *hydrocarbons*. See *Bergius process*; *Fischer-Tropsch process*.

hydrogenation of oils Artificial hardening of *liquid* animal and *vegetable oils* by the action of hydrogen. Liquid *fats and oils* contain a high percentage of liquid *triolein*, $C_{57}H_{104}O_6$, which may be converted into a solid *tristearin*, $C_{57}H_{110}O_6$, by the action of hydrogen in the presence of a finely divided nickel *catalyst*; the result being a hard fat of higher *melting point*. This is the basis of the process for manufacturing *margarine*.

hydrogen azide Hydrazoic acid, azoimide. HN_3. A colourless poisonous *explosive liquid*, b.p. 37°C., that forms explosive *salts* with heavy *metals*. The salts are called *azides*.

hydrogen bomb See *nuclear weapons*.

hydrogen bond A weak *electrostatic bond* that forms between covalently bonded (see *chemical bond*) hydrogen *atoms* and a strongly *electronegative* atom with a *lone pair of electrons* (e.g. oxygen, nitrogen, fluorine). Hydrogen bonds have about one tenth of the strength of a covalent bond. *Ice crystals* are held together by this type of bond, a *tetrahedral* structure being built up as in Fig. 20 (where the dotted lines represent hydrogen bonds). Because the oxygen atom has two lone pairs of electrons it can make two hydrogen bonds, as shown above. When ice melts this structure breaks down but some hydrogen bonds continue to exist, and liquid *water* consists of groups of water *molecules* held together by hydrogen bonds. The hydrogen bond is of enormous importance in biochemical processes, especially the N–H - - - N bond, which enables complex *proteins* and *nucleic acids* to be built up. Life would be impossible without this type of linkage.

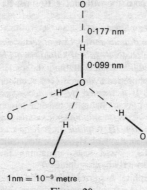

1nm = 10^{-9} metre

Figure 20.

hydrogen bromide See *hydrobromic acid*.

hydrogencarbonate Bicarbonate. A *salt* of *carbonic acid* in which one hydrogen atom has been replaced, i.e. a salt containing the *ion* HCO_3^-.

hydrogen chloride See *hydrochloric acid*.

hydrogen cyanide See *hydrocyanic acid*.

hydrogen electrode A *half cell* used as a standard for measuring *electrode potentials*, for which purpose it is assigned a potential of zero. It consists of a platinum electrode, over which hydrogen is bubbled, immersed in a dilute acid. The effective reaction is:

$$H_2 \rightarrow 2H^+ + 2e$$

This arrangement is designed to produce a standard concentration of hydrogen ions. See also *redox reaction*.

hydrogen fluoride See *hydrofluoric acid*.

hydrogen iodide See *hydroiodic acid*.

hydrogen ion A positively charged hydrogen *atom*; a *proton*. The general properties of *acids* in aqueous *solution* are due to the presence of *hydroxonium ions*, H_3O^+. See also *hydrogen ion concentration*.

hydrogen ion concentration The number of moles of *hydrogen ions* per cubic decimetre of *solution*. It is useful as a measure of the *acidity* of a solution and in this context is usually expressed in terms of pH = $\log_{10}1/[H^+]$, where $[H^+]$ is the hydrogen ion concentration. As pure *water* at ordinary *temperatures* dissociates slightly into hydrogen ions and *hydroxyl ions* ($H_2O = H^+ + OH^-$), the concentration of each type of ion being 10^{-7} *mole* dm^{-3}, the pH of pure water will be $\log_{10}1/10^{-7} = 7$; this figure is accordingly taken to represent neutrality on the pH scale. If *acid* is added to water its hydrogen ion concentration will increase and its pH will therefore decrease. Thus a pH below 7 indicates acidity and similarly a pH in excess of 7 indicates alkalinity.

hydrogen peroxide H_2O_2. A thick syrupy *liquid*, b.p. 150.2°C.; the usual form in which it is sold is a *solution* of the pure compound in *water*. It gives off oxygen readily, and is used as a *disinfectant*, *bleaching* agent, and as the *oxidant* in rockets. Strength of solution is usually given in terms of 'volume strength'; thus, 10 volume hydrogen peroxide will evolve 10 times its own volume of oxygen *gas*.

hydrogen phosphide See *phosphine*.

hydrogen spectrum See *Balmer series*; *Lyman series*; *Paschen series*.

hydrogensulphate Bisulphate. A *salt* of *sulphuric acid* in which one hydrogen atom has been replaced, i.e. a salt containing the *ion* HSO_4^-.

hydrogen sulphide Sulphuretted hydrogen. H_2S. A colourless poisonous *gas* with a smell of bad eggs. It is formed by the *decomposition* of organic matter containing sulphur and occurs naturally in some mineral waters. It is prepared by the action of dilute *acids* on *sulphides* of *metals* (e.g. by the action of dilute *hydrochloric acid* on *iron(II) sulphide* in a *Kipp's apparatus*); it is used in chemical analysis. A solution of H_2S in water, called hydrosulphuric acid, contains the *ion* HS^-. Acid salts containing this ion are known as hydrogensulphides (formerly hydrosulphides).

hydrogensulphite Bisulphite. A *salt* of *sulphurous acid* in which one hydrogen atom has been replaced, i.e. a salt containing the *ion* HSO_3^-.

hydroiodic acid HI. A *solution* of the colourless *gas*, hydrogen iodide, in water.

hydrolases A class of *enzymes* that control *hydrolysis*; e.g. *esterases*, *proteases*.

hydrolith Calcium hydride. CaH_2. A substance that is *decomposed* by *water* and used for the production of hydrogen, according to the equation:
$$CaH_2 + 2H_2O = Ca(OH)_2 + 2H_2.$$

hydrological cycle Water cycle. The cycle of events by which water evaporates from the seas into the atmosphere, where it forms clouds, from which it falls back onto the land as rain, snow, etc. Some of the rain water falling on the land evaporates back into the atmosphere but some is drained back into the seas by rivers and some soaks into the earth to form ground water stores beneath the surface.

hydrology The study of *water* with reference to its occurrence and properties in the *hydrosphere* and *atmosphere*.

hydrolysis The chemical *decomposition* of a substance by *water*, the water itself being also decomposed; the reaction is of the type:
$$AB + H_2O = A(OH) + HB$$
Salts of weak *acids*, weak *bases*, or both, are partially hydrolyzed in *solution*; *esters* may be hydrolyzed to form an *alcohol* and acid. See *saponification*.

hydrometer An instrument for measuring the *density* or *relative density* of *liquids*. The common type consists of a weighted bulb with a graduated slender stem; the apparatus floats vertically in the liquid being tested. In liquids of high density a greater length of stem is exposed than in liquids of low density.

hydronium ion Former name for the *hydroxonium ion*, H_3O^+.

hydrophilic Having an affinity for *water*.

hydrophobic Having no affinity for *water*; water-repellent.

hydroponics Cultivation of plants without the use of *soil*, using instead *solutions* of those mineral *salts* that a plant normally extracts from the soil.

hydroquinone See *benzene-1,4-diol*.

hydrosol A *colloidal solution*, as distinct from a *hydrogel*, *water* being the *solvent*.

hydrosphere The watery portion of the *Earth's* crust, comprising the oceans, seas, and all other waters. About 74% of the Earth's surface consists of water. Composition by *mass* is oxygen 85.8%, hydrogen 10.7%, chlorine 2.1%, sodium 1.1%, magnesium 0.14%, not more than 0.05% of any other *element* being present. The chief constituents are *water*, H_2O (approx. 10^{21} kg), *sodium chloride*, NaCl, and *magnesium chloride*, $MgCl_2$.

hydrostatic equation The relationship between atmosphere pressure, p, and the height, h, at which it is measured, i.e.
$$dp/dh = -g\rho,$$
where g is the acceleration of free fall and ρ is the *density* of the atmosphere. A simple aircraft *altimeter* consists of an *aneroid barometer* calibrated in metres (or feet).

hydrostatics The mathematical study of *forces* and *pressures* in *liquids* at rest.

hydrosulphate A *salt* formed when an *organic base* (e.g. an *alkaloid*) combines with *sulphuric acid*. The salt so formed is usually more *soluble* than the base.

hydrosulphide See *hydrogen sulphide*.

hydrosulphuric acid See *hydrogen sulphide*.

hydrous Containing *water*.

hydroxide A *compound* derived from *water*, H_2O, by the replacement of one of the hydrogen *atoms* in the *molecule* by some other atom or group; either a compound containing the hydroxide ion, OH^-, or a compound containing the *hydroxyl group*, –OH. Hydroxides of metals are *alkalis*.

hydroxonium ion The *ion* H_3O^+ formed in *aqueous solutions* of *acids*. See also *oxonium ion*.

hydroxy acid An *organic acid* containing *hydroxyl* groups in addition to *carboxyl* in its *molecule*; e.g. *lactic acid*, $CH_3CH(OH)COOH$.

3-hydroxy-2-butanone See *acetoin*.

2-hydroxy-1,2-diphenylethanone Benzoin. $C_6H_5CHOH.CO.C_6H_5$. An *optically active* crystalline substance, m.p. 133–7°C., used in organic synthesis.

hydroxyl group The *univalent* –OH group. It is present in covalently bonded form in *alcohols*.

2-hydroxypropanoic acid See *lactic acid*.

hygro- Prefix denoting moisture, humidity. E.g. *hygrometer*.

hygrodeik A *wet and dry bulb hygrometer* with a chart attached, which enables the *relative humidity* to be obtained directly from the readings of the two *thermometers*.

hygrometer Any instrument designed to measure the *relative humidity* of the *atmosphere*. The simplest type is the hair hygrometer, which uses the expansion and contraction of a human hair to operate a needle. In electric hygrometers, the resistance of a *hygroscopic* substance is used as a measure of humidity. See also *dew-point hygrometer, wet and dry bulb hygrometer*.

hygroscope An instrument for showing variations of *relative humidity* of the air.

hygroscopic Having a tendency to absorb moisture.

hyoscine Scopolamine. $C_{17}H_{21}NO_4$. A colourless crystalline *alkaloid*, m.p. 82°C., used in the form of its *hydrobromide* as a *sedative* and *narcotic*.

hyoscyamine $C_{17}H_{23}NO_3$. A poisonous crystalline *alkaloid*, m.p. 106°C., obtained from henbane, and used in the form of its *hydrobromide* or *hydrosulphate* as a *sedative* and antispasmodic.

hypabyssal rock See *igneous rock*.

hyper- Prefix denoting over, above, beyond.

hyperbola A curve traced out by a point that moves so that its distance from a fixed point, the *focus*, always bears a constant ratio greater than unity to its distance from a fixed straight line, the *directrix*. The curve has two branches and is formed by a *plane* cutting a right circular *cone* when the angle the plane makes with the base is greater than the angle formed by the cone's side (see *conic sections*).

hyperbolic functions Six mathematical functions analogous to the *trigonometrical ratios*. The hyperbolic functions are sinh, cosh, tanh, cosech, sech, and coth. Sinh x is defined as

$$\tfrac{1}{2}\,(e^x - e^{-x})$$

and cosh x as

$$\tfrac{1}{2}\,(e^x + e^{-x}).$$

(See *exponential*.) The remaining functions are derived from sinh and cosh, on the same basis as the related trigonometrical ratios.

hypercharge A quantized property of certain *elementary particles*; it is equal to the particle's *baryon* number added to its *strangeness*. This property is not conserved in *weak interactions* but it is in *strong* and electromagnetic *interactions*.

hyperfine structure of spectrum lines The very *fine structure* of certain *spectrum* lines observed when they are examined under very high resolution. The lines are caused either (a) by the presence of different *isotopes* of the *element* emitting the spectrum, or (b) if the atomic *nuclei* of the element possess a *spin*, and therefore a resultant *magnetic moment*.

hypergolic Denoting constituents of *rocket fuels* that ignite spontaneously upon contact with some other specific constituent.

hypermetropia Long sight. A defect of vision in which the subject is unable to see near objects distinctly. It is corrected by the use of *convex* spectacle *lenses*.

Hyperol* Trade name of a crystalline *compound* of *urea* and *hydrogen peroxide*; $CO(NH_2)_2.H_2O_2$. It evolves *hydrogen peroxide* by the action of *water*.

hyperons A group of *elementary particles*, belonging to the class called *baryons*, which have greater *mass* than the *neutron* but very short lives. All baryons that are not *nucleons* are known as hyperons, but as all hyperons *decay* into nucleons they can be regarded as *excited* nucleons. For each hyperon there is a corresponding *antiparticle*.

hypersonic Having a speed in excess of Mach 5. See *Mach number*.

hypertonic A *solution* is said to be hypertonic with respect to another if it has a greater *osmotic pressure*.

hyphae The fine threads forming the network (mycelium) constituting the main body of a *fungus*; their cell walls consist of *chitin* or *cellulose*; they absorb nutrients from their surroundings.

hypnotic (chem.) A substance producing sleep. A *sedative*.

hypo- Prefix denoting under, below.

hypo (phot.) See *sodium thiosulphate*, $Na_2S_2O_3.5H_2O$. It was formerly incorrectly called 'sodium hyposulphite'. See *fixing*.

hypochlorite See *chlorate*.

hypochlorous acid See *chloric acids*.

hypocycloid The figure traced by a point on the circumference of a *circle* that rolls, without slipping, round the inside of a larger fixed circle.

hypophosphorous acid Phosphinic acid, HPH_2O_2. A colourless *deliquescent* crystalline substance, m.p. 26.5°C. It decomposes on heating into *phosphoric(V) acid* and *phosphine*. It is used as a *reducing agent*.

hypotenuse The side opposite the right angle (i.e. the longest side) in a right-angled *triangle*.

hypothesis A supposition put forward in explanation of observed facts.

hypotonic A *solution* is said to be hypotonic with respect to another if it has a smaller *osmotic pressure*.

hypsometer 'Height-measurer'. An apparatus for the determination of the *boiling point* of a *liquid*. Since the boiling points of liquids depend upon the *pressure*, and the atmospheric pressure varies with the altitude, the apparatus may be used for the determination of altitude above sea-level.

hysteresis A physical phenomenon chiefly met in the elastic and magnetic behaviour of materials. When a body is stressed, the *strain* produced is a function of the *stress*. On releasing the stress, the strain lags behind; i.e. the strain for a given value of stress is greater when the stress is decreasing than when it is increasing. On removing the stress completely, a residual strain remains. This lagging of effect behind cause is called *hysteresis*. It also occurs in induced *magnetism*. See *hysteresis cycle*.

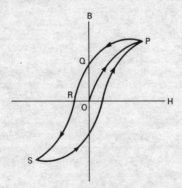

Figure 21.

hysteresis cycle A cycle of magnetizing field variations to which an initially demagnetized *ferromagnetic* substance is subjected. The magnetizing field is periodically reversed in direction until a steady state is reached in which the *magnetic flux density* in the specimen at any instant is a *function* only of the magnitude of the magnetizing field and the sign of its rate of change at that instant. When this condition has been reached, a plot of magnetic flux density (B) against magnetizing field (H) gives a 'hysteresis loop' or curve. See Fig. 21. When H is reduced to zero along PQ, there is a residual magnetic flux density in the substance, OQ, which is called the *remanence*. Reversing the *polarity* of H along QRS reduces the value of B to zero at R. The strength of field, OR, required to reduce B to zero is called the *coercive force*. Rapidly reversing the field of an *electromagnet* causes *energy* to be lost by heating up the core. This 'hysteresis loss' is proportional to the area of the hysteresis loop, which should therefore be as small as possible.

I

IAA See *indole-3-acetic acid.*

IAT See *International Atomic Time.*

iatrochemistry Medieval medical *chemistry*; early attempts at the application of drugs to medicine.

ice *Water*, H_2O, in the *solid* state, formed at the *freezing point* of water, 0°C. As it is less dense than water, water expands (See *water, expansion of*) on freezing, and ice floats on water. See *hydrogen bond.*

ice age Any of a number of periods in the Earth's history when the polar icecaps spread towards the equator and there was a general lowering of *temperature*. The last of these occurred some 10 000 years ago. The cause of these ice ages is unknown.

ice point The *temperature* of equilibrium between *ice* and *water* under normal atmospheric pressure (see *atmosphere*); i.e. the *melting point* of ice. The ice point is assigned the value of 0°C., in the *Celsius temperature* scale.

iconoscope A form of camera tube (see *camera, television*) in which an *electron* beam scans a *mosaic* of small globules of *photoemissive* material. When light from the image falls on the mosaic, the illuminated globules emit electrons leaving a pattern of positively charged globules in the shape of the image. The scanning electron beam discharges the positively charged globule *capacitors* as it passes, onto a backing *electrode*. This creates an electrical signal containing information regarding the image.

icosahedron A *polyhedron* with twenty faces.

-ide A chemical suffix denoting a *binary compound* of the two named *elements* or *radicals*; e.g. *hydrogen sulphide*, a compound of hydrogen and sulphur only.

ideal crystal A *crystal* whose *lattice* is perfectly regular and contains no foreign *atoms* or *ions* or other *defects* or imperfections.

ideal gas See *perfect gas.*

ideal solution A *dilute solution* that obeys *Raoult's law* exactly.

identity (math.) A statement of equality between known or unknown quantities that holds true for all values of the unknown quantities. E.g. $3x \equiv 2x + x$, irrespective of what value is assigned to x.

IGFET See *field-effect transistor.*

igneous rock A rock formed from molten silicates (*magma*). The types of igneous rock depend on the depths at which they solidified. Plutonic rocks, the coarsest igneous rocks, formed at the greatest depths; the medium-grained hypabyssal rocks formed relatively close to the Earth's surface; and volcanic rocks formed from magma poured onto the Earth's surface from *volcanoes*. Compare *metamorphic rocks*; *sedimentary rocks.*

ignis fatuus Will-o'-wisp. A pale *flame* sometimes seen over marshy ground, probably caused by the *combustion* of *methane*, CH_4, or other inflammable *gases* ignited by some chance event.

ignition The action of setting fire to something. Initiating *combustion* by raising the *temperature* of the reactants to the *ignition temperature*; particularly the process or means of firing the explosive mixture in an *internal-combustion engine* by an *electric spark*.

ignition temperature Ignition point. **1.** The *temperature* to which a substance must be heated before *combustion* can take place in air. **2.** The temperature to which a *plasma* has to be raised to enable *nuclear fusion* to occur.

illuminance E_v. The *luminous flux* falling on a surface per second. The derived *SI unit* of illuminance is the *lux* (*lumen* per square *metre*).

ilmenite Natural iron(II) titanate, $FeTiO_3$. An *ore* of *titanium*.

image converter A device for converting an image formed by non-visible radiation (such as *infrared* or *ultraviolet radiation*) into a visible image. It usually consists of a *photocathode* onto which the non-visible radiation is focused and a *fluorescent* screen, which is activated by the *electrons* emitted by the photocathode. This device is used in *fluoroscopes*, *infrared telescopes*, *ultraviolet microscopes*, etc.

image, real (phys.) An image formed by a *mirror* or *lens* at a point through which the *rays* of *light* entering the observer's eye actually pass. Such an image can be obtained on a screen. See Fig. 25 under *lens*.

image, virtual (phys.) An image seen at a point from which the *rays* of *light* appear to come to the observer, but do not actually do so; e.g. the image seen in a plane *mirror* or through a *diverging lens*. Such an image cannot be obtained on a screen placed at its apparent position, since the rays of light do not pass through that point. See Fig. 25 under *lens*.

imaginary numbers Numbers with negative *squares*; thus $\sqrt{-1}$ is an imaginary number, denoted by i; $i^2 = -1$. See also *complex number*.

imbibition The absorption of water by an insoluble substance, such as *cellulose*, *starch*, some *proteins*, and other biological materials. It causes the substance to swell.

imidazole Iminazole, glyoxaline. $C_3H_4N_2$. A colourless *soluble heterocyclic* crystalline substance, m.p. 90°C., used in organic synthesis.

imide Imido compound. A *compound* containing the imido group, –CO.NH.CO–.

imine Imino compound. A *compound*, derived from *ammonia*, containing the imino group, NH=, in which the two hydrogen *atoms* of ammonia are replaced by non-acidic *organic* groups.

immersion objective Oil-immersion lens. A type of *objective* used in high-power *microscopes*, the lowest *lens* of the objective lens system being immersed in a drop of cedar-wood oil placed upon the slide to be examined. Such an arrangement causes more *light* to enter the system than if the oil were absent.

immiscible Incapable of being mixed to form a *homogeneous* substance; it is usually applied to *liquids*, e.g. *oil* and *water* are immiscible.

immune response See *antibody*.

immunization The stimulation of an immune response (see *antibody*) in a person by giving them a *vaccine*, either orally or by injection, to enable them to produce the appropriate *antibodies*. This is known as active immunization. In passive immunization the actual antibodies themselves are injected.

immunoglobulins See *globulins*.

impact The collision of bodies. See *conservation of momentum*.

216

impedance Z. The quantity that determines the *amplitude* of the current for a given *voltage* in an *alternating current* circuit. It is determined by the *resistance* R and the *reactance* X, according to the relationship:
$$Z^2 = R^2 + X^2.$$
For a circuit containing *self-inductance* L and *capacitance* C connected in series, the impedance of the circuit is given by the expression
$$Z = [R^2 + (L\omega - 1/C\omega)^2]^{1/2},$$
where ω is the *angular frequency* of the alternating current. $\omega = 2\pi f$, where f is the *frequency* of the current.

Imperial units A British system of units based on the *pound*, *yard*, and *gallon*. It is being replaced by *metric units* for many general purposes and has been replaced by *SI units* for scientific purposes.

impermeable Not permitting the passage of *fluids*.

impfing See *seeding*.

implantation Nidation. The process in which a fertilized *ovum* of a mammal embeds itself in the wall of a female's uterus (womb).

implicit function A variable quantity, x, is said to be an implicit function of y, when x and y are connected by a relation that is not explicit. See *explicit function*.

implosion The inward collapse of an evacuated vessel.

improper fraction See *proper fraction*.

impulse (bio.) Nerve impulse. A momentary change in the concentrations of *ions* in a fibre of a *neuron*, due to a transient change in the *permeability* of the cell *membrane*; the impulse sets up ionic *currents* and voltage changes, resulting in its propagation along the nerve fibre. The propagation of impulses along nerve fibres and their transfer and interaction at *synapses* result in the movement of muscles and the action of other effector organs (such as glands) and thus result in the behaviour of animals.

impulse (phys.) A *force* acting for a very short time; it is given (for a constant force) by the product of the magnitude of the force and the time for which it acts; it is equal to the total change of *momentum* produced by it, i.e. impulse
$$J = m(v_1 - v_0),$$
where m is the constant *mass* during a change of *velocity* from v_0 to v_1. If the force is not constant,
$$J = \int_{t_1}^{t_0} F \, dt.$$

incandescence The emission of *light* caused by high *temperatures*; white or bright-red heat.

incidence, angle of The angle between a *ray* of *light* meeting a surface and the *normal* to the surface at that point. See Fig. 35 under *refraction, angle of*.

inclination See *magnetic dip*.

inclinometer 1. See *dip circle*. **2.** An instrument for measuring the angle of inclination that an aircraft makes with the horizontal.

incubator A box designed to maintain a constant internal *temperature* by the use of a *thermostat*; it is used for such purposes as rearing chickens and prematurely born infants and growing microorganisms in *culture*.

indefinite integral See *integration*.

indene $C_6H_4.C_3H_4$. A colourless *liquid aromatic hydrocarbon*, b.p. 182.2°C., obtained from *coal-tar* and used in organic synthesis.

indeterminancy principle See *uncertainty principle*.

index (math.) The *exponent* of a quantity raised to a *power*; the number indicating the power to which the quantity is raised. E.g. the index of a in $4a^5$ is 5.

indicator (chem.) A substance that, by a sharp *colour* change, indicates the completion of a *chemical reaction*. It is frequently used in *volumetric analysis*. Indicators for *titrations* of *acids* and *alkalis* are usually weak *organic acids* or *bases*, yielding *ions* of a different colour from the un-ionized *molecules* (see *ionization*). E.g. *phenophthalein* is a weak acid which dissociates slightly in solution; because in an acid solution its molecules are largely un-ionized it remains colourless, whereas in an alkaline solution it dissociates into ions, the negative ion having a strong red colour. See *end point*.

indigo $C_{16}H_{10}N_2O_2$. An important blue *vat dye*, formerly extracted from plants of the genus Indigofera, in which it occurs as indican, a *glucoside*. It is now manufactured artificially on a large scale.

indium In. Element. R.a.m. 114.82. At. No. 49. A soft silvery-white *metal*, r.d. 7.31, m.p. 156.4°C., b.p. 2080°C. *Compounds* are rare but some, such as InSb and InAs, are used in *semiconductors*. The metal itself is used in some special electroplates and dental *alloys*.

indole C_8H_7N. A yellow *soluble* substance, m.p. 52.5°C., that occurs in oil of jasmin and is a *decomposition* product of *proteins* in animal intestines. Despite its unpleasant smell it is used in perfumes.

indole-3-acetic acid IAA, indole-3-ethanoic acid $C_{10}H_9NO_2$. A white crystalline substance, m.p. 168–170°C., that promotes plant growth. See *auxins*.

induced current See *induction, electromagnetic*.

induced radioactivity Artificial radioactivity. *Radioactivity* induced by bombarding substances with *neutrons* or other high *energy* particles (or *photons*).

inductance 1. The property of an electric *circuit* as a result of which an *electromotive force* is generated by a change in the *current* flowing through the circuit (see *self-induction*), or by a change in the current of a neighbouring circuit with which it is magnetically linked (see *mutual induction*). The derived *SI unit* of inductance is the *henry*. **2.** A device or circuit having this property.

induction, charging by A process of electrically charging an insulated *conductor*, using the *force* due to another nearby charge to separate the positive and negative charges existing on the conductor.

induction, electromagnetic When the *magnetic flux* through a circuit changes, an *electromotive force* is induced in the circuit. This phenomenon is called electromagnetic induction. The induced EMF is equal to the rate of change of magnetic flux through the circuit. These two statements are known as Faraday's laws of electromagnetic induction. If the circuit is closed, this EMF gives rise to an induced current, and the phenomenon forms the basis of the *dynamo*, *transformer*, etc. The induced current is in such a direction that its *magnetic field* tends to neutralize the change in magnetic flux producing it (*Lenz's Law*).

induction, magnetic See *magnetic induction*.

induction coil An instrument for producing a high *electromotive force* from a supply of low EMF. Essentially it consists of a cylindrical soft-iron core, usually laminated to prevent losses due to *eddy currents*, round which are wound two coils, the primary and the secondary. The primary coil consists of a few hundred turns; rapid variation of an *electric current* in this coil, produced by a repeated interruption or break in the circuit by a mechanism similar to that in the electric *bell*, produces an induced EMF (see *induction, electromagnetic*) in the secondary coil, which contains a very large number of turns of thin wire. Induction coils are widely used in *internal-combustion engines* to provide the electric sparks.

induction heating A form of heating in which electrically conducting material is heated as a result of the *eddy currents* induced in it by an alternating *magnetic field*.

induction motor A type of *electric motor* in which an *alternating current* supply fed to the primary winding of the *stator* sets up a flux causing *electrical currents* to be induced in the secondary winding of the *rotor*. The interaction between these currents and the flux causes the rotor to rotate.

inductometer A calibrated variable *inductance*.

inelastic collision A collision between bodies in which there is a loss of total *kinetic energy* and a corresponding rise in the *internal energy* of one or both of the colliding bodies. Referring to *nuclear physics*, an inelastic collision is one in which an incoming particle causes *excitation* or breaking up of the struck *nucleus*.

inelastic cross-section See *cross-section*.

inelastic scattering See *scattering*.

inert Not easily changed by *chemical reaction*.

inert gases See *noble gases*.

inertia (phys.) The tendency of a body to preserve its state of rest or uniform motion in a straight line. Its *mass* is a measure of its inertia. See *inertial mass*; *Mach principle*.

inertial guidance A method of automatic control used in *guided missiles* that depends upon *inertia*. The *velocities* or distances covered, computed from the *acceleration* measured within the missile, are compared with data stored before launching.

inertial mass The *mass* of a body as determined by its *momentum* (in accordance with the law of *conservation of momentum*), as opposed to 'gravitational mass', which is determined by the extent to which it responds to the *force* of *gravity*. The *acceleration* of a falling body increases in proportion to its gravitational mass and decreases in proportion to its inertial mass. Since all falling bodies have the same constant acceleration it follows that the two types of mass must be equal.

inertial system Inertial frame. A *frame of reference* in which bodies are not accelerated, i.e. remain at rest or move with constant *velocity*, unless acted upon by external *forces*. *Newtonian mechanics* is valid in such a system.

infinitesimal A quantity smaller than any assignable quantity; the concept is obtained by imagining a quantity decreasing indefinitely without actually becoming zero.

infinity ∞. A quantity that is greater than any assignable quantity.

inflationary cosmology A variation on the *big-bang theory* of the origin of the Universe, in which in its early stages the Universe is thought to have passed through a stage of rapid expansion (inflation) as the result of a change of *phase*.

inflection A point on a *curve* at which the *tangent* to the curve changes its direction of rotation. In the curve $y = f(x)$, a point of inflection has the characteristic that $d^2y/dx^2 = 0$.

information theory A branch of *cybernetics* that attempts to define the amount of information required to control a process of given complexity. See *bit*, *noise*, *redundancy*, *equivocation*, *channel capacity*.

infrared astronomy The use of *infrared telescopes* to study the emissions of *infrared stars* and other sources of *infrared radiation* from space.

infrared radiation IR. *Electromagnetic radiation* possessing *wavelengths* between those of visible *light* and *radio* waves, i.e. from approximately 0.8μm to 1 mm. IR is produced by the natural vibrations of atoms and molecules and the rotation of some *gas* molecules. Infrared radiation has the power of penetrating *fog* or haze, which would scatter ordinary visible light; thus photographs taken on a plate made sensitive to infrared radiation may often disclose detail invisible on an ordinary plate or to the naked eye.

infrared spectroscopy A form of chemical analysis based on the *infrared* emissions produced as a result of molecular vibrations. Organic *functional groups* have characteristic absorption *frequencies* in the infrared region ($2.5 - 16 \mu$m), which enables them to be identified by this technique. An infrared spectrometer consists of an infrared source split into two beams of equal intensity. One is a reference beam and the other is passed through the sample (which can be solid, liquid, or gas). The spectrograph shows absorption peaks at particular wavelengths, enabling many molecular structures to be identified.

infrared stars Celestial bodies whose principal emission is *infrared radiation*. They are believed to consist of *stars* surrounded by dust clouds. In some cases the *light* from the central star penetrates the dust so that it can be seen with *optical telescopes*.

infrared telescope A form of reflecting *telescope* used to study *infrared radiation* from space (refractors cannot be used as glass is opaque to most infrared radiation). As the eye is not sensitive to infrared and photographic film is only sensitive to short infrared wavelengths (up to 1.2 μm), other forms of detector have to be used. These are devices based on the *photovoltaic* and *photoconductive effects*; *bolometers* are also used. To reduce *noise* from the instrument, the detectors need to be cooled with liquid helium. As water vapour in the Earth's atmosphere, and other greenhouse gases (see *greenhouse effect*) can absorb infrared radiation, *infrared astronomy* is best carried out at high altitudes or from spacecraft.

infrasonic Having a *frequency* below the frequency of audible *sound waves*, i.e. a frequency of less than about 20 *hertz*.

infrasound Vibrations or pressure waves with a *frequency* below that of *sound*, i.e. below about 20 *hertz*.

infusible Difficult to melt; having a very high *melting point*.

infusorial earth See *kieselguhr*.

inhibitor A substance that reduces the rate of a catalysed *reaction*. In biochemical reactions catalyzed by an *enzyme*, the inhibitor may bind onto the enzyme so blocking the *substrate*.

injection moulding A process by which *thermoplastic* articles are moulded. The thermoplastic material is softened in a heated chamber and then injected under pressure through an orifice into a cool closed mould.

inorganic (chem.) Of *mineral* origin; not belonging to the large class of *organic* carbon *compounds*.

inorganic chemistry The study of the *elements* and their *compounds*. Inorganic chemistry usually includes the study of elemental carbon, its *oxides*, *metal carbonates*, and *sulphides*, while all other carbon compounds belong to the study of *organic chemistry*.

inositol Hexahydroxycyclohexane. $C_6H_6(OH)_6$. An *optically active* white crystalline solid, m.p. $228-248°C$.; it is sometimes regarded as a member of the *vitamin B* complex, although some animals can synthesize it.

insecticide A substance used for killing insect pests.

insolation 1. Exposure to the *rays* of the *Sun*. 2. The *electromagnetic energy* from the Sun per unit area of the Earth's surface on which it falls. It is measured in *joules* per square *metre*.

insoluble Not capable of forming a *solution* (in *water*, unless some other *solvent* is specified). It is a relative term, since most substances dissolve in water to some extent.

instantaneous frequency The rate of change of *phase* of an oscillation, expressed in *radians* per second divided by 2π.

insulation The prevention of the passage of an *electric* current, or *heat*, by *conduction*.

insulator A non-conductor of *electricity* or *heat*.

insulin A *hormone* produced in the pancreas that controls *sugar metabolism* in the body. When injected, it reduces the *blood* sugar level and so relieves the symptoms of diabetes mellitus. Insulin was the first *protein* to have its *amino acid* sequence and structure worked out in detail.

integer A whole number.

integral 1. Consisting of whole numbers or *integers*. 2. A mathematical *function* obtained by the process of *integration*. See Appendix, Table 9.

integral calculus The branch of the *calculus* making use of the processes of *integration*. It is used for calculating *areas* and *volumes* and for other problems concerned with summation of infinitesimally small elements.

integrand A mathematical *expression* that is to be subjected to *integration*.

integrated circuit Microcircuit. A *microelectronic* circuit incorporated into a chip of *semiconductor*, usually crystalline silicon (a silicon chip). Integrated circuits consist of whole systems rather than single components, and are used in *computers*, calculators, and watches. They are also used in other industries (e.g. cars, radios, etc.) in which small reliable electronic control circuits are required.

integration A mathematical process used in the *calculus*; the inverse process of *differentiation*. It gives a method of finding the area enclosed by curves, and of finding solutions to other problems involving the summation of *infinitesimals*. The integration of a function of a variable x, is written
$$\int f(x)dx + C.$$
This is an indefinite integral and C is the constant of integration. If the interval over which the integration is to take place is specified, the integral becomes a definite integral, written
$$\int_a^b f(x)dx,$$
i.e. the function has to be integrated between $x = a$ and $x = b$. See Appendix, Table 9.

intensifier (phot.) A substance used to increase the *density* or contrast of an image on a photographic *film* or plate. It is usually a *compound* from which a *metal* (e.g. silver, lead, uranium, etc.) can be deposited.

intensity 1. See *electric field.* **2.** See *magnetic field strength.* **3.** See *luminous intensity.* **4.** See *sound intensity.*

inter- Prefix denoting between, among.

interaction Mutual action between bodies, particles or systems. In *nuclear physics* the word is often used to mean the *force* between interacting particles. See *fundamental interactions.*

intercalation compound A compound in which *atoms*, *ions*, or *molecules* are trapped between the layers of a *crystal lattice.* The mineral muscovite (see *mica*) is an example.

interface The surface that separates two chemical *phases*, or more generally a boundary between two parts of any system.

interference of wave motions (phys.) The addition or combination of waves; if the crest of one wave meets the trough of another of equal *amplitude* travelling in the same direction, the wave is destroyed at that point; conversely, the superposition of one crest upon another leads to an increased effect (see also *Huygens' principle of superposition*). The *colour* effects of thin films are due to interference of *light* waves; *beats* produced by two notes of similar *frequency* are the result of the interference of *sound* waves. Interference provides evidence for the *wave theory of light.*

interferometer Any instrument that divides a *beam* of *light* into a number of beams and re-unites them to produce *interference.* Uses include the accurate determination of *wavelengths* of light, the testing of *prisms* and *lenses*, the examination of the *hyperfine structure of spectrum lines*, measurement of the diameters of *stars* and the determination of the number of light waves of a certain wavelength in the standard *metre.* See also *radio interferometer.*

interferon A number of *proteins* produced in many animals *cells* as the result of the presence of *viruses* (either active or inactive) in the cell; it acts as a form of protection against these viruses.

intergalactic space The *space* between *galaxies*, in which intergalactic *matter* may occur.

intermediate (chem.) A compound used in an intermediate step in the manufacture of a final product by chemical *synthesis.*

intermediate frequency In *superheterodyne radio* receivers, the *carrier wave frequency* of the incoming radio wave is changed to a fixed intermediate frequency by *heterodyne* action, for ease of amplification before detection.

intermediate neutrons *Neutrons* with *kinetic energies* between those of *epithermal* and *fast neutrons*, i.e. between 100 eV and 0.1 MeV.

intermediate vector boson W or Z *particle.* The *virtual particles* that are exchanged in *weak interactions.*

intermetallic compound A *compound* in which two or more *metals* are held together by metallic bonds. They occur in some *alloys.*

internal-combustion engine A *heat engine* in which *energy* supplied by a burning *fuel* is directly transformed into *mechanical energy* by the controlled *combustion* of the fuel in an enclosed cylinder behind a piston. The *Otto engine* and the

Diesel engine are the common types of internal-combustion engine. See also *gas turbine*; *Wankel rotary engine*.

internal conversion A process in which an *excited nucleus* decays to the *ground state*, the energy released being used to eject a *conversion electron* from an inner shell of the atom. The excited *ion* so formed may emit a *photon* (*X-ray*) or an Auger electron (see *Auger effect*).

internal energy Thermodynamic energy. U. The sum of the *kinetic energies* of random motions and the *potential energies* of interactions of the constituent particles within a system, which cannot itself usually be determined. However the change in the internal energy of a system, ΔU, is a useful *thermodynamic* quantity, and is defined by $\Delta U = Q - W$ where Q is the *heat* abstracted by the system from its surroundings and W is the *work* done simultaneously on the surroundings.

internal resistance The resistance inside an *electric cell* or other source of *current*. It is equal to $(E - V)/I$, where E is the EMF of the cell, V the *potential difference* between its *terminals*, and I the current supplied.

internal standard line In *spectrographic analysis* an internal standard line is a line within the *line spectrum* of the material being analysed, due to a known amount of an *element* present in, or added to, the material. See also *homologous pair*.

internal stress The *stress* within a solid material, e.g. *metal*, *glass*, etc., as a result of heat treatment, cold working, or non-uniform molecular structure.

International Atomic Time IAT. The internationally agreed scale of *time*, the fundamental unit of which is the *second*, as defined in *SI units*. It uses *atomic clocks* at the Bureau Internationale de l'Heure in Paris as a basis for civil timekeeping.

international candle The former unit of *luminous intensity*. A point source emitting *light* uniformly in all directions at one-tenth of the rate of the Harcourt pentane lamp burning under specified conditions. It has now been replaced by the *candela*.

international date line An imaginary line on the surface of the *Earth* joining the North and South poles, approximately following the 180° *meridian* through the Pacific Ocean. This line is used to mark the internationally agreed start of a calendar day. Crossing from east to west a traveller changes the day to the next day, and crossing it from west to east goes one day back.

International Practical Temperature Scale This temperature scale, first devised in 1968, superseded all previous practical scales. It consisted of a practical scale of *temperature* defined so that it conformed as closely as possible to the *thermodynamic temperature*. The unit of temperature is the *kelvin* (symbol K). The eleven fixed points on the 1968 scale have now been replaced by the 16 fixed points of the most recent version of the scale introduced in 1990 and known as IPTS-90 (see table). Between these points *interpolation* is made with a defining formula using a platinum *resistance thermometer* or a *thermocouple*. Above the *freezing point* of silver a radiation *pyrometer* is based on *Planck's law of radiation*.

interplanetary space The *space* between the *planets* within the *solar system*.

interpolation The process of filling in intermediate values or terms of a series between known values or terms.

IPTS-90

	T/K
Triple point of hydrogen	13.8033
B.p. of hydrogen (33 321.3 Pa)	17.035
B.p. of hydrogen (101 292 Pa)	20.27
Triple point of neon	24.5561
Triple point of oxygen	54.3584
Triple point of argon	83.8058
Triple point of mercury	234.3156
Triple point of water	273.16
M.p. of gallium	302.9146
F.p. of indium	429.7485
F.p. of tin	505.078
F.p. of zinc	692.677
F.p. of aluminium	933.473
F.p. of silver	1234.93
F.p. of gold	1337.33
F.p. of copper	1357.77

interrupted continuous waves ICW. A *continuous wave electromagnetic radiation* switched on and off at an *audiofrequency*.

interstellar matter Clouds of hydrogen *atoms* or *molecules*, mixed with a small proportion of dust, that exist between *stars*. The *density* of these clouds is very low, ranging between some 10^6 and 10^9 atoms m^{-3} (compared to about 10^{25} molecules m^{-3} for a *perfect gas* at *s.t.p.*).

interstellar space The *space* between *stars* within a *galaxy*, in which *interstellar matter* may occur.

interstitial An additional *atom* or *ion* situated between the normal sites in a *crystal lattice*, causing a *defect*.

interstitial compound See *compound, interstitial*.

intra- Prefix denoting within; e.g. intra-molecular *forces* are forces within the *molecule*, while *inter*-molecular forces are forces between molecules.

intrinsic semiconductor See *extrinsic semiconductor*.

inulin $(C_6H_{10}O_5)_6.H_2O$. A *soluble polysaccharide* consisting of *fructose* units; it occurs in many plant roots or tubers as a stored food.

Invar* An *alloy* containing 63.8% iron, 36% nickel, 0.2% carbon that has a very low coefficient of *expansion*. It is used for balance wheels of watches and in other accurate instruments, which would otherwise be affected by *temperature* changes.

inverse Compton effect See *Compton effect*.

inverse square law The intensity of an effect at a point B due to a source at A varies inversely as the *square* of the distance AB. Examples include the *illumination* of a surface, *gravitational field*, field due to an *electric charge*, etc. Thus, the illumination of a surface 1 *metre* away from a source will be 9 times as great as that of a surface 3 metres away.

inverse trigonometrical functions If $y = \sin x$ (see *trigonometrical ratios*), then the inverse trigonometrical function of x is $\sin^{-1}y$ (or arc sin y), where $\sin^{-1}y$ is the *angle* whose sine is y. Similar inverse functions exist for the other trigonometrical and *hyperbolic* ratios.

inverse variation One quantity is said to vary inversely as another, or to be inversely proportional to another, if the *product* of the two is a constant.

inversion layer See *field-effect transistor*.

inversion of cane sugar The conversion of *cane sugar* (*sucrose*, $C_{12}H_{22}O_{11}$) into a mixture of equal amounts of *glucose* and *fructose*, two isomeric sugars (see *isomerism*) having the formula $C_6H_{12}O_6$. The action is one of *hydrolysis* and may be carried out by the action of the *enzyme invertase*, or by boiling with dilute *acids*. The resulting mixture is *laevorotatory*, while a solution of cane sugar is *dextrorotatory*, inversion of the *optical rotation* being thus obtained.

inversion temperature See *Joule-Thomson effect*.

invertase Sucrase. An *enzyme* contained in *yeast* that converts *cane sugar* into *glucose* and *fructose*. See *inversion of cane sugar*.

inverter A device for converting *direct current* into *alternating current*.

invert sugar A *mixture* of *glucose* and *fructose* in equal proportions, obtained by the *inversion of cane sugar*.

in vitro Denoting experiments involving biological or biochemical processes that are carried out in 'glass' (i.e. after the *cells* or *tissues* in which the processes occur have been removed from the *organism* to which they belong) rather than in the living organism, when they are said to take place 'in vivo'.

in vivo See *in vitro*.

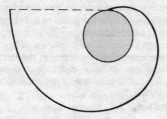

INVOLUTE OF A CIRCLE

Figure 22.

involute The curve formed when a piece of string is unwound from, or wound on to, another curve (the *evolute*). See Fig. 22.

iodate A *salt* of *iodic acid*.

iodic(V) acid HIO_3. A colourless or yellow *soluble* powder formed by the *oxidation* of iodine with *nitric acid*, or *hydrogen peroxide*. It is a powerful *oxidizing agent*.

iodic(VII) acid Periodic acid. H_5IO_6. A white hygroscopic solid made from the electrolytic *oxidation* of *iodic(V) acid*.

iodide A *binary compound* with iodine; *salt* of *hydroiodic acid*, HI.

iodine I. Element. R.a.m. 126.9045. At. No. 53. A blackish-grey crystalline *solid*. R.d. 4.953, m.p. 113.6°C., b.p. 184°C. It is very *volatile*, giving off a violet vapour; iodine is slightly soluble in water but readily soluble in *alcohol* (giving 'tincture' of iodine') and in *potassium iodide* solution, KI. *Compounds* occur in seaweed; sodium iodate, $NaIO_3$, occurs in crude *Chile saltpetre*. It is essential to the functioning of the thyroid gland; lack of iodine in the diet is a cause of goitre. It is used in medicine, chemical *analysis*, and *photography*. The *radioisotope* iodine-131 (*half-life* 8.6 days) is used in the treatment and diagnosis of disorders of the thyroid gland.

iodine number Iodine value. Hübl number. A measure of the degree of unsaturation (content of double *bonds*) of a product, such as an *oil* or fat; it is expressed in grams of iodine absorbed by 100 g of the given substance under standard conditions.

iodoform Tri-iodomethane. CHI_3. A yellow crystalline *solid* with a peculiar odour, m.p. 119°C. It is used as an *antiseptic*.

ion An electrically charged *atom* or group of atoms. Positively charged ions (*cations*) have fewer *electrons* than is necessary for the atom or group to be electrically neutral; negative ions (*anions*) have more. Thus, the *proton*, the hydrogen atom without its circumnuclear electron, is a *hydrogen ion*; the *alpha-particle* is a helium ion. Gaseous ions can be produced in *gases* by *electric sparks*, the passage of energetic charged particles, *X-rays*, *gamma-rays*, *ultra-violet radiation*, etc. (see *ionizing radiation*). Ions in *solution* are due to the *ionization* of the dissolved substance.

ion engine A type of *reaction propulsion* engine for propelling *rockets* in *space*, the exhaust jet of which consists of a stream of positive *ions* accelerated to a high *velocity* by an *electric field*. A beam of *electrons* must also be ejected to enable recombination to occur and to avoid the engine becoming charged.

ion exchange Certain substances have the power of acting on *solutions* containing *ions*, such as solution of *salts*, and replacing some of the ions by others; e.g. in a typical *cation* exchange ('base exchange') action, when *hard water* is passed through a suitable ion exchange resin or a *zeolite*, the calcium ions in the water are replaced by sodium ions. In *anion* exchange *acid radicals* or anions are exchanged similarly. Ion exchange has many important industrial uses including water softening and *desalination*.

ionic bond Electrovalent bond. See *chemical bond*.

ionic crystal Electrovalent *crystal*. A crystal in which the component *ions* are held in their positions in the *lattice* by ionic bonds (see *chemical bond*). *Sodium chloride* is a typical example.

ionic strength I. A measure of the *charge* of the *ions* in a *solution* of an *electrolyte*. It is defined as the sum of the *molality* of each ion multiplied by the square of its *charge*.

ionization The formation of *ions*. See *ionizing radiation*; *photoionization*; *thermal ionization*.

ionization chamber A device for measuring the amount of *ionizing radiation*. It consists of a gas-filled chamber containing two *electrodes* (one of which may be the chamber wall) between which a *potential difference* is maintained. The radiation ionizes gas in the chamber and an instrument connected to one electrode measures the *ionization current* produced as the ions are driven to the appropriate electrode by the *electric field*.

ionization current The *electric current* produced by the movement of *ions* or *electrons* in an *electric field* as a result of *ionizing radiation*.

ionization gauge A type of *pressure gauge* for monitoring low gas *pressures* (of the order 10^{-6} Pa). It consists essentially of a *triode thermionic valve* in which some *electrons* from the *cathode* pass through the grid causing *ionization* of any gas molecules with which they collide. The positive ions so formed are attracted to the negatively charged *anode*; the current produced is a measure of the number of molecules present and therefore of the pressure.

ionization potential Ionization energy. *I*. The *work* that must be done, measured in *electronvolts*, to remove an *electron* from an *atom*. (See *atom, structure of.*) It was formerly defined as the potential through which an electron would need to fall to cause an atom to become ionized, hence the misleading name. More work is required to remove the second electron from an atom and each subsequent electron requires additional work. Table 7 in the Appendix gives the first five ionization potentials for the commonest atoms.

ionizing radiation *Radiation* (either *electromagnetic* or corpuscular) that is capable of causing *ionization*, either directly or indirectly. *Electrons* and *alpha particles* are considerably more effective in this respect than *neutrons* or *gamma-rays*. Ionizing radiation can cause damage to biological tissue as a result of the reactive transient species produced when water molecules are exposed to it (e.g. H_2O^+, H_3O^+, H, OH, etc.). See *dose*.

ion microprobe A very sensitive method of analysing the surface of a solid, enabling a few parts per million of a substance to be detected and its chemical and isotopic composition determined. The surface is bombarded by a narrow beam of *ions* and the ions *sputtered* from the surface are detected by a *mass spectrometer*.

ion mobility The *velocity* of an *ion* in a unit *electric field*.

Ionol* BHT, butylated hydroxytoluene. 2,6-Di-*tert*-butyl-4-methylphenol. A white crystalline substance, m.p. 70°C., used as an antioxidant.

ionomer resins *Synthetic resins* cross-linked (see *cross-linkage*) through ionized *carboxyl groups* in their *macromolecules*. Although they have the usual properties of cross-linked *polymers*, they can be processed like *thermoplastic resins*.

ionone $C_{13}H_{20}O$. A yellow *optically active soluble liquid ketone*, b.p. 140–146°C., used in perfumes.

ionosphere The region of the Earth's *upper atmosphere* in which *free electrons* rising from *ionization* occur, mainly as a result of *ultraviolet radiation* and *X-rays* from the *Sun*. The ionosphere is useful in that it enables intercontinental *radio* transmission round the curved surface of the Earth to be achieved, as a result of its property of reflecting *electromagnetic radiations* of *radio frequencies* (see *sky wave*); but it is an obstacle to *radio astronomy* because it reflects a large proportion of the radiation that arrives from extra-terrestrial sources. The ionosphere is usually divided into three regions: the D-region between 50 and 90 kilometres above the Earth, the E-region (the *Heaviside-Kennelly layer*) between 90 and 150 km, and the F-region (the *Appleton layer*) above 150 km. See Fig. 44 under *upper atmosphere*. At night the electron concentration in the E-region falls off due to recombination with *ions*, but the F-region remains substantially ionized owing to the lower *density* of ions and their consequent infrequency of collisions with electrons. With the advent of artificial Earth *satellites* it is now possible to study the electron density of the different regions of the ionosphere from the top side.

ionospheric wave See *sky wave*.

ion pump A *vacuum pump* in which gas is removed from a system by ionizing the atoms or molecules and adsorbing the resulting *ions* on a surface, usually of a metal. This device will enable a vacuum as low as 10^{-9} pascal to be obtained. However, the metal surface soon becomes saturated, which limits its effectiveness. In the sputter-ion pump, fresh metal surface is continuously created by *sputtering*.

IPTS-90 See *International Practical Temperature Scale*.

IR See *infrared radiation*.

iridium Ir. Element. R.a.m. 192.22. At. No. 77. A rare *metal* resembling, and occurring together with, platinum, r.d. 22.42, m.p. 2410°C., b.p. 4130°C. It is extremely hard and resistant to chemical action. *Alloys* of platinum and iridium are used for fountain-pen nib-tips, *crucibles* for fine analytical work, and numerous other purposes where extreme hardness and a high *melting point* are required.

iris 1. The coloured part of the eye of vertebrates. **2.** A diaphragm forming an adjustable opening over a *lens* in an optical instrument. It consists of a number of overlapping crescent-shaped discs which when moved can produce a central aperture of varying diameter.

iron Fe. (Ferrum.) Element. R.a.m. 55.847. At. No. 26. A white magnetic *metal*, r.d. 7.86, m.p. 1535°C., b.p. 2750°C. Physical properties are greatly modified by the presence of small amounts of other metals and of carbon. It occurs as *magnetite*, Fe_3O_4; *haematite*, Fe_2O_3; *siderite*, $FeCO_3$; *limonite, hydrated* Fe_2O_3; and as *pyrites* in combination with sulphur. It exists in three crystalline forms (see *alpha-iron*; *gamma-iron*; *delta-iron*). Iron is extracted from its ores by a *blast furnace*. According to the method and conditions of working and cooling, the carbon in iron and *steel* may be present in various forms, upon which the particular properties of the metal depend. *Compounds* of iron are essential to the higher forms of life. See *pig iron*; *cast iron*; *wrought iron*.

iron alum See *ferric alum*.

iron(II) chloride Ferrous chloride. $FeCl_2$. A greenish *deliquescent* solid, m.p. 670°C., which forms the hydrates $FeCl_2.2H_2O$ and $FeCl_2.4H_2O$.

iron(III) chloride Ferric chloride. $FeCl_3.6H_2O$. A brown-yellow deliquescent crystalline salt, m.p. 306°C. It is used as a *mordant* and in medicine.

iron(II) oxide Ferrous oxide. FeO. A black solid, m.p. 369°C., that is readily *soluble* in dilute acids. It is a non-*stoichiometric* compound with a simple cubic structure that is deficient in iron ions. A mixture of iron(II) and iron(III) oxides forms the ore *magnetite*, Fe_3O_4.

iron(III) oxide Ferric oxide. Fe_2O_3. A red-brown insoluble substance that occurs naturally as *haematite*, m.p. 1565°C. It is used as a *mordant* and a *pigment*.

iron pyrites See *pyrites*.

iron(II) sulphate Ferrous sulphate, green vitriol, copperas. $FeSO_4.7H_2O$. A pale green *soluble salt*, m.p. 64°C., made by dissolving scrap iron in dilute *sulphuric acid*. It is used in dyeing and *tanning*.

iron(III) sulphate Ferric sulphate. $Fe_2(SO_4)_3$. A yellow *hygroscopic* compound obtained by heating acidified *iron(II) sulphate* with *hydrogen peroxide*.

irradiance E. The *radiant flux* striking unit area of surface. Measured in W m^{-2}, it refers to all forms of *electromagnetic* radiation, whereas *illuminance* refers only to *light*.

irradiation Exposure to *radiation* of any kind. Artificial *radioisotopes* are made by irradiation of stable isotopes with *neutrons* in a *nuclear reactor*. Intense irradiation can alter the physical and chemical properties of *solids*, but even small doses may be used for sterilization of food owing to the sensitivity of biological *cells* to irradiation by *ionizing radiation*.

irrational number A number that cannot be expressed exactly as the ratio of two *integers*. It may be a *surd* (e.g. $\sqrt{2}$) or a *transcendental* (e.g. e). Compare *rational number*.

irreversible process Any process, except one that is a completely *reversible process*.

irreversible reaction A *chemical reaction* that proceeds to completion; the resulting products do not react to form the original substances. See *chemical equilibrium*.

irritability The property of living *organisms* that enables them to respond to external stimuli.

isatin $C_8H_5NO_2$. An orange *soluble* crystalline substance, m.p. 203–5°C., used in the manufacture of *dyes*.

isenthalpic Of equal *enthalpy*.

isentropic Of equal *entropy*.

isinglass A product containing about 90% *gelatin*, made from the swimming bladders of fish. It is used for clarifying alcoholic beverages.

iso- 1. Prefix denoting equal. **2.** (chem.) Prefix denoting an *isomer* with a branched chain.

isobar 1. A line on a map or chart connecting points having equal (atmospheric) *pressure*. **2.** One of two or more *nuclides* of different elements that have different *atomic numbers* but identical *mass numbers*. E.g. the tin isotope, $^{115}_{50}Sn$, and the indium isotope, $^{115}_{49}In$, are isobars, 115 being the mass number and 50 and 49 the atomic numbers. Isobars have the same number of *nucleons*, but different numbers of *protons* in their *nuclei*.

isobaric spin See *isotopic spin*.

isobaric surface A surface of equal (atmospheric) *pressure*. An altimeter will record constant height when moving along such a surface. The intersection of an isobaric surface with the ground is along an *isobar*.

isochore A line that graphically represents the relationship between the *pressure* and the *temperature* of a *liquid* or *gas*, the *volume* of the system being kept constant.

isochromatic film See *orthochromatic film*.

isocline A line on a map or chart connecting points of equal angle of *magnetic dip*.

isocyanate A *salt* or *ester* of *isocyanic acid*; a *compound* containing the –NCO group.

isocyanic acid HN=C=O. An unstable *isomer* of *cyanic acid*, which forms stable *salts* called *isocyanates*.

isocyanide See *isonitrile*.

isodiapheres *Nuclides* in which the difference between the number of *neutrons* and *protons* is the same, e.g. a nuclide and its *decay* product after it has emitted an *alpha-particle* are isodiapheres.

isodimorphism The phenomenon of a *dimorphous* substance being isomorphous (see *isomorphism*) with another dimorphous substance in both its forms.

229

isodynamic line A line on a map or chart passing through points of equal *horizontal intensity* of the Earth's magnetic field (see *magnetism, terrestrial*).

isoelectric point The pH value (see *hydrogen ion concentration*) at which a substance or system (e.g. a *protein solution*) is electrically *neutral*; at this value *electrophoresis* does not occur when a direct *electric current* is applied.

isoelectronic Denoting compounds that have equal numbers of *valence electrons*.

isogonal line A line on a map or chart passing through points of equal *magnetic declination*.

isogonism (chem.) A type of *isomorphism* in which two substances having little or no chemical resemblance have the same crystalline form.

isokom A line on a *phase diagram* joining points of equal *viscosity*.

isoleucine A colourless crystalline *amino acid*. See Appendix, Table 5.

isomegethic solutions *Solutions* formed of *solute molecules* of the same size.

isomeric Exhibiting *isomerism*.

isomerism 1. The existence of two or more chemical *compounds* with the same *molecular formula* but having different properties owing to a different arrangement of *atoms* within the *molecule*. In structural isomerism, the molecular structure is different, e.g. ammonium cyanate, NH_4CNO, and *urea* $CO(NH_2)_2$ are structural isomers because although the molecules contain the same atoms, each has a different structure. See also *stereoisomerism*; *cis-trans isomerism*; *optical isomerism*; *tautomerism*. **2.** In *nuclear physics*, *nuclei* having the same *atomic number* and the same *mass number*, but which exist in different *energy* states, are said to be isomeric. E.g. a nucleus in its *ground state* and a nucleus in a *metastable* excited state are isomers.

isomers See *isomerism*.

isometric 1. Referring to a system of *crystallization* in which the *axes* are at right angles to each other. **2.** A method of projecting a drawing (isometric projection) in which the three axes are equally inclined to the surface of the drawing, and all lines are drawn to scale. **3.** A line on a *graph* (isometric line) showing change of *temperature* with *pressure*, when the *volume* is kept constant.

isomorphism Similarity or identity of crystalline form, usually indicating similar or analogous chemical composition; e.g. the *alums* are isomorphous.

isonitrile Isocyanide, carbylamine. A compound containing the group –NC.

isooctane 2,2,4-trimethylpentane. $(CH_3)_3CCH_2CH(CH_3)_2$. The *isomer* of *octane* used in defining *octane numbers*.

isophthalic acid Benzene-1,2-dicarboxylic acid. $C_6H_4(COOH)_2$. The *meta-isomer* of *phthalic acid*, m.p. 345 – 7°C., used in the manufacture of synthetic *resins* and *plasticizers*.

isopoly compound See *cluster compound*.

isoprene 2-methylbuta-1,3-diene. $CH_2:CH.C(CH_3):CH_2$. A colourless *liquid*, b.p. 34°C. Natural *rubber* consists mainly of a polymer of isoprene and it is used in making synthetic rubbers. See *polymerization*; *terpenes*.

isosceles triangle A *triangle* having two of its sides equal.

isospin See *isotopic spin*.

isosterism The phenomenon of substances having *molecules* with the same number of *atoms* and the same total number of *electrons*; this leads to similarity in physical properties. E.g. *carbon dioxide*, CO_2, and *nitrous oxide*, N_2O.

isotactic polymer See *atactic polymer*.

isotherm 1. Isothermal line. A line on a graph, map, or chart connecting points at an equal *temperature*. **2.** A relationship (graphical or mathematical) between two *variables* at constant temperature.

isothermal change Isothermal process. A change or process (e.g. an isothermal reaction) that takes place at constant *temperature*. E.g. the isothermal *expansion of a gas*. Compare *adiabatic*.

isotones *Nuclides* containing the same number of *neutrons* but a different number of *protons*.

isotonic solutions *Solutions* having the same *osmotic pressure*, being of the same *molar concentration*.

isotopes Types of *atoms* of the same *element* (i.e. having the same *atomic number*) that differ in *mass number*. The isotopes of an element are identical in chemical properties, and in all physical properties except those determined by the *mass* of the atom. The different isotopes of an element contain different numbers of *neutrons* in their *nuclei*. Nearly all elements found in nature are mixtures of several isotopes. See *atom*, *structure of*.

isotopes, separation of As the *isotopes* of an *element* have identical chemical properties but some slightly different physical properties, their separation depends upon physical operations. The following methods are used: *diffusion* (either *gaseous* or *thermal*); *distillation*; centrifuging of *gases* or *liquids*; *electrolysis* (depending upon different rates of discharge or *ionic mobility* of isotopic *ions*); electromagnetic or electrostatic methods (depending upon different mass-to-charge ratios between isotopic ions and their consequent separation in a steady *magnetic field* or an *electric field* varied at *radio frequencies*). *Lasers* have also been used to excite one isotope, which can then be separated electromagnetically.

isotopic mass The *relative atomic mass* of an individual *nuclide*. Isotopic masses are very nearly *integral* (whole numbers), the *integer* being called the *mass number* of the isotope concerned.

isotopic number Neutron excess. The difference between the number of *neutrons* in a *nuclide* and the number of *protons*.

isotopic spin Isospin. Isobaric spin. A *quantum number*, I, used to work out the properties of groups of *hadrons* when the members of the group are identical in all respects except that of *electric charge*. E.g. the *nucleon* has isotopic spin, $I = \frac{1}{2}$, and its two states, the *proton* and the *neutron* are then described as different orientations of that spin in a fictitious 'isotopic space'. The word 'spin' is not intended to imply any conventional image of rotation in this context, it is used in analogy to *angular momentum* to which the concept of isotopic spin bears a close formal resemblance. Isotopic spin is conserved in all *strong nuclear interactions*.

isotropic Exhibiting uniform properties throughout, in all directions. Compare *anisotropic*.

-ite A suffix formerly denoting a *salt* of the corresponding -ous *acid*; e.g. *sulphite* from sulphurous acid.

ivory black A form of carbon obtained from *animal charcoal*, by dissolving out *inorganic compounds*, such as *calcium phosphate*, by means of *hydrochloric acid*.

J

jade A hard semiprecious stone consisting of either the rare jadeite, $NaAlSi_2O_6$, usually green but occasionally brown or orange, or the more common nephrite, $Ca_2(Mg,Fe)_5Si_8O_{22}(OH,F)_2$, which is also usually green, but occasionally other colours. It is used for jewellery and carved ornaments.

jadeite See *jade*.

Jansky noise High-frequency radio disturbance of cosmic origin. Named after K. F. Jansky (1905–50).

jasper A coloured impure form of natural *silica*, SiO_2.

javelle water Eau de Javelle. A *solution* containing potassium hypochlorite, KOCl; made by the action of chlorine on a cold solution of *potassium hydroxide*, KOH. It is used for *bleaching* and as a *disinfectant*.

jet A very hard lustrous form of natural carbon, allied to *coal*, that is used in jewellery.

JET Joint European Torus. See *thermonuclear reaction*.

jet engine A *gas turbine* that produces a stream of hot *gas* enabling an aircraft to be propelled through the air by *reaction propulsion*. Air taken in at the front of the engine is compressed by a radial compressor. The compressed air then enters the combustion chambers providing the *oxidant* for the combustion of the *liquid fuel*. The *energy* released expands the gas and accelerates it rearwards, some of the energy of the gas being used to drive a *turbine*, which in turn operates the compressor. After leaving the turbine the gas passes to the rear jet nozzle producing forward *thrust* by reaction on the structure of the jet tube.

jet propulsion See *reaction propulsion*.

Josephson effects Effects that occur when two superconductors (see *superconductivity*) are separated by a thin insulated layer, such as an *oxide* film 10^{-8} m thick. At normal *temperatures* a small current can flow between the *conductors* as a result of the *tunnel effect*. At superconducting temperatures, the insulating barrier can have zero *resistance*, but if the current exceeds a critical value this superconductivity disappears. Moreover, if a *magnetic field* is applied to the junction when the current is less than the critical value, the current will depend on the strength of the field; as the field increases, the current increases to a maximum, decreases to zero, and then increases again, and so on. Above a critical value for the field this phenomenon ceases. If a *potential difference* is applied across the junction, a high-frequency *alternating current* flows through it, the frequency of which is related to the potential difference.

   The Josephson junction has proved useful as a research tool and also as a *logic* component and special purpose high-speed switch. Named after B. D. Josephson (born 1940).

Joint European Torus JET. See *thermonuclear reaction*.

joule The derived *SI unit* of *work* or *energy*. The work done when the point of application of a *force* of one *newton* is displaced through a distance of 1 *metre* in the direction of the force. The joule is also the work done per second by a *current* of 1 *ampere* flowing through a *resistance* of 1 *ohm*. Symbol J. 1 joule = 10^7 *ergs*. Named after James Prescott Joule (1818–89).

Joule heating See *electric current, heating effect of; Joule's laws* (2).

Joule's equivalent See *mechanical equivalent of heat*.

Joule's laws 1. The *internal energy* of a *gas* at constant *temperature* is independent of its *volume*. Joule's law is obeyed strictly only by a *perfect gas*, real gases show deviations from it as a result of intermolecular forces, which would cause a change in the internal energy when a change of volume occurred. **2.** The *heat* produced (sometimes known as Joule heating) by an *electric current I*, passing through a *conductor* of *resistance R*, for a time *t*, is equal to I^2Rt. If *I* is in *amperes*, *R* in *ohms*, and *t* in *seconds*, the heat produced will be in *joules*.

Joule-Thomson effect Joule-Kelvin effect. When a *gas* expands through a porous plug, a change of *temperature* occurs, proportional to the *pressure* difference across the plug. The temperature change is due partly to a departure of the gas from *Joule's law*, the gas performing internal work in overcoming the mutual attractions of its *molecules* and thus cooling itself; and partly to deviation of the gas from *Boyle's law*. The latter effect can give rise to either cooling or heating, depending upon the initial temperature and pressure difference used. For a given mean pressure, the temperature at which the two effects balance, resulting in no alteration of temperature, is called the 'inversion temperature'. Gases expanding through a porous plug below their inversion temperature are cooled, otherwise they are heated. Named after J. P. Joule and Sir William Thomson (Lord Kelvin) (1824–1907).

J particle See *psi particle*.

JUGFET See *field-effect transistor*.

junction detector A detector of *ionizing radiation* making use of a *semiconductor junction*. It produces a current *pulse*, which is proportional to the energy falling on the *depletion region* of the junction, when it is reverse biased. They usually consist of gold-silicon devices and are used in medicine and space research.

junction rectifier A *rectifier* based upon a *semiconductor junction*.

junction transistor A *transistor* having a *base electrode* and two or more electrodes connected to *semiconductor junctions*.

Jupiter (astr.) A *planet*, having sixteen small *satellites*, with its *orbit* between those of *Mars* and *Saturn*. It is the largest of the planets, diameter 142 700 kilometres. Mean distance from the *Sun* 778.34 million kilometres. *Sidereal period* ('year') = 11.86 years. *Mass* approximately 317.89 times that of the *Earth*. The largest satellite, Io, has a diameter of 3242 km. It also has a planetary ring of rocks.

juvenile hormone A hormone secreted by *endocrine glands* in insects that inhibits their *metamorphosis* and keeps them in the larval stage.

K

kainite A double *salt* of *magnesium sulphate* and *potassium chloride*, $MgSO_4.KCl.3H_2O$, that occurs naturally in Poland and in the *Stassfurt deposits*. It is a valuable source of potassium salts.

kalium See *potassium*.

kaolin See *china clay*.

kaon A K-*meson*. See *elementary particles*; *meson*.

karyo- A prefix denoting the *nucleus* of a *cell* or its contents; e.g. 'karyotype', the sum of the morphological characteristics of the *chromosomes* of a cell.

katharometer A device for comparing the *thermal conductivities* of two gases. It can be used to detect the presence of impurities in air and as a detector in *gas chromatography*.

K-capture The process in which an atomic *nucleus* absorbs an *electron* from the inner shell (the K shell) of its orbital electrons, thus transforming into another *nuclide*, e.g. ${}^{7}_{4}Be \rightarrow {}^{7}_{3}Li$. During K-capture the atom is raised to an excited state, and usually decays by emitting an *X-ray photon*.

keepers of magnets Short bars of soft iron used to prevent permanent *magnets* from losing their *magnetism*.

Kekulé forumula The graphic representation of *benzene* first suggested by F. A. Kekulé von Stradonitz (1829–96). See *benzene ring*.

kelp Sea-weed or its *ash*, used as a source of iodine.

kelvin The *SI unit* of *thermodynamic temperature* defined as the fraction 1/273.16 of thermodynamic temperature of the *triple point* of *water*, i.e. the triple point of water contains exactly 273.16 kelvins. The units of kelvin and *Celsius* (*centigrade*) *temperature* interval are identical. A temperature expressed in degrees *celsius* is equal to the temperature in kelvins less 273.15°C. This is true both for thermodynamic temperatures and on the *International Practical Temperature Scale*. Symbol K. Named after Lord Kelvin (1824–1907).

Kelvin effect See *Thomson effect*.

Kelvin temperature *Temperature* expressed in *kelvins*. The same as the thermodynamic temperature. Symbol K.

Kepler's laws 1. The *planets* move about the *Sun* in *ellipses*, at one focus of which the Sun is situated. 2. The *radius vector* joining each planet with the Sun describes equal areas in equal times. 3. The ratio of the *square* of the planet's year to the cube of the planet's mean distance from the Sun is the same for all planets. Named after Johannes Kepler (1571–1630).

keratin A tough sulphur-containing *protein* forming the principal constituent of wool, hair, horns, hoofs, and the cells in the epidermis of skin.

kerma Kinetic energy released in matter. *K*. The sum of the initial *kinetic energies* of all the charged particles produced by the indirect effect of *ionizing radiation* in a small volume divided by the mass of the substance in that volume. It is measured in *grays*.

kernite An *ore* of boron consisting of *disodium borate*.

kerogen See *oil shale*.

kerosine Kerosene. See *paraffin oil*.

Kerr cell A transparent cell (based on the *Kerr effect*) filled with a *liquid*, such as *nitrobenzene*, which contains two *electrodes* placed between two polarizing media. *Light* can only pass through the cell if the two planes of *polarization* are parallel. As the Kerr effect occurs in time intervals as short as 10^{-8} *second*, the cell may be used as a high-speed shutter, and also as a means of modulating a *laser* beam.

Kerr effect When *plane-polarized light* is reflected from a highly polished *pole* of an *electromagnet* the light becomes *elliptically polarized*. Similarly, if a *beam* of light is passed through certain transparent *liquids* or *solids* to which a *potential difference* is applied, the plane of polarization of the light is rotated through an angle that depends upon the magnitude of the applied potential difference. This effect is made use of in the *Kerr cell*. Named after John Kerr (1824–1907).

ketal An *organic* compound formed from a *ketone* and an *alcohol*; it has the general formula RR′C(OR″)(OR‴).

ketene $CH_2:C:CO$. A colourless gas, b.p. $-56°C$., used as an acetylating agent in the manufacture of *cellulose acetate* and aspirin. It is the first member of the ketene series, which has the general formula $R:C:CO$, where R represents a *bivalent radical* or two *univalent* radicals.

keto-enol tautomerism The type of *tautomerism* that occurs in *ketones* as the result of the migration of a hydrogen *atom* from an *alkyl group* to the *carbonyl group*. Thus *propanone* (acetone) contains in addition to ketone *molecules* ($CH_3.CO.CH_3$, the keto-form) a small proportion of molecules having the structure of an *unsaturated alcohol* ($CH_2:COH.CH_3$, the enol-form).

ketones A series of *organic compounds* having the general formula RR′C:O, where R and R′ are *univalent hydrocarbon* groups. E.g. *propanone* (acetone), $(CH_3)_2CO$. The names of ketones end in -one.

ketose A *monosaccharide* that contains a *ketone* group.

kicksorter See *pulse height analyser*.

kieselguhr Diatomaceous earth, infusorial earth. A mass of *hydrated silica* (SiO_2) formed from skeletons of minute plants known as diatoms. It is a very porous and absorbent material, used for filtering and absorbing various *liquids*, in the manufacture of *dynamite* and in other industries.

killed spirits of salts A *solution* of *zinc chloride*, $ZnCl_2$, made by reacting zinc with *hydrochloric acid*. It is used in soldering.

kilo- Prefix denoting a thousand times; 10^3. Symbol k, e.g. $kV = 10^3$ *volts*.

kilogram Kilogramme. 1000 *grams*. The *SI unit* of *mass* defined in terms of the international prototype in the custody of the Bureau International des Poids et Mesures at Sèvres near Paris. It is equal to 2.204 62 lbs. Symbol kg.

kilohertz kHz. 1000 *hertz*. A measure of *frequency* equal to 1000 *cycles* per *second*.

kilometre 1000 *metres*. Equal to 1094 yards or 0.6214 mile.

kiloton bomb A *nuclear weapon* with an explosive power equivalent to one thousand tons of *TNT* (approximately 4×10^{12} *joules*).

kilowatt kW. A unit of *power* equal to 1000 *watts*.

kilowatt-hour kWhr. Board of Trade unit. A practical unit of *work*. The work done when a rate of work of 1000 *watts* is maintained for 1 hour.

kinematic equations See *motion, equations of*.

kinematics The branch of *mechanics* concerned with the phenomena of motion without reference to *mass* or *force*. Kinematics deals with motion from the standpoint of measurement and precise description, while *dynamics* is concerned with the causes or laws of motion (see *motion, equations of*).

kinematic viscosity. $v = \eta/\rho$. The ratio of the coefficient of *viscosity* to the *density* of a *fluid*. Measured in square *metres* per *second* (*SI units*) or *stokes*. 1 centistoke = 10^{-6} m^2/s.

kinetic energy The *energy* a body possesses by virtue of its motion, i.e. it is the amount of work the body will do if it is brought to rest. The kinetic energy of a rigid body of constant *mass m*, moving with a speed v, is $\frac{1}{2}mv^2$. The energy will be in *joules* if m is in *kilograms* and v is in *metres* per second (in *c.g.s. units* it will be in *ergs*). The kinetic energy of rotation of a body whose *moment of inertia* about an *axis* is I, and whose *angular velocity* about this axis is ω, is $\frac{1}{2}I\omega^2$. Again the energy will be in joules if I is in kg m^2 and ω is in *radians* per second (in c.g.s. units it will be in ergs).

kinetics Chemical kinetics, molecular kinetics. The branch of *physical chemistry* concerned with the study of the rates at which *chemical reactions* proceed.

kinetic theory of gases A mathematical explanation of the behaviour of *gases* on the assumption that gases consist of *molecules* in ceaseless motion in space, the *kinetic energy* of the molecules depending upon the *temperature* of the gas; the molecules are considered to be on average elastic particles that collide with each other and with the walls of the containing vessel (see *elastic collision*). The *pressure* exerted by a gas on the walls of the vessel is due to the collisions of the molecules with it. Thus the pressure, p, of one *mole* of a *perfect gas* contained in a vessel of volume V is given by: $p = N_A m \bar{c}^2/3V$, where N_A is the *Avogadro number*, m is the mass of each molecule, and $\bar{c}$ is the mean square speed of the molecules. The *gas laws* may be shown to be in full agreement with this theory.

kink instability In a *thermonuclear reaction* experiment, an instability in the magnetically confined *plasma* resulting from a local deformation of the plasma. The kink tends to grow because the *magnetic lines of forces* of the self-induced confining field are crowded on the *concave* side of the kink.

Kipp's apparatus A device used in laboratories for the production of a supply of any *gas* that can be evolved by the action of a *liquid* on a *solid* without heating (e.g. *hydrogen sulphide* by the action of dilute *hydrochloric acid* on iron(II) sulphide). The simplest form is illustrated in Fig. 23. Opening the tap T allows the liquid in C to reach the solid in B. A reaction occurs and gas is produced. When the tap is closed, gas production continues until the liquid is forced back into B. Named after Petrus Jacobus Kipp (1808–64).

Kirchhoff's laws 1. In any network of wires the algebraic sum of the *electric currents* that meet at a point is zero. 2. The algebraic sum of the *electromotive forces* in any closed circuit or mesh is equal to the algebraic sum of the products of the *resistances* of each portion of the circuit and the currents flowing through them. Named after Gustav Robert Kirchhoff (1824–87).

kish A variety of *graphite* occasionally formed in iron *smelting* furnaces.

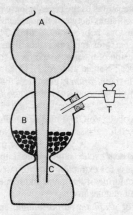

Figure 23.

Kjeldahl flask A round-bottomed glass flask with a long wide neck, used in the estimation of nitrogen by the *Kjeldahl method.*

Kjeldahl's method An analytical method of determining the nitrogen content of an *organic* compound. The compound is decomposed with concentrated *sulphuric acid* to convert the nitrogen into *ammonium sulphate.* The sulphate is estimated by adding excess *alkali*, distilling the ammonia into a standard acid solution, and measuring the excess acid by *titration.* Named after Johan Kjeldahl (1849–1900).

klystron An *electron* tube used to generate or amplify *electromagnetic radiation* in the *microwave* region, by *velocity modulation.* It consists of two or more *resonant cavities* in which the electrons, from an *electron gun*, are concentrated into 'bunches'.

knocking Violent *explosions* in the cylinder of a petrol engine, often due to over-compression of the *mixture* of air and petrol vapour ahead of the flame front. The result is a *shock wave* that causes overheating, plug damage, and loss of power. It is overcome by the use of high-octane fuels (see *octane number*) and the use of antiknock agents, such as *tetraethyllead.* It can also be avoided by an engine design that increases turbulence in the cylinder head. The use of tetraethyllead is now deprecated because it increases the level of lead in the atmosphere. Lead-free petrol is now sold at a lower price than leaded petrol in many countries.

knock-on collision (phys.) A process in which an *elementary particle* or *nucleus* is set in motion by being struck by another high-*energy* particle (or *photon*). The term is also used in relation to collisions as a result of which an *electron* is knocked out of its atomic *orbit* by some other particle. The 'knock-on' particle is the particle set in motion as the result of the collision.

knot A unit of speed equal to 1 *nautical mile* per hour. (Approximately 1.15 statute miles per hour.)

Knudsen flow Molecular flow. The flow of a *gas* through a pipe at pressures that are sufficiently low for the *mean free path* of the gas *molecules* to be large compared to the dimensions of the pipe. In these circumstances the rate of flow is determined by collisions between the molecules and the walls of the tube rather than by collisions between molecules. Thus the flow depends on the *relative molecular mass* of the gas rather than its *viscosity*. The ratio of a characteristic dimension of the pipe to the mean free path of the gas is known as the 'Knudsen number'. Named after M. H. C. Knudsen (1871–1949).

Kohlrausch's law When *ionization* is complete, the *conductivity* of an *electrolyte* is equal to the sum of the conductivities of the *ions* into which the substance dissociates.

Kovar* An *alloy* of cobalt, iron, and nickel, which has an *expansivity* similar to that of *glass*. It is used for glass-to-metal seals, particularly in *thermionic valves* and *transistors*.

Krebs cycle See *citric acid cycle*. Named after Hans Adolf Krebs (1900–81).

Kroll process A process for extracting titanium or zirconium from their ores by producing the tetrachloride of the metal and reducing it under reduced pressure or by reacting it with magnesium.

Kryptol* A *mixture* of *graphite*, *carborundum*, and *clay*, used as an electrical *resistance* in electric furnaces.

krypton Kr. Element. R.a.m. 83.80. At No. 36. A *noble gas*, which occurs in the *atmosphere* (1 part in 670 000); b.p. –152.3°C. It is used in some *lasers*.

Kundt's tube A device used to measure the speed of *sound*. It consists of a closed glass tube with a sound source at one end and a light powder sprinkled along its length. The gas column is adjusted by means of a piston so that it is exactly an integral number of half wavelengths long. The resulting *standing waves* cause the powder to form rings at the *nodes*, enabling the distance between nodes to be measured. Named after August Kundt (1839–94).

Kupfer-nickel Natural nickel arsenide, NiAs. An important *ore* of nickel.

kurchatovium A name proposed for the element with *atomic number* 104. See *transactinides*.

L

labelled compound A *compound* in which a stable *atom* is replaced by a *radioactive isotope* of that atom. The path taken through a mechanical or biological system by such a labelled compound can be traced by the *radiation* emitted by the 'labelled atom'. In some cases a non-radioactive isotope is used as the labelled atom and in this case its presence is observed using a *mass spectrometer*. See also *radioactive tracing*; *tritiated compound*.

labile Prone to undergo change or displacement; *unstable*. A labile chemical compound is usually a *coordination compound* in which *ligands* are easily replaced by other ligands.

lachrymator See *tear-gas*.

lactams A group of *organic* ring *compounds* in which –NH–CO– appears in the ring. The *tautomer* –N=C(OH)–, called a lactim, also occurs. Lactams are formed by the combination of an –NH_2 group and a –COOH group in the same molecule.

lactase An *enzyme* that catalyses the conversion of *lactose* into *glucose* and *galactose*. It is present in the digestive juices of mammals.

lactate 1. (chem.) A *salt* or *ester* of *lactic acid*. **2.** (bio.) To produce milk from the mammary glands.

lactic acid 2-hydroxypropanoic acid. $CH_3CH(OH)COOH$. A colourless crystalline *organic acid*, occurring in three stereoisomeric forms (see *stereoisomerism*), m.p. 18°C. *dl*-lactic acid, a mixture of equal amounts of (*dextrorotatory*) *d*-acid and (*laevorotatory*) *l*-acid, is formed by the action of certain *bacteria* on the *lactose* of milk during souring. The *d*-form, sarcolactic acid, occurs in muscle tissue. The optically inactive *dl*-form is used in dyeing and *tanning*.

lactim See *lactam*.

lactones A group of *organic* ring *compounds* in which –CO–O– appears in the ring. They are formed by the combination of an –OH group and a –COOH group in the same molecule.

lactoprotein Any of the *proteins* present in milk.

lactose Milk sugar. $C_{12}H_{22}O_{11}$. A hard gritty crystalline *soluble disaccharide*, m.p. 203°C., less sweet than *cane-sugar*, that occurs in the milk of all mammals. *Hydrolysis* gives a mixture of *glucose* and *galactose*. In the action of certain *bacteria* on milk ('lactic acid fermentation') lactose is converted into *lactic acid*.

laevorotatory Rotating or deviating the plane of vibration of *polarized light* to the left (observer looking into the oncoming light). See *dextrorotatory*; *optical activity*.

laevulose See *fructose*.

Lagrangian *L*. A *function* used in Lagrangian dynamics to define a system in terms of the *scalar* generalized coordinates, $q_i, \dot{q}_i$, *kinetic energy T*, and *potential energy V*, such that

$$L = T(q_i, \dot{q}_i) - V(q_i, \dot{q}_i).$$

239

In certain circumstances this can provide a valuable simplification as it avoids such *vectors* as *force* and *acceleration*. Named after J. L. Lagrange (1736–1813).

lake In dyeing, a coloured *insoluble* substance formed by the chemical combination of a *soluble dye* with a *mordant*.

Lamarckism The theory of *evolution* proposed by Jean-Baptiste Lamarck (1744–1829) in 1809, which postulated the inheritance of characteristics acquired during the lifetime of an organism. The theory finds little acceptance today, but enjoyed a revival in Soviet genetics in the 1930s with the backing of T. D. Lysenko (1898–1976). Compare *Darwin's theory*; *neo-Darwinism*.

lambda particle An *elementary particle*, classified as a *hyperon*, that has no charge and is 2183 times heavier than an *electron*.

lambda point λ. The *temperature* below which helium becomes *superfluid*. $\lambda = 2.186$ K.

lambert A former unit of *luminance*. The luminance of a uniform diffuser of *light* that emits one *lumen* per sq cm. 1 lambert = 3180 candela per square metre. Named after J. H. Lambert (1728–77).

Lambert's law of illumination 1. The *illuminance* of a surface upon which the *light* falls normally from a *point source* is inversely proportional to the *square* of the distance between the surface and the source. If the *normal* to the surface makes an angle θ with the direction of the *rays*, then the *illuminance* is proportional to cos θ. **2.** See also *linear absorption coefficient*.

Lamb shift A small difference in the *energy levels* $^2S_{1/2}$ and $^2P_{1/2}$ of the *hydrogen spectrum* resulting from the quantization of the interaction between the atomic *electron* and the *electromagnetic field*. Named after W. E. Lamb (born 1913).

lamina A thin sheet.

laminar flow The steady flow of a *fluid* in parallel layers (laminae) that closely follow the shape of a streamlined surface without turbulence. Compare *turbulent flow*.

laminated iron Thin sheets of iron (or, more frequently *Stalloy**) used for cores of *transformers* instead of solid iron cores. The sheets are varnished or oxidized to give a high resistance between them, in order to reduce losses due to *eddy currents*.

lamp-black Soot; an *allotropic* form of powdered *carbon*. It is used as a *pigment* and *filler*. It is made by burning organic matter without sufficient oxygen for complete *combustion*.

LAN See *local area network*.

lanolin A *wax*-like *emulsion* obtained from wool-grease and containing *cholesterol*, $C_{27}H_{45}OH$, and other complex *organic* substances in water. It is readily absorbed by the skin and is used in ointments and cosmetics.

lanthanides Lanthanons, lanthanoids, rare earths. A group of rare metallic *elements* with *atomic numbers* from 57 to 71 inclusive. The properties of these *metals* are all very similar and resemble those of aluminium. The elements occur in *monazite* and other rare minerals. See Appendix, Table 8.

lanthanum La. Element. R.a.m. 138.9055. At. No. 57. A silvery metal, r.d. 6.146, m.p. 920°C., b.p. 3464°C., used in *pyrophoric alloys* and as a *catalyst* in oil *cracking*.

lapis lazuli Sodium aluminium silicate containing sulphur. A rare mineral of beautiful blue colour.

Laplace operator Laplacian. ∇^2. The *differential* operator that gives the sum of the *partial derivatives* of second *order* with respect to each *variable*, i.e.
$$\nabla^2 u = \partial^2 u/\partial x^2 + \partial^2 u/\partial y^2 + \partial^2 u/\partial z^2 = 0$$
This *equation* is known as the 'Laplace equation'. Named after Pierre Simon Laplace (1749–1827).

large calorie Kilogram-calorie, Calorie. 1000 *calories*.

Larmor precession The orbital motion of the *electrons* about the *nucleus* of an *atom* usually gives the atom a *resultant angular momentum* and a *magnetic moment*. These two properties cause the atom to precess (see *precessional motion*) about the direction of any applied *magnetic field*. This is Larmor precession; the *frequency* of this precession, known as the Larmor frequency, is equal to $eH/4\pi mv$, where e and m are the electronic charge and mass, H is the *magnetic field strength*, and v is the velocity of the electron. Named after Sir Joseph Larmor (1857–1942).

laser Light Amplification by Stimulated Emission of Radiation. An optical *maser*. The laser produces a powerful, highly directional, *monochromatic*, and *coherent* beam of electromagnet radiation in the *infrared*, visible, and *ultraviolet* regions of the *spectrum*. It works on essentially the same principle as the maser, except that the 'active medium' in the simplest type consists of, or is contained in, an optically transparent cylinder with a reflecting surface at one end and a partially reflecting surface at the other. The stimulated waves make repeated passages up and down the cylinder, some of them emerging as light through the partially reflecting end. In the *ruby* laser, the chromium *atoms* of a cylindrical shaped ruby *crystal* are optically pumped to an excited state (see *excitation*) by a flash lamp, and it can then be made to emit *pulses* of highly coherent light (see *population inversion*). Lasers have also been constructed using a mixture of *inert gases* (helium and neon) to produce a continuous beam. Another type of laser consists of a cube of specially treated gallium arsenide, which is capable of emitting *infrared radiation* when a current is passed through it. The uses of lasers include eye surgery, *holography*, cutting metals, printing, and communications.

laser printer An output device from a *computer*, which can produce up to 150 pages per minute of high quality print. The beam from a *laser* transfers the image to be printed onto the surface of a drum or band in the form of a pattern of *electric charges*. The pattern is transferred to paper on which a permanent visible image is formed as in *xerography*.

latent heat *L*. The quantity of *heat* absorbed or released in an *isothermal* transformation of *phase*. The *specific* latent heat of *fusion* is the *heat* required to convert unit *mass* of a *solid* to a *liquid* at the same *temperature*. The specific latent heat of vaporization is the heat required to convert unit mass of liquid to *vapour* at the same temperature. It is thus the energy a substance must absorb from its environment in order to overcome the forces between its molecules so that it can change from a liquid to a gas, plus the energy it must absorb in order to do work against the environment to accommodate its expansion from a liquid to a gas at the same temperature. In thermodynamic terms, $L = \Delta H = \Delta U + p\Delta V$. *L* is therefore the change in *enthalpy H*, as a result of the change in the *internal energy U*, and the change in volume *V*, at constant pressure, *p*. Specific latent heat is measured in *joules* per *kilogram*. The corresponding *molar* latent heats are measured in joules per *mole*. At the *melting* and *boiling points* of a substance, the addition of heat causes no rise in temperature until the change of state is complete.

lateral In a sideways direction.

lateral inversion The inversion produced by a plane *mirror*. It is seen when the image of a printed page is observed in a mirror.

lateral velocity The component of a celestial body's *velocity* perpendicular to the *line of sight velocity*.

latex 1. A milky fluid produced by certain plants; the most important is that obtained from the rubber tree (*Hevea brasiliensis*), consisting mainly of a colloidal *suspension* of *rubber* globules in a watery liquid **2.** An analogous *emulsion* or *suspension* of a synthetic rubber or similar polymer.

latitude The *angular distance* of a point from the *equator* measured upon the curved surface of the Earth. In *astronomy* it is the *coordinate* of a celestial body from a fixed *plane*. The 'galactic latitude' is the *angular distance* from the plane of the *Milky Way*. The 'celestial latitude' is the angular distance between the celestial body and the *ecliptic*.

latitude, lines of Parallels of latitude. Circles parallel to the *equator*, joining points of equal *latitude*; the equator itself is latitude $0°$, while the poles are latitude $90°$. Compare *longitude, lines of.*

lattice 1. The regular network of fixed points about which *molecules*, *atoms*, or *ions* vibrate in a *crystal*. **2.** In a *nuclear reactor*, a structure consisting of discrete bodies of *fissile* and non-fissile material (especially *moderator*), arranged in a regular geometrical pattern.

lattice energy The *energy* required to separate the *ions* of a *crystal* to an infinite distance from each other.

lattice point One of the fixed points in a crystal *lattice* about which *molecules*, *atoms*, or *ions* vibrate.

laudanum An alcoholic *tincture* of *opium*.

laughing gas Dinitrogen oxide. See *nitrogen oxides*.

launch window The *period* during which a space vehicle can be launched in order to comply with various *parameters* and achieve its planned *orbit* or objective.

lauric acid See *dodecanoic acid*.

lauroyl The *univalent* group $CH_3(CH_2)_{10}CO-$.

lauryl alcohol See *dodecanol*.

lava See *magma*.

lawrencium Lr. *Transactinide element*. At. No. 103. The only known *nuclide*, lawrencium-260, has a *half-life* of only 8 secs. The name unniltrium is sometimes used.

Lawson criterion A criterion for a *thermonuclear reactor* to be a source of energy. It is the product of the *density* of the fusing particles and the *containment time* required in order that they will react sufficiently to raise the temperature of the *plasma* to the *ignition temperature*. For a mixture (50:50) of deuterium and tritium the Lawson criterion has the value 10^{15} s cm^{-3}.

LD$_{50}$ See *median lethal dose*.

L-D process See *basic-oxygen process*.

leaching Washing out a *soluble* constituent from a mixture of solids.

lead Pb (Plumbum). Element. R.a.m. 207.2, At. No. 82. A soft bluish-white *metal*, r.d. 11.34, m.p. 327.5°C., b.p. 1760°C. It occurs chiefly as *galena*, PbS, and is extracted by roasting the *ore* in a *reverberatory furnace*. *Compounds* are poisonous. The metal is used in the lead *accumulator*, in *alloys* and in plumbing; compounds are used in *paint* manufacture and in petrol additives (see *tetraethyllead*).

lead accumulator See *accumulator*.

lead acetate See *lead ethanoates*.

lead arsenate $Pb_3(AsO_4)_2$. A white crystalline substance, m.p. 1042°C., used as an *insecticide*.

lead carbonate Normal lead(II) carbonate, $PbCO_3$, is a white powder that occurs naturally as cerussite. Basic lead carbonate, or lead(II) carbonate hydroxide, $2PbCO_3$. $Pb(OH)_2$, is known as white lead and was formerly widely used as a *pigment*.

lead-chamber process The manufacture of *sulphuric acid* by the action of *nitrogen dioxide*, NO_2, on *sulphur dioxide*, SO_2, to give *nitrogen monoxide*, NO, and *sulphur trioxide*, SO_3. The former reacts with oxygen from the air to give NO_2 again; the SO_3 combines with *water* to give sulphuric acid, the process being carried out in large lead chambers. The process is obsolete and was replaced by the *contact process*.

lead ethanoates Two *ethanoates* (acetates) of lead. **1.** Lead(II) ethanoate, sugar of lead. $Pb(CH_3COO)_2$. A white crystalline solid that exists in the anhydrous and trihydrate forms. It is soluble in water, m.p. 280°C., and has a sweet taste. It is used as a *mordant* and as a drier in *paints*. **2.** Lead(IV) ethanoate, lead tetraacetate. $Pb(CH_3COO)_4$. A colourless solid that decomposes in water, m.p. 175°C.

lead oxides Three *oxides* of lead. **1.** Lead(II) oxide, lead monoxide, litharge. PbO. A yellow crystalline substance, m.p. 888°C, which is insoluble in water and is used in *glass*, *paints*, and *glazes*. **2.** Lead(IV) oxide, lead dioxide, lead peroxide. PbO_2. A dark-brown *amorphous* powder, which was formerly used in safety matches. **3.** Dilead(II) lead(IV) oxide, red lead, minium. Pb_3O_4. A bright red powder formerly used as a *pigment* and *oxidizing agent*. When it is heated it is black. It is a nonstoichiometric compound containing less oxygen than implied by the formula.

lead tetraethyl See *tetraethyllead*.

Leblanc process Salt-cake process. An obsolete process for the manufacture of *sodium carbonate*, Na_2CO_3. *Common salt* is converted into *sodium sulphate*, Na_2SO_4 ('salt-cake') by heating with *sulphuric acid*. This is heated with *coal* and *limestone*; the sodium sulphate is reduced by the carbon to sodium sulphide, which then reacts with the limestone to give sodium carbonate and calcium sulphide. It has been replaced by the *Solvay process* and by conversion of *trona*. Named after Nicolas Leblanc (1742–1806).

Le Chatelier Principle If a system in *equilibrium* is subjected to a disturbance the system tends to react in such a way as to oppose the effect of the disturbance. Named after Henri-Louis Le Chatelier (1850–1936).

lecithins Naturally occurring complex *phospholipids* in which *choline* is bound to the *phosphoric acid* group. They are found in *cell membranes*, egg yolk, etc.

Leclanché cell A *primary cell* with a positive *electrode* or pole of carbon surrounded by a mixture of *manganese dioxide* and powdered carbon in a porous pot. This stands in a *solution* of *ammonium chloride*, the *electrolyte*, in a jar, which also contains the negative electrode of zinc. When the external circuit is completed, a current flows, chlorine *ions* in the electrolyte moving towards the zinc and ammonium ions toward the carbon electrode. The chlorine ions react with the zinc to form *zinc chloride*, and the ammonium ions decompose at the positive electrode to give *ammonia* and hydrogen. The hydrogen liberated tends to cause *polarization* of the cell. This tendency is partly counteracted by the manganese dioxide, which oxidizes the hydrogen. The *E.M.F.* is approximately 1.5 *volts*. The common *dry cell* is a special form of Leclanché cell. Named after Georges Leclanché (1839–82).

LED See *light-emitting diode*.

lens Any device that causes a *beam* of *rays* to converge or diverge on passing

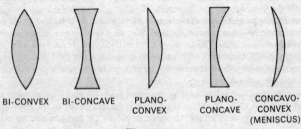

BI-CONVEX   BI-CONCAVE   PLANO-CONVEX   PLANO-CONCAVE   CONCAVO-CONVEX (MENISCUS)

Figure 24.

CONVERGING AND DIVERGING LENSES

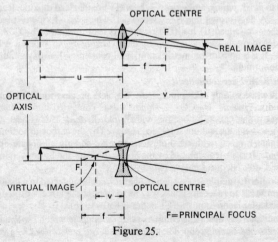

Figure 25.

244

through it. The optical lens is a portion of a *transparent* refracting medium (see *refraction of light*), usually *glass*, bounded by two surfaces, generally curved. Such lenses are classified according to the nature of the surfaces into *biconcave*, *biconvex*, planoconvex, etc. (See Fig. 24.) The centres of the *spheres* of which the lens surfaces are considered to form a part are termed the *centres of curvature*; the line joining these is the *optical axis*, the *optical centre* is a point on the axis within the lens; all rays passing through this point emerge without deviation. A parallel beam of *light* incident on a lens is made to converge (convex lens) or diverge (concave lens). The point of divergence or convergence is called a principal focus (see Fig. 25). Regarding all distances as being measured from the optical centre, and taking all distances as positive when measured in a direction opposite to that of the incident light, the distances of the object and image from the lens are given by the formula:

$$1/v - 1/u = 1/f,$$

where u and v are the distances from the lens of object and image respectively, and f is the *focal length*, i.e. the distance of the focus from the lens (see Fig. 25). Electrostatic and electromagnetic lenses, for converging beams of *electrons* and other elementary charged particles, are also of importance, e.g. in the *electron microscope*. See *electron lens*.

lenticular Pertaining to a *lens*, especially a *biconvex* lens, or resembling such a lens in shape.

Lenz's law When there is a change in the *magnetic flux* linked with a circuit, the *electric current* induced in the circuit will have a magnetic field opposing the change producing it. See *induction, electromagnetic*. Named after Heinrich Lenz (1804–65).

lepton A class of *elementary particles* that react by the *electromagnetic interaction* and the *weak interaction* but are insensitive to the *strong interaction*. They include the *electron, muon, tauon*, and *neutrino*. The electron, muon, and tauon all have a charge of −1 and a corresponding *antiparticle* of charge +1. The neutrinos, of which there are three (one associated with each of the other types of lepton), all have a charge of zero. The neutrinos are massless (or almost massless). The muon has a mass 200 times that of the electron and the tauon is 3500 times heavier than the electron. The number of leptons minus the number of corresponding antileptons taking part in a process is called the 'lepton number'; a quantity that appears to be conserved in all processes. All leptons have *spin* ½. Unlike *hadrons*, leptons have no internal structure.

LET See *linear energy transfer*.

leucine A white *soluble amino acid*, m.p. 293–295°C., essential to mammals. See Appendix, Table 5.

leucocytes White *blood cells*. The cells of the *blood* that contain no *haemoglobin*. There are several types of leucocytes, the main function of which is the combating of infection, including the production of *antibodies*. Human blood contains between 5000 and 10 000 leucocytes per cubic millimetre, many of which (lymphocytes) are produced in *lymph* nodes.

lever A rigid bar that may be turned freely about a fixed point of support, the fulcrum. The *mechanical advantage* of a lever is given by the ratio of the perpendicular distance of the line of action of the effort from the fulcrum, to the perpendicular distance of the line of action of the resistance from the fulcrum.

Lewis acids and bases A concept of *acids* and *bases* put forward by G. N. Lewis

(1875–1946) in which an acid is defined as a substance that forms a covalent bond (see *chemical bond*) with a base by accepting from it a *lone pair of electrons*. A base is defined as a substance that forms a covalent bond with an acid by donating to it a lone pair of electrons.

lewisite 1. β-Chlorovinyldichloroarsine. $ClCH:CHAsCl_2$. An oily liquid, b.p. 190°C., developed as a war gas and having *vesicant* and other lethal properties. **2.** The mineral calcium titanium antimonate, $5CaO.2TiO_2.3Sb_2O_5$.

Leyden jar A form of electrostatic *capacitor* of historical interest. It was invented in 1745 in the Dutch town of Leyden and consists of a glass jar with both internal and external layers of metal foil. Contact is made with the inner foil by means of a loose chain hanging inside the jar.

LF See *low frequency*.

libration An oscillation of the *Moon's* face from side to side. Due to libration about 58% of the Moon's surface can be seen from the Earth.

Liebig condenser See *condenser* (chem.). Named after Baron von Liebig (1803–73).

life The attribute that enables a discrete quantity of matter (*organism*) to take in energy from its environment to maintain its growth, increase its complexity, and convert some of itself into new organisms (offspring). See *cell*.

lifetime 1. See *mean life*. **2.** The mean time between the generation of a charge *carrier* and its *recombination*.

ligand 1. A single *molecule* or *ion* that donates a *lone pair of electrons* to a metal atom or ion to form a *coordination compound* (see also *chemical bond*). In a ligand one, two, or more atoms may be attached to the central atom, and it is referred to correspondingly as a uni-, bi-, or multi-dentate ligand. **2.** A molecule that binds to another large molecule, especially to a *protein* molecule, by a number of weak bonds.

light The agency by means of which a viewed object influences the observer's eye. It consists of *electromagnetic radiation* within the *wavelength* range 4×10^{-7} metre to 7.7×10^{-7} metre approximately; variations in the wavelength produce different sensations in the eye, corresponding to different *colours*. See *colour vision*.

The nature of light is a problem that has long perplexed scientists. It is now accepted that it is a phenomenon that has both *wave*-like properties and particle-like properties. On the principle of *complementarity*, light will appear in some circumstances to behave as a wave (see *interference*; *wave theory of light*) and in others as particles called *photons* (see *corpuscular theory*; *photoelectric effect*).

light, speed of c. The mean value is $2.997\ 924\ 58 \times 10^8$ m s^{-1} exactly, or approximately 186 281 miles per second. The special significance of the speed of light in the *Universe* was revealed by the special theory of *relativity*. According to this, the speed of light is absolute (i.e. independent of the speed of the observer) and represents a limiting speed in that the speed of no body can exceed it. The special significance of the speed of light is that it forms the 'connecting link' between *mass* and *energy* in the *mass-energy equation*.

light-emitting diode (LED) A device used to display figures (*digital display*), etc., in calculators and other equipment giving a visual display. It consists essentially of a *semiconductor diode*, made from such materials as gallium arsenide, in which light is emitted at a p-n junction when *electrons* and *holes* recombine. When the junction is forward biased light emitted is proportional to the bias current and its colour depends on the type of material used.

lightness See *colour*.

lightning A luminous electric *discharge* in the form of a flash between two charged clouds, between two parts of the same cloud (usually the upper part of a cloud is positively charged and the lower part is negatively charged), or between a cloud and the *Earth*. In a discharge from cloud to Earth, a leader stroke with a current of some 10 000 amperes is followed by a return stroke up the pre-ionized path of the leader with a current of approximately twice this value. Cloud to cloud discharges also have leader and return strokes.

lightning conductor A *conductor* of *electricity* connected to earth and ending in one or more sharp points attached to a high part of a building. It provides a direct path of low resistance to earth.

light pen An input/output *computer* device used with a *visual display unit*. When pointed at a spot on a *cathode-ray tube* it can sense whether or not the spot is illuminated.

light quantum See *photon*.

light-year An astronomical measure of distance; the distance travelled by *light* (see *light, speed of*) in one year: equal to $9.460\,528 \times 10^{15}$ metres or $0.306\,5949$ *parsec*.

lignin A complex *polymer*, which includes *polysaccharides* and *phenols* and is the major constituent of wood; lignin is associated with *cellulose* in the *walls of cells* of woody plant tissue, conferring toughness and rigidity. Removing the lignin from cellulose is an important step in the manufacture of pulp for the *paper* and *rayon* industries.

lignite Brown coal. A brownish-black natural deposit resembling *coal*, which contains a higher percentage of *hydrocarbons* than ordinary coal; it is probably of more recent origin.

ligroin A mixture of *hydrocarbons* similar to *benzine*, but boiling in a higher temperature range (80–130°C.).

lime Quicklime, see *calcium oxide*; or slaked lime, see *calcium hydroxide*. The term is sometimes loosely applied to calcium *salts* in general.

limestone Natural *calcium carbonate*, $CaCO_3$.

lime-water A solution of *calcium hydroxide*, $Ca(OH)_2$, in *water*. It turns milky by the action of *carbon dioxide*, CO_2, owing to the formation of *insoluble calcium carbonate*, $CaCO_3$.

limit Limiting value (math.). A *function* of a variable quantity x, written $f(x)$, approaches a limiting value k as x approaches a value a, if the difference $k - f(a + \delta)$ may be made smaller than any assignable value of making δ sufficiently small.

limit of proportionality See *elasticity*; *yield point*.

limit of spectral series The lines appearing in the *line spectrum* of any *element* can be grouped into definite series. The shortest *wavelength* of any such series is called the limit of the series. At this series limit, the lines crowd closer and closer together from the long wavelength side.

limonene Dipentene. $C_{10}H_{16}$. An *optically active* liquid *terpene*, b.p. 176–178°C., which occurs in some *essential oils*; it is used as a *solvent* and in the manufacture of *resins* and *surface-active agents*.

limonite A natural *hydrated* form of *iron(III) oxide*, Fe_2O_3. An *ore* of iron.

linac See *linear accelerator*.

linalool $C_{10}H_{17}OH$. A colourless *liquid terpene alcohol*, b.p. 198–200°C., occurring in certain *essential oils* and used in perfumes.

linalyl ethanoate Bergamol. $C_{10}H_{17}COOCH_3$. A colourless *liquid*, b.p. 220°C., with a pleasant odour; it is used in perfumes and *soaps*.

Linde process A process for producing *liquid air*, based on the *Joule-Thomson effect*. Air is compressed (to about 150 atmospheres) and expanded through a nozzle, which causes it to cool. The cool air is passed through a counter-current *heat exchanger* to reduce the temperature of the incoming air. Eventually the temperature is reduced sufficiently to liquefy the air. The process is also used to liquefy other gases, such as hydrogen and helium. Named after Carl von Linde (1842–1934).

linear 1. Arranged in a line. **2.** Having only one dimension. **3.** (of a mathematical expression or *equation*) Having only first *degree* terms (see also *linear relationship*). **4.** (of a component, *circuit*, or piece of *electronic* equipment) Having an output directly proportional to the input.

linear absorption coefficient *a*. A measure of a medium's ability to absorb radiation, but not to scatter or diffuse it (compare *linear attenuation coefficient*). It is given by $\phi_x/\phi_0 = e^{-ax}$, where ϕ_0 is the initial *radiant flux* or *luminous flux* and ϕ_x is the flux after it has travelled a distance x through the medium. This relationship is sometimes known as *Lambert's law* or as Bougner's law.

linear accelerator Linac. An apparatus for accelerating *ions* or *electrons* to high *energies*. It consists of a row of cylindrical *electrodes* separated by small gaps and having a common axis. Alternate electrodes are connected to each other and a *high-frequency potential* is applied between the two sets of electrodes. The frequency, and the lengths of the different electrodes, are such that the particles are accelerated each time they cross a gap between two electrodes. In an ion linac, this implies that the electrodes increase in length along the machine but in an electron linac the particles travel at *relativistic speeds* and the increase of speed with energy is not sufficient for this to be necessary. A typical rate of energy gain in an electron linac is 7 MeV m^{-1}.

linear attenuation coefficient Linear extinction coefficient. μ. A measure of a medium's ability both to absorb and to diffuse radiation (compare *linear absorption coefficient*). If a *luminous flux* or a *radiant flux*, ϕ_0, passes perpendicularly through a section of the attenuating medium, x, being reduced to ϕ_x, then the linear attenuation coefficient, μ is given by $\phi_x/\phi_0 = e^{-\mu x}$.

linear energy transfer LET. The average *energy* transferred by a charged particle of specified energy to the *atoms* and *molecules* of the medium through which it is passing in unit path length. The LET is proportional to the *charge* on the particle and decreases as the speed of the particle increases. It is an important parameter in living tissue as it modifies the effect of a particular *dose* of radiation.

linear motor A form of *induction motor*, in which the *stator* and *rotor* are linear instead of cylindrical, and parallel instead of *coaxial*. Linear motors have been used with *magnetic levitation* to drive experimental monorail vehicles. The *super conducting* magnets in the track function both to support the vehicle and to act as the stator in the linear motor.

linear relationship A relationship existing between two *variable* quantities such that a *graph* representing the manner in which they vary with each other will be a straight line. If the straight line passes through the *origin* the quantities are directly proportional and the equation of the line is $y = m\,x$, where m is the *gradient* of the line. In the more general case, the equation of a linear relationship is $y = mx + C$, where C is the intercept of the line with the y-axis, i.e. it does not pass through the origin.

line defect See *defect*; *dislocation*.

line-of-sight velocity Radial velocity. The *velocity* at which a heavenly body approaches, or recedes from, the Earth. It is measured spectroscopically by observing the shift of the spectral lines (see *spectrum*) of *elements* within the body, relative to those of the same elements on Earth. See *Doppler effect*.

line pair In *spectrographic analysis* a line pair consists of the particular spectral line (see *line spectrum*) utilized in the determination of the *concentration* of an *element* and the *internal standard line* with which it is compared.

line printer An output device from a *computer*, which prints a line of *characters* at a rate of between 300 and 3000 lines per minute. Compare *laser printer*.

lines of force See *electrical lines of force*; *magnetic lines of force*.

line spectrum A *spectrum* (*emission* or *absorption*) consisting of definite single lines, each corresponding to a particular *wavelength*; it is characteristic of an *element* in the atomic state.

Linnaean system See *binomial nomenclature*. Named after Carolus Linnaeus (1707 – 78).

Linnz-Donnewitz process See *basic-oxygen process*.

linoleic acid $C_{17}H_{31}COOH$. A yellow oily liquid *unsaturated fatty acid*, b.p. 229°C., which occurs in various *vegetable oils*, particularly *linseed oil*. Once known as *vitamin F*, its function in this capacity is now discredited.

linseed oil A *vegetable oil* extracted from the seeds of flax plants. It contains *glycerides* of *oleic acid* and other *unsaturated fatty acids*. Being easily oxidized and polymerized it is widely used in the *paint* and varnish industries and for manufacturing linoleum.

lipase An *enzyme* with the power of hydrolyzing (see *hydrolysis*) *fats*.

lipids Lipoids. A group of *organic compounds* that are *esters* of *fatty acids* and are characterized by being *insoluble* in *water* but *soluble* in many organic *solvents*. They are usually divided into three groups: (1) 'Simple lipids', which include *fats and oils* as well as *waxes*; (2) 'Compound lipids', which include *phospholipids* and *glycolipids*; (3) 'Derived lipids', of which the most important are the *steroids*.

lipoic acid $C_8H_{14}O_2S_2$. A *vitamin* of the B complex, it is one of the *coenzymes* in the decarboxylation of *carbohydrates*, which is necessary before they can enter the *citric-acid cycle*. It occurs in liver and yeast.

lipolytic Lipoclastic. Fat-splitting; denoting *enzymes* having the power of hydrolysing (see *hydrolysis*) *fats* into *fatty acids* and *glycerol*; e.g. *lipase*. Most animals store energy in the form of *lipids*, especially as triglycerides in *fats and oils*. Lypolysis is the process by which these animals use lypolytic enzymes to split these triglycerides into glycerol and fatty acids which are transported in the blood to the tissues, where they are oxidized to provide energy.

lipoprotein A *protein* that includes a *lipid* in its structure. Lipoproteins are the main structural components of *cell membranes* and they occur in *blood* and *lymph*.

Lipowitz' alloy A fusible *alloy*, m.p. 65 – 70°C., consisting of 50%, bismuth, 27% lead, 13% tin, 10% cadmium.

liquation The separation of a *solid mixture* by heating until one of the constituents melts and can be drained away.

liquefaction of gases A *gas* possessing a *critical temperature* above room temperature may be liquefied merely by increasing the *pressure* on it. Otherwise, the gas must first be cooled to below its critical temperature and then compressed; or, if desired, cooled directly to its *boiling point* under normal pressure. The methods of cooling are (1) by *evaporation* under reduced pressure, as in the *cascade liquefier*; (2) by using the principle of the *Joule–Thomson effect* (see *Linde process*); (3) by causing the gas to expand against an external pressure; in so doing the gas does *work*, thereby cooling itself. This principle is used in the *Claude process*.

liquefied natural gas LNG. *Natural gas*, principally *methane*, that has been liquefied for convenience of shipping or for use as a *liquid* engine *fuel*. Unlike *liquid petroleum gas* it cannot be liquefied by pressure alone owing to its low *critical temperature* (190 K). It must therefore first be cooled to below this temperature and stored in insulated containers.

liquefied petroleum gas LPG. A mixture of *petroleum* gases, usually *propane* and *butane*, that is stored under *pressure* as a *liquid* and used as an engine *fuel*. It burns clearly causing little atmospheric pollution or engine deposits.

liquid A state of *matter* intermediate between a *solid* and a *gas*, in which the *molecules* are relatively free to move with respect to each other but are restricted by cohesive *forces* to the extent that the liquid maintains a fixed *volume*. Liquids assume the shape of the vessel containing them, but are only slightly compressible. However, no comprehensive theory of the liquid state exists, although it is clear that the cohesive forces do keep bundles of atoms, molecules, or ions in short-range ordered arrays. It is in the absence of a long-range order that they differ from solids.

liquid air A pale blue *liquid*, containing mainly liquid oxygen, b.p. – 182.9°C., and liquid nitrogen, b.p. – 195.7°C.

liquid-crystal display A *digital display* in an electronic calculator, etc., based on *liquid-crystal* cells that change their reflectivity in an applied *electric field*. The most common display uses a twisted *nematic crystal* placed between two polarizers. Polarized light entering such a liquid-crystal cell follows the twist in the nematic crystal and is rotated through 90°, enabling it to pass through the second polarizer. When an electric field is applied to the cell no light is transmitted as the molecular alignment is changed by the field. A mirror placed behind the second polarizer will cause the display, shaped in the form of a digit, to appear black when a voltage is applied.

liquid crystals Relatively large regions of regularly aligned *molecules* in *liquids* that are analogous to crystals (exhibiting *cybotaxis*) and sufficiently distinct from the bulk liquid to constitute identifiable 'mesophases'. Under the influence of an *electric field*, these phases undergo realignments leading to optical effects. See *liquid-crystal display*; *cholesteric crystals*; *nematic crystals*; *smectic crystals*.

liquid drop model of the nucleus A hypothetical model of the atomic *nucleus* in which its properties are compared to those of a drop of *liquid*.

l-isomer See *optical activity*.

Lissajous figure The *locus* of the *resultant* displacement of a point on which two or more simple periodic motions are impressed. In the common case, two periodic motions are at right angles and are of the same *frequency*. The Lissajous figures then become, in general, a series of *ellipses* corresponding to the possible differences of *phase* between the two motions. Named after Jules Lissajous (1822–80).

litharge See *lead oxides*.

lithia See *lithium oxide*.

lithium Li. Element. R.a.m. 6.941. At. No. 3. A light silvery-white *alkali metal*, m.p. 180.6°C., b.p. 1360°C., r.d. 0.533. It is the lightest *solid* known. Chemically it resembles sodium, but is less active. It is used in *alloys* and its *salts* have various uses.

lithium carbonate Li_2CO_3. A white crystalline *solid*, m.p. 735°C, used in the treatment of endogenous depression. It is also used in some *glazes*.

lithium chloride LiCl. A white *soluble deliquescent* substance, m.p. 614°C., used as a *flux* and in mineral waters.

lithium hydride LiH. A white crystalline substance, m.p. 680°C., used in organic synthesis as a *reducing agent*. The deuteride, in which some of the hydrogen is replaced with *deuterium*, is used as a deuterating compound.

lithium oxide Lithia. LiO_2. A white crystalline *solid*, m.p. 1700°C., used in greases, *refractories*, *fluxes*, and in some *accumulators*.

lithopone A *mixture* of *zinc sulphide*, ZnS, and *barium sulphate*, $BaSO_4$. Used in *paints* as a non-poisonous substitute for *white lead*.

lithosphere See *Earth*.

litmus A *soluble* purple substance of vegetable origin; it is turned red by *acids* and blue by *alkalis*. It is used as a rough *indicator*, especially in the form of litmus paper, an absorbent paper soaked in a solution of litmus.

litre A unit of *volume* in the *metric system*. Formerly defined as the volume of 1 *kilogram* of pure *water* at 4°C. and 760 mm pressure (which is equivalent to $1.000\ 028\ dm^3$). However, in *SI units* the litre is a special name for the cubic decimetre, but is not used for high precision measurements. For approximate purposes 1 litre = 1000 cc, and the symbol ml (for millilitre) is often used synonymously with cc, though this practice is now deprecated. For accurate measurements the SI symbol dm^3 should be used.

liver of sulphur A *mixture* of *sulphides* and other sulphur *compounds* of potassium, obtained by fusing *potassium carbonate*, K_2CO_3, with sulphur. It is used as an *insecticide* and *fungicide* in gardening.

lixiviation The extraction of *soluble* material from a *mixture* by washing with *water*.

LNG See *liquid natural gas*.

loaded concrete Normal *concrete* to which has been added some material containing *elements* of high *atomic number* (e.g. iron or lead). It is used in the shielding of *nuclear reactors*.

local area network LAN. A system for linking *computers* (usually *microcomputers*) in a single site (office, university, factory, etc.) so that they can share data files and *databases*, as well as such expensive resources as hard disks, *laser printers*, etc.

local group of galaxies The cluster of *galaxies* to which the *Galaxy* belongs. Distant clusters of galaxies are receding from the local group. See *expansion of the Universe*.

local oscillator The *oscillator* in a *heterodyne* or *superheterodyne radio* receiver that produces the *radio frequency* oscillation with which the received wave is combined.

locus (math.) The locus of a point is the line that can be drawn through adjacent positions of the point, thus tracing out the path of the point in space.

lodestone A magnetic variety of natural *iron oxide*. Fe_3O_4, *magnetite*.

logarithmic scale A scale of measurement in which an increase of one unit represents a tenfold increase in the quantity measured (for common *logarithms*).

logarithms If a number, *a*, is expressed as a *power* of another number, *b*, i.e. if $a = b^n$, then *n* is said to be the logarithm of *a* to *base b*, written $\log_b a$. Common logarithms are to base 10. Multiplication, division, and other computations are shortened by the use of common logarithms; the addition of logarithms of numbers gives the logarithm of the *product* of the numbers; similarly *division* can be performed by subtraction of the logarithms. Logarithms corresponding to ordinary numbers have been tabulated, and calculations are carried out by the use of such tables. Natural or Napierian logarithms are to the base 'e' (which has the value 2.71828). $\log_e a$, also written $\ln a$, is equal to $2.303 \log_{10} a$. See also *characteristic*; *mantissa*; *exponential*.

logic In an automatic data processing system, the systematic scheme that defines the interactions of the physical entities representing data. In digital *computers* the logic circuits are the basic switching circuits formed in *microchips* and *integrated circuits*. In digital computers the *binary notation* is used to express data, which can then manipulated in logic circuits by using 0 to represent a switch in the off position and 1 in the on position. Thus the basic AND logic circuit functions by giving an output current if all its input circuits have currents; an OR circuit gives an output if at least 1 input has a current; and the NOT circuit inverts the input. Any logical procedure can be carried out using combinations of such logic circuits (also called logic gates).

lone pair of electrons A pair of unshared *valence electrons* that are responsible for the formation of coordinate bonds (see *chemical bond*). They occupy the same *orbital* but have opposite *spins*. For example, in the water molecule, of the six outer electrons of the oxygen atom, two form single covalent bonds with the two hydrogen atoms and the remaining four electrons consist of two lone pairs. The shape of many molecules is determined by a repulsion between chemical bonds and lone pairs.

longitude The angle that the terrestrial *meridian* through the geographical poles and a point on the Earth's surface makes with a standard meridian (through Greenwich) is the longitude of the point. In astronomy, the 'celestial longitude' is the *angular distance* of a celestial body from the vernal *equinox* along the *ecliptic*, measured through 360° towards the East.

longitude, lines of Imaginary *meridians* on the Earth's surface, referred to a standard meridian through Greenwich; they are *great circles* of the Earth intersecting at the poles. Compare *latitude, lines of.*

longitudinal Lengthwise; in a line with the length of the object under consideration.

longitudinal waves Waves in which the vibration or displacement takes place in the direction of propagation of the waves; e.g. *sound* waves. See also *transverse waves*.

long sight See *hypermetropia*; *presbyopia*.

Lorentz-Fitzgerald contraction A contraction in the length of a moving object believed to occur as a result of the absolute motion of a body through the *ether*; it was postulated by H. A. Lorentz and G. F. Fitzgerald (1851–1901) to account for the negative result of the *Michelson-Morley experiment*. The contraction was given a theoretical explanation by Einstein in his special theory of *relativity*. In special relativity an object at rest, of length l_o, in one frame of reference, will appear to an observer in another frame of reference to have a length $l_o(1 - v^2/c^2)^{1/2}$, where v is the speed of one frame of reference relative to the other and c is the *speed of light*. This formula shows that the apparent contraction is negligible unless the relative speed of the two frames of reference is comparable to the speed of light.

Lorentz transformation A set of *equations* for correlating *space* and *time coordinates* in two *frames of reference*, especially at *relativistic velocities*. Named after Hendrik Lorentz (1853–1928).

Loschmidt's constant N_L. The number of particles per unit volume of a *perfect gas* at *s.t.p*; equal to $2.686\ 763 \times 10^{25}$ m^{-3}. N_L is equal to the ratio of the *Avogadro constant* to the *molar volume*.

loudness of sound The magnitude of the physiological response of the ear to *sound*. As the ear responds differently to different *frequencies*, the loudness of a sound will depend to a certain extent on its frequency. However, loudness can be roughly correlated with the *cube root* of the intensity of sound, and different levels can be conveniently compared by the units *decibel* and *phon*.

loudspeaker A device for converting *electric currents* into *sounds* loud enough to be heard at a distance. The commonest type consists of an *electromagnetically* operated *moving-coil* device vibrating a paper cone.

Lovibond tintometer* A *colorimeter* in which the *colour* of a liquid, surface, powder, or light source is compared with a series of glass slides of standardized colours.

low frequency LF. A *radio frequency* in the range 30–300 *kilohertz*.

Lowry-Brønsted theory A theory of *acids* and *bases* in which *acids* are regarded as *proton* donors and bases as proton acceptors. Water, in this definition, can act as both acid and base. Named after T. M. Lowry (1874–1936) and Johannes Brønsted (1879–1947), who arrived at the same conclusion independently in 1923.

LPG See *liquid petroleum gas*.

L-series See *optical activity*.

lubrication See *tribology*.

lumen (phys.) The derived *SI unit* of *luminous flux*. The amount of *light* emitted per second in unit *solid angle* of one *steradian* by a uniform *point source* of one *candela* intensity; i.e. the amount of light falling per second on unit area placed at unit distance from such a source. Symbol lm.

lumen (bio.) The space inside a hollow tube, duct, vessel, etc. For example, the space inside an intestine or blood vessel.

luminance *L*. The *luminous intensity* of any surface in a given direction per unit of orthogonally projected area of that surface, on a plane perpendicular to the given direction. It is given by:

$$L = dI/(dA\cos\theta),$$

where *I* is the luminous intensity, *A* is the surface area, and θ is the angle between the surface and the given direction. It is measured in *candela* per square *metre*.

luminescence The emission of *light* from a body from any cause other than high *temperature*. It is caused by the emission of *photons* when an *excited* atom returns to the *ground state*. *Fluorescence* and *phosphorescence* are particular cases of luminescence.

luminosity 1. The property of emitting *light*. **2.** The amount of light emitted by a *star*, irrespective of its distance from the *Earth*, usually expressed as a *magnitude*.

luminous flux Φ_v. The luminous flux through an area is the amount of *light* passing through that area in one *second*. The derived *SI unit* of luminous flux is the *lumen*. Compare *radiant flux*.

luminous intensity I_v. The amount of *light* emitted per second in unit *solid angle* by a *point source*, in a given direction. The *SI unit* of luminous intensity is the *candela*. The term is restricted to point sources. Compare *radiant intensity*.

luminous paint *Paint* prepared from phosphorescent *compounds* such as *calcium sulphide*, etc., which glows after exposure to *light*. See *phosphorescence*.

lunar caustic *Silver nitrate*, $AgNO_3$, usually fused and cast into sticks.

lunation Synodic month. The time between one new moon (see *phases of the moon*) and the next; equal to 29 days 12 hours and 44 minutes.

lutetium Cassiopeium. Lu. Element. R.a.m. 174.967. At. No. 71. A very rare silvery metal; r.d. 9.842, m.p. 1700°C., b.p. 3400°C. See *lanthanides*.

lux Metre candle. The derived *SI unit* of *illuminance*; one *lumen* per square *metre*. Symbol lx.

lyddite An *explosive* consisting of *picric acid* (trinitrophenol, $C_6H_2OH(NO_2)_3$, mixed with 10% *nitrobenzene* and 3% *Vaseline**.

Lyman series A series of lines that occurs in the *ultraviolet* region of the *spectrum* of hydrogen. Named after T. Lyman (1874–1954).

lymph A clear colourless liquid resembling *blood plasma* and consisting of water with dissolved *salts* and *proteins* that occurs in the lymphatic system of vertebrates. Derived from the blood, it bathes the *cells* of the body, supplying them with nutrients and absorbing their waste products. The lymph nodes, through which the lymph passes, filter out bacteria and other foreign particles and also produce the lymphocytes (see *leucocytes*).

lyophilic colloid 'Solvent-loving colloid'. See *colloidal solutions*.

lyophobic colloid 'Solvent-hating colloid'. See *colloidal solutions*.

lysergic acid $C_{15}H_{15}N_2COOH$. A crystalline substance obtained from ergot and used in the manufacture of the hallucinogen LSD (lysergic acid diethylamide).

lysine An essential crystalline soluble *amino acid*, m.p. 224°C. See Appendix, Table 5.

lysis The dissolution or destruction of *blood cells* or *bacteria* by a class of *antibodies* called lysins.

M

machine A device for overcoming resistance at one point by the application of a *force*, usually at some other point. Physics recognizes six "simple machines", the *lever*, wedge, inclined plane, screw, pulley, and wheel and axle. More complex machines are usually an arrangement for the purpose of taking in some definite form of *energy*, modifying it, and delivering it in a form more suitable for the desired purpose.

machmeter An instrument for measuring the speed of an aircraft relative to the speed of *sound*. See *Mach number*.

Mach number The ratio of the speed of a *fluid* or body to the local speed of *sound*. The speed of a fluid or body is therefore said to be *supersonic* if its Mach number is greater than unity. See also *hypersonic*. Named afer Ernst Mach (1838–1916).

Mach's principle The *inertia* of a body is a result of its interaction with the rest of the *Universe*. An isolated body would have no inertia.

macro- Prefix denoting large, in contrast to *micro*, small.

macrocyclic Containing a ring structure consisting of more than twelve *atoms* in the *molecule*.

macromolecular Consisting of or pertaining to *macromolecules*; having a very high *relative molecular mass*.

macromolecule A very large *molecule*, generally of a *polymer*. See *polymerization*.

Magellanic clouds Two small patches of *light* that appear, from the southern hemisphere, to be detached from the main bright band of *stars* constituting the *Milky Way*. These objects are separate *galaxies*, being two of the smaller members of the *Local Group* to which our *Galaxy* belongs. Named after Ferdinand Magellan (1480–1521).

magenta Fuchsine. $C_{20}H_{22}N_3OCl$. A red dye, prepared from *aniline* and *toluidine*.

magic numbers The numbers 2, 8, 20, 28, 50, 82, and 126. Atomic *nuclei* containing these numbers of *neutrons* or *protons* have exceptional stability.

maglev See *magnetic levitation*.

magma Molten material, consisting of *silicates* with occluded *gases* and other substances, that forms in the mantle or crust of the *Earth* and solidifies into *igneous rocks*. Magma extruded from *volcanoes* is called lava.

Magnadur* A *ferrite* used for making permanent *magnets*. It consists of sintered iron oxide and barium oxide.

Magnalium* A light *alloy*, r.d. 2–2.5; it consists of aluminium with from 5% to 30% magnesium and 1% to 2% of copper. It is highly reflective.

magnesia See *magnesium oxide* or *magnesium hydroxide*; 'magnesia alba' of pharmacy is *basic magnesium carbonate*; 'fluid magnesia' is a *solution* of magnesium hydrogencarbonate.

magnesite Natural *magnesium carbonate*, $MgCO_3$, which occurs in white masses; it is used in the manufacture of *refractories* and *fertilizers*.

magnesium Mg. Element. R.a.m. 24.305. At. No. 12. A light silvery-white *metal*, r.d. 1.74, m.p. 650°C., b.p. 1100°C., that tarnishes easily in air. It burns with an intense white *flame* to form *magnesium oxide*, MgO. Magnesium occurs as *magnesite*, $MgCO_3$; *dolomite*, $MgCO_3.CaCO_3$; *carnallite*, $KCl.MgCl_2.6H_2O$, and in many other *compounds*; it is prepared by *electrolysis* of fused carnallite. It is used in lightweight *alloys*, in *photography* and compounds are used in medicine. It is essential to life as it occurs in *chlorophyll*.

magnesium carbonate *Magnesite*. $MgCO_3$. A white solid that exists in the anhydrous, trihydrate, and pentahydrate forms. Basic magnesium carbonate, $MgCO_3.Mg(OH)_2.3H_2O$ or $3MgCO_3.Mg(OH)_2.3H_2O$, also occurs. Magnesium carbonate is used as a drying agent in table salt and as an *antacid*.

magnesium chloride $MgCl_2$. A white *deliquescent* substance, m.p. 708°C., that occurs in *sea-water* and also as *carnallite*. A *concentrated solution* mixed to a paste with *magnesium oxide* sets to a stone-like mass owing to the formation of the oxychloride, Mg_2OCl_2 (Sorel's cement).

magnesium hydroxide *Magnesia*. $Mg(OH)_2$. A white crystalline substance, used as an *antacid* in 'milk of magnesia'.

magnesium oxide *Magnesia*. MgO. A white tasteless substance, m.p. 2800°C., used as an *antacid* and a laxative and as a *refractory*.

magnesium sulphate Epsom salts. $MgSO_4.7H_2O$. A white crystalline *soluble salt*, used in medicine and in leather processing.

magnesium trisilicate Dimagnesium trisilicate. $2MgO.3SiO_2.nH_2O$. A white tastless powder used as an *antacid* and to absorb odours.

magnesothermic reduction *Reduction* of *oxides* to the corresponding metals at high temperatures with the aid of metallic *magnesium*. It is analogous to *aluminothermic reduction*.

magnet, permanent A *ferromagnetic substance* that has a permanent *magnetic field* and *magnetic moment* associated with it. See also *electromagnet*; *magnetic domains*.

magnetic amplifier A device for the amplification of small *direct currents* and of low *frequency alternating currents*. It depends upon the fact that the output from the secondary coil of a *transformer* due to an alternating current in the primary coil is also a function of a direct current (the signal to be amplified) in a third winding on the transformer core.

magnetic bottle Any configuration of *magnetic fields* used in the *containment* of a *plasma* during controlled *thermonuclear reaction* experiments. As all known materials vaporize at the temperatures of thermonuclear reactions, magnetic bottles are the only useful means of containing a plasma.

magnetic bubble A *computer* memory element consisting of a small magnetized region in a material, such as *garnet*, that is easily magnetized in one direction but not in the perpendicular direction. A magnetic chip consists of a thin film of this material deposited on a nonmagnetic substrate, and may measure some $15-25$ mm^2. When a *magnetic field* is applied to a chip, cylindrical *domains*, called magnetic bubbles, form. These bubbles consist of tiny regions of one magnetic polarity in an environment of the opposite polarity, each chip being able to store as many as one million bubbles. Information is represented in *binary notation* as the presence or absence of a bubble in a specified place on the chip. A rotating magnetic field is used to recover the information.

magnetic circuit A closed path following the *lines of force* of a *magnetic field*.

magnetic constant Permeability of free space. μ_0. The fundamental constant that has the value $4\pi \times 10^{-7}$ henry per metre. It arises as the constant of proportionality in *Ampere's law*, its value depending on the choice of units. See also *magnetic permeability*.

magnetic containment See *thermonuclear reaction*.

magnetic declination Magnetic variation. The *angle* between the *planes* of the geographic and *magnetic meridian*. See *magnetic elements*; *magnetism, terrestrial*.

magnetic dip Angle of dip, inclination. The *angle* between the direction of the Earth's magnetic field (see *magnetism, terrestrial*) and the horizontal; i.e. the angle through which a magnetic needle will 'dip' from the horizontal when suspended so that it is free to swing in a vertical plane in the *magnetic meridian*. See *dip circle*; *magnetic elements*; *magnetic equator*.

magnetic dipole See *dipole*; *magnetic moment*.

magnetic disk A metal disc coated with magnetic *iron oxide* for use as a *computer* memory. Data is recorded onto the rotating disk in concentric tracks and retrieved by means of a play-back head. The standard disk usually forms part of a ten-disk pack. The floppy disk is a similar, but smaller, device consisting of a plastic disk used in a stiff envelope in a *microcomputer*.

magnetic domains Regions, some $1-0.1$ mm across, within a *ferromagnetic* material, in which the atomic *magnetic moments* are aligned when the material is magnetized. The application of an external *magnetic field* increases the number and size of the aligned domains. In a very strong field all the magnetic moments of the domains are aligned with the field and the material becomes a saturated permanent *magnet*. If the material becomes demagnetized the domains cease to be aligned.

magnetic elements The three quantities, *magnetic declination*, *magnetic dip*, and *horizontal component*, which define completely the Earth's magnetic field (see *magnetism, terrestrial*) at any point.

magnetic equator Aclinic line. A line of zero *magnetic dip* lying fairly near the geographical *equator*, but passing North of it in Africa and the Indian Ocean, and South of it in America and the Eastern Pacific.

magnetic field A *field* of *force* that is said to exist at any point if a small coil of wire carrying an *electric current* experiences a *couple* when placed at that point. A magnetic field may exist at a point as a result of the presence of either a permanent *magnet* or a circuit carrying an *electric current*, in the neighbourhood of the point. The strength and direction of the field can be expressed either in terms of the *magnetic flux density*, B, or the *magnetic field strength*, H, both of which are *vector* quantities related by $B = H\mu$, where μ is the *magnetic permeability* of the medium.

magnetic field of electric current A wire or coil carrying an *electric current* is surrounded by a *magnetic field*. The direction of the field relative to the current may be determined by the following corkscrew rule: If a corkscrew, held in the right hand, is turned along the conductor in the direction of the current, the movement of the thumb indicates the direction of the magnetic field produced.

The strength of the magnetic field at the centre of a circular coil of wire of radius r, consisting of n turns, in which a current of I amperes is flowing, is $nI/2\pi r$ amperes per metre in SI units or $2\pi nI/10r$ oersted in c.g.s. units.

magnetic field strength Magnetic intensity. H. The *vector* quantity giving strength of a *magnetic field* measured in *amperes* per *metre* (*SI units*) or *oersteds* (*c.g.s. units*). It is given by $H = B/\mu_0 - M$, where B (also a vector quantity) is the *magnetic flux* density, M is the *magnetization*, and μ_0 is the *magnetic constant*. See *magnetic field of an electric current*.

magnetic flux Φ The strength of a *magnetic field* through an area, given by the product of the *magnetic flux density* and the area. The *c.g.s. unit* of magnetic flux is the *maxwell*. The derived *SI unit* of magnetic flux is the *weber*.

magnetic flux density Magnetic induction. B. The *vector* quantity giving *magnetic flux* passing through unit area of a *magnetic field* in a direction at right angles to the *magnetic force*. If a *charge q* experiences a *force F* when travelling through the field at a velocity v at an angle θ to the field, then $B = F/qv\sin\theta$. The derived *SI unit* of magnetic flux density is the *tesla* (*weber* per square *metre*). The *c.g.s unit* is the *gauss*.

magnetic force The *force* exerted by a *magnetic field* on a *magnetic pole* or an *electric charge*.

magnetic induction 1. The induction of *magnetism* in a body by an external *magnetic field*. **2.** See *magnetic flux density*.

magnetic intensity See *magnetic field strength*.

magnetic iron ore See *magnetite*.

magnetic levitation Maglev. Supporting an object by means of an upward *magnetic field* in a downward *gravitational field*. Maglev has been used to raise a vehicle above its track or tracks so that its horizontal motion is almost frictionless. Experimental maglev trains (in the UK and Japan) use systems in which the vehicle starts to move on wheels but is supported by maglev over a monorail as it gathers speed as a result of the interaction between *superconducting* magnets in the track and those in the train. The train is moved forward using a *linear motor*. The system is fast and comfortable, but very expensive.

magnetic line of force An imaginary line whose direction at each point is that of the *magnetic field* at that point.

magnetic meridian See *meridian, magnetic*.

magnetic mirrors The regions of high field strength at the end of an externally generated *magnetic field* used in the *containment* of a *plasma* in controlled *thermonuclear reaction* experiments. *Ions* that enter these regions of high field strength reverse their direction of motion (are reflected) and return to the central region of the plasma in which they become trapped.

magnetic moment 1. The maximum torque experienced by a magnetic *dipole* in a field of unit *magnetic field strength* perpendicular to it. It is measured in weber metres. This is also called the magnetic dipole moment. **2.** The product IA, where I is the current flowing through a small loop of wire of area A. It is measured in ampere metres squared ($A\ m^2$). This is called the electromagnetic moment. An *electron* (charge q) in an atomic orbit has a magnetic moment IA, where A is the orbital area and I is the equivalent current, given by $I = q\omega/2\pi$, where ω is the *angular velocity*.

magnetic monopole A hypothetical unit of magnetic 'charge' analogous to *elec-*

tric charge. No evidence has been found for the existence of a separate *magnetic pole*, they are always found in pairs. Some *gauge field theories* predict the existence of very heavy monopoles.

magnetic permeability μ. The ratio of the *magnetic flux density* in a medium to the external *magnetic field strength* that induces it. The 'relative permeability', μ_r, is the ratio of the permeability of a substance to the permeability of free space (see *magnetic constant*). For most substances μ_r has a constant small value. When μ_r is less than 1, the material is said to be *diamagnetic*; if μ_r is greater than 1, it is *paramagnetic*. A few substances, notably iron, have very large values of μ, which tend to fall as the field strength increases so that the magnetic flux density tends to a limiting value called the saturation value. Such substances are said to be *ferromagnetic*.

magnetic pole 1. A *magnet* appears to have its *magnetism* concentrated at two points termed the poles. If a bar magnet is suspended to swing freely, one of these, the North-seeking, North, or positive pole, will point North, and the other South. Unlike poles attract, and like poles repel each other. The *force* of attraction or repulsion between two poles varies inversely as the *square* of the distance between them (see *inverse square law*). The strength of a magnetic pole was formerly expressed in terms of a 'unit magnetic pole', to which the inverse square law was applied. Thus, the force between two poles m_1 and m_2, separated by a distance d in a vacuum, was given by $m_1 m_2 / d^2$. In modern practice the magnetic dipole moment is used (see *magnet moment*; *dipole*). **2.** The points on the Earth's surface close to the North and South poles, which appear to function as the magnetic poles of the Earth's *magnetic field*. See *magnetism, terrestrial*.

magnetic potential See *magnetomotive force*.

magnetic quantum numbers See *quantum numbers*.

magnetic resistance See *reluctance*.

magnetic storm A sudden disturbance in the Earth's magnetic field (see *magnetism, terrestrial*) associated with *sunspot* activity, which affects *compasses* and *radio* transmission.

magnetic susceptibility χ_m. The ratio of the *magnetization* (M) produced in a substance to the *magnetic field strength* (H) to which it is subjected, i.e. $\chi_m = M/H$. The susceptibility is related to the relative permeability, μ_r, (see *magnetic permeability*) by $\chi_m = \mu_r - 1$. Ferromagnetic materials have high positive values of χ_m.

magnetic tape *Plastic* tape coated with a *ferromagnetic* powder, used in tape recordings, video recordings, computers, etc. The tape is passed over the gap in a *magnetic circuit*, which is modulated in accordance with information to be recorded. The tape retains a record of the *modulation*, which can be 'played back' through a suitable circuit. Magnetic tape is used in the *backing storage* of *computers*.

magnetic variation See *magnetic declination*.

magnetism The branch of physics concerned with *magnets* and *magnetic fields*. See *diamagnetism*; *paramagnetism*; *ferromagnetism*; *ferrimagnetism*.

magnetism, terrestrial Geomagnetism. The Earth's magnetism. The Earth possesses a *magnetic field*, the strength of which varies with time and locality. The field is similar to that which would be produced by a powerful *magnet* situated at the centre of the Earth and pointing approximately North and South. A magnetized needle suspended to swing freely in all planes will set itself pointing to the Earth's magnetic North and South poles, at an angle to the horizontal (see *magntic dip*). The vertical plane through the *axis* of such a needle is termed the *magnetic meridian*, defined as the vertical plane that contains the direction of the Earth's magnetic field. At any point on the Earth's surface, terrestrial magnetism is defined by the three *magnetic elements*: the horizontal component B_0 of the *magnetic flux density* at that point; the angle of dip (the angle between B_0 and the resultant magnetic flux density); and the declination (the angle between B_0 and the geographic true north. See *magnetic declination*).

The cause of the Earth's magnetism is not definitely known but is believed to be associated with movements in the Earth's liquid core. The variations of the Earth's magnetic field with time are of two types, the 'secular' and the 'diurnal'. The secular variations are slow changes in the same sense, but at different rates, as a result of which the Earth's magnetic field has decreased by some 5% over the last hundred years. Moreover the evidence of *palaeomagnetism* suggests that the Earth's magnetic field has reversed in direction several times in the distant past. The cause of these variations is unknown. The diurnal variations are much smaller and more rapid variations which have been shown to be associated with changes in the *ionosphere* related to *sunspot* activity.

magnetite Magnetic iron ore. Natural black *iron oxide*, Fe_3O_4.

magnetization M. The magnetic moment per unit volume of a magnetized body. It is equal to $B/\mu_0 - H$, where B is the *magnetic flux density*, μ_0 is the *magnetic constant*, and H is the *magnetic field strength*. It is measured in amperes per metre.

magneto A small *alternating-current dynamo* provided with a secondary winding to produce a high voltage to enable a *spark* to jump between the *electrodes* of a sparking plug in the ignition of a petrol engine. Most magnetos consist of a permanent magnet *rotor* and a primary low-voltage *stator*, around which a high-voltage secondary coil is wound.

magnetohydrodynamics MHD. **1.** The study of the behaviour of moving electrically conducting fluids in magnetic fields. **2.** A method of generating an *electric current* by subjecting the *free electrons* in a high velocity *flame* or *plasma* to a strong *magnetic field*. The free-electron concentration in the flame is increased by the *thermal ionization* of added substances of low *ionization potential* (e.g. containing sodium or potassium). These electrons constitute a current when they flow between *electrodes* within the flame, under the influence of the external magnetic field. MHD devices are used to top up the output of some power stations at peak periods. The power output per unit of fluid volume of the unit is proportional to $\sigma v^2 B^2$, where σ is the conductivity of the fluid or flame whose velocity is v; B is the *magnetic flux density*.

magnetometer A instrument for comparing strengths of *magnetic fields*, and *magnetic moments*. The deflection magnetometer consists of a short *magnet* with a long, non-magnetic pointer at right angles across it, pivoted at the junction. The pointer swings along a circular scale, thus enabling deflections of the short magnet to be measured.

magnetomotive force MMF. Formerly called the magnetic potential. A quantity analagous to the *electromotive force*. It is defined as the circular integral of the *magnetic field strength* around a closed path, i.e.

$$\oint H \mathrm{d}x$$

It is measured in ampere turns.

magneton A *unit* for measuring the *magnetic moments* of atomic particles. The Bohr magneton, μ_B, is equal to
$$eh/4\pi m_e = 9.274 \times 10^{-24} \text{ A m}^2,$$
where e and m_e are the charge and mass of the *electron* and h is the *Planck constant*. The nuclear magneton, μ_N, is equal to
$$\mu_B . m_e/m_p = 5.05 \times 10^{-27} \text{ A m}^2,$$
where m_p is the mass of the *proton*. The symbols $\boldsymbol{m}_B$ and $\boldsymbol{m}_N$ are sometimes used for the Bohr magneton and the nuclear magneton respectively.

magnetosphere The *space* surrounding the *Earth*, or any celestial body, in which there is a *magnetic field* associated with that body. It includes the *Van Allen radiation belts*. In the solar system the magnetospheres of the magnetic planets, including the Earth, are comet-shaped regions in which the charged particles of the *solar wind* are controlled by the planet's magnetic field rather than the Sun's magnetic field.

magnetostriction A change in the dimensions of *ferromagnetic substances* on magnetization. It arises as a result in changes in the sizes of the *magnetic domains*.

magnetostriction oscillator A device consisting of a central rod (usually of nickel or iron) magnetized by a coil carrying a *direct current* on which an *alternating current* is superimposed. If the a.c. *frequency* coincides with the natural frequency of the rod, the resulting *magnetostriction* causes a mechanical vibration of the rod of considerable *amplitude*. Oscillators of this kind are used to generate *acoustic* and *ultrasonic* waves for a variety of purposes. If a *tuned circuit* is included the frequency of the oscillation is maintained constant.

magnetron A *thermionic valve* capable of producing high power oscillations in the *microwave* region. It consists of a heater, a central *cathode*, and an *anode* with a number of radial segments, all enclosed in an evacuated container, which is situated in the gap of an external *magnet*. The movement of the *electrons* is controlled by a combination of crossed *electric* and *magnetic fields*. It is used extensively in *radar*.

magnification (Of a *microscope* or other optical instrument). The ratio of the linear dimensions of the final image to the linear dimensions of the object. It is also important to take into account the *resolving power* of the instrument.

magnifying glass A *convex lens*. See *microscope, simple*.

magnifying power of a compound microscope The ratio of the *angle* subtended at the eye by the final image to the angle subtended by the object placed at the distance of most distinct vision.

magnifying power of a lens The ratio of the *angle* subtended at the eye by the *virtual image* to the angle subtended by the object when placed at the distance of most distinct vision; this latter is generally taken to be 0.25 metres.

magnitude of stars The apparent magnitude is a measure of the relative apparent brightness of *stars*. A star of any one magnitude is approximately 2.51 times brighter than a star of the next magnitude. E.g. a star of the first magnitude is $(2.51)^3$ times as bright as a star of the fourth magnitude. The absolute magnitude is defined as the apparent magnitude a given star would have at the standard distance of 10 *parsecs*.

Magnox* A group of magnesium *alloys* used for sheathing uranium *fuel elements* in certain types of *nuclear reactor*. See *gas-cooled reactor*.

main-sequence stars See *Hertzsprung-Russell diagram*.

major axis The *axis* of an *ellipse* that passes through both *foci*. See Fig. 15, under *ellipse*.

majority carriers In a *semiconductor*, the type of *carrier* that constitutes more than half the total number of carriers.

Maksutov telescope An astronomical telescope developed by D. D. Maksutov in 1944. It consists of a *concave* spherical *mirror*, the *aberration* of which is reduced by a *meniscus lens*.

malachite Natural *basic* copper carbonate, $CuCO_3.Cu(OH)_2$. A bright green *mineral* used as a gemstone and as a copper *ore*.

malate A *salt* or *ester* or *malic acid*.

maleate A *salt* or *ester* of *maleic acid*.

maleic acid See *butenedioic acid*.

malic acid 2-hydroxybutanedioic acid. $HOOCCH(OH)CH_2COOH$. A white crystalline *organic acid*, m.p. 98 – 99°C. It occurs in unripe apples and other fruits.

malleability The ability to be hammered out into thin sheets.

malonic acid See *propanedioic acid*.

malonyl The *bivalent* group $-OCCH_2CO-$, derived from *propanedioic* (malonic) *acid*.

malonylurea See *barbituric acid*.

malt Grain (usually barley) that has been allowed to germinate and then heated and dried. See *brewing*.

maltase An *enzyme* occurring in *yeast* and other *organisms* that hydrolyzes (see *hydrolysis*) *maltose* into *glucose*.

maltose Malt sugar, maltobiose. $C_{12}H_{22}O_{11}$. A hard crystalline *soluble disaccharide*, less sweet than *cane-sugar*. It is formed in *malt* by the action of the *enzyme amylase* on *starch*.

malt sugar See *maltose*.

mandelic acid $C_6H_5CHOHCOOH$. A white crystalline *optically active* substance, the *racemic form* of which has a m.p. of 120.5°C.; it is used as an *antiseptic*.

manganates 1. Manganate(VI). A salt containing the *ion* $MnO_4{}^{2-}$. They are dark green. **2.** Manganate(VII), permanganate. A salt containing the ion $MnO_4{}^-$. They are purple and are strong *oxidizing agents*.

manganese Mn. Element. R.a.m. 54.938. At. No. 25. A reddish-white hard brittle *metal*, r.d. 7.473, m.p. 1255°C, b.p. 2120°C. It occurs as *pyrolusite*, MnO_2, from which it is extracted by *reduction* with carbon or aluminium. It is used in numerous *alloys*.

manganese bronze Manganese brass. A copper-zinc *alloy* containing up to 4% manganese.

manganese dioxide Manganese(IV) oxide. MnO_2. A heavy black powder that occurs naturally as *pyrolusite*. It is used as a source of manganese metal, as an *oxidizing agent*, in *glass* manufacture, in *Leclanché cells*, as a *catalyst* in the laboratory preparation of oxygen, etc.

manganese steel A very hard variety of *steel* containing up to 13% manganese.

manganin An *alloy* containing 83% copper, 13%–18% manganese, 1%–4%

nickel. As its electrical *resistance* is affected only slightly by change in *temperature* it is used for resistance coils.

mannitol $HOCH_2(CHOH)_4CH_2OH$. A white crystalline *optically active polyhydric alcohol*, the *racemic form* of which has a m.p. of 168°C.; it is found in *kelp* and is used in the manufacture of synthetic *resins* and *plasticizers*.

mannitol hexanitrate $C_6H_8(ONO_2)_6$. A colourless *insoluble* substance, m.p. 112°C., used as an *explosive* and in medicine.

manometer Any instrument used for measuring differences in gaseous *pressure*, especially a U-tube containing mercury.

mantissa The decimal, always positive, portion of a common *logarithm*.

mantle See *Earth*.

map projections The way in which the *Earth*'s spherical surface is projected onto a plane surface. The parallels of *latitude* and the meridians of *longitude* form a grid of intersecting lines on the plane surface. All projections involve some distortion. There are three basic forms of projection. In the 'cylindrical projections', the globe is projected onto a cylinder that touches the *equator*. The common Mercator projection is of this type, parallels of latitude are represented as straight lines of equal length to the equator and the meridians of longitude are equally spaced. It was devised in 1564 by Gerardus Mercator. The more recent (1973) Peters' projection is a modified equal-area cylindrical projection that places greater prominence on third world countries than on Europe. Named after Arno Peters. In the 'conic projections' the parallels and meridians are projected onto a cone. The result is that the parallels appear as concentric arcs from the centre of which the meridians radiate. The 'azimuthal projections' are constructed as if a plane is placed at a tangent to the Earth's surface and the portion of the Earth covered is transferred onto that plane. On this projection all points have their true compass bearings. The appropriate map projection is chosen for the purpose to which the map will be put and also on the part of the Earth to be represented.

marble A *metamorphic rock* consisting of *calcite* or *dolomite*. Pure marble is white but impurities, such as silica, can make it coloured. Commercially, marble is any dolomite or *limestone* that can be cut and polished.

margarine A butter substitute prepared from hydrogenated *oils* (see *hydrogenation of oils*). Milk powder is emulsified with the hydrogenated oil; bacterial action in the milk produces a butter-like flavour; vitamins A and D (see *vitamins*) and suitable colouring materials are added.

Markovnikov rule In the addition of a hydrogen *halide* to an asymmetric *alkene*, the *halogen atom* becomes attached to the carbon atom with the fewer hydrogen atoms. Named after V. V. Markovnikov (1838 – 1904).

Mariotte's law The name often given to *Boyle's law* in France.

Mars (astr.) A *planet*, with two small *satellites* (Phobos and Deimos), having its *orbit* between those of the *Earth* and *Jupiter*. Mean distance from the *Sun* 227.94 million kilometres. *Sidereal period* ('year') = 686.98 days. Mass 0.107 that of the Earth, diameter 6790 kilometres. The atmosphere, composed mainly of *carbon dioxide*, has a pressure of only about 0.01 atmosphere. The polar ice caps are solid carbon dioxide. The day temperature at the equator is about −25°C., dropping to about − 120°C. at night.

marsh gas See *methane*, CH_4.

Marsh's test A sensitive test for arsenic that depends upon the formation of *arsine* when arsenic or its *compounds* are present in a *solution* evolving hydrogen. This is achieved by adding hydrochloric acid and zinc to the sample to produce hydrogen. The arsine so formed is passed through a narrow heated tube, where it decomposes leaving a deposit of metallic arsenic.

martensite The hard and brittle constituent of *steel* produced when the material is cooled from its hardening *temperature* at a greater rate than its critical cooling rate.

mascon A local concentration of high *mass*, below the surface of the *Moon*. They occur in lunar maria as a result of flooding of the basins with *basalt* or some other high-density substance from the lunar mantle.

maser Microwave Amplification by Stimulated Emission of Radiation. A class of *amplifiers* and *oscillators* that makes use of the internal *energy* of *atoms* and *molecules* to obtain low noise-level amplification and *microwave* oscillations of precisely determined *frequencies*. Stimulated emission, which is the basic principle on which these devices work, is the emission by an *atom* in an excited *quantum* state (see *excitation*) of a *photon*, as the result of the impact of a photon from outside of exactly equal energy. Thus the stimulating photon, or wave, is augmented by the one emitted by the excited atom. A maser consists of an 'active medium' (either in the gaseous or *solid state*), in which most of the atoms can be optically pumped to an excited state by subjecting the system to *electromagnetic radiation* of different frequencies to that of the stimulating frequency (see *population inversion*). The active medium is enclosed in a *resonant cavity* so that a wave is built up with only one mode of oscillation, which is equivalent to a single output frequency. Masers can also be made to operate at optical frequencies, when they are referred to as optical masers or *lasers*.

mass m. A characteristic of a material body that can be defined in either of two ways. The *inertial mass* of a body is the constant of proportionality in the relationship $F \propto a$, where a is the acceleration produced when the body is acted on by a force F. The gravitational mass is determined by Newton's law of *gravitation*. The standard of mass, the International Prototype kilogram, is a cylinder of platinum-iridium; it is defined in terms of the inertial mass, but because the inertial mass is equal to the gravitational mass, it is the gravitational mass that is used in practice to determine the *weight* of a body. See also *rest mass*; *relativistic mass*; *mass-energy equation*.

mass action law The rate of a chemical change is proportional to the *active masses* (*molar concentrations*) of the reacting substances. This law applies accurately only to *perfect gases*. In real cases the *activity* can be used in place of the molar concentration.

mass decrement The difference between the *rest mass* of a radioactive *nuclide* and the rest masses of its *decay* products.

mass defect The difference between the *rest mass* of a *nucleus* and the sum of the rest masses of its constituent *nucleons*. The *energy* equivalent of the mass defect, on the basis of the *mass-energy equation*, must be supplied to a nucleus to split it into its component nucleons.

mass-energy equation A quantity of *energy*, E, always has a mass, m, such that $E = mc^2$, where c is the speed of *light* in m s^{-1}, E is in *joules*, and m is in *kilograms*. This is a consequence of Einstein's special theory of *relativity*. In this theory it became clear that the mass of a body is a measure of its total energy content; if the energy increases, e.g. by an increase in its speed or temperature the mass increases (see *relativistic mass*). See also *annihilation radiation*; *conservation of mass and energy*.

massicot A yellow powder consisting of unfused *lead oxide*, PbO.

mass number Nucleon number. *A*. The *integer* nearest to the relative *atomic mass* of an *isotope*, i.e. the number of *nucleons* in the *nucleus* of an *atom*.

mass spectroscopy A method of determining *relative atomic masses* and the relative *abundance* of isotopes in a sample; it is also used in chemical analysis. In the mass spectrometer a sample of the material to be examined, usually in a gaseous form, is ionized. The positive *ions* so produced are accelerated into a region of high vacuum, where they are subjected to both an *electric field* and a *magnetic field*, which focus the ions onto a detector. The mass spectrum so produced consists of a series of peaks as a result of the separation by the fields of ions having different values of *m/e*, where *m* is the mass of the ion and *e* is its charge. Different molecules can be identified by their characteristic mass spectra. Relatively portable instruments avoid the use of heavy magnets by using alternating electric fields to select ions of differing mass by means of their times of flight.

mass spectrum See *mass spectroscopy*.

mast cells Large *cells* found in the connective tissue of vertebrates. They produce *histamine* in certain conditions as well as the anticoagulant *heparin*.

masurium Former name of *element* of At. No. 43; it was replaced in 1949 by the name *technetium*.

matches The heads of safety matches usually contain *antimony trisulphide*, *oxidizing agents* such as *potassium chlorate*, and some sulphur or *charcoal*; while the striking surface contains red phosphorus. Ordinary non-safety match-heads contain phosphorus sulphide, P_4S_3; very rarely red phosphorus.

matrix 1. A mould for shaping a cast. **2.** (math.) An arrangement of mathematical elements into rows and columns according to algebraic rules, in order to solve a set of *linear* equations. **3.** (computers) An array of components for translating from one code to another. **4.** (metallurgy) The crystalline *phase* in an *alloy*, in which the other phases are contained.

matte A mixture of the *sulphides* of iron and copper obtained as an intermediate stage in the *smelting* of copper.

matter A substance that has the attributes of *mass* and extension in *space* and time.

mauve Mauveine, aniline violet. A reddish-violet *dye*; a complex *organic compound*, it was the first organic dye to be prepared artificially.

maximum (math.) A *function* $y = f(x)$ has a maximum value at $x = a$ if $f(a)$ is greater than the values of the function immediately preceding and immediately following $x = a$. The function has a minimum value at $x = b$ if $f(b)$ is less than the value of the function immediately preceding and immediately following $x = b$.

maximum and minimum thermometer See *thermometer*.

maximum permissible dose (or level) See *dose*.

maxwell The *c.g.s. unit of magnetic flux*. The flux through 1 square *centimetre normal* to a *magnetic field* of strength 1 *gauss*. 1 maxwell = 10^{-8} *weber*. Named after James Clerk Maxwell (1831–79).

Maxwell-Boltzmann distribution A statistical *equation* giving the distribution of *velocities* or positions of the *molecules* in a *gas*; it is based on the assumption that each particle has an equal *probability* of appearing in a particular region. Named after James Clerk Maxwell (1831–79) and Ludwig Boltzmann (1844–1906).

Maxwell's equations A series of differential equations that describe the *electromagnetic field* in terms of *electric field strength* and *magnetic flux density*. They form the basis of classical *electrodynamics* and enabled Maxwell to conclude that *light* is a form of electromagnetic radiation and to calculate the *speed of light*, which agreed well with the measured value at that time.

McLeod gauge A vacuum pressure gauge in which a large volume of low-pressure gas is compressed so that at the higher pressure it will support a column of mercury the height of which can be measured accurately. On the basis of *Boyle's law* this enables the lower pressure to be calculated. It is accurate down to 10^{-3} pascal, providing that condensable vapours are not used.

mean (math.) Average. **1.** See *arithmetic mean*. **2.** See *geometric mean*.

mean free path λ. **1.** The average, or *mean*, distance travelled by a particle, *atom*, or *molecule* between collisions. In a *gas*, the mean free path between molecules is inversely proportional to the *pressure*. If n is the number of particles of mean diameter d (assuming that they are spherical):
$$\lambda = 1/\sqrt{2}\pi n d^2.$$
See *kinetic theory of gases*. **2.** In particle physics, the mean distance a particle travels before undergoing an interaction.

mean free time The average, or *mean*, time that elapses between two collisions of a particle, *atom*, or *molecule*.

mean life τ. **1.** The reciprocal of the *disintegration constant*. **2.** The average survival time of an elementary particle, ion, etc., in a particular medium. **3.** The average survival time of an *electron* or *hole* in a *semiconductor*.

mean solar day See *solar day*.

mechanical advantage Force ratio. In a *machine*, the ratio of the actual load raised to the *force* required to maintain the machine at constant speed.

mechanical equivalent of heat If H units of *heat* are completely converted into W units of *work* then $W = JH$, where J is a constant formerly called the mechanical equivalent of heat, or Joule's equivalent. It has the value 4.185×10^7 ergs/calorie. In *SI units*, W and H would both be measured in *joules*, and J would therefore equal 1. Thus, in modern units the concept ceases to be of value.

mechanics The branch of physical science dealing with the behaviour of *matter* under the action of *force*. See *dynamics*; *fluid mechanics*; *statics*; *kinematics*. See also *Newtonian mechanics*; *quantum mechanics*.

mechanistic theory The view that all biological phenomena may be explained in mechanical, physical, and chemical terms, in opposition to the *vitalistic theory*.

median 1. A line joining a vertex of a *triangle* to the mid-point of the opposite side. **2.** The middle number in a sequence of numbers.

median lethal dose (M)LD_{50}. The *dose* of *ionizing radiation* or toxic chemicals that would kill 50% of a large batch of *organisms* within a specified period.

medical physics The application of *physics* to medicine. It includes *radiotherapy*, *nuclear medicine*, *dosimetry*, medical electronics, and the design and maintenance of a wide range of medical equipment using advanced physical techniques.

medium frequency MF. *Radio frequencies* in the range 300–3000 *kilohertz*.

medulla The central part of certain bodily organs, such as the adrenal glands, in which the medulla functions separately from the cortex, which surrounds it.

meerschaum Natural *hydrated* magnesium silicate, $Mg_2Si_3O_8.2H_2O$. It is a white *solid* used for tobacco pipes.

mega- Prefix denoting one million times; 10^6. Symbol M, e.g. MW = 10^6 *watts*. More loosely, denoting 'very large'.

megahertz MHz. 1 million *hertz*. A measure of *frequency* equal to 10^6 *cycles* per *second*.

megaton bomb A *nuclear weapon* with an explosive power equivalent to one million tons of *TNT* (approximately 4×10^{15} *joules*).

megohm One million *ohms*.

meiosis The process by which the *nucleus* of a *diploid* reproductive *cell* divides to produce four *haploid* nuclei – often *gamete* nuclei. Meiosis is divided into two stages, each of which occurs in four phases (prophase, metaphase, anaphase, and telophase). In the first stage the *chromosomes* become associated in pairs and may exchange genetic material, after which they separate into two daughter nuclei, each with half the number of chromosomes of the parent nucleus. In the second stage the chromosomes of these two haploid chromosome sets undergo a *mitosis*-like separation to produce four haploid sets, each of which may be genetically unique.

Meissner effect The reduction in the *magnetic flux* within a *superconductor* when it is cooled to below its *critical temperature* in a *magnetic field*. It was first noted by Walther Meissner.

mel A subjective unit of *pitch* in which a sound judged to have a pitch n times that of a tone of pitch 1 mel is said to have a pitch of n mels. The standard of 1000 mels is set by a tone with a *frequency* of 1000 Hz at a loudness of 40 *decibels* above the threshold of hearing.

melamine $C_3H_6N_6$. Triaminotriazine. A white crystalline cyclic compound with a six membered ring of alternate carbon and nitrogen atoms, m.p. 354°C., that forms a *thermosetting* resin with *methanal*.

melanin $C_{17}H_{98}O_{33}N_{14}S$. A dark brown pigment produced in the skin *cells* called melanocytes. Skin and hair colours in many animals, including man, are due to melanin. The number of melanocytes a person has is genetically determined. The *Sun* stimulates the production of melanin in melanocytes, and the function of the melanin is to absorb the Sun's harmful radiations.

melting point The constant *temperature* at which the *solid* and *liquid phase* of a substance are in equilibrium at a given *pressure*. Melting points are normally quoted for standard atmospheric pressure.

membrane A thin tissue that encloses or lines biological *cells*, organs, or other structures. It consists of a double layer of *lipids* (mostly *lecithin* and cephalin) with *protein* molecules between the two layers. Membranes are permeable to water and fat-soluble substances but not to such *polar molecules* as *sugars*. See also *carrier (bio.)*.

membrane carrier See *carrier (bio.)*.

memory Store. A device in which data and *programs* are stored in a *computer*. The main store (main memory) is closely associated with the central processor and is used for programs and data waiting to be used by the processor. Main stores consist of *semiconductor memories* and may be random-access memories (see *RAM*) or read-only memories (see *ROM*); main stores provide very fast access. The slower but larger capacity memory of a computer is the *backing store*, usually magnetic disks or tapes, on which vast amounts of data and programs are stored before being transferred to the main memory for use by the processor.

Mendeleev's law See *periodic law*. Named after Dimitri Ivanovich Mendeleev (1834–1907).

mendelevium Md. *Transuranic element*, At. No. 101. The most stable *isotope*, mendelevium–258, has a *half-life* of 60 days. The alternative name unnilunium is sometimes used (see *unnil*).

Mendelism The theoretical basis of classical *genetics*. It is named after Gregor Mendel (1822–84) who formulated two laws encapsulating the concept that an *organism*'s characteristics were determined by inherited factors (now called *alleles*).

meniscus 1. The curved surface of a *liquid* in a vessel. If the *contact angle* between the liquid and the wall of the vessel is less than 90°, the meniscus is *concave*; if greater, the meniscus is *convex*. **2.** A *concavo-convex lens*. See Fig. 24, under *lens*.

menstrual cycle The monthly cycle that is associated with *ovulation* in most primates (including humans). It replaces the *oestrous cycle*, which occurs in other mammals. In the first part of the cycle, the lining of the uterus becomes thickened with additional blood vessels. Ovulation occurs in the middle of the cycle; if the *ovum* produced is fertilized by a *spermatozoon* and if *implantation* takes place, the cycle ends and the *zygote* develops into a new individual. If neither fertilization nor implantation occur, the lining of the uterus breaks down and is discharged in the process of menstruation. In an approximately 28-day cycle, women are only fertile for about five days (11th–15th days).

mensuration The measurement of lengths, areas, and volumes.

menthol $C_{10}H_{19}OH$. One of a series of *organic compounds* of the *camphor* group. It is white crystalline *terpene alcohol* that occurs in natural *oils*, m.p. 42°C., with a characteristic smell. It is used in medicine.

mercaptans See *thiols*.

mercaptide See *thiolates*.

Mercator projection See *map projections*.

mercuric A *compound* of mercury in its +2 *oxidation state*, e.g. mercury(II) chloride, mercuric chloride.

mercurous A *compound* of mercury in its +1 *oxidation state*, e.g. mercury(I) chloride, mercurous chloride.

Mercury (astr.) The *planet* with its orbit nearest the *Sun*. Mean distance from the Sun 57.91 million kilometers. Sidereal period ('year') = 87.969 days. Mass 0.054 that of the *Earth*, diameter 4878 kilometres. It has no atmosphere and a day temperature of about 400°C.

mercury Quicksilver, hydrargyrum. Hg. Element. R.a.m. 200.59. At. No. 80. A *liquid*, silvery-white *metal*, r.d. 13.546, m.p. –39°C., b.p. 357°C., which occurs as *cinnabar*, HgS. It is extracted by roasting the *ore* in a current of air. It is used in *thermometers*, *barometers*, *manometers*, and other scientific apparatus; *alloys* (called *amalgams*) are used in dentistry. *Compounds* are poisonous; some are used in medicine.

mercury cell A *primary cell* consisting of a zinc *anode*, a *cathode* of *mercury(II) oxide* (HgO) mixed with *graphite* (about 5%), and an *electrolyte* of *potassium hydroxide* (KOH) saturated with *zinc oxide* (ZnO). The *EMF* is about 1.3 *volts* and by suitable design the cell can be made to deliver about 0.3 *ampere-hour* per cm^3.

mercury chlorides 1. Mercury(I) chloride, mercurous chloride, calomel. Hg_2Cl_2. A white *insoluble* powder, m.p. 3°C., used in medicine and as a *fungicide*. **2.** Mercury(II) chloride, mercuric chloride, corrosive sublimate. $HgCl_2$. A poisonous white *soluble salt*, m.p. 276°C., used as an antiseptic and to make other mercury compounds.

mercury cyanate Fulminate of mercury, mercuric fulminate. $Hg(ONC)_2$. A white crystalline substance that explodes on being struck and is therefore used in detonators to initiate explosions.

mercury(II) oxide HgO. A *soluble* poisonous powder that occurs as either yellow or red *crystals*; it is used as a pigment and as an *antiseptic*.

mercury(II) sulphide HgS. An *insoluble* substance that occurs naturally as *cinnabar*. The pure compound is a red powder, m.p. 583.5°C., which is used as a pigment, known as *vermilion*.

mercury-vapour lamp A lamp emitting a strong bluish *light* by the passage of an *electric current* through mercury *vapour* in a bulb. The light is rich in *ultraviolet radiations*; used in artificial sun-ray treatment and in street lighting. See also *fluorescent lamp*.

meridian, celestial The imaginary *great circle* of the *celestial sphere* passing through the *zenith* and the celestial poles, meeting the horizon at points called the North and South points.

meridian, magnetic An imaginary *great circle* on the Earth's surface that passes through both *magnetic poles*. A compass needle on the Earth's surface that was influenced only by the Earth's *magnetic field* would point along a magnetic meridian.

meridian, terrestrial An imaginary *great circle* drawn round the *Earth* that passes through both poles. See *longitude, lines of*.

mescaline $C_{11}H_{17}NO_3$. A white *soluble* crystalline powder, m.p. 35–36°C., obtained from the mescal cactus and used as a hallucinogen.

mesitylene $C_6H_3(CH_3)_3$. 1,3,5-trimethylbenzene. A colourless *aromatic liquid hydrocarbon*, b.p. 164.7°C., that occurs in *coal-tar* and is used in organic synthesis.

meso- 1. (phys.) A prefix indicating that a substance is optically inactive due to intramolecular compensation. **2.** (bio.) A prefix meaning 'middle'.

mesomerism See *resonance*.

mesons A group of *elementary particles* belonging to the class called *hadrons*. They consist of a *quark* and its antiquark. See also *elementary particles*. Positive, negative, and neutral mesons exist; when charged the magnitude of the charge is equal to that of the electron; they are *bosons* with zero or integral *spin*. Mesons are found in *cosmic rays* and are emitted by *nuclei* under bombardment by high *energy* particles. Mesons include the *kaon* and the *pion*. *Muons* were originally called μ-mesons, but they are now classified as *leptons* rather than mesons.

mesophases *Phases* intermediate between crystalline and liquid phases (see *liquid*

270

crystals; *cybotaxis*). Three different types are recognized: *smectic*, *nematic*, and *cholesteric crystals*, in accordance with the different arrangements of the *molecules* in them.

mesosphere 1. The region of the Earth's *atmosphere* between the *ionosphere* and the *exosphere*, extending from about 400 kilometres to 1000 kilometres above the Earth's surface. It is sometimes considered to be part of the exosphere. **2.** The region of the Earth's atmosphere between the *stratosphere* and the *thermosphere*, extending from some 40 kilometres to 80 kilometres above the Earth's surface.

messenger RNA See *ribonucleic acid.*

mesyl Methylsulphonyl. The *univalent* group $CH_3.SO_2-$.

meta 1. Denoting positions separated by one atom in a hexagonal ring of atoms, particularly the *benzene ring*. Abbreviated to *m-* as a prefix in naming a compound; e.g. *m*-dichlorobenzene is 1,3-dichlorobenzene. Compare *ortho*; *para*. **2.** A prefix indicating an *inorganic acid* (or a corresponding *salt*) of a lower degree of hydration; e.g. *metaphosphoric acid*, HPO_3, as compared with *orthophosphoric acid*, H_3PO_4.

metabolism The chemical processes associated with living *organisms*. It is usually divided into two parts: *catabolism*, as a result of which complex substances are decomposed into simple ones, sometimes with the release of *energy*, which becomes available for the organism's activities; and *anabolism*, which comprises the building up of complex substances with the absorption or storage of energy. Metabolic reactions are usually under the control of *enzymes*, which are consequently of immense importance in the chemistry of life. Metabolism usually proceeds in a number of steps (e.g. the *citric-acid cycle*), the sequence of which is known as a metabolic *pathway*.

metabolite Any substance that takes part in the process of *metabolism*.

metal A substance having a 'metallic' lustre and being malleable, ductile, of high *relative density*, and a good *conductor* of *heat* and *electricity*. *Elements* having such physical properties to a greater or less degree are generally *electropositive* and usually combine with oxygen to give *bases* (although some form *amphoteric* oxides); their *chlorides* are stable towards *water*. A number of elements normally regarded as metals have only some of the above properties. See *metalloid*.

metaldehyde Meta. A white, *volatile*, flammable poisonous solid *polymer* of *ethanal* (acetaldehyde) CH_3CHO. It is used as *fuel* in small heaters and as a slug killer.

metallic bond See *metallic crystal*.

metallic crystal The type of *crystal* formed by most *metals*, in which the outer *electrons* of the metallic *atoms* are shared by the crystal as a whole. Thus, the positively charged metal *ions* in the crystal *lattice* are surrounded by a 'gas' of *free electrons*. The ions are held in position by the *electrostatic* force (sometimes called the metallic bond) between the ions and electrons. These free electrons account for the fact that most metals are good conductors of *heat* and *electricity*.

metallic soap An *insoluble salt* formed by a *metal* and a *fatty acid* (especially salts of lead and aluminium). It is used for waterproofing textiles and as a drier for *paints*.

metallized dyes See *acid dyes*.

metallocene See *ferrocene*.

metallography The microscopic study of the crystalline structure of *metals* and *alloys*.

metalloid Semimetal. An *element* having some properties characteristic of *metals* and others of non-metals. Many metalloids give rise to an *amphoteric oxide* (e.g. *arsenic* or *antimony*) and many are *semiconductors*.

metallurgy The science and technology of *metals*; in particular, the extraction of metals from their *ores*, their heat treatment, and the compounding of *alloys*.

metamerism 1. (chem.) A type of *isomerism* exhibited by *organic compounds* of the same chemical class or type; it is caused by the attachment of different *radicals* to the same central *atom* or group. E.g. diethyl *ether* $(C_2H_5)_2O$, and methyl propyl ether, $CH_3OC_3H_7$. **2.** (bio.) Segmentation. The division of the body of an animal into similar units, as in the earthworm.

metamict minerals *Minerals* in which the crystalline structure has been disrupted by *alpha particles* produced by *radioactive nuclei* within the minerals.

metamorphic rock Rock, such as *marble* or *slate*, formed from existing rock, such as *limestone* and *shale* respectively, by heat, pressure, or chemical fluids. Compare *igneous rock*; *sedimentary rock*.

metamorphosis The change from a larva to an adult form during the life cycle of an invertebrate or amphibian (e.g. the change from a tadpole to a frog).

metaphosphoric acid See *phosphoric acids*.

metastable state (chem.) The state of supercooled *water* (see *supercooling*) or of supersaturated *solutions* (see *supersaturation*) in which the *phase* that is normally *stable* under the given conditions does not form unless a small amount of the normally stable phase is already present. Thus supercooled water will remain as *liquid* water below 0°C. until a small *crystal* of *ice* is introduced.

metastable state (phys.) An excited state (see *excitation*) of an *atom* or *nucleus* that has an appreciable life-time.

metathesis (chem.) See *double decomposition*.

meteor A *solid* body from outer *space*. A meteor becomes incandescent ('shooting star') on entering the Earth's atmosphere owing to the frictional *forces* set up at its surface. Small meteors burn up completely in the atmosphere, but some of the larger ones survive and fall to Earth as meteorites. Meteorites are of two kinds, those that are predominantly stone and those predominantly iron. The largest meteorites can weigh up to 100 tons. Every day some million meteors enter the Earth's atmosphere and some 10 tons of meteorite material are added to the planet's surface. See also *micrometeorite*.

meteorite See *meteor*.

meteorology The science of the weather; the study of such conditions as atmospheric *pressure*, *temperature*, *wind* strength, *humidity*, etc., from which conclusions as to the forthcoming weather are drawn. Modern meteorology is based on the information provided by weather stations and *satellites*.

meteor showers Exceptionally heavy falls of *meteors* (about 20 times greater than the average) that enter the Earth's *atmosphere* when the Earth's *orbit* crosses the orbit of a *comet*, i.e. an orbit that contains either the material of which comets are made or into which they disintegrate.

-meter Suffix denoting measurer; e.g. *voltmeter*.

methacrylate A *salt* or *ester* of *methacrylic acid*.

methacrylic acid 2-methylpropenoic acid. $CH_2:C(CH_3)COOH$. A corrosive *liquid*, m.p. 15°C., b.p. 163°C. The *polymer* of its methyl *ester*, methyl methacrylate, is an important *plastic* (Perspex*).

methanal Formaldehyde. HCHO. A *gas* with an irritating smell, b.p. −19°C. The simplest *aldehyde*, it is made by oxidizing *methanol* at 500°C with air over a silver *catalyst*. A *trimer*, methanal trimer, $C_3O_3H_6$, consisting of alternate -O- and -CH_2- groups in a six-membered ring, forms when an acidic solution of methanal is distilled. A solid polymer, polymethanal (see *paraformaldehyde*), is formed by evaporating an aqueous solution of methanal. See also *formalin*.

methanal trimer See *methanal*.

methane Marsh gas, fire-damp. CH_4. The first *hydrocarbon* of the *alkane series*. An odourless, flammable *gas*, b.p. −161.5°C., that forms an explosive mixture with air. It is formed from decaying *organic* matter and in coalmines; it is the main constituent of *natural gas* and an important starting material for producing other organic compounds.

methanoate Formate. A *salt* or *ester* of *methanoic acid* (formic acid).

methanoic acid Formic acid. HCOOH. A colourless, corrosive fuming liquid with a pungent smell, m.p. 8.4°C., b.p. 100.5°C. It occurs in various plants and in ants; it is made industrially from sodium methanoate, HCOONa, which is produced by the action of carbon monoxide on sodium hydroxide. The simplest of the *carboxylic acids*, it is used in dyeing, tanning, and electroplating.

methanol Methyl alcohol, wood spirit. CH_3OH. A colourless, poisonous liquid, b.p. 64.6°C., formerly obtained as wood naphtha by the *destructive distillation* of wood. Now made by the catalytic *oxidation* of *methane*, it is used to *denature methylated spirit*, as a *solvent*, and in the chemical industry (largely as a starting material for making *urea-formaldehyde resins*).

methionine An *amino acid* found in *casein*, wool, and other *proteins*, used in the treatment of certain liver diseases. See Appendix, Table 5.

methoxy The *univalent* group $CH_3O–$.

methoxybenzene Anisole, methyl phenyl ether. $CH_3OC_6H_5$. A colourless liquid with an aromatic odour, b.p. 155.4°C., used in perfumes and as a vermicide.

methyl The *univalent* organic group $CH_3–$.

methyl alcohol See *methanol*.

methylamine CH_3NH_2. A *gas* with an odour of *ammonia*, b.p. −6.3°C.

methylated spirit A *liquid fuel* consisting, by *volume*, of 90% *ethanol*, 9.5% *methanol*, 0.5% *pyridine*, together with small amounts of *petroleum* and methyl violet dye. Industrial methylated spirit (IMS) is free from *pyridine*; it consists of *ethanol* with 5% *methanol*.

methylbenzene Toluene, toluol. $C_6H_5CH_3$. A hydrocarbon of the *benzene* series. A colourless flammable *liquid* with a characteristic odour, b.p. 110°C. It occurs in *coal-tar* or can be made from methylcyclohexane from *petroleum* and is used as a solvent and in the manufacture of *TNT*, *saccharin*, and drugs.

methyl chloride See *chloromethane*.

methyl cyanide See *acetonitrile*.

methylcyclohexanol $CH_3C_6H_{10}OH$. A colourless *viscous liquid* consisting of a mixture of *isomers* with b.p. in the range 167–174°C; it is obtained from *methylphenol* and used as a *solvent* for *rubber* and *cellulose*.

methylene See *carbene*.

methylene blue $C_{16}H_{18}N_{13}SCl$. A *soluble*, intense blue *dye*. Used as a dyestuff, in medicine, and as a stain in *biology*.

methyl methacrylate See *methacrylic acid*; *polymethyl methacrylate*.

methylol Hydroxymethyl. The *univalent* group $HO.CH_2-$.

methyl orange $C_{14}H_{14}N_3NaO_3S$. An orange *indicator*, used in acid-base titrations. It is red below a *pH* of 3.1 and yellow above 4.4.

methylphenols Cresols. $CH_3C_6H_4OH$. A *liquid aromatic* mixture of *compounds* obtained from *coal-tar*. Consisting of three *isomers*, which boil in the range $191-203°C.$, they are used in the *plastics*, explosives, and *dye* industries, and as a disinfectant (Lysol).

2-methylpropenoic acid See *methacrylic acid*.

methylpyridine See *picoline*.

methyl red $C_{15}H_{15}N_3O_2$. A dark red *indicator*, used in acid-base titrations. It is red below a *pH* of 4.4 and yellow above 6.0.

methyl salicylate Oil of wintergreen, methyl 2-hydroxybenzenecarboxylate. $OH.C_6H_4COOCH_3$. A colourless oil, b.p. $223.3°C.$, used in flavours, perfumes, and medicine.

metol 4-methylaminophenol. $CH_3NH.C_6H_4OH$. A white crystalline *compound*, m.p. $87°C$. It is used as a developer in *photography*. The same name is often applied to the *sulphate* of the compound.

metre m. The *SI unit* of length, defined since 1983 as the length of the path travelled by *light* during 1/299 792 458 of a *second*. The unit was introduced in France in 1791 with the *metric system* and was intended to be 1/10 000 000 of the quadrant of the Earth's *meridian* passing through Paris. The surveys attempting to establish this standard proved unworkable and in 1793 the original metre bar, made of platinum and known as the *mètre des archives*, was made. This was replaced in 1960 by a definition based on 1 650 763.73 wavelengths of the radiation emitted in the transition $2p_{10}$ to $5d_5$ of the *nuclide* krypton-86.

metre bridge See *Wheatstone bridge*.

metre-candle See *lux*.

metric system (units) A system of *weights* and measures originally devised by the French Academy in 1791 and based upon the *metre*. See *c.g.s. system*; *m.k.s. system*; *SI units*.

metric ton Tonne. 1000 *kilograms*; 2204.61 lb, 0.9842 ton.

metrology The scientific study of weights and measures.

MeV Million *electronvolts*.

MF See *medium frequency*.

MHD See *magnetohydrodynamics*.

mho Reciprocal ohm. The unit of electrical *conductance* now known as the *siemens*.

mica A group of *minerals*, the most important of which are muscovite, $H_2KAl_3(SiO_4)_3$, and phlogopite, $H_2KMg_3Al(SiO_4)_3$. Naturally occurring mica can be split along its cleavages into small thick pieces ('blocks') or thin sheets ('splittings'). Being an excellent insulator and being resistant to high *temperatures*, mica is used as a *dielectric* in *capacitors*, as a support for heating elements in irons, etc. As mica is also transparent it is used for inspection windows of furnaces. Micanite* sheet is manufactured by bonding mica splittings with *shellac* or synthetic *resins*.

micelle A cluster or group of associated (see *association*) *molecules*, especially in a *colloidal solution*.

Michelson-Morley experiment An experiment carried out in 1887 to measure the *velocity* of the Earth through the *ether*, by measuring the effect that such a velocity would have upon the speed of *light*. It was hoped to be able to detect a shift in the *interference* fringes when the *interferometer* used was rotated through 90°. No such motion of the Earth relative to the ether was detected: a result of the greatest importance for the theory of *relativity*. It also led to the abandonment of the ether concept. The result was explained by the *Lorentz-Fitzgerald contraction*. Named after Albert A. Michelson (1852–1931) and Edward Morley (1838–1923).

micro- 1. Prefix denoting one-millionth; 10^{-6}. Symbol μ, e.g. $\mu A = 10^{-6}$ *ampere*. **2.** Prefix meaning 'very small'; on a small scale. Compare *macro-*.

microbalance A *balance* for weighing objects of very small *mass*, i.e. down to the order of 10^{-8} g.

microbiology The branch of *biology* concerned with the structure and behaviour of *microorganisms*.

microcomputer A small personal computer using a *microprocessor* as its control unit. It usually includes a *VDU* and a printer, and an input keyboard. *Memory* is usually on floppy disks or cassettes. See also *word processor*.

microcosmic salt See *ammonium sodium hydrogen orthophosphate*.

microelectronics The design, manufacture, and use of *electronic* units using extremely small *solid-state* components, especially those based on *integrated circuits* and silicon *chips*.

microfarad μF. One-millionth of a *farad*.

micrometeorite A *meteor* with a diameter of less than 1 mm. They survive atmospheric friction and reach the surface of the Earth because their small *mass* compared to their relatively large surface area enables them to radiate away the *heat* produced by *friction* before they vaporize.

micrometer An instrument for the accurate measurement of small distances or *angles*.

microminiaturization The techniques or the devices used in *microelectronics*.

micron One-millionth of a *metre*. The former name for a micrometre.

microorganism An *organism* that can only be seen with the aid of a *microscope*. Examples are *bacteria*, some *fungi*, *viruses*, and unicellular plants and animals. Microorganisms are important contributors to the Earth's *ecosystem*; they produce more than 80% of the *organic* carbon compounds in the oceans and on land some 30% of organic nitrogen occurs in fungi and bacteria that are invisible to the human eye.

microphone A device for converting *sound* waves into an electric signal, which may then be reconverted into sound after transmission by wire or *radio*. One common type consists of a diaphragm in contact with, or close to, loosely packed carbon granules. The vibration of the diaphragm set up by sound disturbs the packing of the carbon granules and alters the electrical *resistance* of the carbon. Thus an *electric current* flowing through the carbon will vary in a manner that depends upon the *frequency* and intensity of the vibrations produced by the sound on the diaphragm. See also *capacitor microphone*; *crystal microphone*; *ribbon microphone*.

microphotometer A special form of *densitometer* enabling density variations over a very small area of the image to be measured.

microprocessor A *semiconductor chip* or set of chips that functions as the central processing unit of a *microcomputer*. They were introduced in the 1970s and their peformance has steadily improved since then.

PRINCIPLE OF COMPOUND MICROSCOPE

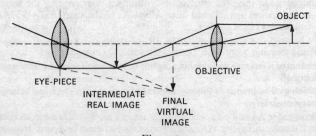

Figure 26.

microscope, compound An instrument consisting essentially of two *converging lenses* or systems of lenses called the *objective* and the *eyepiece* respectively. The objective, which is nearest the viewed object, forms a real inverted magnified *image* of the object just inside the focal distance (see *focal length*) of the eyepiece. This image is viewed through the eyepiece, which then acts as a *simple microscope* producing an inverted further magnified *virtual image*. See Fig. 26. The useful *magnification* obtainable with an optical microscope is limited by the *wavelength* of visible *light* as two points on a microscopic specimen cannot be distinguished from each other if they are not as far apart as half the wavelength of the light used to illuminate them. Thus for magnifications in excess of about 1500, an *ultraviolet microscope* or an *electron microscope* must be used. The binocular microscope has two eyepieces, light from the objective being split into two beams by means of a *prism*. It has the advantage of being more comfortable to use and gives greater perception of depth. The stereoscopic microscope has two eyepieces and two objectives and provides stereoscopic vision. See also *atomic force microscope*; *field-emission microscope*; *field-ionization microscope*; *phase-contrast microscope*; *scanning tunnelling microscope*.

microscope, simple Magnifying glass. A *convex lens* used to produce a virtual *image* larger than the viewed object. In Fig. 26, the *eyepiece* is used as a simple microscope.

microtome An apparatus for cutting thin (3 – 5 micrometres) sections of material, for microscopical examination.

microwave background *Electromagnetic radiation* of *wavelengths* in the *microwave* range that occur in *space* and are thought to have originated in the big bang with which the *Universe* began (see *big-bang theory*). The *energy density* of this radiation in space is of the order 10^{-14} J m^{-3}.

microwaves *Electromagnetic radiation* with *wavelengths* ranging from very short *radio* waves almost to the *infrared* region; i.e. wavelengths from 30 cm to 1 mm.

microwave spectroscopy The measurement of the absorption or emission of *electromagnetic radiation* in the *waveband* 0.1 mm to 10 cm by atomic or molecular systems. It is used to study molecular structure and is based on the changes that *microwaves* cause to the rotational energy levels of molecules and their consequent absorption at characteristic *frequencies*. See *electron spin resonance*.

migration 1. (phys.) The movement of *ions* in an *electrolyte* when a *current* is passed through it. Because the *cations* and *anions* can migrate at different speeds, they can transport different fractions of the current through the electrolyte (see *transport number*). **2.** (chem.) The movement of an *atom*, group of atoms, or a *double bond* from one part of a *molecule* to another. **3.** (bio.) The movement of a group or population of animals from one environment to another, usually on a seasonal basis as a response to temperature changes causing a depletion in the supply of food.

mil One thousandth of an inch.

milk The fluid secreted by female mammals to provide food for their offspring. Cows' milk contains 3.3% *protein*, 3.6% *lipids*, 4.7% *lactose*, and up to 88% *water*, with some *vitamins* and *minerals*. The composition varies from species to species; human milk contains more lactose and less protein than cows' milk.

milk of lime A *suspension* of *lime* in *water*.

milk of magnesia See *magnesium hydroxide*.

milk sugar See *lactose*.

Milky Way Originally the luminous band of *stars* encircling the heavens. It is now known that these stars are members of the *Galaxy* to which the *Solar system* belongs, and the Galaxy is therefore often referred to as the Milky Way.

milli- Prefix denoting one thousandth; 10^{-3}. Symbol m, e.g. mA = 10^{-3} *ampere*.

milliammeter A sensitive *ammeter* graduated to measure *milliamperes*.

milliampere mA. 1/1000 *ampere*.

millibar A unit of atmospheric pressure, used in *meteorology*. 1000 *dynes* per square centimetre or 100 *pascals* approximately equal to 1/32 inch of mercury. See *pressure*.

milligram mg. 1/1000 *gram*; 0.0154 *grain*.

millilitre ml. See *litre*.

millimetre mm. 1/1000 *metre*; 0.0393701 inch.

mineral 1. A natural *inorganic* substance having a chemical composition in a characteristic range and specific properties. See also *rock*. The names of many minerals end in -ite. **2.** A natural *organic* mixture of substances having no single chemical formula, e.g. *coal*, oil (see *petroleum*), and *natural gas*. In this sense a mineral is often taken to mean any material taken from the Earth, especially by mining.

mineralocorticoid See *corticosteroid*.

mineralogy The study of *inorganic minerals*. It is a branch of *geology*.

mineral oil See *paraffin oil*; *petroleum*.

minicomputer Originally a *computer* that could be contained in a single desk-top cabinet. The term was introduced before the term *microcomputer*, which now encompasses the devices formerly called minicomputers.

minim British fluid measure; 1/60 of a fluid drachm; 0.0591 cm^3. See *apothecaries' fluid measure*.

minimum (math.) See *maximum*.

minium See *lead oxides*.

Minkowski's geometry A four-dimensional geometry devised by Hermann Minkowski (1864–1909) in 1907 to give a mathematical formulation of the special theory of *relativity*. It is based on the concept that an event is specified by three spatial coordinates and one time coordinate. The motion of a body can then be described by a curve in this *space-time* coordinate system, which is sometimes called Minkowski's space-time.

minority carriers In a *semiconductor*, the type of charge *carrier* that constitutes less than half the total number of carriers.

minor planets See *asteroids*.

minute 1. A unit of time equal to 60 *seconds*. Although not an *SI unit* it can be used with SI units. **2.** A unit of angle equal to 60 *seconds* of arc, i.e. 1/60 *degree* or 0.291 milliradian.

mirror A surface that reflects regularly most of the *light* falling upon it, thus forming *images*. See *mirrors, spherical*; *reflection*.

mirror image An *image* of an object as viewed in a *mirror*; it is reversed in such a way that the image bears to the object the same relation as a right hand to a left.

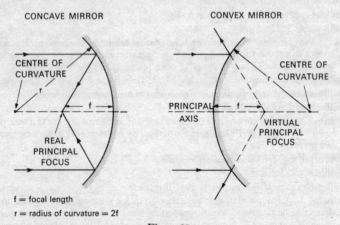

CONCAVE MIRROR   CONVEX MIRROR

CENTRE OF CURVATURE

CENTRE OF CURVATURE

PRINCIPAL AXIS

VIRTUAL PRINCIPAL FOCUS

REAL PRINCIPAL FOCUS

f = focal length
r = radius of curvature = 2f

Figure 27.

mirrors, spherical *Mirrors* the reflecting surfaces of which form a portion of a *sphere*. The surface of such a mirror may be regarded as being made up of an infinitely large number of very small plane mirrors, each at a *tangent to the curve* of the mirror. Thus a *ray* of incident *light* would be reflected at any point as if from such a small plane mirror. Spherical mirrors may be *convex*, with the reflecting surface on the outside of the sphere, or *concave*. The centre and radius of the sphere of which the mirror is considered to form a part, are called the *centre* and *radius of curvature*; the centre of the mirror is the pole, and the line joining the centre of curvature to the pole is the axis. The principal focus (see *focus*) is at a point halfway between the pole and the centre of curvature. Regarding all distances as measured from the mirror and taking all distances in the direction opposite to that of the incident light as positive, the following relationship holds for spherical mirrors:

$$1/v + 1/u = 1/f = 2/r,$$

where u and v are the distances of object and image from the mirror, r the radius of curvature, and f the focal length. See Fig. 27. Spherical mirrors suffer from spherical *aberration* but not chromatic aberration.

misch metal An *alloy* of cerium with small amounts of other *rare earth metals*. It is used for 'flints' in automatic lighters and in various non-ferrous alloys.

miscible Capable of being mixed to form a *homogenous* substance; it is usually applied to *liquids*, e.g. *water* and *alcohol* are completely miscible.

mispickel Arsenical pyrites. A natural *sulphide* of iron and arsenic, FeAsS.

missing mass The *mass* of the hypothetical quantity of *matter* in the *Universe* that cannot be observed. The existence of this dark matter or hidden matter (matter that has mass but that emits no *radiation*) has been postulated on the grounds that the density of the Universe appears to exceed the value that would be obtained on the basis of the visible matter. In particular, it has been found that the mass of some large clusters of *galaxies* is considerably too small to account for the *gravitational forces* that bind them together. The missing mass and dark matter has not been accounted for but suggestions include the existence of many planetary bodies throughout the Universe and of various undetected massive particles, such as weakly interacting massive particles (WIMPS) and massive *neutrinos*.

mist Droplets of *water*, formed by the *condensation* of *water-vapour* on dust particles.

mitochondria (singular **mitochondrion**) *Membrane*-bounded *organelles* in the *cytoplasm* of most eukaryotic cells (see *eukaryote*); they are rounded and elongate, $1-2\ \mu$m in diameter. Mitochondria contain many of the *enzymes* of the cell, particularly those that produce *energy* from *aerobic respiration* (the *citric-acid cycle* and *electron transport chain*), hence they have been called the 'powerhouse of the cell'.

mitosis The process by which eukaryotic *cells* (see *eukaryote*) reproduce. It is divided into four stages. 1. Prophase, during which the chromosomes appear in the *nucleus* as duplicated threads, which become shorter and thicker. 2. Metaphase, during which the nuclear membrane dissolves and a spindle forms, to the centre of which the chromosomes attach themselves. 3. Anaphase, during which the duplicates of the chromosomes separate and migrate to the ends of the spindle. 4. Telophase, during which two nuclear membranes form, each enclosing one set of chromosomes. The *cytoplasm* also divides in this stage, so that two new cells are formed, each with a nucleus containing a set of chromosomes identical to that of the parent cell. Compare *meiosis*.

mixed crystals See *solid solutions*.

mixtures Mechanical mixtures. Mixtures differ from chemical *compounds* in the following respects: 1. The constituents may be separated by suitable physical or mechanical means. 2. Most mixtures may be made in all proportions; in the case of *solutions*, which may be regarded as molecular mixtures, there are often limits of *solubility*. 3. No *heat* effect (except in the case of solutions) is produced on formation; the formation of chemical compounds is invariably accompanied by the evolution or absorption of *energy* in the form of heat. 4. The properties of a mixture are an aggregate of the properties of the constituents, whereas a compound has individual properties, often quite unlike those of the component *elements*.

m.k.s. system Giorgi system. A system of *units* derived from the *metre, kilogram*, and *second*. The ampere was used with m.k.s. units; later the system was rationalized and known as the MKSA system. In the rationalized system the *magnetic constant* was given the value $4\pi \times 10^{-7}$ henry per metre. *SI units*, which are based on the MKSA system, are now used for all scientific purposes.

MMF See *magnetomotive force*.

mmHg A unit of pressure equal to one millimetre of mercury. 1 mmHg = 133.322 *pascals*.

modem Modulator-demodulator. A device that converts the signals from one type of electronic system into signals that are suitable for another type. Modems are required to convert the *digital* signal from a *computer* into an *analogue* signal for transmission over a telephone line – and for converting incoming analogue signals into digital form.

moderator A substance used in *nuclear reactors* to reduce the speed of *fast neutrons* produced by *nuclear fission*. These substances consist of *atoms* of light *elements* (e.g. *deuterium* in *heavy water, graphite, beryllium*) to which the neutrons are able to impart some of their *energy* on collision, without being captured. Neutrons which have been slowed down in this way are called *thermal neutrons* and are much more likely to cause new fissions of uranium-235 than they are to be captured by uranium-238.

modulation The process of varying some characteristic of one wave (usually a *radio frequency carrier wave*) in accordance with some characteristic of another wave. The main types are *amplitude, frequency*, and *phase modulation*. See also *pulse modulation; velocity modulation*.

module 1. A *unit* used as a standard, especially in architecture. **2.** A detachable section of a *spacecraft*. **3.** A detachable unit in a *computer* system.

modulus 1. A constant *factor* or multiplier for the conversion of *units* from one system to another. See also *elastic modulus*. **2.** See *Argand diagram*. **3.** See *absolute value*.

Moebius strip A rectangular ribbon-shaped strip of paper or material one end of which has been twisted through 180 degrees before attaching it to the other end. This forms a single continuous surface, bounded by a continuous curve. Named after A. F. Moebius (1790–1868).

Moho Mohorovicic discontinuity. The discontinuity between the *Earth's* crust and its underlying mantle. It lies some 30–40 kilometres below the surface of the land and some 5–12 kilometres below the ocean floor. Earthquake waves suffer an abrupt increase of speed at this discontinuity. Named after Andrija Mohorovicic (1857–1936).

Mohs scale of hardness A scale in which each *mineral* listed is softer than (i.e. is scratched by) all those below it. 1. *Talc*. 2. *Gypsum*. 3. *Calcite*. 4. *Fluorite*. 5.

Apatite. 6. *Orthoclase*. 7. *Quartz*. 8. *Topaz*. 9. *Corundum*. 10. *Diamond*. Named after Friedrich Mohs (1773–1839).

molality Molal concentration. A method of expressing the strength of a *solution* (see also *concentration*): the number of *moles* of *solute* per *kilogram* of *solvent*.

molar When the adjective 'molar' is used before the name of an extensive physical property, it implies 'divided by the amount of substance'. This usually, but not always, means 'per *mole*'. It is often denoted by the use of the subscript m, e.g. V_m for *molar volume*. In some exceptional cases 'molar' is used to mean 'divided by *concentration*'.

molar concentration The *concentration* of a *solution* expressed in *moles* per unit *volume*, usually mol dm^{-3}.

molar gas constant See *gas constant*.

molar heat capacity C_m. The *heat capacity* of a substance, divided by the amount of substance. The amount of *heat* required to raise the *temperature* of 1 *mole* of a substance by 1 *kelvin*. Expressed in *joules* per mole per kelvin (*SI units*), or *calories* per *gram-molecule* per °C. (*c.g.s. units*).

molarity A former word for *concentration* expressed in *moles* of *solute* per cubic decimetre of *solvent*. However, owing to its confusion with *molality* its use for this purpose is now deprecated.

molar solution An obsolete expression for a *solution* with a *concentration* of 1 *mole* per dm^3.

molar volume Molecular volume. V_m. The volume occupied by 1 *mole* of a substance. All gases have approximately equal molar volumes under the same conditions of *temperature* and *pressure*. At 760 mmHg and 0°C., the molar volume of a *perfect gas* is 22.413 837 dm^3 per mole.

mole The basic *SI unit* of amount of substance. The amount of substance that contains as many elementary units as there are *atoms* in 0.012 kg of carbon-12. The elementary units must be specified and may be an atom, *molecule, ion, radical, electron*, etc., or a specified group of such entities. For example, 1 mole of HCl has a mass of 36.46 g: i.e. 1 mole of a compound has a mass equal to its *relative molecular mass* in grams. 1 mole of electrons has a mass of 5.486×10^{-4} g, i.e. $m_e \times N_A$ (see Appendix, Table 2).

The mole has replaced such former units as the *gram-atom, gram-molecule, gram-ion*, and *gram-equivalent*. Symbol *mol*.

molecular biology The study of the structure of the *molecules*, such as *proteins* and *nucleic acids*, that are of importance in *biology*. See also *molecular genetics*.

molecular compounds Chemical *compounds* formed by the chemical combination of two or more complete *molecules*. E.g. the *hydrates* of *salts*.

molecular distillation The *evaporation* of *molecules* from a surface, at *pressures* of about 10^{-2} mmHg, and their subsequent *condensation* under such conditions that their *mean free path* is of the same order as the distance between the heated and cooled surfaces, i.e. so that the liquid molecules have a low probability of having collisions with other molecules. It is used for *isotope separation* and distilling heat-sensitive *organic compounds*.

molecular flow See *Knudsen flow*.

molecular formula A *formula* of a chemical *compound*, showing the kind and the number of *atoms* present in the *molecule*, but not their arrangement. See *structural formula*.

molecular gauge A type of *pressure gauge* for measuring a low gas *pressure* ($10^{-1} - 10^{-5}$ Pa). A common type uses a small propellor disc turning at constant rate to turn a second propellor disc mounted close to it by means of the viscous drag of the intervening gas. The *couple* acting on the second disc is a measure of the gas pressure, being related to the number of *molecules* per unit volume of the gas. Such gauges are usually calibrated using a *McLeod gauge*.

molecular genetics The branch of *molecular biology* concerned with analysing the structure of *genes* and with *gene sequencing*.

molecular orbital See *orbital*.

molecular sieves See *zeolites*.

molecular spectrum The *absorption spectrum* of *molecules*. It is caused by transitions between different states of molecular rotation, vibration, etc.

molecular volume See *molar volume*.

molecular weight See *relative molecular mass*.

molecule The smallest portion of a chemical *compound* capable of existing independently and retaining the properties of the original substance.

mole fraction Mol fraction. The ratio of the number of *moles* of a particular component of a *mixture*, to the total number of moles present in the mixture.

molybdate A *salt* of *molybdic acid*.

molybdenum Mo. Element. R.a.m. 95.94, At. No. 42. A hard white *metal* resembling iron, r.d. 10.222, m.p. 2623°C., b.p. 4630°C., that occurs as molybdenite, MoS_2. It is extracted by roasting the *ore* and reducing the *oxide* so formed in an electric furnace with carbon. It is used for special *steels* and *alloys* and the sulphide is used as a lubricant.

molybdenum trioxide Molybdic anhydride MoO_3. A yellow crystalline substance, m.p. 795°C., used in the manufacture of molybdenum *compounds*.

molybdic acid H_2MoO_4. A yellow crystalline substance that loses a *molecule* of *water* at 70°C. to form *molybdenum trioxide* (molybdic anhydride).

moment, magnetic See *magnetic moment*.

moment of force A measure of the tendency of a *force* to cause angular acceleration of the body to which it is applied. It is measured by multiplying the magnitude of the force by the perpendicular distance from the line of action of the force to the *axis* of rotation.

moment of inertia The moment of inertia I of a body about any *axis* is the sum of the products of the *mass*, dm, of each element of the body and the *square* of r, its distance from the axis. $I = \Sigma r^2 dm$. If the body is subjected to a *torque* T, giving it an angular *acceleration* α, then $I = T/\alpha$.

momentum 1. p The linear momentum is the product of the *mass* and the *velocity* of a body. For speeds approaching that of *light*, the variation of mass with velocity must be taken into account, and the value of m appropriate to the velocity of the body must be used in the expression for the momentum. See *relativistic mass*. **2.** See *angular momentum*.

momentum, conservation of See *conservation of momentum*.

monad An *element* having a *valence* of one.

monatomic molecule A *molecule* of an *element*, consisting of a single *atom* of the element. E.g. the molecules of the *inert gases*.

monazite A *mineral* containing *phosphates* of cerium, thorium, and other *rare earths*, with some occluded helium.

Mond process The extraction of nickel by the action of *carbon monoxide*, CO, on the impure *metal*. This gives *nickel carbonyl*, $Ni(CO)_4$, a *gas* that decomposes when heated to 200°C. into pure nickel and carbon monoxide, the latter being used again. Named after Ludwig Mond (1839–1909).

Monel metal* An *alloy* of copper (25%–35%), nickel (60%–70%) and small amounts of iron, manganese, silicon, and carbon. It is used as an *acid*-resisting material in chemical industry.

mono- Prefix denoting one, single.

monobasic acid An *acid* having one *atom* of *acidic* hydrogen in a *molecule*; an acid giving rise to only one series of *salts*. E.g. *nitric acid*, HNO_3.

monochromatic radiation *Radiation* consisting of vibrations of the same or nearly the same *frequency*; especially light of one *colour*. Compare *polychromatic radiation*.

monoclinic Relating to *crystals* that have three unequal axes with one oblique intersection.

monoclonal antibody A specific *antibody* that can be produced artificially in large quantities and used to distinguish the major blood groups and to produce highly specific *vaccines*. The antibody is produced by a single *clone* of identical *cells* that are derived from a single parent cell obtained by fusing a normal antibody-producing lymphocyte (see *leucocyte*) with a cell from a malignant tumour in the lymphoid tissue of a mouse. These hybrid cells multiply rapidly in the appropriate culture and yield large quantities of antibody.

monohydrate Containing one *molecule* of *water*.

monohydric Containing one *hydroxyl group* in a *molecule*.

monolayer Monomolecular layer. A layer or film one molecule thick.

monomer A chemical *compound* consisting of single basic groups of *atoms*, as opposed to a *polymer*, the molecules of which are built up by the repeated union of monomer molecules. See *polymerization*.

monosaccharides Simple sugars. A group of *carbohydrates* consisting chiefly of *sugars* having a *molecular formula*, $C_6H_{12}O_6$ (*hexoses*) or $C_5H_{10}O_5$ (*pentoses*). Monosaccharides are either *aldoses*, containing a –CHO group, or *ketoses*, containing a –CO– group. They exhibit *optical activity*. See also *pyranose*.

monosodium glutamate See *sodium hydrogen glutamate*.

monotropic Existing in only one *stable* physical form, any other form obtainable being unstable under all conditions.

monovalent Univalent. Having a *valence* of one.

monozygotic Denoting twins that are formed from one fertilized egg (*zygote*), i.e. identical twins.

month The 'solar month' is one twelfth of a solar *year*. The 'calendar month' is any of the twelve divisions of the year according to the Gregorian calendar. The 'lunar month' is the time taken for the *Moon* to complete one *orbit* of the *Earth*. This may be measured in various ways. The 'synodic month' is the period between two successive *phases of the Moon*, equal to 29.5306 days. The 'sidereal month' is the Moon's period with respect to successive conjunctions with

MOON

a *star*, equal to 27.3217 days. The 'anomalistic month' is the Moon's period between two successive *perigees*, equal to 27.5546 days. The 'Draconic month' is the Moon's period with respect to two successive similar *nodes*, equal to 27.2122 days.

Moon The only *satellite* of the *Earth*. Mean distance from the Earth 384 400 kilometres; synodic *month* 29.5306 days, sidereal month 27.3217 days. Mass 0.0123 that of the Earth; diameter 3476 kilometres. It is devoid of *water* or an *atmosphere*. Man first set foot on the Moon in July 1969.

mordants Substances used in dyeing, especially fabrics of plant origin. The fabric is first impregnated with the mordant, which is generally a *basic metal hydroxide* for *acidic dyes*, or an acidic substance for basic dyes. The dye then reacts chemically with the mordant forming an insoluble *lake*, which is firmly attached to the fabric.

morphine $C_{17}H_{19}O_3N$. A white crystalline *alkaloid* that occurs in *opium*, m.p. 253°C. It is a powerful *narcotic*, used medically in the form of its *sulphate* or *hydrochloride* for relieving pain, but it is habit-forming and its misuse can be dangerous.

morpholine $O(CH_2CH_2)_2NH$. A colourless *hygroscopic liquid*, b.p. 128°C., used as a *solvent* for *resins* and *waxes*.

morphology The study of the shape, form, and structure of *organisms*.

mortar A building material consisting mainly of *lime* and *sand* that hardens on exposure through chemical action between the ingredients and atmospheric *carbon dioxide*.

mosaic 1. In televison cameras (see *camera*, *television*), a device for the electrical storage of the optical image. It usually consists of a sheet of *mica* one side of which is covered with mutally insulated particles of a *photoemissive* material, each of which is capacitively coupled through the mica to a conducting coating on the reverse side. This conducting coating, called the signal plate, is the output *electrode* from which the electrical signal representing the optical image is obtained. **2.** In *nuclear physics*, a *photomicrograph* of a track in an *emulsion*, prepared from a number of photographs of consecutive fields of view and reconstructed as though the track lay in one *plane*. **3.** In biology, an *organism* composed of two or more genetically distinct *clones* of *cells*.

mosaic gold Crystalline *tin(IV) sulphide*, SnS_2, consisting of shining, golden-yellow scales.

MOSFET See *field-effect transistor*.

Mössbauer effect An effect observed when certain *nuclei* emit or absorb *gamma radiation*. The *momentum* of the gamma-ray *photon* is hf/c, where h is the *Planck constant*, f is the *frequency*, and c is the *speed of light*. When a stationary nucleus emits or absorbs such a photon, it must recoil to conserve momentum. The *kinetic energy* of this recoil, for a free nucleus of mass m, is $(hf/c)^2/2m$. Mössbauer showed that for low quantum energies an atom bound in a *crystal* may remain in its lattice when emitting or absorbing radiation because the momentum of recoil is shared by the whole lattice (up to 10^{20} atoms), making the recoil energy negligible (m in the above expression will become relatively large). This process is called recoilless.

This effect is used in Mössbauer *spectroscopy* in which a gamma-ray source is mounted so that it can be moved at varying speeds. A similar sample is held stationary nearby and a detector measures gamma-rays scattered by the sample. The source is moved towards the sample at a series of different speeds, so that the frequency of the gamma radiation is varied by the *Doppler effect*. Resonance absorption by the sample is indicated by a sharp fall in the detector signal at particular frequencies. This effect provides valuable information regarding nuclear energy levels and the *chemical bonds* and structure of compounds. Named after R. L. Mössbauer (born 1929).

mother-liquor A *solution* from which substances are crystallized.

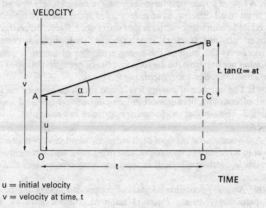

u = initial velocity
v = velocity at time, t

Figure 28.

motion, equations of Kinematic equations that apply to bodies moving with uniform *acceleration*, a; the equations numbered (1)–(4) below. In the velocity-time *graph* (Fig. 28):
Gradient of AB = $\tan \alpha = a$
therefore BC = $t.\tan \alpha = at$
and BD = v = final *velocity*
hence $v = u + at$...(1) where u is the initial velocity.
The area under AB equals the distance covered, s,
therefore s = area ACDO + area ABC
i.e. $s = ut + \frac{1}{2}at^2$...(2)
Also, using (1); $s = ut + \frac{1}{2}(v - u)t$
or $s = (u + v)t/2$...(3)
Combining (1) & (2), $v^2 = u^2 + 2as$...(4).

motion, laws of See *Newton's laws of motion*.

motor A device for converting other forms of *energy* into mechanical energy. The most common forms are the *internal-combustion engine* and the *electric motor*.

moving-coil ammeter See *ammeter*.

moving-iron ammeter See *ammeter*.

mucopolysaccharide Former name for *glycosaminoglycan*.

mucoproteins Glycoproteins. *Proteins* that contain a *carbohydrate* group.

285

multicellular (Of an *organism*) Consisting of more than one *cell*.

multimeter An electrical measuring instrument that functions as a *voltmeter* and *ammeter* over various ranges. It usually also has an internal *dry battery* to enable it to function as an ohmmeter. Usually a moving-coil instrument, it has a number of switches to incorporate series or parallel *resistors* in the circuits.

multiple proportions, law of See *chemical combination, laws of.*

multiple star A system of *stars* consisting of three or more components held together by *gravitation.*

multiplet 1. A line in a *spectrum* formed by two or more closely spaced lines and resulting from small differences of *energy level* in the atoms or molecules. **2.** A group of related elementary particles that differ only in electric *charge.*

multiplication constant (factor) The 'effective' multiplication constant of a *nuclear reactor* is the ratio of the average number of *neutrons* produced by *nuclear fission* per unit time, to the total number of neutrons absorbed or leaking out in the same time. See *subcritical; supercritical.*

multiplicity 1. The number of *energy levels* into which an *atom* or *nucleus* splits as a result of coupling between orbital *angular momentum* and *spin* angular momentum. **2.** The number of *elementary particles* in a *multiplet.*

multiwire chamber Multiwire proportional chamber. See *wire chamber.*

Mumetal* A *ferromagnetic alloy* of high *magnetic permeability* containing up to 78% nickel in addition to iron, copper, and manganese and some chromium and molybdenum in modern alloys. It is used for *transformer cores* and magnetic shielding.

Muntz metal* An *alloy* containing 3 parts of copper and 2 parts of zinc. It is used where a particularly strong *brass* is required. Named after G. F. Muntz (1794–1857).

muon μ-meson. An *elementary particle* with a *mass* 207 times that of an *electron*; it exists in negatively and positively charged forms. It was originally so called as it was classified as a *meson*. However as these particles have *spin* ½, they are now classified as *leptons*. The negative muon, μ^-, decays into an *electron*, a *neutrino*, and an antineutrino, with a mean life of 2.197 microseconds.

muriate Obsolete term for *chloride.*

muriatic acid Obsolete term for *hydrochloric acid.*

muscovite See *mica.*

musical scale See *temperament.*

mustard gas Dichlorodiethyl sulphide. $(CH_2CH_2Cl)_2S$. An oily *liquid* that has been used as a war gas. It is destroyed by *oxidizing agents*, e.g. *bleaching powder.*

mutagen A substance that produces *mutations*, e.g. *mustard gas, colchicine.*

mutarotation A change in the *optical rotation* of a substance.

mutation A change in the chemical constitution of the *deoxyribonucleic acid* (DNA) in the *chromosomes* of an *organism*: the changes are normally restricted to individual *genes*, but occasionally involve serious alteration to whole chromosomes. When a mutation occurs in *gametes* or *gametocytes* an inherited change may be produced in the characteristics of the organisms that develop from them. Mutation is one of the ways in which genetic variation is produced in organisms (see *meiosis; natural selection*). A *somatic* mutation is one that occurs to a body *cell*, and is consequently passed on to all the cells derived from it by *mitosis*. Natural mutations, at this stage of biological evolution, when they occur in the cells of higher animals, almost always produce deleterious characteristics. Both natural and artificial mutations can be brought about by *ionizing radiation* (hence the genetic and *carcinogenic* dangers of *nuclear weapons*) and by certain chemical substances called *mutagens*.

mutual conductance The ratio of the change of *anode current* to the change in *control-grid voltage*, when a small change is made to the control-grid voltage in a *thermionic valve*. It is used as a measure of the valve's performance. See also *transconductance*.

mutual induction The induction of an *EMF* in a *circuit* due to a changing current in a separate circuit with which it is magnetically linked. The induced EMF (E) is proportional to the rate of change of the current (I) in the second circuit, the constant of proportionality being called the coefficient of mutual induction, or the mutual inductance (M), i.e. $E = -M.dI/dt$. The derived *SI unit* of mutual inductance is the *henry*.

mycology The study of *fungi*.

mydriatic A substance used to dilate the pupil of the eye.

myocelium A network of *hyphae*, forming the main body of a *fungus*.

myoglobin A *respiratory pigment* consisting of a conjugated *protein*, similar to *haemoglobin* (but monomeric). It occurs in vertebrate muscle fibres.

myopia Short sight. A defect of vision in which the subject is unable to see distant objects distinctly. It is corrected by the use of *concave* spectacle *lenses*.

myosin One of the two major *proteins* of muscle *cells*. See *actomyosin*.

N

NAD See *nicotinamide adenine dinucleotide*.

nadir (astr.) The lowest point; the point opposite the *zenith* on the *celestial sphere*. See Fig. 2 under *azimuth*.

NADP See *nicotinamide adenine dinucleotide*.

nano- Prefix indicating one thousand millionth; 10^{-9}. Symbol n, e.g. ns $= 10^{-9}$ *second*.

naphtha A *mixture* of *hydrocarbons* in various proportions, obtained from *paraffin oil*, *coal-tar*, etc. Wood naphtha is impure *methanol*, CH_3OH, produced by the *destructive distillation* of wood.

naphthalene $C_{10}H_8$. A white crystalline *cyclic hydrocarbon* with a penetrating odour that occurs in *petroleum* and *coal-tar*. M.p. 80.2°C., b.p. 218°C. It is used in the manufacture of organic *dyes* and in moth-balls.

naphthol $C_{10}H_7OH$. Two *isomeric derivatives* of *naphthalene*, both of which darken in colour on exposure to *light*: naphthalen-1-ol (α-naphthol) is a yellow crystalline substance, m.p. 93.3°C., used in the manufacture of *dyes* and perfumes; naphthalen-2-ol (β-naphthol) is a white crystalline substance, m.p. 122°C., used as an *antiseptic* and in the manufacture of dyes, *drugs*, and perfumes.

naphthoyl The *univalent* group $C_{10}H_7.CO-$ (from naphthoic acid, $C_{10}H_7.COOH$).

naphthyl The *univalent* group $C_{10}H_7-$ (from *naphthalene*, $C_{10}H_8$).

Napierian logarithm See *logarithm*. Named after John Napier (1550–1617).

narceine $C_{23}H_{27}NO_8.3H_2O$. A white crystalline *alkaloid* that occurs in *opium*, m.p. 176°C.; it is used as a muscle relaxant.

narcotic Producing sleep, stupor, or insensibility. A drug used to control pain. Most of these drugs can cause dependence and their medical use is controlled.

nascent state Certain *elements*, notably hydrogen, are more active when being set free in a *chemical reaction* than in their ordinary state; such 'nascent' elements were formerly thought to owe their activity to being composed of single *atoms* instead of *molecules* but they are now thought to be excited molecules, which are highly reactive before they revert to the *ground state*.

natrium See *sodium*.

natron Natural sodium sesquicarbonate, $Na_2CO_3.NaHCO_3.2H_2O$.

natural (chem.) Occurring in nature; not artificially prepared.

natural abundance The *abundance* of each different *isotope* in an *element* as it is normally found in nature.

natural frequency The *frequency* of free oscillation of any system.

natural gas A mixture of gaseous *hydrocarbons*, predominantly *methane* (85%), ethane (10%), and propane (3%), often containing other *gases*, issuing from the Earth in some localities, more particularly near deposits of *mineral oil*. Like *petroleum* it originated from the decomposition of organic matter; it is used as a fuel (having largely replaced *coal-gas* for this purpose) and as a source of *intermediates* for organic synthesis. See also *liquefied natural gas*.

natural logarithm See *logarithm*.

natural selection The theory, first proposed by Charles Darwin and A. R. Wallace, that explains the mechanism of biological evolution (see *Darwin's theory* of evolution). According to this theory, the life-forms best adapted to their environment will tend to survive longer and so leave more offspring, thus passing on any *genes* responsible for their adaptiveness with higher frequency.

nautical mile Defined in the U.K. as 6080 ft, but internationally as 1852 *metres*, 1 U.K. nautical mile therefore equals 1.000 64 international nautical miles. 1 international nautical mile equals 1.150 78 miles.

near infrared or ultraviolet The shortest *infrared* or the longest *ultraviolet wavelengths*; i.e. those wavelengths of these two types of *radiation* that are 'nearest' in magnitude to those of visible *light*.

near point The nearest point at which an object can be placed in front of the eye and still be clearly seen. The near point increases with age.

nebula (astr.) A cloudy luminous patch in the heavens that consists of a *galaxy* of *stars*, or of materials from which such galaxies are being formed.

Néel temperature The *temperature* above which an *antiferromagnetic* substance becomes *paramagnetic*. The transition was discovered in 1930 by L. E. F. Néel (born 1904).

negative (math. and phys.) In any convention of signs, regarded as being counted in the minus, or negative direction, as opposed to positive.

negative, photographic See *photography*.

negative feedback See *feedback*.

negative pole The south-seeking pole of a *magnet*. See *magnetic pole*.

negatron Negaton. See *electron*.

nekton The animals that actively swim in the sea or other aqueous environment. Compare *plankton*.

nematic crystals *Liquid crystals* in which the molecules are not arranged in layers but all their axes are parallel. See also *cholesteric crystals*; *smectic crystals*.

neo-Darwinism The modern version of *Darwin's theory* of evolution, which is widely accepted now. It combines Darwin's theory of *natural selection* with subsequent discoveries in *genetics* that explain the source of the variation on which his theory is based.

neodymium Nd. Element. R.a.m. 144.24. At. No. 60. A soft silvery metal, r.d. 7.0, m.p. 1024°C., b.p. 3100°C., used to colour *glass* purple and in *misch metal*. See *lanthanides*.

neon Ne. Element. R.a.m. 20.179. At. No. 10. A colourless odourless invisible *noble gas*, m.p. −248.67°C., b.p. −246.05°C., that occurs in the *atmosphere* (1 part in 55 000). It is obtained by the *fractional distillation of liquid air*. A discharge of *electricity* through neon at low *pressures* produces an intense orange-red glow (see *neon tube*).

neon tube A *discharge* tube containing neon at a low pressure. As a result of the potential difference maintained between the *cathode* and the *anode*, electrons are accelerated towards the anode and in colliding with neon atoms ionize them and excite the positive ions. When these decay to a lower energy level they emit the characteristic pink light.

Small neon lamps, which use very little current, are used as indicator lights to show that a device is live.

neoplasm New growth of abnormal *tissue* in plants or animals; a tumour, which may be either benign or malignant.

Neoprene* *trans*-Polychloroprene. $(CH_2:CH.CCl:CH_2)_n$. A synthetic *rubber* having a high *tensile strength* and better heat and *ozone* resistance than natural rubber. It is made by polymerizing 2-chlorobuta-1,2-diene.

neper A unit for expressing the *ratio* of two values (e.g. *currents*, *voltages*, etc.) equal to the natural *logarithm* of the ratio of the quantities. 1 neper = 8.686 *decibels*. Named after John Napier (1550–1617).

nephelometer An instrument for measuring turbidity of *liquids*, or *scattering of light* by particles in *suspensions*.

nephoscope A grid-like instrument for determining the speed of celestial objects (including clouds) by observation of time of transit.

nephrite See *jade*.

Neptune (astr.) A *planet* with two main *satellites* (Triton and Nereid); six smaller satellites were discovered in 1989 by the US Voyager II probe. Its *orbit* usually lies between those of *Uranus* and *Pluto* but because of the high *eccentricity* of Pluto's orbit, between 1979 and 1999 Pluto is nearer to the *Sun* than Neptune. Mean distance from the Sun 4496.7 million kilometres; *sidereal period* ('year') 164.8 years; mass 17.46 times that of the *Earth*; diameter 48 600 kilometres. The surface temperature is about $-200°C$. and the dense atmosphere consists mainly of *methane* and hydrogen.

neptunium Np. *Transuranic element*. At. No. 93. Most stable *isotope*, neptunium-237, has a *half-life* of 2.2×10^6 years. A *metal* of silvery appearance, r.d. 20.45, m.p. 640°C., b.p. 3900°C., produced as a *by-product* by *nuclear reactors* in the manufacture of plutonium.

neritic zone The sea above the continental shelf, which is less than 200 metres deep, the maximum depth for *photosynthesis* to occur. Compare *oceanic zone*.

Nernst effect If a *temperature* gradient is maintained across an electrical *conductor* or *semiconductor* that is placed in a transverse *magnetic field*, a *potential difference* will be produced across the conductor. Named after Walter Nernst (1864–1941).

Nernst heat theorem The *entropy* change for *chemical reactions* involving crystalline *solids* is zero at the *absolute zero* of temperature. See also *thermodynamics, laws of*.

nerol $C_{10}H_{17}OH$. A colourless *liquid unsaturated alcohol*, isomeric with *geraniol*, b.p. 224°C. It is used in perfumes and obtained from *neroli oil*.

neroli oil An *essential oil* obtained from the flowers of orange trees.

nerve cell See *neurone*.

nerve fibre An *axon* or *dendrite*.

nerve gas A wargas that attacks the nervous system, especially the nerves controlling respiration. Most nerve gases are *derivatives* of *phosphoric acid*.

nerve impulse See *impulse*.

Nessler's solution A *solution* of potassium mercury(II) iodide, $KHgI_3$, in *potassium hydroxide* solution. It is used as a test for *ammonia*, with which it forms a brown coloration or precipitate. Named after Julius Nessler (1827–1905).

neuron(e) Nerve cell. A special type of biological *cell*, being the unit of which the nervous systems of animals are composed. It consists of a *nucleus* surrounded by a *cytoplasm* from which thread-like fibres project. In most neurones *impulses* are received by numerous short fibres called *dendrites* and carried away from the cell by a single long fibre called an *axon*. Transfer of impulses from neurone to neurone takes place at junctions between axons and dendrites, which are called *synapses*.

neurotoxin A poison that attacks the nervous system.

neurotransmitter A substance that transmits a nerve *impulse* across a *synapse*, being released by the tip of the *axon* into the synaptic space. Substances that function in this way include the *catecholamines adrenaline* and *noradrenaline*, and acetylcholine.

neutral (chem.) Neither *acid* nor *alkaline*. Containing equal numbers of *hydroxyl* and *hydrogen ions* and having a pH of 7.

neutral (phys.) Having neither negative nor positive net *electric charge*.

neutralization (chem.) The addition of *acid* to *alkali*, or vice versa, until neither is in excess and the *solution* is *neutral*, consisting of a *salt* and *water*.

neutral temperature The temperature of the hot junction of a *thermocouple* at which the EMF round the circuit is a maximum and the rate of change of EMF with *temperature* is zero.

neutrino A stable *elementary particle* with no *electric charge* or *rest mass*, but with *spin* $\frac{1}{2}$. It was originally postulated to preserve the laws of *conservation of mass and energy* and *conservation of momentum*. The existence of the particle has since been established experimentally, and it is known to exist in three forms: one associated with the *beta decay* process, one with the *tau particle*, and the other with the *muon*. All forms have antiparticles. Neutrinos travel at the *speed of light* and are classified as *leptons*. The existence of some massive neutrinos has been postulated in certain *grand unified theories*.

neutron An *elementary particle* that is a constituent of all atomic *nuclei* except that of normal hydrogen. The neutron has no *electric charge* and a *mass* only very slightly greater than that of the *proton* ($1.674\,929 \times 10^{-27}$ *kilogram*). Outside a nucleus a neutron decays, with a *half-life* of 12 minutes, into a proton, an *electron*, and an antineutrino. It is classified as a *hadron*.

neutron activation analysis See *activation analysis*.

neutron diffraction A technique used to determine *crystal* structure by irradiating a solid with a beam of *neutrons* and analysing the diffraction patterns in a similar manner to the technique of *electron diffraction*. Usually a flux of neutrons from a *thermal reactor* is used, with average energy of 4×10^{-21} J (0.025 eV), equivalent to a *wavelength* of 0.1 nm, an order suitable for studying interatomic interference. With neutron diffraction there are two types of interaction, one between the neutrons and the atomic nuclei, the other between the *magnetic moments* of the neutrons and the spin and magnetic moments of the nuclei.

neutron excess See *isotopic number*.

neutron flux A measure of the number of free *neutrons* passing through unit volume multiplied by their mean speed.

neutron number N. The number of *neutrons* in an atomic *nucleus*; it is equal to the *mass number* minus the *atomic number*.

neutron star A hypothetical state of a *star* at the end of its evolutionary process (see *stellar evolution*) when it has consumed all its *nuclear fuel* and no longer has a source of internal *energy*. The star would then become highly compressed by *gravitational collapse* and apart from a thin outer shell would consist only of *neutrons*. Such a star would be expected to have a density some 10^7 times greater than a *white dwarf*. No neutron stars have been identified with certainty, although it is thought that *pulsars* may be this type of star.

neutron temperature The *energies* possessed by *neutrons* in thermal equilibrium with their surroundings may be expressed in terms of a *temperature*, if it is assumed that they behave as a *monatomic gas*. Under these conditions, the neutron temperature on the *Kelvin* scale, T, is given by: $E = 3kT/2$, where E is the neutron energy and k is *Boltzmann's constant*.

new candle See *candela*.

newton The derived *SI unit* of *force*. The force required to give a *mass* of one *kilogram* an *acceleration* of one *metre* per second per second. Symbol N. Named after Sir Isaac Newton (1642–1727).

Newtonian fluid A fluid that obeys Newton's law of viscosity, i.e. the viscosity is independent of the rate of shear or the velocity gradient. The tangential force, F, between two parallel layers of fluid is given by
$$F = \eta A.dv/dx$$
where A is the area of the fluid layers, dx is the distance between them, and dv is their velocity. η is a constant called the coefficient of viscosity. A large number of liquids obey Newton's law. Compare *non-Newtonian fluid*.

Newtonian mechanics A system of *mechanics* developed from *Newton's laws of motion*. It provides an accurate means of determining the motions of bodies travelling at ordinary *speeds*. The motions of particles having very high speeds (comparable to the *speed of light*) must be treated by relativistic mechanics, i.e. a system of mechanics based on the theory of *relativity*, as the change of *mass* of a particle with its speed becomes important under such conditions.

NEWTONIAN TELESCOPE

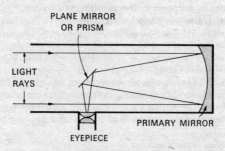

Figure 29.

Newtonian telescope Newtonian reflector. A form of astronomical reflecting *telescope* consisting of a large *concave* focusing *mirror* on the *axis* of which is

292

mounted a small plane mirror or reflecting *prism*, enabling the image to be viewed through an *eyepiece*, which is perpendicular to the axis of the main mirror. See Fig. 29.

Newton's law of cooling The rate at which a body loses *heat* to its surroundings is proportional to the *temperature* difference between the body and its surroundings. It is an *empirical* law, true only for small differences of temperature.

Newton's law of gravitation Every particle in the *Universe* attracts every other particle with a *force* directly proportional to the product of the *masses* of the particles and inversely proportional to the square of the distance between them. Thus, the force of attraction between two masses m_1 and m_2, separated by a distance s, is given by:

$$F = Gm_1m_2/s^2,$$

where G is the *gravitational constant*.

Newton's laws of motion The fundamental laws on which classical *dynamics* is based. 1. Every body continues in its state of rest or uniform motion in a straight line except in so far as it is compelled by external *forces* to change that state. 2. Rate of change of *momentum* is equal to the applied force, and takes place in the direction in which the force acts, i.e. $F = d(mv)/dt$, where F is the applied force, m is the *mass* of the body, and v is its velocity. If the mass remains constant, $F = mdv/dt$ or $F = ma$, where a is the acceleration. 3. If body A exerts a force on body B, then B will simultaneously exert an equal force on A (sometimes called the reaction), in the opposite direction in the same straight line.

Newton's rings Coloured rings that may be observed round the point of contact of a *convex lens* and a *plane* reflecting surface. They are caused by the *interference* effects that occur between *light* waves reflected at the upper and lower surfaces of the air film separating the lens and the flat surface.

niacin See *nicotinic acid*.

Nichrome* Trade name for a nickel-chromium *alloy* (usually 80% nickel) used for wire in electrical devices owing to its high *resistance* and its ability to withstand high *temperatures*.

nickel Ni. Element. R.a.m. 58.69. At. No. 28. A silvery-white magnetic *metal*, resembling iron, that resists corrosion; r.d. 8.90, m.p. 1455°C., b.p. 2900°C. It occurs combined with sulphur or arsenic in pentlandite, *kupfer-nickel*, smaltite, and other *ores*. The ore is roasted to form the *oxide*, which is reduced to the metal by hydrogen, and the metal is then purified by the *Mond process*. It is also obtained by *electrolysis*. It is used for *nickel-plating*, in coinage, for such *alloys* as *Invar*, *nickel steel*, *nickel silver*, *platinoid*, *Mumetal**, *constantan*, and *Nichrome**, and as a *catalyst* (see *Raney nickel*).

nickel carbonyl $Ni(CO)_4$. A colourless *volatile liquid*, b.p. 43°C., that decomposes at 200°C. into nickel and *carbon monoxide*. See *Mond process*.

nickel ethanoate $(CH_3COO)_2Ni.4H_2O$. A green crystalline *soluble* substance, used in *nickel plating*.

nickel-iron accumulator See *accumulator*.

nickel oxides 1. Nickel(II) oxide, nickel monoxide, nickelous oxide. NiO. A green insoluble powder, m.p. 1990°C., used as a pigment and in the manufacture of nickel compounds. **2.** Nickel(III) oxide, nickelic oxide. Ni_2O_3. A black or grey powder that decomposes into nickel(II) oxide at 600°C. It is used in nickel-iron *accumulators*.

nickel plating Depositing a thin layer of metallic nickel by an electrolytic process. See *electrolysis*.

nickel silver German silver. A group of *alloys* of copper, nickel, and zinc in varying proportions, containing up to 30% nickel. A typical composition is 60% copper, 20% nickel, 20% zinc.

nickel steel *Steel* containing up to 6% nickel.

Nicol prism An optical device, constructed from two crystals of *calcite* stuck together with *Canada balsam*, used for obtaining plane polarized light. The extraordinary ray passes through the device but the ordinary ray is totally reflected at the interface between the two crystals. Named after William Nicol (1768–1851). See *polarization of light*.

nicotinamide Niacinamide. $C_5H_4NCONH_2$. See *nicotinic acid*.

nicotinamide adenine dinucleotide NAD. A *coenzyme* derived from *nicotinic acid* that takes part in many biological dehydrogenation reactions. It can accept one hydrogen atom and two *electrons* to form NADH, which is produced during the *oxidation* of food. This releases two electrons and a proton to the *electron transport chain*, reverting to NAD^+. NADP, which possesses a *phosphate* group, functions in the same way as NAD. Enzymes tend to be specific to either NAD or NADP. NADP is important in *photosynthesis*.

nicotine $C_{10}H_{14}N_2$. A colourless intensely poisonous oily *liquid alkaloid*, b.p. 247.3°C., that occurs in tobacco leaves.

nicotinic acid Pyridine-3-carboxylic acid, niacin. $C_5H_4N.COOH$. *Vitamin* of the B complex. A colourless crystalline *solid*, m.p. 235°C., that occurs in meat and *yeast*; deficiency causes pellagra. It can be made by plants and animals from the *amino acid* tryptophan and it occurs in liver and groundnuts. The amide derivative, nicotinamide, is a component of the *coenzymes* NAD and NADP, which are important in many metabolic processes (see *nicotinamide adenine dinucleotide*).

nielsbohrium A name proposed for the element with *atomic number* 105. See *transactinides*.

Ni-Fe* accumulator See *accumulator*.

niobium Columbium. Nb. Element. R.a.m. 92.9064. At. No. 41. A rare grey *metal*, r.d. 8.578, m.p. 2477°C., b.p. 4900°C. Small quantities in *stainless steel* preserve the steel's *corrosion* resistance at high *temperatures*. It is also used in *superconducting alloys*.

nit A unit of *luminance* equal to one *candela* per square *metre*.

niton An obsolete name for *radon*.

nitrate A *salt* or *ester* of *nitric acid*.

nitration Introduction of the *nitro group*, $-NO_2$, into *organic compounds* by the use of *nitric acid*. It is of importance in the production of *explosives*, many nitro derivatives of organic compounds being chemically *unstable*.

nitre Saltpetre. See *potassium nitrate*.

nitric acid Aqua fortis. HNO_3. A colourless corrosive *acid liquid*, b.p. 86°C., that is a powerful *oxidizing agent*. It attacks most *metals* and many other substances with evolution of brown fumes of *nitrogen dioxide*, NO_2. It is manufactured by the action of concentrated *sulphuric acid*, H_2SO_4, on *sodium* or *potassium nitrate*, and by the *oxidation* of *ammonia*, NH_3, by passing a mixture of ammonia and air over heated platinum, which acts as a *catalyst*. It is widely used in chemical industry.

nitric oxide See *nitrogen oxides*.

nitrides *Binary compounds* of nitrogen.

nitrification 1. The treatment of a substance with *nitric acid*. **2.** The process of conversion, by the action of *bacteria*, of nitrogen *compounds* from animal and plant waste and decay, into *nitrates* in the *soil*. See *nitrogen cycle*.

nitrile An organic compound containing the cyanide group -CN.

nitrile rubbers A group of synthetic *rubbers* that are copolymers (see *polymerization*) of *butadiene* and *acrylonitrile*. These materials, which can be vulcanized in a similar manner to natural rubber, have a high resistance to *oil*, *fuels*, and *aromatic solvents*. Their properties can be modified by varying the proportions of the constituents; increasing the acrylonitrile content results in greater oil resistance.

nitrite A *salt* or *ester* of nitrous acid, HNO_2.

nitro The *univalent* group O_2N-.

nitrobenzene $C_6H_5NO_2$. A pale yellow oily poisonous *liquid*, b.p. 211°C., with an odour of bitter almonds. It is produced by the action of *nitric acid* on *benzene*; reduction of nitrobenzene yields *aniline*.

nitrocellulose See *cellulose nitrate*. Although the term nitrocellulose is chemically incorrect for this compound, it is extensively used.

nitrochalk A *mixture* of *calcium carbonate*, $CaCO_3$, and *ammonium nitrate*, NH_4NO_3, used as a *fertilizer*.

nitrogen N. Element. R.a.m. 14.0067. At. No. 7. An odourless invisible chemically inactive *gas*, m.p. -210°C., b.p. -195.8°C., forming approximately 4/5 of the *atmosphere*, from which it is obtained by *fractional distillation*. The chief natural *compound* is *Chile saltpetre*. Compounds are used as *fertilizers* and the manufacture of *nitric acid*. The element is vital to living *organisms*, forming an essential part of *proteins* and *nucleic acids*. See *fixation of atmospheric nitrogen*; *nitrogen cycle*.

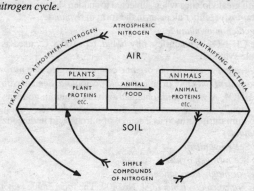

Figure 30.

nitrogen cycle The circulation of nitrogen *compounds* in nature through the various *organisms* to which nitrogen is essential. *Inorganic* nitrogen compounds in the *soil* are taken in by plants, and are combined by the plants with other *elements* to form *nucleic acids* and *proteins*, the latter being the form in which nitrogen can be utilized by the higher animals. The result of animal waste and decay is to bring the nitrogen that the animals had absorbed back into the soil in the form of simpler nitrogen compounds. Bacterial action of various kinds converts these into compounds suitable for use by plants again. In addition to this main circulation, a certain amount of atmospheric nitrogen is 'fixed' (i.e. combined) by the action of free-living *bacteria* and blue-green *algae* as well as by the *bacteria* associated with the roots of leguminous plants, and by the action of atmospheric *electricity*; while some combined nitrogen is set free by the action of *denitrifying bacteria*. See Fig. 30.

nitrogen dioxide See *nitrogen oxides*.

nitrogen fixation See *fixation of atmospheric nitrogen*.

nitrogen monoxide See *nitrogen oxides*.

nitrogen oxides 1. Nitrogen monoxide, nitric oxide. NO. A colourless *gas*, m.p. $-163.6°C$., b.p. $-151.8°C$., that reacts with oxygen to form nitrogen dioxide. **2.** Dinitrogen oxide, nitrous oxide, laughing gas. N_2O. A colourless gas, m.p. $-90.8°C$., b.p. $-88.5°C$., used as a mild *anaesthetic* in dentistry. **3.** Nitrogen dioxide, nitrogen peroxide. NO_2. A compound consisting of two forms, the *monomer*, NO_2, and the *dimer*, N_2O_4 (dinitrogen tetroxide). The degree of dissociation of the colourless N_2O_4, b.p. $21.15°C$., increases with *temperature*. As more of the brown gas NO_2 forms, so the colour darkens. At $150°C$., the vapour is black. It is formed by *reduction* of *nitric acid* and by heating some nitrates. It is used as an *oxidant* (e.g. for *rocket fuels*) and for *nitration*. N_2O_4 is produced by internal-combustion engines and is a serious atmospheric pollutant, thought to be responsible in part, for the depletion of the *ozone layer*.

nitroglycerin Glyceryl trinitrate. $CH_2(NO_3)CH(NO_3)CH_3(NO_3)$ A pale yellow heavy oily *liquid*, made by reacting *glycerol* with *nitric acid* and *sulphuric acid*. It is an *ester* of nitric acid, despite its name, and not a *nitro* compound. It explodes with great violence when subjected to sudden shock or detonation. It is used as an *explosive*, either alone or in the form of *dynamite*.

nitrolime See *calcium cyanamide*.

nitromethane CH_3NO_2. A colourless oily *liquid*, b.p. $100.8°C$., used as a *solvent* and in organic synthesis.

nitroso The *univalent* group ON– in *organic compounds*. See also *nitrosyl*.

nitrosyl The *univalent ion* ON^+ in an *inorganic compound*. See also *nitroso*.

nitrous acid HNO_2. A weak *acid*, obtained only in solution; aqueous solutions decompose rapidly to give *nitric acid* and *nitrogen dioxide*. Among its *salts* (*nitrites*), sodium nitrite is used as a source of nitrous acid in diazotization (see *diazo compounds*).

nitrous ether See *ethyl nitrite*.

nitrous oxide See *nitrogen oxides*.

NMR See *nuclear magnetic resonance*.

nobelium No. *Transuranic element*. At. No. 102. The most stable *isotope*, nobelium-259, has a *half-life* of 3 minutes. The name unnilbium is sometimes used (see *unnil*).

noble gases Rare gases, inert gases. The *elements* helium, neon, argon, krypton, xenon, radon forming group 0 of the *periodic table*. They are all chemically inactive, although some *compounds* of xenon and krypton are known (e.g. XeF_2, XeO_3, $XePtF_6$, KrF_2). For this reason the former name 'inert gases' is rarely now used. Argon occurs in appreciable amounts (0.8%) in the air; the others, with the exception of radon, occur in the air in very minute amounts.

noble metals *Metals* such as silver, gold, and platinum, that do not corrode or tarnish in air or *water*, and are not easily attacked by *acids*. Unreactive metals are low in the *electromotive series*.

nodal points Two points on the axis of a *lens* system, such that if the incident *ray* passes through one, travelling in a given direction, the emergent ray passes through the other in a parallel direction.

nodes 1. Points of zero displacement in a system of *standing waves*. See also *antinodes*. **2.** (astr.) Two points at which the *orbit* of a celestial body intersects the *ecliptic*. **3.** (math.) Points on a curve or surface that can have more than one *tangent*. **4.** (biol.) The points on the stem of a plant or similarly shaped organism from which leaves or other outgrowths arise.

noise (elec.) **1.** An effect observed in amplifying circuits due to the amplification, together with the input signal, of spurious *voltages* arising from such causes as the vibration of certain components, the random motion of the *electrons* constituting the *current* in the *conductors*, etc. **2.** In *information theory*, a disturbance that does not represent any part of a message from a specified source.

noise level A measure of the intensity level of *noise* (in *decibels*) or its *loudness* (in *phons*).

nomogram Nomograph. An alignment chart arranged so that the value of a *variable* can be found, without calculation, from the values of one or two other variables which are known.

nonanoic acid Pelargonic acid, $CH_3(CH_2)_7COOH$. A colourless oily *liquid*, b.p. 253–5°C., used in the manufacture of lacquers and *plastics*.

non-conservation of parity See *parity*.

non-cyclic photophosphorylation A process by which *adenosine triphosphate* (ATP) and reduced *nicotinamide adenine dinucleotide* phosphate (NADP) are generated in the light stage of *photosynthesis*. *Electrons* from *chlorophyll* excited by *light* pass along an *electron carrier system*, then with hydrogen *ions* (from the *photolysis* of water) reduce NADP.

non-electrolytes Substances that do not yield *ions* in *solution* and therefore form solutions of low electrical *conductivity*. See *electrolysis*.

non-ferrous metal Any *metal* other than *iron* or *steel*.

non-metallic elements Chemical *elements* not possessing the properties of the *metals*.

non-Newtonian fluid A fluid that does not obey Newton's law of viscosity, i.e. the viscosity is not independent of the rate of shear or the velocity gradient. In *colloids* and other fluids consisting of more than one *phase* the viscosity usually diminishes as the velocity gradient increases (see *dilatancy*). Compare *Newtonian fluid*.

non-polar compound A *covalent* compound the molecules of which have no permanent *dipole moment*.

non-stoichiometric compound See *Berthollide compound*; *stoichiometric compound*.

nor- (chem.) A combining form of *normal*. The prefix is also used to indicate the loss of a *methyl group*, e.g. *noradrenaline*, or the loss of a *methylene group* from a chain.

noradrenaline Norepinephrine. $C_8H_{10}NO_3$. A *catecholamine hormone*, similar in structure to *adrenaline*, and produced by the *medulla* of the adrenal glands. It functions as a *neurotransmitter* of the sympathetic nervous system.

Nordhausen sulphuric acid See *sulphuric acids*.

normal 1. (math.) A line *perpendicular* to a surface. **2.** (chem.) A prefix denoting either a *normal solution* (abbrev. N-) or an *isomer* with an unbranched chain (abbrev. *n*-).

normality (chem.) An obsolete method of expressing *concentrations* of *solutions*; the number of *gram-equivalents* of *reagent* per *litre* of solution. Thus, a solution containing 2 gram-equivalents per litre is a twice-*normal* or 2N solution.

normalizing A *heat* treatment applied to *steel* in order to relieve internal *stresses*. It involves heating above a critical *temperature* and cooling in air.

normal solution (chem.) A *solution* containing 1 *gram-equivalent* of *solute* per *litre* of solution. See *normality*.

normal state of atom See *ground state*.

notation The representation of numbers, quantities, or other entities by symbols; a system of symbols for such a purpose.

NOT circuit See *logic*.

nova A *star* that ejects a small part of its material in the form of a *gas* cloud. During the process the star becomes 5000 to 10 000 times more luminous than it was before the outburst. 'Dwarf' novae increase their *luminosity* by a factor of only $10-100$. Novae appear to one of a pair of *binary stars* in which a *red giant* transfers some of its matter to a *white dwarf*. See also *supernova*.

***n-p-n* transistor** See *transistor*.

N.T.P Normal *temperature* and *pressure*. See *s.t.p.*

n-type conductivity The *conductivity* in a *semiconductor* caused by a flow of *electrons*, whereas p-type conductivity is caused by a flow of *holes*.

nuclear barrier Potential barrier. The region of high *potential energy* through which a charged particle must pass on entering or leaving an atomic *nucleus*.

nuclear battery A *cell* or *battery* of cells in which *energy* from a *beta emitter*, such as strontium-90, is converted into electric energy by collecting the emitted *electrons* on a suitable *electrode*. In the low-voltage devices the primary electrons from the beta emitter are used to ionize a *gas* in an *electric field*, thus increasing the number of electrons up to 200 times; such devices can deliver nanoamperes at about 1.5 volts. These devices have various uses, especially in space technology.

nuclear charge The positive *electric charge* on the *nucleus* of an *atom*. When expressed in units equal to the charge on the *electron*, this is numerically equal to the *atomic number* of the *element*, to the number of *protons* in the nucleus, and to the number of electrons surrounding the nucleus in the *neutral* atom. See *atom, structure of*.

nuclear cross-section See *cross-section*.

nuclear energy Atomic energy. *Energy* released during a *nuclear reaction* in accordance with the *mass-energy equation*. Nuclear energy is released in *nuclear reactors* and *nuclear weapons*.

nuclear fission A *nuclear reaction* in which a heavy atomic *nucleus* (e.g. uranium) splits into two approximately equal parts, at the same time emitting *neutrons* and releasing very large amounts of *nuclear energy*. The energy released when a slow neutron causes fission of a uranium-235 nucleus is approximately 3×10^{-11} joules. 1 kg of U-235 therefore produces the energy equivalent of some 3 million tons of *coal*. Fission can be spontaneous or it may be caused by the absorption of a neutron (see *chain reaction*), an energetic charged particle, or a *photon* (*photofission*). See also *nuclear reactor*; *nuclear weapon*.

nuclear force The attractive *force* that acts between *nucleons* when they are extremely close together (of the order 10^{-15} m). The nuclear force is much stronger than the repulsive *electromagnetic interaction* between *protons* at such proximities and holds the nucleons together in the atomic *nucleus*. See *strong interaction*.

nuclear fuel A substance that undergoes *nuclear fission* or *nuclear fusion* in a *nuclear reactor*, a *nuclear weapon*, or a *star*.

nuclear fusion A *nuclear reaction* between light atomic *nuclei* as a result of which a heavier nucleus is formed and a large quantity of *nuclear energy* is released. E.g. the fusion of two *deuterium* nuclei to form a *tritium* nucleus and a *proton* is accompanied by an energy release of 4 *MeV* or 6.4×10^{-13} J (D + D = T + p + 4 *MeV*). For fusion to be possible the reacting nuclei must possess sufficient *kinetic energy* to overcome the *electrostatic field* that surrounds them. The *temperatures* associated with fusion reactions are therefore extremely high (of the order of 10^8 K). Fusion reactions occur on *Earth* during the explosion of a hydrogen bomb (see *nuclear weapons*); see also *thermonuclear reactions*. Fusion reactions are the source of the energy of the *stars* (including the *Sun*).

In recent years a considerable amount of work has been devoted to attaining 'cold fusion', especially using an electrolytic method. It has been suggested that the *electrolysis* of D_2O using a palladium *cathode* can result in a fusion reaction at room temperature. The principle is that D^+ *ions* are absorbed into the *lattice* of the palladium *electrode* and are forced together sufficiently to overcome the electrostatic repulsion. Claims to have created a fusion reaction by this method have not, however, been confirmed. Another cold fusion method is based on replacing the *electrons* in the deuterium with *muons*, which being 207 times heavier than electrons so reduce their size that they can approach each other sufficiently closely for the fusion reaction to occur. The muon is then released to form another muonic atom, so that the muon catalyses the fusion reaction. This approach is being investigated.

nuclear isomers *Nuclides* of the same *mass* but possessing different rates of *radioactive decay* as a result of being in different quantum states.

nuclear magnetic resonance NMR. All atomic *nuclei*, except *even-even nuclei*, have *magnetic moments* associated with them, which tend to be aligned by an externally applied *magnetic field*, but because nuclei possess *angular momentum*, they precess (see *precessional motion*) about the direction of the applied field. The *energy* of the interaction between the applied and the nuclear magnetic fields is *quantized* (see *quantum mechanics*), so that only certain orientations of the nucleus relative to the applied field are permitted: a transition from one orientation to another involves the absorption or emission of a *quantum* of *electromagnetic radiation*, the *frequency* of which can be shown to equal the precessional frequency. With the *magnetic flux densities* customarily used (up to about 2 *tesla* or 20 kilo*gauss*) the energies involved are small, and the radiations fall in the *radio frequency* band, i.e. $1-100$ *megahertz*. Transitions from one *energy level* to another can be induced by applying a second magnetic field, at right angles to the first, which rotates in *phase* with the nuclear precession. NMR spectroscopy (also called 'radio frequency spectroscopy') consists of observing the point of *resonance* at which such transitions are induced. Data obtained in this way provide valuable information concerning nuclear properties. As the *orbital electrons* 'shield' the nucleus to a certain extent from the applied magnetic field, at a given frequency nuclei in different electronic (i.e. chemical) environments will resonate at slightly different values of the applied field. This phenomenon, known as the 'chemical shift', enables NMR spectroscopy to be of great value in working out the structure of complex *molecules*. It also enables an image of soft tissue to be formed showing variations in *proton* density and is used medically to diagnose some forms of brain abnormalities, vascular diseases, and cancers.

nuclear medicine The application of techniques involving *radioactive nuclei* to medical diagnosis and therapy. *Radioisotopes* follow the same route in the body as stable isotopes and will tend to accumulate in the same tissues. Measuring the radioactivity in certain tissues after injection of particular radioisotopes can indicate the presence of overactive (cancer) cells. Similarly, a radionuclide that emits *gamma radiation* can be used with a *gamma camera* and a *computer* to form an image of the heart; e.g. using an injection of thallium-201 to scan the blood flow through the heart (thallium scan). See also *radiotherapy*.

nuclear physics The study of the *physics* of the atomic *nucleus* and of sub-atomic particles, especially with reference to *nuclear energy*.

nuclear power Electric or motive power produced from a unit in which the primary *energy* source is a *nuclear reactor*.

nuclear reaction Any reaction that involves a change in the *nucleus* of an *atom*, as distinct from a *chemical reaction*, which only involves the *orbital electrons*. Such reactions occur naturally, on the *Earth* in *radioactive elements*, and in *stars* as *thermonuclear reactions*. They are also produced artificially in *nuclear reactors*, *nuclear weapons*, and controlled thermonuclear reactions. See also *nuclear fission*; *nuclear fusion*.

Nuclear reactions are represented by enclosing within a bracket the symbols for the incoming and outgoing particles or *quanta* (separated by a comma), the initial and final *nuclides* being shown outside the bracket. Thus the reaction:
$$^{14}_{7}N + ^{4}_{2}He \rightarrow ^{17}_{8}O + ^{1}_{1}H$$
is represented: $^{14}N(\alpha,p)^{17}O$.

nuclear reactor An assembly in which a *nuclear fission chain reaction* is maintained and controlled for the production of *nuclear energy*, *radioactive nuclides*, or artificial *elements*. The *nuclear fuel* used in a reactor consists of a *fissile* material (e.g. uranium-235), which undergoes fission, as a consequence of which two *nuclides* of approximately equal *mass* are produced together with between two and three *neutrons* and a considerable quantity of *energy*. These neutrons cause further fissions so that a chain reaction develops; in order that the reaction should not get out of control, its progress is regulated by neutron absorbers (see *control rods*), only sufficient free neutrons being allowed to exist in the reactor core to maintain the reaction at a constant level. The fissile material is usually mixed with a *moderator*, which slows down (see *thermalize*) the *fast neutrons* emitted during fission, so that they are more likely to cause further fissions of the fissile material than they are to be captured by the uranium-235 *nuclides*. In a 'heterogeneous reactor' the fuel and the moderator are separated in a geometric pattern called a *lattice*. In a 'homogeneous reactor' the fuel and the moderator are mixed so that they present a uniform medium to the neutrons (e.g. the fuel, in the form of a uranium *salt*, may be dissolved in the moderator).

Besides this classification, reactors may be described in a number of ways. They may be described in terms of neutron energy (see *fast reactor; thermal reactor*) or in terms of function, e.g. a 'power reactor' for generating useful *electric power*, a 'production reactor' for manufacturing fissile material (see also *breeder reactor* and *converter reactor*), and a 'propulsion reactor' for supplying motive power to ships or submarines. Reactors are also described in terms of their moderator (e.g. 'graphite-moderated reactor') or their coolant (e.g. *boiling-water reactor, gas-cooled reactor*).

All current reactors used to feed electricity to their national grids are thermal reactors, which raise *steam*, usually in a counter-current *heat exchanger*, to drive a steam *turbine*. This turbine, in turn, drives an electric *generator*. The most widely used type is now the *pressurized-water reactor*, which is replacing the earlier *gas-cooled reactors. Fast reactors* are more complex than thermal reactors and require liquid sodium as the *coolant*, as a result they are not expected to become commercially viable until early in the 21st century.

In 1992, 30% of Europe's electricity came from *nuclear power* (20% in the UK, 73% in France), while the USA generated 22% of its electricity in this way, although the 1986 disaster at Chernobyl (USSR) caused widespread concern over nuclear safety and several countries postponed their plans for nuclear expansion.

The term nuclear reactor may also be applied to a device in which a controlled *thermonuclear reaction* takes place, in which case it is also referred to as a 'fusion reactor'. The first nuclear reactor, built by Enrico Fermi (1901–54) at Chicago University in 1942, was called an atomic pile.

nuclear transmutations The changing of *atoms* of one *element* into those of another by suitable *nuclear reactions*.

nuclear waste See *radioactive waste*.

nuclear weapons Weapons in which the explosive power is derived from *nuclear fission* or a combination of nuclear fission and *nuclear fusion*. The fission bomb (atom[ic] bomb or A-bomb) consists essentially of two or more *masses* of a suitable *fissile* material (e.g. uranium-235 or plutonium-239) each of which is less than the *critical mass*. When the bomb is detonated the subcritical masses are brought rapidly together to form a supercritical assembly, so that a single fission at the instant of contact sets off an uncontrolled *chain reaction*. The resulting release of *nuclear energy* produces a devastating *explosion* the effect of which is comparable to the explosion of tens of *kilotons* of *TNT*. The fusion bomb (thermonuclear bomb, hydrogen bomb, or H-bomb) consists of a fission bomb surrounded by a layer of hydrogenous material (e.g. lithium deuteride). At the *temperature* resulting from the explosion of the fission bomb, fusion of the hydrogen nuclei to form helium nuclei takes place (see *thermonuclear reaction*) with the evolution of even greater quantities of energy. The explosive effect of a fusion bomb (or fission-fusion bomb) is comparable to the explosion of tens of *megatons* of TNT. See also *fall-out*.

nucleases A group of *enzymes* that break down *nucleic acids*.

nucleation (chem.) The formation of *nuclei*, e.g. preceding crystallization from solutions or in seeding rain clouds.

nucleic acids Large *polymer molecules* consisting of chains of *nucleotides*. They are present in all living matter and are responsible for storing and transferring *genetic* information. See *deoxyribonucleic acid*; *ribonucleic acid*.

nucleoid The region of a *prokaryote* cell in which the *deoxyribonucleic acid* (DNA) lies.

nucleolus A small dense body in the *nucleus* of *eukaryote cells*, in which the ribosomal RNA (see *ribonucleic acid*) is *transcribed* and combined with *protein* to form the major subunits of *ribosomes*.

nucleon A constituent of the atomic *nucleus*, i.e. a *proton* or a *neutron*.

nucleonics The practical applications of *nuclear physics*, and the techniques associated with these applications.

nucleon number See *mass number*.

nucleophilic reagents *Reagents* that react at centres of low *electron* density (e.g. *hydroxyl ions*). Nucleophilic reagents behave as electron pair donors, either transferring electrons or sharing their electrons with outside *atoms* or *ions*. Compare *electrophilic reagents*.

nucleoproteins Associations of *nucleic acids* and *proteins* found in *cell nuclei* principally in the form of *chromosomes*. *Viruses* consist almost entirely of nucleoproteins.

nucleoside A *compound* formed from a nitrogenous base (*purine* or *pyrimidine*) and a *pentose sugar*, e.g. adenosine, which consists of *adenine* and D-ribofuranose. The phosphorylated *derivative* of a nucleoside is called a *nucleotide*.

nucleotide A most important type of *compound* found in all living matter. Nucleotides consist of a nitrogenous base (*purine* or *pyrimidine*), a *pentose sugar* and a phosphate group. They are found free in *cells* as *adenosine triphosphate* and as part of various *coenzymes*; they also occur in the form of *polynucleotide* chains as *nucleic acids*.

nucleus 1. A vital central point, especially a particle of matter that acts as a centre for the *condensation* of water vapour in mist or as a centre for the formation of crystals. **2.** (chem.) A characteristic ring of *atoms* in a molecule that retains its identity in chemical changes; e.g. the *benzene* nucleus of six carbon atoms in the benzene ring.

nucleus, atomic The positively charged core of an *atom*, consisting of one or more *protons* and, except in the case of hydrogen, one or more *neutrons*. The number of protons in the nucleus is given by the *atomic number* and the number of neutrons by the difference between the *mass number* and the *atomic number* (i.e. the *neutron number*). Nearly the whole of the *mass* of an atom is concentrated in its nucleus, which occupies only a tiny fraction of its *volume*. See *atom, structure of*.

nucleus of cell A *membrane*-bounded body found within the *cytoplasm* of most biological *cells* of both plants and animals. The nucleus contains *deoxyribonucleic acid* (DNA) in the form of *chromosomes*, which become visible under a *microscope* during *mitosis* or *meiosis*. The nucleus is therefore the repository of the molecular information that controls the characteristics of cells and their progeny.

nuclide 1. The *nucleus* of a type of *atom*, characterized by its *atomic number* and *mass number*. **2.** The type of atom to which such a nucleus belongs.

numerator The number above the line in a *vulgar fraction*. E.g. 3 in 3/16.

nutation An oscillation of the *Earth's* poles about the mean position.

nylon Officially defined as 'a generic term applied to any long-chain synthetic *polyamide* that has recurring *amide* groups as an integral part of the main *polymer* chain and is capable of being formed into a filament in which the structural elements are oriented in the direction of the axis'. The familiar commercial form of nylon (e.g. Brinylon*) is a substance formed by the *condensation polymerization* of *hexanedioic acid* (adipic acid) with 1,6-*diaminohexane*. The *solid* polymer is melted and forced through fine jets to make filaments, which are then collected in the form of yarn.

nystatin Fungicidin. A yellow *insoluble antibiotic* obtained from *Streptomyces noursei* and other *Streptomyces* species. It is used to treat infections caused by fungi.

O

objective (phys.) A *lens* or system of lenses nearest the object in a *telescope* or compound *microscope*.

oblate spheroid See *spheroid*.

obtuse angle An *angle* greater than 90°.

occlusion Absorption of a gas into the bulk of a solid.

occultation The cutting off of the *light* or *radio* emission from one celestial body when another is interposed between it and the observer. E.g. a *star* may become invisible to an *optical* or *radio telescope* when it is hidden behind the *Moon*.

oceanic zone The sea beyond the continental shelf where the depth exceeds 200 metres. Compare *neritic zone*.

oceanography The study of the oceans, sea floor, sea waters, and their tides and currents.

ochre A natural *hydrated* form of *iron(III) oxide*, Fe_2O_3, containing various impurities. It is used as a red or yellow *pigment*.

octa-, octo- Prefix denoting eight, eightfold.

octacalcium phosphate OCP. $Ca_2H(PO_4)_3.2\frac{1}{2}H_2O$. A crystalline substance of importance in the chemistry of bones, teeth, and precipitated *calcium phosphates*.

octadecanoic acid Stearic acid. $C_{17}H_{35}COOH$. A white solid *fatty acid*, m.p. 69°C., that occurs as *glycerides* in many fats. It is used in the manufacture of soaps and cosmetics.

octagon An eight-sided *polygon*. The angle between the sides of a regular octagon is 135°.

octahedron A *polyhedron* having eight faces.

octane $CH_3(CH_2)_6CH_3$. A *hydrocarbon* of the *alkane series*. It is a colourless *liquid*, b.p. 126°C., r.d. 0.704, that occurs in *petroleum*. The isomer 2,2,4-trimethylpentane (iso-octane) also occurs in *petroleum*.

octane number of a fuel The percentage by *volume* of iso-*octane* in a mixture of iso-octane and *normal heptane*, C_7H_{16} that is equal to the *fuel* in knock characteristics (see *knocking*) under specified test conditions.

octanoic acid Caprylic acid. $CH_3(CH_2)_6COOH$. A colourless oily liquid, b.p. 237°C., used as an *intermediate* in the manufacture of *dyes* and perfumes.

octanol Octyl alcohol. $C_8H_{17}OH$. A group of *isomeric alcohols* of which the most important is *l*-octanol, a colourless *liquid*, m.p. 16.7°C., b.p. 194–5°C., used as a *solvent*.

octant The portion of a *circle* cut off by an arc and two radii at 45°; one-eighth of the area of a circle.

octavalent Having a *valence* of eight.

octave The interval between two musical notes, the fundamental components (see *quality of sound*) of which have *frequencies* in the ratio two to one. This use of the word has been extended to include the interval between two frequencies of any type of oscillation that are in the ratio two to one.

octaves, law of An incomplete statement of the *periodic law* made by J. A. R. Newlands (1837–1898) in 1864 independently of Dimitr Mendeleev (1834–1907), who published his paper on the *periodic table* in 1869.

octet A stable group of eight *electrons* that constitutes the outer electron *shell* of an *atom* of a *noble gas* (except helium whose only electron shell contains two electrons). When the atoms of the *elements* (except hydrogen) combine to form *compounds*, they do so by donating or sharing electrons so that each combining atom has a completed octet in its outer shell. See *valence*.

octyl The *univalent* group C_8H_{17}–.

odd-even nucleus A *nucleus* that contains an odd number of *protons* and an even number of *neutrons*.

odd-odd nucleus A *nucleus* that contains an odd number of both *protons* and *neutrons*.

oersted The unit of *magnetic field strength* or *magnetic intensity* in *c.g.s. electromagnetic units*, defined as the strength of a magnetic field that would cause a unit *magnetic pole* to experience a *force* of 1 *dyne* in a vacuum. It is equivalent to $1/4\pi \times 10^3$ *amperes* per *metre*. Symbol Oe. Named after Hans Christian Oersted (1777–1851).

oestrogens Female *sex hormones* of which the most important are the *sterols* oestradiol ($C_{18}H_{24}O_2$), oestrone ($C_{18}H_{22}O_2$), and oestriol ($C_{18}H_{24}O_3$). They are produced by the ovaries, in particularly high levels during *ovulation*, and control the *oestrous cycle*. They are also used in *oral contraceptives*.

oestrous cycle The cycle of reproductive activity of most sexually mature animals (except most primates; see *menstrual cycle*). The length of the cycle depends on the species. 'Monoestrous mammals' have one breeding season during the year, whereas polyoestrous species have two or more. The males of some species also have cycles of sexual activity.

ohm The derived *SI unit* of *resistance* defined as the resistance between two points of a *conductor* when a constant difference of potential of 1 *volt*, applied between these two points, produces in the conductor a *current* of 1 *ampere*. The former 'international ohm' was defined as the resistance, at 0°C., of a column of mercury 106.3 cm in length, of *mass* 14.4521 g, and of uniform cross-sectional area. 1 'international ohm' = 1.000 49 'absolute' SI ohms. Symbol Ω. Named after Georg Ohm (1787–1854).

ohmic contact An electrical contact across which the *potential difference* is proportional to the current flowing through it.

ohmmeter An instrument for measuring *resistance* in *ohms*, e.g. a *multimeter*.

Ohm's law The ratio of the *potential difference* between the ends of a *conductor* and the *current* flowing in the conductor is constant. This ratio is termed the *resistance* of the conductor. For a potential difference of *V volts* and a current of *I amperes*, the resistance, R, in *ohms* is equal to V/I.

-oic See *carboxylic acid*.

oil cake A mass of oilseeds (e.g. linseed, cottonseed) from which the *oil* has been expelled in a press (expellers) or extracted by a *solvent* (extractions); used as cattle food.

oil-immersion lens See *immersion objective*.

305

oil of vitriol See *sulphuric acids*.

oil of wintergreen See *methyl salicylate*.

oils Viscous *liquids*, which are usually immiscible with water. They may be plant or animal products (see *fats and oils*; *essential oils*) or mineral oils (see *petroleum*). See also *oil, synthetic*.

oil sand Tar sand. *Sandstone* or loose sand impregnated with a viscous *hydrocarbon* mineral oil. They occur as tar sands or bituminous sands in Alberta (Canada) and as asphalt lakes in Trinidad and elsewhere. They are a potential source of oil, depending on the cost of extraction compared to the price of *petroleum*.

oil shale A fine-grained *sedimentary rock* containing kerogen, a form of organic matter that decomposes on heating to yield mineral oil. Although there are very large deposits of oil shale throughout the world, the commerical extraction of oil would only be economic if other sources reached a high price.

oil, synthetic Natural *mineral oils* are composed of various *hydrocarbons*. Similar products can be made artificially from *coal*, etc., by combining carbon or *carbon monoxide* with hydrogen. See *Bergius process*; *Fischer-Tropsch process*.

Olbers' paradox If the Universe contains an infinite number of uniformly distributed stars the night sky should be uniformly bright. In fact it is not: this is explained by the *expansion of the Universe* and the recession of the galaxies, which causes a *redshift* making the most distant galaxies no longer visible. Named after Heinrich Wilhelm Olbers (1758–1840).

oleate A *salt* or *ester* of *oleic acid*.

olefiant gas An obsolete name for *ethene* (ethylene).

olefins Olefines. See *alkenes*.

oleic acid cis-octadec-9-enoic acid. $C_{17}H_{33}COOH$. An *unsaturated liquid organic acid*, m.p. 15°C., that occurs in the form of *glycerides* in many *fats and oils*. A high proportion of *triolein*, the glyceride of oleic acid, in a fat or oil makes it more liquid.

olein See *triolein*.

oleoyl The *univalent unsaturated* group $C_{17}H_{33}CO-$ (from *oleic acid*).

oleum Fuming sulphuric acid. See *sulphuric acids*.

oleyl alcohol $CH_3(CH_2)_7CH:CH(CH_2)_7CH_2OH$. An *unsaturated liquid alcohol*, b.p. 205°C., used in organic synthesis.

olfactory Pertaining to the sense of smell.

oligomer A *polymer* having comparatively few *monomer* units in the molecule.

olivine $(Mg,Fe)_2SiO_4$. A mineral *silicate* of magnesium and iron. The transparent form is used as a gem.

omega-minus Ω^-. A negatively charged *elementary particle*, classified as a *hyperon* and having a mass 3276 times that of the *electron*.

omegatron An instrument in which *ions* are caused to move in spiral paths by the application of an *electric field* at right angles to a constant *magnetic field*. As the *angular frequency* of rotation of the ions depends upon their *charge* to *mass* ratio, it is possible by this means to separate ions of different *isotopes*. The instrument may be used for the absolute determination of atomic *masses* and for isotopic and chemical analysis.

oncogenic Causing cancer. Some chemical substances and some environmental factors are oncogenic. Oncogenic viruses are able to transform a normal host cell into a cancerous cell.

oncology The study of cancer, its causes and treatment.

onium ion An *ion* formed by the addition of a *proton* to a neutral *molecule*; e.g. the *ammonium ion*, hydroxonium ion (see *oxonium ion*).

on-line working The use of a device that is connected directly to a *computer* so that it becomes a *peripheral* device. In 'off-line working', the device produces information in readable form for subsequent processing by a computer.

ontogeny (bio.) The history of the development of an individual member of a species, from *zygote* to death (often with special reference to its development as an embryo), as opposed to 'phylogeny', which is the history of the evolution of the species (or other biological group).

oocyte A female *gametocyte* that undergoes *meiosis* to form an *ovum*.

opacity The extent to which a medium is *opaque*. It is the *reciprocal* of the *transmittance*, i.e. the ratio of the *radiant flux* falling on it to the flux transmitted.

opal Hydrated *amorphous silica*, SiO_2, some forms of which are used as gems. The milky white variety is sometimes coloured by impurities. Opalescence, a characteristic rainbow effect within the stone, is caused by *interference* from internal cracks and cavities.

opaque Not permitting a *wave motion* (e.g. *light*, *sound*, *X-rays*) to pass. It is usually applied to *electromagnetic radiation* (especially light); not *transparent* or *translucent*. A medium that does not permit *X-rays* or *gamma-rays* to pass is said to be *radio-opaque*. See *opacity*.

open-chain compounds *Organic compounds* not derived from ring compounds; *aliphatic compounds*.

open clusters Clusters of *stars* that have a common motion through *space*. The open clusters are much less densely populated with stars than the *globular clusters*, containing only some hundreds of stars interspersed with *gas* and dust clouds.

open-hearth process Siemens-Martin process. A process for *steel* manufacture. *Pig-iron* and steel scrap or iron *ore* in calculated amounts are heated together by *producer gas* on a hearth in a furnace.

opera glasses See *binocular field glasses*.

operand See *operator*.

operator A symbol representing a mathematical operation to be carried out on a particular operand, e.g. in $\sqrt{3}$, $\sqrt{}$ is the operator and 3 is the operand.

operon A length of *deoxyribonucleic acid* (DNA) forming a group of *genes* whose function is to control the *synthesis* of the individual *enzymes* that act together as one enzyme system. One of the genes in an operon, known as the 'operator gene', is involved in control of the binding of the enzyme that initiates activation of the entire operon.

ophthalmology The study of the eye and its diseases and the correction of impaired vision.

ophthalmoscope An instrument for examining the eye. A powerful light and *lens* system, combined with the lens of the eye, enable the *retina* and blood vessels of the eye to be examined at high magnification.

opiate See *opium*.

opium The dried, milky juice from unripe fruits of the opium poppy, *Papaver somniferum*. It contains several *alkaloids*, including *morphine*, *narceine*, and *codeine*. The group of drugs called opiates includes these alkaloids and their synthetic analogues, such as pethidine and methadone. They are extensively used as powerful pain killers in medicine. They are also widely abused by unfortunate people who have become addicted to them.

opposition 1. (astr.) A *planet* having its *orbit* outside that of the *Earth* is in opposition when the Earth is in a line between the *Sun* and the planet. **2.** Two periodic quantities having the same *frequency* and *waveform* are said to be in opposition if they differ in *phase* by 180°.

optical activity Optical rotation. The property possessed by some substances and their *solutions* of rotating the plane of vibration of polarized light (see *polarization of light*). It occurs with asymmetric *molecules* that can exist in two different forms, called optical isomers or enantiomorphs, one being a *mirror image* of the other. One form rotates the light in one direction, the other rotates it equally in the other direction. The *dextrorotatory* form, or *d*-isomer, rotates it to the right (observer looking against the incoming light) and the *laevorotatory* form, or *l*-isomer rotates it to the left. An equimolecular mixture of the two forms, called a *racemic mixture*, is optically inactive and designated *dl*-. Some natural compounds exhibit optical isomerism and in such cases usually only one isomer occurs in nature. For example *d*-glucose occurs in nature but *l*-glucose cannot be made by living organisms (although it can be synthesized in vitro).

The absolute configuration of an optical isomer is referred to *d*-glyceraldehyde, which is taken as a reference structure, designated D-glyceraldehyde. Any compound containing an asymmetric carbon atom having an analogous configuration to this compound is said to belong to the D-series. Compounds belonging to the opposite configuration are members of the L-series. However, not all D-series compounds are dextrorotatory; for example, D-glyceric acid is laevorotatory, i.e. is *l*-glyceric acid. The prefixes D- and *d*- are not the same.

optical axis Principal axis. The line passing through the *optical centre* and the *centre of curvature* of a spherical *mirror* or *lens*.

optical centre A point, situated for all practical purposes at the geometrical centre of a thin *lens*, through which an incident *ray* passes without being deviated.

optical density See *density, optical*.

optical fibres A *glass* fibre that functions as a *waveguide* for light. They are used in medical instruments (called fibrescopes) to examine internal organs (stomach, bladder, uterus, etc.). They are also used in short-range *telecommunications*. The step-index fibre consists of a glass core with a *coaxial* glass or plastic cladding of lower *refractive index* so that *total internal reflection* takes place at the interface between the core and the cladding. In graded-index fibres, the fibre is structured, each layer of glass having a lower refractive index than the one inside it. Optical fibres are used individually or in bunches.

optical glass *Glass* for use in optical instruments. High quality *crown glasses* usually contain potassium or barium in place of sodium and have a *refractive index* of 1.50 to 1.54. *Flint glasses* have a refractive index of 1.57 to 1.72. Lanthanum crown glasses and flint glasses, containing oxides of the lanthanides (rare earths), have higher refractive indexes.

optical isomerism Enantiomorphism. The occurrence of a compound in two different forms, one a mirror image of the other. The two forms have similar properties in all respects except for their *optical activity*.

optically flat A surface is said to be optically flat if the irregularities do not exceed the *wavelength* of light. This is a requirement for many optical devices.

optical maser See *laser*.

optical pumping See *population inversion; laser; maser*.

optical pyrometer See *pyrometer*.

optical rotation See *optical activity*.

optical telescope An astronomical *telescope* used to observe celestial bodies by the *light* that they emit, as compared to a *radio telescope*, which is used to observe their *radio frequency* emissions.

optical temperature The *radiation temperature* of a celestial body as calculated from its *light radiation*.

optic axis The direction in a doubly refracting *crystal* in which *light* is propagated without *double refraction*.

optics The study of *light*. Geometrical optics is built up on the laws of *reflection* and *refraction*, and assumes the *rectilinear propagation of light*; it involves no consideration of the physical nature of light. It is mainly concerned with the formation of images by *mirrors* and *lenses*. Physical optics is concerned with light as *electromagnetic waves*, and the phenomena associated with this aspect.

oral contraceptive A pill taken daily by a woman to prevent conception. The commonest form contains the *sex hormones oestrogen* and *progestogen*, which suppress *ovulation*. In addition the progestogen causes changes in the lining of the uterus which makes implantation less likely should ovulation and fertilization occur. The mini-pill has fewer undesirable side-effects as it contains only progestogen.

orbit 1. The path of one heavenly body around another as a result of their mutual gravitational attraction. Particularly the path of the *planets* around the *Sun*, or the *Moon* (or *artificial satellites*) around the *Earth*. **2.** The path of an *electron* around the *nucleus* of an *atom*. See *orbital electron; atom, structure of*.

orbital The space containing all the points in an *atom* or *molecule* at which the *wave function* of an *electron* (two electrons may be present if they have opposite *spins*) has an appreciable magnitude. It is so called in modern atomic theory by analogy to its counterpart (orbit) in *Bohr's theory*. An atomic orbital (AO), i.e. one associated with a single atomic *nucleus*, has an energy and a shape determined by its *quantum numbers*, and various types (*s, p, d,* etc.) of AO can be distinguished accordingly. Relative to the nucleus an *s* orbital is spherically symmetrical, whereas a *p* orbital is dumbbell-shaped with a definite orientation in space. In the formation of a *covalent bond* (see also *chemical bond*) between two atoms, a molecular orbital (MO) containing two electrons and associated with both nuclei is formed. In the formation of a single carbon-carbon bond, as in *ethane*, the MO arises by the overlapping of two AOs, and it surrounds the two nuclei and is centred on the line joining them; the bond is called a σ (sigma) bond. In a double carbon-carbon bond, as in *ethene*, the second bond is formed by the overlapping of two *p* AOs and is called a π (pi) bond; the overlapping of the two dumbbells results in the formation of two sausage-like spaces of *electron density* at some distance on each side of the line joining the nuclei. In *benzene*, represented as a ring containing alternating single- and double-bonds, a *p* orbital concerned in the formation of a double bond will overlap with the *p* orbital of one adjacent carbon atom as much as with that on the other. The result is two *torus*-shaped MOs, one on each side of the benzene ring, which thus becomes a symmetrical structure with six identical carbon-carbon bonds. This MO treatment is an alternative to the *resonance* (valence-bond) treatment of molecular structure.

orbital electron Planetary electron. An *electron* contained within an *atom*; it may be thought of as orbiting around the *nucleus*, in a manner analogous to the *orbit* of a *planet* around the *Sun*. See *atom, structure of*; *Bohr theory*; *orbital*.

orbital quantum number See *quantum number*.

orbital velocity The *velocity* of a *satellite* or spacecraft that enables it to *orbit* round the *Earth* or other celestial body. A *synchronous orbit* round the Earth requires an orbital velocity of about 3200 *metres* per *second* (7200 miles per hour).

OR circuit See *logic*.

orcinol $CH_3C_6H_3(OH)_2.H_2O$. A white crystalline substance, m.p. $107-8°C.$, that reddens on exposure to *air*. It is used in the analytical detection of *carbohydrates*.

order (bio.) A group (*taxon*) of similar *families*. See *classification*.

order (chem.) A measure of the rate of a chemical reaction in terms of the *concentrations* of the *reactants*. If the rate, *R*, of a chemical reaction:
$$X + Y = Z,$$
is given by $R = K[X]^2[Y]$, then the rate of the reaction would be third order, or first order in Y and second order in X. If the reaction rate is independent of the concentration, it is said to be zero order.

order (math.) The number of times a *function* has been differentiated to give a particular *derivative*: the *degree* of the highest derivative in a *differential equation*.

order of magnitude A magnitude expressed to the nearest power of 10.

ordinary ray When a *ray* of *light* is incident upon a *crystal* that exhibits *double refraction* so that the direction of the ray makes an angle with the *optic axis* of the crystal, the ray splits into two rays. One of these obeys the ordinary laws of *refraction* and is called the ordinary ray. The other is the 'extraordinary ray'.

ordinate In *analytical geometry*, the ordinate of a point is the perpendicular distance of the point from the *x*-axis. See Fig. 5 at *Cartesian coordinates*.

ore A naturally occurring mineral material from which a desired product (usually a *metal*) can be extracted; e.g. *bauxite* is an ore of aluminium. See *beneficiation*.

organ A distinct part of an *organism*, such as an eye, lung, or leaf, that is specialized to perform one or more functions. Organs are composed of many different *tissues*.

organelle A discrete part of a biological *cell* having a particular function, often bounded by *membrane*, e.g. *mitochondria*; *nucleus*; *plastids*.

organic acid An *organic compound* that is able to give up a *proton* to a *base*; i.e. one that contains one or more *carboxyl groups* or in some cases *hydroxyl groups* (e.g. *phenol*).

organic base A *molecule* or *ion* possessing a *lone pair of electrons* that can be used for coordination (see *chemical bond*) with a *proton*. The common *organic* *compounds* that fulfil this condition owe their basic character to an oxygen or nitrogen *atom.*

organic chemistry The *chemistry* of the *organic compounds*; the chemistry of carbon compounds excluding the *metal carbonates* and the *oxides* and *sulphides* of carbon. Originally, it was the chemistry of substances produced by living *organisms*, as distinct from the *inorganic chemistry* of substances of *mineral* origin.

organic compounds Chemical *compounds* containing carbon combined with hydrogen, and often also with oxygen, nitrogen, and other *elements.* The *molecules* of organic compounds are often very complex, and contain a large number of *atoms.* They are not usually ionized in *solution* (see *dissociation*), and frequently show the phenomenon of *isomerism.*

organism A living being, animal, plant, or *unicell*, capable of absorbing *energy* from its environment and of replicating itself.

organometallic compound An *organic compound* in whose molecule a carbon atom is linked directly to a metal atom; e.g. methylsodium, CH_3Na.

organosilicon compounds Chemical compounds in which *silicon atoms* play the part of carbon atoms in organic compounds; e.g. *silanes* (general formula Si_nH_{2n+2}) are the organosilicon analogues of *alkanes.*

origin (math.) The point of intersection of two or more axes (see *Cartesian coordinates* and *polar coordinates*).

ormulu An *alloy* of copper, zinc, and tin in various proportions; generally containing at least 50% copper.

orogenesis The building of a mountain range. See *plate tectonics.*

orpiment Natural *arsenic trisulphide*, As_2S_3. A yellow mineral.

Orsat apparatus A portable apparatus for determining the amount of *carbon dioxide*, oxygen, and *carbon monoxide* in flue or exhaust *gases.* A measured *volume* of the gas is successively passed through three tubes, the first of which contains *potassium hydroxide* to absorb the CO_2, the second alkaline *pyrogallol* (benzene-1,2,3-triol) to absorb O_2, and the third copper(I) chloride in *hydrochloric acid* to absorb the CO. The diminution of volume after the gas has been passed through each tube indicates the quantity of each constituent gas.

ortho- 1. Prefix denoting right, straight, correct. **2.** Prefix denoting adjacency in position in a hexagonal ring of atoms, particularly the *benzene ring.* Abbreviated to *o-* as a prefix in naming a compound, e.g. *o*-dichlorobenzene is 1,2-dichlorobenzene. Compare *meta*; *para*. **3.** Prefix formerly indicating an *inorganic acid* (or a corresponding *salt*) of a higher degree of hydration; e.g. *orthophosphoric acid*, H_3PO_4, as compared with *metaphosphoric acid*, HPO_3.

orthoclase feldspar Natural potassium aluminium silicate, $K_2O.Al_2O_3.6SiO_2$. A constituent of *granite.*

orthogonal 1. (math.) Rectangular, or involving right angles. **2.** (of *crystals*) Having a set of mutually *perpendicular axes.*

orthohydrogen Hydrogen *molecules* in which the *spins* of the two constituent *nuclei* are parallel. Compare *parahydrogen.*

orthophosphoric acid H_3PO_4. See *phosphoric acids.*

oscillator 1. A device for producing sonic or *ultrasonic* pressure waves in a medium. **2.** A device with no rotating parts for converting *direct current* into *alternating current* usually of a relatively high *frequency*; it usually consists of

311

a *transistor* coupled with a suitable *resonant circuit*. A *sinusoidal* (or harmonic) oscillator produces an output current or voltage of a specific frequency that has the form of a *sine wave*. A 'relaxation oscillator' produces an output in which there is an abrupt change of voltage from one level to another, e.g. a *sawtooth waveform* or *square wave*.

oscilloscope See *cathode-ray oscilloscope*.

osmic acid Osmium(IV) oxide, osmium tetroxide, OsO_4. A colourless crystalline *solid*, m.p. $40°C$.: its *solution* is used in *electron microscopy* as an electron-opaque stain of *cell* components.

osmiridium A natural *alloy* of osmium (up to 48%) and iridium, with smaller amounts of platinum, rhodium, and ruthenium. Hard and resistant to *corrosion*, it is used for tipping pen-nibs.

osmium Os. Element. R.a.m. 190.2. At. No. 76. A hard white crystalline *metal*. The heaviest substance known; r.d. 22.58, m.p. $3030°C$., b.p. $5000°C$. It occurs together with platinum (see *osmiridium*) and is used in *alloys* with platinum and iridium.

osmometer An instrument for measuring *osmotic pressures*.

osmoregulation The control of the water content and concentration of *salts* in the body of an animal.

osmosis The net diffusion of *water* (or other *solvent*) through a *semipermeable membrane*; i.e. a membrane that will permit the passage of the solvent but not of dissolved substances. There is a tendency for *solutions* separated by such a membrane to become equal in *molecular concentration*; thus water will flow from a weaker to a stronger solution, the solutions tending to become more nearly equal in concentration. See *osmotic pressure*.

osmotic pressure The *pressure* that must be applied to a *solution* in order to prevent the flow of *solvent* through a *semipermeable membrane* separating the solution and the pure solvent. When a solvent is allowed to flow through such a membrane into a vessel or cell containing a solution, the solvent will flow into the cell (see *osmosis*) until such a pressure is set up as to balance the pressure of the solvent flowing in. The osmotic pressure of a dilute solution is analogous to gaseous pressure; a substance in solution, if not dissociated (see *dissociation*), exerts the same osmotic pressure as the gaseous pressure it would exert if it were a *gas* at the same *temperature*, and occupying the same *volume*. The osmotic pressure, temperature, and volume of a dilute solution of a *non-electrolyte* are connected by laws exactly similar to the *gas laws*, i.e. $\Pi V = RT$, where Π is the osmotic pressure and R is the *gas constant*. Osmotic pressure is a *colligative property*, i.e. it depends on the number of particles in the solution, not on their nature.

Ostwald's dilution law A law relating the dissociation constant, K (see *dissociation*), and the degree of dissociation (or *ionization*), α, of a weak *electrolyte* of *concentration c moles* per *litre*. This law states that for a binary electrolyte
$$K = c\alpha^2/(1 - \alpha),$$
an equation that applies with a fair degree of accuracy to weak *organic acids* and *bases*. Named after Wilhelm Ostwald (1853–1932).

Ostwald viscometer See *viscometer*.

OTEC Ocean Thermal Energy Conversion. An experimental method of obtaining energy from deep stretches of ocean, where there is a marked difference in temperature between layers of water at the top and bottom of the sea. See also *heat pump*.

Otto engine A four-stroke *internal-combustion engine* invented by Nikolaus Otto (1832–91) in 1876. In the ideal Otto cycle, on which it is based, combustion takes place at constant volume, whereas in the later *Diesel engine* combustion takes place at constant pressure.

ouabain $C_{29}H_{44}O_{12}$. A white crystalline *glycoside*, m.p. 200°C., obtained from wood; it is used as a heart stimulant.

ounce 1. (avoirdupois) 437.5 grains or 28.3 grams. **2.** (fluid) 8 fluid drachms or 28.41 cm³. **3.** (Troy) 480 grains or 31.1 grams.

overtones Notes of lesser intensity and higher *pitch* (i.e. of higher *frequency*) than the *fundamental note*, and superimposed upon the latter to give a note of characteristic *quality*. Overtones are the upper *harmonics of a wave motion* as applied to *sound* waves. Sometimes overtone is used synonymously with harmonic in this context, but sometimes the fundamental is regarded as the first harmonic, in which case the first overtone is the second harmonic.

Ovshinsky device Ovonic device. A device consisting of a special *glass*, which incorporates selenium and tellurium, the *resistance* of which drops rapidly when a suitable *voltage* is applied across it. These devices are used as special purpose switches in *electronic circuits*. The type that stays 'on' after the voltage has been removed is called a 'memory switch'.

ovulation The release of an *ovum* by an ovary.

ovum Egg cell. A *gamete* characteristic of female animals produced by *meiosis* from an *oocyte*.

oxalate A *salt* or *ester* of *oxalic acid*.

oxalic acid Ethanedioic acid. $(COOH)_2$. A white crystalline poisonous *soluble solid*, m.p. 189.5°C., whose *salts* occur in wood sorrel and other plants. It is used in dyeing, *bleaching*, ink manufacture, metal polishes, and for removing ink stains.

oxidant The substance that supplies the oxygen in an *oxidation* reaction. The term is frequently used with reference to the substance that supplies the oxygen in a *combustion* process, particularly in a *rocket*. The oxidant used in rockets is usually liquid oxygen, *hydrogen peroxide*, or *nitric acid*.

oxidase An *enzyme* that catalyses *oxidation* of the *substrate*. See *oxidoreductase*.

oxidation The combination of oxygen with a *substance*, or the removal of hydrogen from it. The term is also used more generally to include any reaction in which an *atom* loses *electrons*; e.g. the change of an iron(II) *ion*, Fe^{2+}, to an iron(III) ion, Fe^{3+}. See also *redox reaction*.

oxidation number Oxidation state. The number of *electrons* that must be added to a positive *ion* or removed from a negative ion to produce a neutral atom. Pure elements have an oxidation number of 0. In electrovalent compounds the oxidation state is equal to the charge on the ion, e.g. in $MgBr_2$ the oxidation number of the Mg is +2 and of the Br is –1. In covalent compounds the electrons are notionally assigned to the more electronegative elements. Oxidation numbers are used in naming inorganic compounds, e.g. Fe_2O_3 is known as iron(III) oxide. See also *redox reaction*.

oxidation-reduction reactions See *redox reaction*.

oxidative phosphorylation A process by which *adenosine triphosphate* (ATP) is generated in the *mitochondria* of living *cells*. See *electron transport chain*.

oxide A *binary compound* with oxygen. Oxides of *metals* are usually ionic compounds (see *chemical bond*) and *basic* or *amphoteric*. Oxides of nonmetals are usually covalent, and *acidic* or *neutral*.

oxidizing agent A substance that brings about an *oxidation* reaction. It contains atoms with high *oxidation numbers*, i.e. atoms that have lost electrons. In the process of oxidation they gain electrons and are themselves reduced. See *redox reaction*.

oxidoreductase An *enzyme* that functions both as an *oxidase* and a *reductase*, i.e. one that catalyses a *redox reaction*.

oxime A compound formed from hydroxylamine, (H_2NOH) and an *aldehyde* (aldoximes) or *ketone* (ketoximes), i.e. a compound containing the group $-C:NOH$.

oxirane See *epoxyethane*.

oxo- Prefix denoting the O= group in a *compound*.

oxo acid An *acid* in which the *acidic hydrogen* is bound to an oxygen atom, e.g. H_2SO_4, *sulphuric acid*, which has the structure $(OH)_2SO_2$. Compare *binary acid*.

oxonium ion Formerly called hydronium ion. The *cation* R_3O^+, where R is an organic group or hydrogen; in the latter case it is known as the hydroxonium ion (H_3O^+), which forms when acids dissolve in *water*.

2-oxopropanoic acid See *pyruvic acid*.

oxyacetylene burner A device for obtaining a very high-*temperature flame* (3300°C.) for *welding* ferrous metals, by burning a mixture of oxygen and acetylene (*ethyne*) in a special jet.

oxydiacetic acid See *oxydiethanoic acid*.

oxydiethanoic acid Diglycolic acid, oxydiacetic acid. $O(CH_2COOH)_2$. A white *soluble dibasic organic acid*, m.p. 148°C., used in the manufacture of *plastics* and *plasticizers*.

oxygen O. Element. R.a.m. 15.9994, At. No. 8, m.p. −218.8°C., b.p. −183°C. An odourless, invisible *gas*; the most abundant of all the *elements* in the *Earth's crust* including the seas and the *atmosphere*; it forms approximately one fifth of the atmosphere. It occurs as *diatomic molecules*, O_2, and as the triatomic *allotrope*, O_3, known as *ozone*. Oxygen is chemically very active; *combustion* and *aerobic respiration* both involve combination with oxygen and *compounds* (*oxides*) are very widely distributed. The pure element is made by the *fractional distillation* of *liquid air*. It is used for *welding* and metal-cutting.

oxyhaemoglobin An *unstable compound* formed by the action of oxygen on *haemoglobin* in *blood*.

oxyhydrogen burner A device similar to the *oxyacetylene burner* except that hydrogen instead of acetylene (*ethyne*) is burnt in oxygen; it gives a *flame* temperature of about 2400°C.

ozokerite Earth-wax. A natural *mixture* of *solid hydrocarbons*. A brownish or greyish mass, resembling *paraffin wax*.

ozone O_3. An *allotropic form* of oxygen, containing three *atoms* in the *molecule*. It is a bluish *gas*, very active chemically, and a powerful *oxidizing agent*. Ozone is formed when oxygen or air is subjected to a silent electric discharge. It occurs in ordinary air in very small amounts only; the health-giving effects sometimes attributed to it in sea-air are probably due to other causes. Ozone in the *atmosphere* is mainly present in the *ozone layer*. It is used for purifying *air* and *water*, and in *bleaching*.

ozone layer Ozonosphere. The layer in the *upper atmosphere*, some 15 to 50 kilometres above the *Earth's* surface, in which most of the atmospheric *ozone* is concentrated. Ozone is formed there as a result of the *dissociation* of molecular oxygen by solar *ultraviolet radiation*, and the recombination of some of the single atoms with undissociated molecular oxygen. The ozone formed is responsible for absorbing a large proportion of the Sun's ultraviolet radiation in the *waveband* 230–320 nm. Without this absorption the Earth would be subjected to a degree of ultraviolet radiation that could cause many undesirable *mutations* (ultraviolet radiation is well-established as a *mutagen*). It could also cause the death of many plants. In the 1980s it became apparent that the ozone layer was being depleted, especially over the polar regions. This is thought to be due to complex photochemical reactions between ozone and the highly stable *chlorofluorocarbons* (CFCs) released into the atmosphere by *aerosols*, etc. It is believed that chlorine atoms from CFCs combine with ozone to form chlorine monoxide, which reacts with atomic oxygen to form molecular oxygen and chlorine atoms:

$$CFC \rightarrow Cl + O_3 \rightarrow ClO + O_2$$
$$ClO + O \rightarrow Cl + O_2.$$

The chlorine atoms react again with ozone, each chlorine atom being capable of reacting 20 000 times with ozone, thus causing a drop in the concentration of ozone in this layer of the atmosphere. The reactions in the ozone layer are also thought to involve reactions between ozone and the *nitrogen oxides* produced by jet aircraft. According to the 1990 Montreal Protocol several governments have agreed to limit the use of CFCs. See also *greenhouse effect*.

P

packing density 1. The number of independent devices (e.g. *logic* gates) per unit area of a *silicon chip*. **2.** The amount of information contained in a computer *memory* device, e.g. the number of bits per unit length of *magnetic tape*.

packing fraction The difference between the *relative atomic mass* of a *nuclide* and its *mass number*, divided by the mass number. E.g. a nuclide of one chlorine isotope has a mass of 32.9860 and a mass number of 33, its packing fraction is therefore:

$$(32.9860 - 33.000)/33 = -0.000\ 42$$

Packing fractions are often multiplied by 10^4 for convenience, and in this example the packing fraction would be given as -4.2.

paint A *liquid* containing a coloured material (*pigment*) in *suspension*. The application of the paint to a surface, and the *evaporation* or hardening of the liquid, covers the surface with the pigment in the form of an adhesive skin. The liquid formerly consisted of *linseed oil*, a 'thinner' of *turpentine* or other *volatile* liquid, and a 'drier' to accelerate drying or hardening of the linseed oil. More recent paints use synthetic oils and driers. Paints may also be based on *water* in the form of an *emulsion*, and are then called 'emulsion paints'. Such paints usually consist of an emulsion of *butadiene* and *styrene*, *polyvinyl acetate*, or *acrylic resins* in water.

pair production The creation of a negative *electron* and *positron* as a result of the interaction between a *photon* or a fast particle (usually an electron) and the *field* of an atomic *nucleus* (see also *showers*). 'Internal pair production' occurs as the result of the de-*excitation* of an excited nucleus. Pair production is sometimes extended to mean the creation of any *elementary particle* and its *anti-particle*.

palaeomagnetism The study of the magnetization of iron and iron compounds in rocks. This technique is used to provide a historical survey of the changes in magnitude and direction of the *Earth's magnetic field* (see *magnetism, terrestrial*) since the rocks were formed. It can also be used for dating rocks.

palaeontology The branch of *geology* that is concerned with the study of *fossils* and their relationship to the evolution of the *Earth's* crust and life upon Earth. The main branches are palaeobotany, palaeoecology, and palaeozoology.

palladium Pd. Element. R.a.m. 106.42. At. No. 46. A silvery-white *metal* that occurs with and resembles platinum. R.d. 11.99, m.p. 1554°C., b.p. 3000°C. It is used in *alloys* and as a *catalyst*.

palmitic acid Hexadecanoic acid. $C_{15}H_{31}COOH$. A wax-like *fatty acid*, m.p. 64°C., that occurs in the form of *tripalmitin* in palm oil and many natural *fats*. It is one of the fatty acids whose *salts* form the basis of soap.

palmitin See *tripalmitin*.

palmitoyl The *univalent* group $C_{15}H_{31}CO-$ (from *palmitic acid*).

pantothenic acid $C_9H_{17}NO_5$. A white *insoluble solid* member of the *vitamin* B complex, of importance to many *organisms*. It occurs in rice, bran, and plant and animal *tissues*. It is essential for the growth of *cells*, being a constituent of coenzyme A, which is required in the *oxidation* of *fats* and *carbohydrates*.

papain An *enzyme*, found in the fruit and leaves of the paw-paw tree that is capable of digesting *proteins*. It is used for softening meat for human consumption.

papaverine $C_{20}H_{21}NO_4$. A white *insoluble alkaloid*, m.p. 147°C., obtained from *opium*; it is used in the form of its *hydrochloride* in medicine as an antispasmodic.

paper Paper normally consists of sheets of *cellulose*, mainly obtained from wood pulp from which *lignin* and other non-cellulosic materials have been removed. *Fillers* (e.g. chalk and clay) are added for opacity and sizes (e.g. rosin) are added for water resistance. In the paper-making machine the slurry of pulp and additives is dried and calendered into thin sheets.

paper chromatography A form of *chromatography* in which the mobile phase is liquid and the stationary phase is a strip of porous paper. A drop of the mixture is placed at one edge of the paper and eluted (see *elution*) with the solvent. The components are separated by the rates at which they move across the paper with the solvent. Identification can be by *indicators* or by their *fluorescence* in ultraviolet radiation.

para 1. Prefix denoting beside, beyond; or wrong, irregular. **2.** Denoting positions at opposite apexes in a hexagonal ring of atoms, particularly the *benzene ring*. Abbreviated to *p*- as a prefix in naming a compound; e.g. *p*-dichlorobenzene is 1,4-dichlorobenzene. Compare *ortho*; *meta*.

parabola A curve traced out by a point that moves so that its distance from a fixed point, the *focus*, is equal to its distance from a fixed straight line, the *directrix*. The equation of a parabola with its vertex at the *origin* and its axis along the *x*-axis is $y^2 = 4ax$, where *a* is the distance from the origin to the focus.

parabolic reflector Paraboloid reflector. A *concave* reflector, the section of which is a *parabola*. It is used for producing a parallel *beam* of *electromagnetic radiation* when a source is placed at its *focus*, or for collecting and focusing an incoming parallel beam of radiation. If the radiation is *light* the reflector is usually called a parabolic mirror, but with *microwave* or *radio frequency* radiation (see *radio telescope*) it may be called a 'dish aerial'.

paraboloid of revolution The surface obtained by rotating a *parabola* about its *axis of symmetry*.

paracasein See *caseinogen*.

paraffin (chem.) See *alkanes*.

paraffin oil Kerosine. A mixture of *hydrocarbons* (mostly C_{11} and C_{12} compounds) obtained in the *fractional distillation* of *petroleum*. The boiling range of the kerosines is 150°–300°C. It is used for paraffin lamps, aircraft jet engines, and domestic heating.

paraffin wax A white translucent *solid* melting to a colourless *liquid* in the range 50°–60°C. It consists of a *mixture* of the higher *alkanes*. It is obtained from *petroleum* by solvent *extraction* and used for candles, waxed paper, and polishes.

paraformaldehyde Polymethanal, paraform. $(HCOH)_n$. A solid *polymer* of formaldehyde (see *methanal*), readily converted into formaldehyde on heating. It is used in fumigation.

parahydrogen Hydrogen *molecules* in which the *spins* of the two constituent *nuclei* are anti-parallel. Compare *orthohydrogen*.

paraldehyde Ethanal trimer. $(CH_3CHO)_3$. A liquid *polymer* of acetaldehyde (see *ethanal*), b.p. 124°C. It is used in medicine as a *hypnotic*.

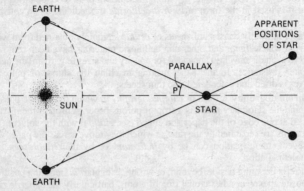

Figure 31.

parallax 1. The difference in direction, or shift in the apparent position, of a body due to a change in position of the observer. **2.** (astr.) The apparent displacement of a celestial body due to the point of observation being either on the *Earth's* surface rather than its centre (diurnal parallax), or on the Earth rather than the centre of the *Sun* (annual parallax). Annual parallax is expressed as the angle P in Fig. 31.

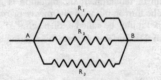

Figure 32.

parallel, conductors in Electrical *conductors* joined in parallel between two points A and B, so that each conductor joins A to B. If R_1, R_2, R_3, etc., are the *resistances* of the separate conductors, the total resistance R between A and B is given by the formula:

$$1/R = 1/R_1 + 1/R_2 + 1/R_3...etc.$$

See Fig 32.

parallel beam of light A *beam* of *light* that neither converges nor diverges. It is a theoretical concept, based on the idea of a beam of light emerging from a source an infinite distance away. *Lasers* are capable of producing nearly parallel beams.

parallelepiped A solid figure having six faces, all *parallelograms*; all opposite pairs of faces being identical and parallel.

parallelogram A plane four-sided *rectilinear* figure having its opposite sides parallel. It may be proved that in all parallelograms the opposite sides and *angles* are equal; the *diagonals* bisect each other; and the diagonals bisect the parallelogram. The *area* of a parallelogram is given by (*a*) the product of the base and the vertical height, and (*b*) the product of two adjacent sides and the *sine* of the angle between them.

parallelogram of forces If a particle is under the action of two *forces*, which are represented in direction and magnitude by the two sides of a *parallelogram* drawn from a point, the *resultant* of the two forces is represented by the *diagonal* of the parallelogram drawn from that point. A parallelogram of forces is a special case of the more general parallelogram of *vectors*.

parallelogram of velocities If a body has two component *velocities*, represented in magnitude and direction by two adjacent sides of a *parallelogram* drawn from a point, the *resultant* velocity of the body is represented by the *diagonal* of the parallelogram drawn from that point. A parallelogram of velocities is a special case of the more general parallelogram of *vectors*.

paramagnetism Substances possessing a *magnetic permeability* slightly greater than unity, i.e. possessing a small positive *magnetic susceptibility*, are said to be paramagnetic. The *atoms* of a paramagnetic substance possess a permanent *magnetic moment* due to unbalanced *electron spins* or unbalanced orbital motions of the electrons around the *nucleus* (see *atom, structure of*). Application of a *magnetic field* to such a substance tends to align the magnetic axes of the atoms in the direction of the field, giving the substance a resultant magnetic moment.

parameter 1. In two-dimensional *analytical geometry* it is often convenient to express the variables (x,y) each in terms of a third variable t, such that x and y are functions of t; $x = f(t)$, $y = g(t)$. The equations are termed parametric equations, and t is a parameter. **2.** A variable that can be kept constant while the effect of other variables is investigated.

parametric amplifier An *amplifier* of *microwaves* that depends on the periodic variation, by an alternating *voltage*, of the *reactance* of a *semiconductor* device.

parasite An *organism* that lives in or on another organism, the host, using the host's ability to acquire *energy* to fuel its own *metabolism*. Examples are roundworms, tapeworms, and the *microorganisms* that cause diseases in humans and their animals and crops.

parasitic capture (phys.) The absorption of a *neutron* by a *nuclide* that does not result in a *nuclear fission* or the production of a useful artificial *element*.

paraxial ray A ray of light that travels close to the *optical axis* of a system.

Paris green Schweinfurt green. A double *salt* of copper(II) ethanoate and arsenate(III), $Cu(CH_3COO)_2.3Cu(AsO_2)_2$. It is used as a *pigment*, *insecticide*, and a wood preservative.

parity Space-reflection symmetry. Mirror symmetry. The principle of space-reflection symmetry, or conservation of parity, states that no fundamental distinction can be made between left and right; that the laws of *physics* are the same in a right-handed system of coordinates as they are in a left-handed system.

If parity is conserved, it is said to be even (or positive) when the *wave function* of a definite state of a system is left unchanged by reversing the sign of all the coordinates; it is said to be odd (or negative) if the sign of the wave function is thereby changed. If parity is not conserved the wave functions bear no simple relation to each other under these circumstances.

Space-reflection symmetry holds for all the phenomena described by *classical physics*, but in 1957 it was shown to be violated by certain interactions between *elementary particles*. For all *strong interactions* and *electromagnetic interactions* parity is conserved, that is to say, if a left-polarized particle exists (i.e. one that *spins* in an opposite sense to its direction of motion) there will be an approximately corresponding number of right-polarized particles. It has been found, however, that for *weak interactions* parity is not conserved. Thus, in a typical weak interaction, such as the *decay* of a *neutron*, the emitted *electron* is always left-polarized. As a result of nonconservation of parity in weak interactions it is now possible to make a fundamental distinction between left and right.

In most instances, symmetry is restored in weak interactions if the parity operation (P) is combined with the operation of *charge conjugation* (C), in which particles are replaced by *antiparticles*. However, in the *decay* of kaons, violation of the combined CP operation also arises, so a complete symmetry for all weak interactions appears to occur only if *time reversal symmetry* (T) is added to CP, to give the combination CPT.

parsec An astronomical unit of distance, corresponding to a *parallax* of one second of arc and equal to 3.26 *light-years* or $3.085\,677 \times 10^{16}$ *metres*.

parthenogenesis The development of an *ovum* into a new individual without *fertilization*. It occurs naturally in some plants (e.g. dandelion) and some animals (e.g. aphids) and can be induced artificially in others.

partial derivative The *derivative* of a *function* with respect to only one of its *variables*, all other variables in the function being taken as *constant*. If z is a function of x and y, i.e. $z = f(x,y)$, the partial derivative of z with respect to x,y being held constant, is written $\partial z / \partial x$.

partial fractions The fractions into which a particular fraction can be separated, so that the sum of partial fractions so obtained equals the original fraction. E.g. the partial fractions of $1/(x^2 - 1)$ are
$$1/2(x - 1) \text{ and } -1/2(x + 1).$$

partial pressures See *Dalton's law* of partial pressures.

particle accelerator See *accelerator*.

particle physics The branch of physics concerned with nuclear structure and the properties of *elementary particles*.

partition coefficient The ratio of the concentrations of a single *solute* in two immiscible *solvents*, at equilibrium. For example, if iodine is dissolved in a mixture of water and benzene, some of the iodine molecules will dissolve in the water layer and some in the benzene layer. At equilibrium, the rate at which iodine molecules cross from the water layer to the benzene layer is equal to the rate of the reverse process. The partition coefficient is the equilibrium constant for the process, usually written so that the concentration of the solute in the more soluble phase is the numerator.

parton A basic particle, such as a *quark*, from which other *elementary particles* are formed.

parylene polymers A series of *polymers* derived from di-1,4-xylylene, $(CH_2C_6H_4CH_2)_2$. They are used for *dielectric* coatings in electronic equipment.

pascal The derived *SI unit* of *pressure*, equal to 1 *newton* per square *metre*. Symbol Pa. Named after Blaise Pascal (1623–62).

Pascal's law of fluid pressures *Pressure* applied anywhere to a body of *fluid* causes a *force* to be transmitted equally in all directions. This force acts at right angles to any surface within, or in contact with, the fluid.

Paschen series A series of lines that occurs in the *infrared* region of the *spectrum* of hydrogen. Named after Friedrich Paschen (1865–1947).

Paschen's law The breakdown or 'sparking potential' for a pair of parallel *electrodes* situated in a *gas*, i.e. the *potential* that must be applied between them for sparking to occur, is a *function* only of the product of the *pressure* of the gas and the separation of the electrodes.

passive 1. Denoting an electronic component, such as a *capacitor*, that does not amplify a signal. **2.** See *satellites, artificial*. Compare *active*.

passivity A state of *metals* in which they become resistant to *corrosion* after treatment with strong *oxidizing agents*. It results from the formation of a surface *oxide* film.

pasteurization Partial sterilization, especially of milk; it involves heating to a *temperature* sufficiently high to kill vegetative *bacteria* (but not necessarily bacterial spores). Milk was originally heated to 63°C. for 30 minutes followed by rapid cooling, now however, it is usual to heat it to a higher temperature for a shorter time. Named after Louis Pasteur (1822–95).

pathogen An *organism* or *virus* that causes disease.

pathway The chain of reactions by which a *chemical change* is brought about. Many biochemical reactions involve complicated pathways. The presence of a *catalyst* can alter the pathway by which a reaction occurs.

patronite Vanadium sulphide, VS_4, a naturally occurring *ore* from which vanadium is extracted.

Pauli exclusion principle Each *electron* moving round the *nucleus* of an *atom* can be characterized by values of four *quantum numbers*. The principle states that no two electrons in an atom can have the same set of four quantum numbers. The principle is of great importance in the theoretical building-up of the *periodic table*. As the principle applies to all *fermions* it also influences the combining of *quarks* to form *elementary particles*. Named after Wolfgang Pauli (1900–58).

PCB See *polychlorinated biphenyl*.

PCM Pulse-code modulation. See *pulse modulation*.

pearl A secretion consisting mainly of *calcium carbonate*, $CaCO_3$, produced by various molluscs.

pearl ash *Potassium carbonate*, K_2CO_3, made from wood *ashes*.

pearlite A microconstituent of iron or *steel* consisting of alternate layers of *ferrite* and *cementite*.

pearl spar See *dolomite*.

peat An early stage in the formation of *coal* from vegetable matter. It is an accumulation of partly decomposed plant material, and is used as *fuel*.

pectins A class of complex *polysaccharides* occurring in plants, particularly fruits.

Solutions have the power of setting to a jelly; this is responsible for the 'setting' of jams.

pedology The scientific study of *soils*, especially their characteristics, structure, and uses.

pelagic Denoting *organisms* that swim or drift in the sea or other aqueous environment. Compare *benthos*. Pelagic organisms are either *plankton* or *nekton*.

pelargonic acid See *nonanoic acid*.

Peltier effect When an *electric current* flows across the junction between two different *metals* or *semiconductors*, a quantity of *heat*, proportional to the total *electric charge* crossing the junction, is evolved or absorbed, depending on the direction of the current. This effect is due to the existence of an *electromotive force* at the junction. Named after Jean Peltier (1785–1845). Compare *Seebeck effect*.

pencil lead A mixture of *graphite* with *clay* in various proportions, to give different degrees of hardness.

pendulum A simple pendulum consists of a weight or 'bob' swinging on the end of a string or wire. In the case of an ideal pendulum, when the *angle* described by the pendulum is small, the string has negligible *mass* and the mass of the pendulum is concentrated at one point, the time of one complete swing, T, is given by the formula $2\pi\sqrt{l/g}$, where l is the length of the string, and g the *acceleration of free fall*. In the compound pendulum, which is a body pivotted about a point within it, the period is given by
$$2\pi\sqrt{[(h^2 + k^2)/hg]},$$
where h is the distance between the pivot and the *centre of mass* and k is the *radius of gyration* about the centre of mass.

penetration factor The *probability* that an incident particle, in a *nuclear reaction*, will pass through the *nuclear barrier*.

penicillin A class of chemically related *antibiotics* produced by the *Penicillium* mould. It is a very powerful agent for preventing the growth of several types of disease *bacteria*, by disrupting the synthesis of the bacterial cell walls. Penicillin and subsequent antibiotics have been responsible for a dramatic change in the incidence of bacterial diseases. However, penicillin-resistance by a wide range of *pathogenic* bacteria has necessitated the use of wide-spectrum antibiotics and the continuing search for new antibiotic drugs.

pennyweight 24 grains, 1/20 Troy ounce. See *Troy weight*.

penta- Prefix denoting five, fivefold.

pentachlorophenol C_6Cl_5OH. A white *insoluble derivative* of *phenol*, m.p. 174°C., used as a *fungicide*.

pentaerythritol $C(CH_2OH)_4$. A white *soluble* powder, m.p. 260°C., used in the manufacture of *plastics*, plasticizers, and *explosives*.

pentagon A five-sided *polygon*: the angle between the sides in a regular pentagon is 108°.

pentane C_5H_{12}. The fifth member of the *alkane* series that exists in three isomeric forms (see *isomerism*). It is contained in light *petroleum*; n-pentane has b.p. 36°C. and r.d. 0.62. It is used as a *solvent*.

pentanoic acid Valeric acid. C_4H_9COOH. A fatty acid that exists in several isomeric forms, the common form being a colourless *liquid*, b.p. 186°C., with a pungent odour. It is used in perfumes.

pentanol Amyl alcohol. $C_5H_{11}OH$. A colourless *liquid*, b.p. 137.3°C., with a characteristic smell. It exists in three isomeric forms (see *isomerism*).

pentavalent Quinquevalent. Having a *valence* of five.

pentlandite A bronze sulphide *mineral*, $(Fe,Ni)_9S_8$, which is the chief *ore* of nickel. It occurs in Ontario, Canada.

pentode A *thermionic valve* containing five *electrodes*: a *cathode*, an *anode* or plate, a *control grid*, and (between the two latter) two other grids called the *screen grid* and the *suppressor grid*.

pentosans *Polysaccharides* that yield *pentoses* on *hydrolysis*.

pentose A *monosaccharide* containing five carbon *atoms* and having the general formula $C_5H_{10}O_5$. The most important pentoses are *ribose* and deoxyribose, which are essential constituents of the *nucleic acids*.

pentyl The *univalent* group $C_5H_{11}-$; it was formerly called the n-*amyl* group.

pentyl ethanoate Amyl acetate, banana oil. $CH_3COOC_5H_{11}$. An ester of *pentanol* and *ethanoic acid*. A colourless liquid, b.p. 148°C., with an odour of pear drops. It is used as a solvent and in perfumes and flavours.

penumbra Half-shadow; it is formed when an object in the path of *rays* from a large source of *light* cuts off a portion of the light. See *shadow*.

pepsin A digestive *enzyme* produced in the stomach that converts *proteins* into *peptones*; it acts only in an *acid* medium.

peptidase An *enzyme* that attacks *peptide* linkages and splits off *amino acids*. See also *proteases*.

peptide A *compound* of two or more (see *polypeptide*) *amino acids* formed by *condensation* of the $-NH_2$ group of one acid and the *carboxyl group* of another. The peptide linkage, $-NH-CO-$, results.

peptones Organic substances produced by the *hydrolysis* of *proteins* by the action of *pepsin* in the stomach. They are *soluble* in *water*, and are absorbed by the body.

per- Prefix denoting, in chemical nomenclature, an excess of the normal amount of an *element* in a *compound*; e.g. *peroxide*.

perborate A *salt* of *perboric acid*, HBO_3.

perboric acid HBO_3. A hypothetical *acid* known only in the form of its *salts*, the *perborates*; e.g. *sodium perborate*.

perchlorate A *salt* of chloric(VII) acid (perchloric acid).

perchloric acid See *chloric acids*.

percussion cap A device used in fire-arms. It consists of a small copper cylinder containing *mercury cyanate* or other violent *explosive* that will explode on being struck, thus initiating the explosion of the main charge.

perdisulphuric acid See *sulphuric acids*.

perfect gas Ideal gas. A theoretical concept of a *gas* that would obey the *gas laws* exactly. Such a gas would consist of perfectly elastic *molecules*, the *volume* occupied by the actual molecules, and the *forces* of attraction between them, being zero or negligible. Compare *real gas*.

perfect solution A *solution* that obeys *Raoult's law* at all concentrations.

peri- Prefix denoting around, about.

periclase Natural *magnesium oxide*, MgO.

pericynthion The time of, or the point of, the nearest approach of a *satellite* in lunar *orbit* to the *Moon's* surface. It is the opposite of *apocynthion*.

perigee The *Moon*, the *Sun*, or an artificial *Earth satellite*, are said to be in perigee when they are at their least distance from the Earth. It is the opposite of *apogee*.

perihelion The time of, or the point of, the nearest approach of a *planet* to the *Sun*. It is the opposite of *aphelion*. See Fig. 1, under *anomaly*.

perimeter The distance all round a *plane* figure; e.g. the perimeter of a *circle* is its circumference.

period 1. (phys.) If any quantity is a *function* of time, and this function repeats itself exactly after constant time intervals T, the quantity is said to be periodic, and T is called the period of the function. $T = 1/f$, where f is the *frequency*. **2.** (chem.) See *periodic table*.

periodate A *salt* of *iodic(VII) acid* (periodic acid).

periodic acid See *iodic(VII) acid*.

periodic law The statement that 'the properties of the *elements* are in periodic dependence upon their *atomic weights*', published by Mendeleev in 1869. The law is brought out clearly when the elements are arranged in the *periodic table*.

periodic table An arrangement of the chemical *elements* in order of their *atomic numbers* to demonstrate the *periodic law*. In such an arrangement elements having similar properties occur at regular intervals and fall into vertical groups of related elements. The horizontal rows in the table (see Appendix, Table 8) are called periods. From the position of an element in the periodic table its properties may be predicted with a fair measure of success; Mendeleev was able to forecast the existence and properties of then undiscovered elements by means of his original table. The periodic law has since been shown to reflect the grouping of *electrons* in the outer *shells* of the *atoms* of the elements. Elements with the same numbers of electrons in their outer shells fall into the same vertical group and have similar chemical properties, as these electrons determine the *valences* of the atoms.

peripherals Peripheral devices. Devices connected to the central processor or the high-speed *store* of a *computer*. Forming part of the *hardware*, they include *backing storage*, input and output devices, *on-line* equipment, *visual display units*, etc.

periphery The external surface or boundary of a body; the circumference or *perimeter* of any closed figure.

periscope A device for viewing objects that are above the eye-level of the observer, or are placed so that direct vision is obstructed. Essentially it consists of a long tube, at each end of which is a right-angled *prism*, so situated that, by *total internal reflection* at the longest faces, *light* is turned through an angle of 90° by each prism. Thus light from a viewed object enters the observer's eye in a direction parallel to, but below, the original direction of the object.

Permalloy* A class of iron-nickel *alloys* with high *magnetic permeability*. They are used in parts of electrical machinery that are subject to alternating *magnetic fields* as they cause only low losses of *energy* due to *hysteresis*. They are also used in *computer* memories.

permanent gas A *gas* that cannot be liquefied by *pressure* alone; a gas above its *critical temperature*.

permanent hardness of water Hardness that is not destroyed by *boiling* the *water*. See *hard water*.

permanent magnet See *magnet, permanent*.

permanganate See *manganates*.

permeability A body is said to be permeable to a substance if it allows the passage of the substance through itself.

permeability, magnetic See *magnetic permeability*.

permeance Λ. The reciprocal of *reluctance*. It is measured in *henries*.

Material	Relative Permittivity	Dielectric Strength V/mm
Air	1	–
Paraffin Wax	2.0–2.5	6.2×10^4
Rubber	2.8–3.0	1.2×10^5
Shellac	3.0–3.7	$3-9 \times 10^4$
Bakelite	4.5–7.5	$2-9 \times 10^4$
Porcelain	6.0–8.0	10^4-10^5
Mica	6.0–8.0	$2-6 \times 10^4$

permittivity ε. **1.** The absolute permittivity of a medium is the ratio of the *electric displacement* to the strength of the *electric field* at the same point. The absolute permittivity of free space, ε_0, is a fundamental constant, called the *electric constant*. In a statement of *Coulomb's law* for the *force*, F, between two charges Q_1 and Q_2, it is given by:
$$F = Q_1 Q_2 / r^2 4 \pi \varepsilon_0,$$
where r is the distance between the charges. ε_0 has the value $8.854\ 185 \times 10^{-12}$ $F\ m^{-1}$. **2.** The relative permittivity, ε_r, also called the dielectric constant, is the ratio of the capacitance of a capacitor with a specified medium (*dielectric*) between the plates, to the capacitance of the same capacitor with free space between the plates, i.e. $\varepsilon_r = \varepsilon / \varepsilon_0$.

The value of the relative permittivity of some common dielectrics at room temperature is given in the table.

permonosulphuric(VI) acid See *sulphuric acids*.

permutation (math.) An arrangement of a specified number of different objects. E.g. the six possible permutations of the digits 123 are 123, 132, 213, 231, 312, 321. The number of possible permutations of n objects if all are taken each time, denoted by ^nP_n, is *factorial n*. The number of permutations of n different objects taken r at a time, ^nP_r, is $n!/(n-r)!$. See also *combination*.

peroxide 1. An *oxide* that yields *hydrogen peroxide* with an *acid*. **2.** An oxide that contains more oxygen than the normal oxide of an *element*.

peroxodisulphuric(VI) acid See *sulphuric acids*.

peroxosulphuric(VI) acid See *sulphuric acids*.

perpendicular At right angles; a straight line making an *angle* of 90° with another line or *plane*.

perpetual motion The concept of a machine that, once set in motion, will go on for ever without receiving *energy*. It is impossible to make a machine that will go on for ever and be able to do *work*, i.e. create energy without receiving energy from outside. To do so would contravene the first law of *thermodynamics*. To create a machine that would give perpetual motion of the second kind, e.g. a ship driven by the *internal energy* of the ocean, would contravene the second law of thermodynamics, because the sea would need to be at a higher temperature than the ship, which could not occur without an external energy source.

persistence of vision The sensation of *light*, as interpreted by the brain, persists for a brief interval after the actual light stimulus is removed; successive images, if they follow one another sufficiently rapidly, produce a continuous impression. Use is made of this in the cine-projector and in *television*.

personal equation The time interval or lag peculiar to a person between the perception and recording of any event. In many physical observations an error is introduced by the time-lag between the actual occurrence of the observed event, its perception by the observer, and its recording.

Perspex* See *polymethyl methacrylate*.

persulphuric acids See *sulphuric acids*.

perturbations Deviations in the motions of the *planets* from their true elliptical *orbits*, as a result of their gravitational attractions for each other.

perturbation theory A method of *successive approximations* used to solve equations if these equations differ only slightly from similar equations to which there are known solutions. The theory was originally used in the solution of complications arising from planetary *perturbations* but has also been used in *quantum mechanics* to calculate *energy levels* in *molecules*. It can also be used with the aid of Feynman diagrams in *quantum electrodynamics*.

pesticides Substances that kill pests; they include *insecticides* and *fungicides*.

peta- Prefix denoting one thousand million million times; 10^{15}. Symbol P, e.g. Pm $= 10^{15}$ *metres*.

Peters' projection See *map projection*.

Petri dish A shallow flat-bottomed circular glass dish, which may have a fitting cover; it is used in laboratories for a variety of purposes, especially for cultivating *microorganisms*. Named after J. R. Petri (1852–1921).

petrifaction The change of an *organic* structure, such as a tree, into a stony or mineral structure. It is generally caused by dissolved *hydrated silica*, SiO_2, penetrating into the pores and gradually losing its *water*.

petrochemicals Chemical substances derived from *petroleum* (or *natural gas*).

petrol Gasoline. A complex *mixture* consisting mainly of *hydrocarbons* in the boiling range 40–180°C., such as *hexane*, *heptane*, and *octane*; other *fuels* and special ingredients, such as the antiknock compound *lead tetraethyl*, are often added. See also *octane number*.

petrolatum Petroleum jelly, Vaseline*. A purified mixture of *hydrocarbons* consisting of a semi-solid whitish or yellowish mass.

petroleum Mineral oil, crude oil. A natural *mixture* of *hydrocarbons* and other *organic compounds* believed to have formed from the remnants of animals and plants that were compacted under raised temperatures and pressures in underground reservoirs formed by impermeable rock. The petroleum, which is often held under pressure under a layer of *natural gas*, may float on a layer of water. The composition of various petroleums varies according to their source; e.g. American petroleum contains a higher proportion of *alkanes* than the Russian variety, which is richer in *cyclic hydrocarbons*. The crude oil is mined and separated by *fractional distillation* into a gas (*refinery gas*), a series of *liquids*, semisolid *petrolatum*, solid *paraffin wax*, and a residue of· *asphalt* and *bitumen*. The liquids include *petrol* (containing C_5 to C_8 hydrocarbons, b.p. 40–180°C.), *paraffin oil* (containing C_{11}–C_{12} hydrocarbons, b.p. 150–300°C.), *gas oil* (C_{13}–C_{25}, b.p. 220–350°C.) and a residual liquid used for lubricating oils. Less desirable products are converted to more desirable products by such processes as *cracking* and *reforming*. Apart from the value of petroleum as a source of *fuels* it is also of enormous value for the *petrochemicals* obtained from it. See also *liquefied petroleum gas*.

petroleum ether A flammable *mixture* of the lower *hydrocarbons* of the *alkane series* consisting mainly of *pentane* and *hexane*, b.p. 30–70°C. It is used as a solvent.

petrology The study of the origin, structure, and composition of rocks.

pewter An *alloy* of tin (80–90%) and lead (20–10%), with small amounts of antimony to harden it or copper to soften it.

pH See *hydrogen ion concentration*.

phage See *bacteriophage*.

phagocyte Any cell that can engulf food by means of pseudopodia into food vacuoles. Phagocytes, such as *leucocytes* in the blood are important in the defence mechanism of many animals.

pharmacology The study of the action of chemical substances upon animals and man.

pharmacophore The portion of a *molecule* of a substance that is regarded as determining the special physiological action of the substance.

pharmacy The study of the preparation and dispensing of *drugs* and medicines.

phase (chem.) A separate part of a *heterogeneous* body or system. E.g. a *mixture* of *ice* and *water* is a two-phase system, while a *solution* of *salt* in water is a system of one phase.

phase (phys.) **1.** Points in the paths of two or more *wave motions* are said to be points of equal phase if they have reached the same fraction of the cycle; i.e. if the wave motion has the same *phase angle* at these points. **2.** One of the circuits in a system or apparatus in which there are two or more alternating *voltages* displaced in phase (meaning as **1**) relative to one another. In a 'two-phase' system the displacement is one quarter of a *period*, in a 'three-phase' system it is one third of a period.

phase angle 1. (phys.) The *angle* between the *vectors* representing two harmonically varying quantities (e.g. *current* and *voltage*) that have the same *frequency*. **2.** (astr.) The *angle*, seen from the *Moon* or a *planet*, between the *Earth* and the *Sun*.

phase-contrast microscope A *microscope* that uses the difference in *phase* of the *light* transmitted or reflected by an object to form an *image* by relative differences in *intensity*.

phase diagram A diagram showing the relations between various *phases* in a chemical system, and the effects of composition and conditions (temperature, pressure) on them.

phase modulation *Modulation* of the *phase angle* of a *sinusoidal carrier wave*. The *phase* of the modulated wave differs from that of the carrier by an amount proportional to the instantaneous value of the modulating wave.

phase rule Gibbs' phase rule. $F + P = C + 2$. For a *heterogeneous* system in equilibrium, the sum of the number of *phases* plus the number of *degrees of freedom* is equal to the number of *components*, plus two. E.g. with *ice*, *water*, and *water vapour* in equilibrium, the number of phases is 3, the number of components 1, and hence the number of degrees of freedom is 0; the system is said to be invariant, since no single variable can be changed without causing the disappearance of one phase from the system. It was discovered by Josiah Gibbs (1839–1903).

phases of the Moon The various shapes of the illuminated surface of the *Moon* as seen from the *Earth* (new Moon, first quarter, full Moon, third quarter); due to variations in the relative positions of Earth, *Sun*, and Moon.

phase speed See *wave motion*.

phasor A rotating *vector* representing a quantity that varies *sinusoidally*. Its length represents the *amplitude* of the quantity and it rotates with an *angular velocity* equal to the quantity's *angular frequency*. The *phase angle* between two quantities can be represented as the angle between two phasors.

phenacetin 1,4-ethoxyphenylethanamide. $CH_3CONHC_6H_4OC_2H_5$. A white crystalline substance, m.p. 134.7°C., used to relieve pain and as an *antipyretic*.

phenazine $C_6H_4N_2C_6H_4$. A yellow crystalline substance, m.p. 171°C., used in the manufacture of *dyes*.

phenetole $C_6H_5OC_2H_5$. Ethoxybenzene. A *volatile aromatic liquid*, b.p. 172°C.

phenobarbitone Phenylethylbarbituric acid. $C_6H_5.C_2H_5.C:(NHCO)_2:CO$. A white crystalline powder, m.p. 174°C.; it is used as a sedative and *hypnotic* drug, usually in the form of the *soluble* sodium *salt*.

phenol Carbolic acid. C_6H_5OH. A white crystalline *solid*, m.p. 41°C., with a characteristic 'carbolic' smell. It is *soluble* in *water*, corrosive, and poisonous. It is made by the *Raschig process*. It is used as a *disinfectant* and in the manufacture of *plastics* and *dyes*.

phenol-formaldehyde resin Phenolic resin. A very widely used type of synthetic *resin* produced by the *condensation* of *phenols* with formaldehyde (see *methanal*); it forms the basis of a *thermosetting* moulding material, and is also used in *paints*, varnishes, and *adhesives*.

phenolphthalein $C_{20}H_{14}O_4$. A colourless crystalline *solid*, m.p. 261°C. A solution in *alcohol* turns a deep purple-red in the presence of *alkalis* (pH greater than 9.6), and is used as an *indicator*. It is also used in *dye* manufacture and as a laxative.

phenols A class of *aromatic organic compounds* containing one or more *hydroxyl groups* attached directly to the *benzene* ring. They correspond to the *alcohols* in the *aliphatic* series, forming *esters* and *ethers*, but they also have weak *acidic* properties and form *salts*. See *phenol*.

phenothiazine Thiodiphenylamine. $C_6H_4NH.S.C_6H_4$. A green *insoluble* substance, m.p. 185.5°C., used as an *insecticide* and in the manufacture of drugs.

phenotype 1. The characteristics possessed by an individual *organism* as a result of the interaction of its inherited characteristics (see *genotype*) with its environment. **2.** A group of organisms having the same phenotype (meaning **1**).

phenyl The *univalent* group C_6H_5-.

phenylalanine A crystalline *soluble amino acid*, m.p. 283°C., obtained from eggs and milk. It is essential to the diet of mammals. See Appendix, Table 5.

phenylamine Aniline $C_6H_5NH_2$. A colourless oily *liquid aromatic*, b.p. 185°C., which turns brown on exposure to sunlight. It is used in drugs, dyes, and the *rubber* industry.

phenylammonium ion The *ion* $C_6H_5NH_3^+$, derived from *phenylamine*.

phenylene The bivalent group $-C_6H_4-$. In modern terminology this has been replaced, e.g. *p*-phenylene diamine would now be called benzene-1,4-diamine.

phenylethanamide Acetanilide. $C_6H_5NHCOCH_3$. A white crystalline solid, m.p. 112°C. It is used in the manufacture of *dyes* and drugs and as an *antipyretic*.

phenylethanone Acetophenone. $C_6H_5COCH_3$. A colourless sweet-smelling *liquid*, b.p. 202.3°C., used in perfumes.

phenylethene See *styrene*.

phenylmethanol Benzyl alcohol. $C_6H_5CH_2OH$. A colourless *aromatic liquid*, b.p. 205.3°C., used as a *solvent* and in the manufacture of perfumes and flavours.

3-phenylpropenoic acid See *cinnamic acid*.

pheromones Chemical substances secreted by an *organism* that elicit a behavioural response from other organisms usually of the same species, especially substances that act as sex attractants.

phlogiston theory A theory of *combustion* that was generally accepted during the eighteenth century until it was refuted by Lavoisier. All combustible substances were supposed to be composed of phlogiston, which escaped on burning, and a calx or ash, which remained. Replacement of phlogiston into the calx would restore the original substance. Lavoisier realized that the calx was in fact the *oxide* formed with oxygen during combustion.

phlogopite See *mica*.

phon A unit of loudness, used in measuring the intensity of *sounds*. The loudness, in phons, of any sound is equal to the intensity in *decibels* of a sound of *frequency* 1000 *hertz* and *root-mean-square* sound pressure of 2×10^{-5} pascal that seems as loud to the ear as the given sound.

phonon The *quantum* of thermal energy in the lattice vibrations of a crystal. If *f* is the vibrational *frequency* the magnitude of the phonon is *hf*, where *h* is the *Planck constant*.

phosgene Carbonyl chloride. $COCl_2$. A colourless poisonous *gas* with a penetrating smell resembling musty hay. It is used as an *intermediate* in organic synthesis and was used in World War I as a war gas.

phosphate A *salt* of *phosphoric(V) acid*, H_3PO_4. Phosphates are used as *fertilizers* to rectify a deficiency of phosphorus in the *soil*.

phosphine PH_3. A colourless flammable poisonous *gas* with an unpleasant smell. It is used for doping *semiconductors*.

phosphinic acid See *hypophosphorous acid*.

phosphite Phosphonate. A *salt* of *phosphorous acid*, H_3PO_3.

phospholipids Phosphatides. Compound *lipids*, *glycerides* in which one of the *fatty acid* chains is replaced by a *phosphoric acid* group (the latter usually with a complex *alcohol* attached). Phospholipids are *amphipathic* and are important components of *cell membranes*.

phosphonate See *phosphorous acid*.

phosphonium ion The *ion* PH_4^+, which is analogous to the ammonium *ion*.

phosphor A substance that is capable of *luminescence*, i.e. storing *energy* (particularly from *ionizing radiation*) and later releasing it in the form of *light*. If the energy is released after only a short delay (between 10^{-10} and 10^{-4} second) the substance is called a 'scintillator'.

phosphor bronze An *alloy* of copper (80%–95%), tin (5%–15%), and phosphorus (0.25%–2.5%) that is hard, tough, and elastic.

phosphorescence A form of *luminescence* in which a substance emits *light* of one *wavelength* after having absorbed *electromagnetic radiation* of a shorter wavelength. Unlike *fluorescence*, phosphorescence may continue for a considerable time after *excitation*.

phosphoric acids 1. Phosphoric(V) acid, orthophosphoric acid. H_3PO_4. A colourless *deliquescent* solid, m.p. 42.5°C. It is used in *fertilizers* and for flavouring drinks. **2.** Metaphosphoric acid. $(HPO_3)_n$. A glossy deliquescent colourless solid polymer derived from *phosphorus(V) oxide*. **3.** Heptaoxodiphosphoric(V) acid, pyrophosphoric acid. $H_4P_2O_7$. A crystalline soluble substance, m.p. 61°C., formed from phosphorus(V) oxide and two molecules of *water*.

phosphorous acid Phosphonic acid. H_3PO_3. A colourless *deliquescent* crystalline substance, m.p. 73.6°C., from which *phosphites* (now often called phosphonates) are obtained.

phosphorus P. Element. R.a.m. 30.97376. At. No. 15. It occurs in several *allotropic forms*, white phosphorus (r.d. 1.82) and red phosphorus (r.d. 2.20) being the commonest. The former is a waxy white, very flammable and poisonous *solid*, m.p. 44°C., b.p. 280°C. Red phosphorus is a non-poisonous dark red powder that is not very flammable. The *element* occurs only in the combined state, mainly as *calcium phosphate*, $Ca_3(PO_4)_2$. It is extracted by heating with *coke* and *silica* (sand) in an electric furnace, and distilling off the phosphorus. Phosphorus is essential to life because of its presence in *nucleic acids* and in *adenosine triphosphate* (ATP); calcium phosphate is also the main constituent of animal bones. Its *compounds* are used as *fertilizers* and *detergents*.

phosphorus chlorides 1. Phosphorus(III) chloride, phosphorus trichloride. PCl_3. A colourless fuming liquid, b.p. 75.5°C., used as a chlorinating agent and in a variety of syntheses. **2.** Phosphorus(V) chloride, phosphorus pentachloride. PCl_5. A yellow crystalline solid, m.p. 148°C., used as a chlorinating agent.

phosphorus oxides 1. Phosphorus(V) oxide, phosphorus pentoxide. P_4O_{10}. A white *deliquescent* crystalline solid that reacts violently with water to give *phosphoric(V) acid*. It is used as a drying agent. **2.** Phosphorus(III) oxide, phosphorus trioxide. P_4O_6. A white waxy solid, m.p. 23.8°C. It is called a trioxide for historical reasons, but the molecule consists of four phosphorus atoms each linked to the other by means of an oxygen bridge.

phosphorus pentachloride See *phosphorus chlorides*.

phosphorus pentoxide See *phosphorus oxides*.

phosphorus trichloride See *phosphorus chlorides*.

phosphorus trioxide See *phosphorus oxides*.

phosphoryl The trivalent group $\equiv PO$.

phosphorylation Conversion into a compound of phosphorus. It is the main mechanism by which the *energy* or *respiration* is stored in *mitochondria* (see *oxidative phosphorylation*). See also *photophosphorylation*.

phot A unit of *illuminance* equal to one *lumen* per square centimetre.

photic zone The upper layer of the sea or other aqueous environment in which there is sufficient sunlight for *photosynthesis* to occur. The depth varies greatly with the condition of the water.

photocathode A *cathode* that emits *electrons* when it is illuminated, i.e. as a result of the *photoelectric effect*.

photocell See *photoelectric cell*.

photochemical reactions *Chemical reactions* initiated, assisted, or accelerated by exposure to *light*. E.g. hydrogen and chlorine combine explosively on exposure to sunlight but only slowly in the dark.

photochemical smog A *smog* produced by a series of complex reactions between *nitrogen oxides* and *hydrocarbons* in the presence of solar *ultraviolet radiation*.

photochemistry The branch of physical chemistry concerned with the effects of radiation on chemical reactions.

photochromism Phototropism. The property of certain *dyes*, or other *compounds*, that undergo a reversible change in the *colours* they absorb when exposed to *light* of different *wavelengths*. Thus some photochromic materials will darken in bright light, but will revert to their original colour when the source of light is removed.

photoconductive effect A *photoelectric effect* in which the electrical *conductivity* of certain substances, notably selenium and *cadmium sulphide*, increases with the intensity of the *light* to which the substance is exposed.

photodiode A *semiconductor diode* in which light from outside is focused on to the *p-n* junction. The diode is usually reverse biased so that the current is a minimum in the dark and increases in proportion to the intensity of the light falling on it. Photodiodes are used as switches to detect light or as cells to measure the intensity of light. They are photovoltaic devices.

photodisintegration 1. Photonuclear reaction. A *nuclear reaction* caused by a *photon* in which the *nucleus* emits charged fragments or *neutrons* (photoneutrons). **2.** See *photodissociation*.

photodissociation Photodisintegration. The *dissociation* of a chemical *compound* as the result of the absorption of *radiant energy*.

photoelasticity When certain materials (e.g. *glass*, *Perspex**, etc.) are stressed they become doubly refracting, which enables the property to be used to detect strain in these transparent materials. If *polarized* white light is passed through a stressed sample into a *polarimeter*, coloured patterns will be visible on the image in the viewing screen in the vicinity of areas of strain.

photoelectric cell Photocell. A device used for the detection and measurement of *light* and other radiations. The cell may depend for its action upon (1) the normal *photoelectric effect*; the cell is then called a *photoemissive* cell; (2) the *photovoltaic effect* (*rectifier* or *barrier layer* cell); or (3) the *photoconductive effect* (conductivity cell). Photoemissive cells consist of two *electrodes*, a plane *cathode* coated with a suitable *photosensitive* material, and an *anode* that is maintained at a positive *potential* with respect to the cathode and that attracts the *photoelectrons* liberated by the latter. These electrodes are arranged in an envelope that is either evacuated, or, for greater sensitivity, contains a *gas* at low *pressure*. The *electric current* passing through the cell is a measure of the *light* intensity incident on the cathode. For rectifier or barrier cells, the *potential difference* developed across the boundary gives rise to a current when the faces of the cell are connected externally. This current can be measured directly by suitable means, such as a *galvanometer*. Photovoltaic cells require no external source of *EMF* and are very convenient for photographic *exposure meters*, etc. (see *photodiode*). They are also used to detect *ultraviolet radiation*. The conductivity cell is simply an arrangement for measuring the *resistance* of a layer of material, usually selenium or *cadmium sulphide*, which shows the photoconductive effect. Photoconductive cells are also used to detect *infrared radiation*.

photoelectric effect In general, any effect arising as a result of a transfer of *energy* from *light* incident on a substance to *electrons* in the substance. The term is normally restricted to the *photoemissive* effect, namely the emission of electrons by substances when irradiated with light of a *frequency* greater than a certain minimum *threshold frequency*. Electrons liberated in this way are called photoelectrons, and constitute a photoelectric current when the system is included in a suitable circuit. The number of electrons emitted is related to the intensity of the radiation; the *kinetic energy* of the emitted electrons depends on the *frequency*, f. The effect is a *quantum* process in which the incident radiation is treated as a stream of *photons*, each having energy hf, where h is the *Planck constant*. If the *work function* of the substance on which the radiation falls is ϕ, then $\phi = hf_0$, where f_0 is threshold frequency at which emission occurs. The maximum kinetic energy, E_m, of the emitted photoelectron is given by the Einstein equation

$$E_m = hf - \phi.$$

The photoelectric effect was important in establishing the corpuscular nature of *light* (see also *complementarity*).

In some contexts the *photoconductive effect* and the *photovoltaic effect* are also included.

photoelectron An *electron* emitted from a surface as a result of illumination, i.e. by the *photoelectric effect* or by *photoionization*.

photoelectron spectroscopy A method of determining *ionization potentials* and examining molecular structure. A *gas* or *vapour* of the substance to be examined is exposed to *ultraviolet radiation*. The photoelectrons produced are directed through a slit into a vacuum region where *electric* and *magnetic fields* deflect them into an energy spectrum, the peaks of which give the ionization potentials of the molecules. If an *X-ray* source is used, electrons ejected from the inner electron shells are subjected to a chemical shift, due to the presence of neighbouring atoms in the molecule, which can be used to provide information regarding the molecular structure.

photoemissive Capable of emitting *electrons* when subjected to *electromagnetic radiation*. The *wavelength* (λ) of the radiation that will provoke such emission depends upon the nature of the substance and its *work function* (ϕ): *light* provokes some *metals* into photoemission, other materials require *ultraviolet radiation* or *X-rays*. For photoemission to take place:

$$\lambda \leq hc/\phi,$$

where h is the *Planck constant* and c is the *speed of light*. See also *photoelectric effect*.

photofission *Nuclear fission* caused by *photons* (of *gamma rays*).

photographic density See *density, photographic*.

photography By means of a system of *lenses* in the *camera* an image of the object to be photographed is thrown for a definite length of time on to a plate or *film* made of *glass*, *celluloid*, or other *transparent* material and covered with an *emulsion* containing *silver bromide*, AgBr, or *silver chloride*, AgCl. The effect of this exposure of the film is to make the silver compound easily reduced (see *reduction*) to metallic silver by the chemical action of *developing*; developers produce a black deposit of fine particles of metallic silver on those portions of the film that had been exposed to *light*, thus giving a negative image. *Fixing* consists of the chemical action of *sodium thiosulphate*, $Na_2S_2O_3$, ('hypo'), and other *reagents* on the unchanged silver *salts* to give a *soluble compound*, which is then washed out with *water*, leaving a negative free of light-sensitive silver salts. By placing the finished negative over a piece of sensitive paper similar to film, and exposing to light, the silver salts in the paper are affected in a similar way to those in the original film; those portions of the negative that were darkest let through least light, and thus give the whitest portions on the developed paper. The negative image is thus again reversed, and a correct image or black-and-white photograph is obtained on the paper, which is then fixed and washed as before. See also *colour photography*.

photoionization The ionization of an *atom* or *molecule* as the result of exposure to radiation. If the *frequency* of the radiation is f, each *photon* will have an energy hf, where h is the *Planck constant*. Photons with energies in excess of the *ionization potential* of the atoms struck will cause ionization to occur.

photoluminescence *Luminescence* caused by *electromagnetic radiation*. The emitted light always has a lower frequency than the radiation absorbed. Whiteners used in *detergents* consist of photoluminescent substances that absorb *ultra-violet radiation* and emit blue light.

photolysis The *decomposition* of a chemical *compound* as the result of *irradiation* by *light* or *ultraviolet radiation*. 'Flash photolysis' is a method of identifying the *free radicals* formed when the *vapour* of a compound at low *pressure* is exposed to an intense, but very brief, flash of radiation. A second flash, following shortly after the first, is used to photograph the *absorption spectrum* of the *gases*, which records the free radicals present. Subsequent flashes at regular intervals may be used to calculate the lifetimes of the radicals so formed.

photomeson A *meson* produced by the interaction between a *photon* and an atomic *nucleus*.

photometer An instrument for comparing the *luminous intensity* of sources of *light*. They originally consisted of devices that enabled a visual source to be compared with a standard source. More modern instruments rely on the *photo-electric effect*. In *astronomy* photoelectric photometers are used to measure the intensity of *light* from distant *stars*.

photometry The study of luminous quantities (relying on the eye) and radiant quantities (relying on photoelectric devices) in the measurement of *luminous intensity*.

photomicrograph A photograph obtained with the aid of an optical *microscope*.

photomultiplier Electron multiplier. A *photoelectric cell* of high sensitivity used for detecting very small quantities of *light* radiation. It consists of a system of *electrodes* suitably arranged in an evacuated envelope. Light falling on the first electrode (the *cathode*) ejects *electrons* from this surface (see *photoelectric effect*). These electrons are accelerated to the second electrode, where they each produce further electrons by the process of *secondary emission*. This process continues until the secondary emission is sufficient to produce a useful *electric current* at the anode, permitting measurement or the operation of a *relay*.

photon A *quantum* of *electromagnetic radiation* that has zero *rest mass*, and *energy* equal to the product of the *frequency* of the radiation and the *Planck constant*. Photons are generated when a particle possessing an *electric charge* changes its *momentum*, in collisions between *nuclei* or *electrons*, and in the *decay* of certain nuclei and particles. In some contexts it is convenient to regard a photon as an *elementary particle*.

photoneutron See *photodisintegration*.

photonuclear reaction See *photodisintegration*.

photophosphorylation *Phosphorylation* induced by *light*, *ultraviolet radiation*, or *infrared radiation*. It is an important part of *photosynthesis*.

photopic vision Vision in which the cones in the eye are the principal receptors. It occurs under normal lighting conditions and colours can be distinguished. Compare *scotopic vision*.

photosensitive 1. Substances are said to be photosensitive if they produce a *photoconductive*, *photoelectric*, *or photovoltaic effect* when subjected to suitable *electromagnetic radiation*. **2.** A substance that experiences a *chemical change* when exposed to *electromagnetic radiation*, e.g. the *emulsion* of a photographic film.

photosphere The visible, intensely luminous portion of the *Sun*, which has an estimated *temperature* of 6000 K. It is several hundred kilometres thick.

photosynthesis The important process by which plants in sunlight manufacture their *carbohydrates* from atmospheric *carbon dioxide* and *water* thereby liberating oxygen. The reaction, which is highly complex in detail, may be summarized by the equation:

$$6CO_2 + 6H_2O = C_6H_{12}O_6 + 6O_2.$$

When *light* falls upon green plants the greater part of the *energy* is absorbed by small particles called *chloroplasts*, which contain a variety of pigments, amongst them *chlorophylls*. The chloroplasts transform the energy of the *light* into *chemical energy* in the form of *adenosine triphospate* (ATP) and reduced NADP (see *nicotinamide adenine dinucleotide*), a process involving the *photolysis* of water (see *photophosphorylation*); in a further series of reactions not requiring light, these compounds are used to reduce carbon dioxide to *sugar molecules*. The process uses the CO_2 produced in *respiration* and releases the oxygen required for *aerobic* respiration. As animals, unlike plants, cannot synthesize their own organic food in this way, they depend for their carbon on the plants (or other animals) that they consume. Photosynthesis is therefore essential to almost all life forms, either directly or indirectly.

phototransistor A bipolar junction *transistor* that is activated by *light* or *ultraviolet radiation* falling on the *emitter-base* junction. The radiation creates new free charge *carriers* in the base region thus increasing the current through the device. Some phototransistors are used as switching devices.

photovoltaic effect A *photoelectric effect* in which *light* falling on a specially prepared boundary between certain pairs of substances (e.g. copper and *copper(I) oxide*) produces a *potential difference* across the boundary. See also *photodiode*.

phthalic acids $C_6H_4(COOH)_2$. Three *isomeric* acids. 1. Phthalic acid, the *ortho* form, 1,2-benzenedicarboxylic acid, is a white crystalline solid, m.p. 207°C., that decomposes into *phthalic anhydride* and water and is used in organic synthesis. 2. The *meta* form, 1,3-benzenedicarboxylic acid. See *isophthalic acid*. 3. The *para* form, 1,4-benzenedicarboxylic acid. See *terephthalic acid*.

phthalic anhydride $C_6H_4(CO)_2O$. The *anhydride* of *o-phthalic acid*, formed from the latter on heating, m.p. 130.8°C. It is made industrially by the *oxidation* of *naphthalene* in the presence of a catalyst and is an important *intermediate* in the production of *dyes*, *polyester resins*, and other organic products.

phthalocyanines Organic colouring matters, usually of outstanding resistance to the action of light and other agencies. The parent compound, phthalocyanine, is a *condensation* product of nitrogen-containing derivatives of *phthalic acid*; its molecule contains a ring of 16 atoms (carbon and nitrogen) similar to that in natural *porphyrins*. Four nitrogen atoms in this ring are positioned to form a small square in the centre of the molecule, and a metal atom, e.g. copper, can occupy a central position in the square becoming bonded to all four nitrogen atoms to form an extremely stable *chelate complex*. For example, copper phthalocyanine is a very stable brilliant-blue pigment.

phylogeny See *ontogeny*.

phylum A major group (*taxon*) of living *organisms*; a group of similar *classes*. See *classification*.

physical change Any change in a body or substance that does not involve an alteration in its chemical composition. It includes *electrochemistry*, *kinetics*, and *thermodynamics*.

physical chemistry The study of the *physical changes* associated with *chemical reactions* and the dependence of physical properties on chemical composition. It includes *electrochemistry*, *kinetics*, and *thermodynamics*.

physical states of matter The physical state in which *matter* exists, at a particular *temperature* and *pressure*, depends upon the *kinetic energy* of, and interaction between, its component *atoms*, *molecules*, or *ions*. In *gases* the distance between the fast moving atoms or molecules is such that the interaction between them is very small (see *Van der Waals' forces*); they are therefore free to move about the space that contains them almost independently of each other (see *kinetic theory of gases*). In the *solid state* the atoms, molecules, or ions have insufficient kinetic energy to overcome the strong *forces* between them, they therefore vibrate about the fixed positions of a *crystal lattice*. *Liquids* represent an intermediate state between gases and solids. Raising the temperature of a solid increases the kinetic energy of its components so that they are able to overcome the forces between them, the solid then becomes a liquid and eventually a gas. Increasing the pressure of a gas increases the number of collisions between the components and thus facilitates their interactions: for this reason increased pressure causes, or assists in, the *liquefaction of gases*. A hot ionized *plasma* has sometimes been referred to as the fourth state of matter.

physics The study of the properties of *matter* and *energy*, traditionally covering the subjects of *mechanics*, *electricity* and *magnetism*, *heat*, *light*, and *sound*. *Quantum theory* and *relativistic mechanics* have given rise to what is usually called modern physics, while the advent of *nuclear reactors*, *nuclear weapons*, and particle *accelerators* have introduced *atomic physics*, *nuclear physics*, and *particle physics*. See also *astrophysics*; *biophysics*; *geophysics*.

physiological saline An *isotonic solution* of *salts* in *distilled water* used for preserving *cells*. Such solutions contain no food for the cells and their survival in them is therefore restricted. An example is *Ringer's fluid*.

physiology The study of the function and chemical reactions within the various organs of plants or animals.

physisorption See *adsorption*.

physostigmine Eserine. $C_{15}H_{21}O_2N_3$. A colourless *alkaloid*, m.p. 105–6°C., used in the treatment of glaucoma.

phytamins See *auxins*.

-phyte, phyto- A suffix or prefix denoting 'plant', e.g. epiphyte (plant growing on other plants), phytocide (a substance that kills plants).

pi π. Symbol for the ratio of the circumference of any *circle* to its diameter. 3.141 59...(Approximately 22/7.)

pi bond π bond. See *orbital*.

pico- Prefix denoting one million millionth; 10^{-12}. Symbol p, e.g. pF = 10^{-12} *farad*.

picoline Methylpyridine. $CH_3C_6H_4N$. A *heterocyclic base* that exists in three *isomeric* forms. All three isomers are found in *coal-tar* and *bone oil*; they are used as *solvents* and as *intermediates* in *organic synthesis*.

picrate A *salt* or *ester* of *picric acid*.

picric acid 2,4,6-trinitrophenol. $C_6H_2(NO_2)_3OH$. A bright yellow crystalline *solid*, m.p. 122°C. Formerly used as an explosive (see *lyddite*), as a *dye*, and (in *solution*) for treating burns.

pie chart A diagrammatic way of showing percentages as slices of a circular pie, i.e. sectors of a *circle*. An example consisting of $x\%$, $y\%$, and $z\%$, where $x + y + z = 100$, would show a circle with three sectors, the central angles of each being $360x/100$, $360y/100$, and $360z/100$, respectively.

pi electron An electron in a pi *orbital*.

piezoelectric effect A property of certain *asymmetric crystals*, such as *Rochelle salt* or *quartz*. When such crystals (called 'piezoelectric crystals') are subjected to a *pressure*, positive and negative *electric charges* are produced on opposing faces; the signs of these charges are reversed if the pressure is replaced by a tension. The *crystal pick-up* and *crystal microphone* make use of the piezoelectric effect by using piezoelectric crystals to convert sound waves into electric currents.

The inverse piezoelectric effect occurs if such crystals are subjected to an *electric potential*, an alteration in size of the crystal taking place. If a quartz crystal is subjected to an alternating *electric field*, this inverse piezoelectric effect makes it expand and contract with the same *frequency* as that of the field. If the field frequency is made equal to the natural elastic frequency of the crystal, this will resonate and so augment the applied field. This is the basis of the *crystal oscillator* and the *quartz clock*.

pig-iron An impure form of iron cast into blocks (pigs) obtained from iron *ores* by the *blast furnace* process. Pig-iron is later converted into *cast iron* or *steel*.

pigment colour Body colour. The *colour* of most natural objects is due to the differential absorption by the substance of the different *wavelengths* (i.e. colours) present in the incident white *light*. The incident light penetrates a small distance into the substance, undergoes this absorption, and is then diffusely reflected out again. The colour the body appears is determined by the wavelengths absorbed the least. Thus, a substance that absorbs chiefly the red and yellow will appear blue. See also *surface colour*.

pigments 1. Materials used generally in the form of insoluble powders for impart-ing various colours to *paints*, *plastics*, etc. **2.** Natural colouring substances in plant or animal tissues.

pile 1. See *voltaic pile*. **2.** See *nuclear reactor* for atomic pile.

pilocarpine $C_{11}H_{16}N_2O_2$. A white crystalline *alkaloid*, m.p. 34°C., used in medi-cine.

pinchbeck An *alloy* of copper and zinc (10–15%) used as an imitation gold. Named after Christopher Pinchbeck (1670–1732).

pinch effect 1. The constriction of a *liquid conductor* of *electricity* (e.g. mercury or molten *metal*) that occurs when a substantial *current* is passed through it. **2.** The constriction of a *plasma* due to the *magnetic field* of a high current within the plasma. In a 'zeta pinch' the current is passed axially through the plasma and the magnetic field forms round it. In a 'theta pinch', current-carrying coils surround the plasma creating an axial magnetic field. Both zeta-pinch and theta-pinch devices are *torroidal*. See *thermonuclear reactions*.

pinene $C_{10}H_{16}$. A *liquid terpene*, b.p. 156.2°C., that is the principal constituent of *turpentine* and is found in other *essential oils*. It is used in the manufacture of *camphor*.

pinking See *knocking*.

pink salt Ammonium chlorostannate. $(NH_4)_2SnCl_6$. It is used as a *mordant* in dyeing.

pint A unit of capacity equal to one eighth of a *gallon*. 1 pint (UK) = 0.5683 *litre*.

pion A pi-*meson*. A type of *meson*. See *elementary particles*.

piperazine Hexahydropyrazine. $C_4H_8(NH)_2$. A colourless *deliquescent heterocy-clic base*, m.p. 108–110°C., used mainly as a *vermifuge*.

piperidine $C_5H_{10}NH$. A colourless *liquid*, b.p. 106°C., used as a *solvent*.

piperine $C_{17}H_{19}NO_3$. A white crystalline *alkaloid*, m.p. 129.5°C., the active constituent of pepper.

pipette A glass tube with the aid of which a definite *volume* of *liquid* may be transferred.

Pirani gauge A type of pressure gauge used to measure low pressures (100–0.01 Pa). An electrically heated wire is placed in the gas, the rate of loss of heat from the wire depending on the gas pressure. It may either be used with a fixed potential difference across the wire, the resistance of which is then a measure of the pressure, or at a fixed resistance so that the p.d. is a measure of the pressure.

pitch Hard dark substances that melt to viscous tarry *liquids*; they may be the residue from the *destructive distillation* of wood, *coal-tar*, *asphalt*, or various *bitumens*, etc.

pitchblende Uraninite. A natural *ore* consisting mainly of uranium oxide, U_3O_8. It occurs in Saxony, Bohemia, East Africa, Canada, and Colorado. Pitchblende contains small amounts of radium, of which it is the principal source.

pitch of a note A measure of the *frequency* of vibration of the source producing the note; a high frequency produces a note of high pitch. 'Concert pitch' is the frequency of the A above middle C to which musical performers tune. By international agreement it is set at 440 hertz. Because pitch is a subjective characteristic of a note that determines its position in the musical scale as experienced by an observer, it is not identical to frequency; it is influenced by loudness. Up to 1000 Hz, increasing the loudness tends to decrease the pitch;

over 3000 Hz an increase in loudness tends to raise the pitch. Between 1000 and 3000 Hz, pitch is independent of loudness. Pitch is measured in *mels*. See *sound*.

pitch of a screw The distance between adjoining crests of the thread, measured parallel to the *axis* of the screw.

Pitot tube An instrument for measuring the speed of a *fluid*; it consists of two tubes, one with an opening facing the moving fluid and the other facing away from it. The difference in *pressure* between the two openings, as measured by a *manometer*, allows the speed of the fluid to be determined according to the relationship

$$\Delta p = \rho V^2 / 2,$$

where Δp is the pressure difference, V is the *velocity* of the undisturbed fluid, and ρ is its *density*. Named after Henri Pitot (1695–1771).

pK A measure of the strength of an *acid*, defined as log $1/K$, where K is the *equilibrium constant* of the *dissociation* of the acid. The higher the value of pK, the weaker the acid.

Planck constant h. The universal constant relating the *frequency* of a *radiation*, v, with its *quantum* of *energy*, E; i.e. E = hv. The Planck constant has the value $6.626\,076 \times 10^{-34}$ *joule second*. The symbol $\hbar$ is often used for $h/2\pi$, which is known as the *rationalized* Planck constant. Named after Max Planck (1858–1947).

Planck's law of radiation. A law giving the *energy* distribution radiated by a *black body*. The energy radiated by unit surface area of a black body at a thermodynamic temperature, T, in unit time is:

$$2\pi hc^2 \lambda^{-5} / (\exp. \ hc/kT\lambda)$$

where h is the *Planck constant*, c is the *speed of light*, k is the *Boltzmann constant*, and λ is the wavelength of the radiation. This law introduced the concept of *quanta* of energy. See *quantum mechanics*.

plane (math.) A flat surface; mathematically, it is defined as a surface containing all the straight lines passing through a fixed point and also intersecting a straight line in space.

plane-polarized light See *polarization of light*.

planetarium 1. A complex system of optical projectors for representing the movements of the *planets* and *stars* on a domed ceiling. **2.** The building that houses such a system.

planetoids See *asteroids*.

planets Heavenly bodies revolving in definite *orbits* about the *Sun*. They are *Mercury*, *Venus*, the *Earth*, *Mars*, *Jupiter*, *Saturn*, *Uranus*, *Neptune*, and *Pluto*. See Appendix, Table 4.

planimeter A mechanical integrating instrument for measuring *plane* areas, consisting of a movable tracing arm the movements of which are recorded on a dial.

plankton The organisms that float and drift in the sea or other aqueous environment. They may be plants (phytoplankton) or animals (zooplankton). Compare *nekton*.

plano- Prefix used in conjunction with the words *concave* and *convex* to describe the shape of a *lens*. See Fig. 24, under *lens*.

plant hormones Plant growth substances. Compounds that affect or regulate the growth of plants. See *auxins*; *gibberellins*; *cytokinins*.

plaque 1. A deposit of dissolved food (mostly *carbohydrates*), *saliva*, and *bacte-*

ria on the exposed enamel surfaces of teeth. The bacteria metabolize the carbohydrates to produce an *acid* that eats into the dental enamel, causing caries (decay). Preventative measures include reducing the intake of carbohydrates (especially sweets), frequent brushing of the teeth to remove the plaque, and the provision of an adequate quantity of fluorine in the food or drinking water to strengthen the dental enamel. **2.** A hole in a growth of bacteria in an *agar* dish caused by a *bacteriophage*.

plasma (bio.) See *blood plasma*.

plasma (phys.) **1.** The region in a *discharge in gases* in which the numbers of positive and negative *ions* are approximately equal. **2.** The very hot ionized gas in which controlled *thermonuclear reaction* experiments are carried out. In such a plasma, which has been described as the fourth state of matter, the *ionization* is virtually complete. Again the numbers of positive ions and electrons are approximately equal and the plasma is therefore virtually electrically *neutral* and highly conducting. See also *containment*.

plasma membrane Plasmalemma. The *membrane* that surrounds and contains the *cytoplasm* of a biological *cell*.

plasmolysis The effect of *osmosis* on *cells* of living *organisms*. A cell placed in a *solution* of a greater *molecular concentration* than (i.e. is *hypertonic* to) the contents of the cell becomes plasmolysed; the *water* in the cell flows out through the cell *membrane* and the cell contents contract.

plaster of Paris Powdered *calcium sulphate* hemihydrate, $CaSO_4.\frac{1}{2}H_2O$, obtained by heating *gypsum* to $120-130°C$. On mixing with water, it sets and hardens.

plasticizer 1. A non-*volatile liquid* added to *paints* and varnishes to prevent brittleness of the dried film. **2.** A liquid or *solid* substance added to synthetic or natural *resins* to modify their flow properties.

plastics Materials that are *stable* in normal use, but at some stage of their manufacture are plastic and can be shaped or moulded by *heat*, *pressure*, or both. Most plastics are polymers (see *polymerization*), and are classified into *thermoplastic* and *thermosetting* materials.

plastid A *membrane*-bounded *organelle* in plant cells that contains a folded inner membrane that may bear complexes of *enzymes*. The most important plastids are *chloroplasts* but they may also store food reserves, e.g. amyloplasts store starch.

platelet See *blood platelet*.

plate tectonics The theory that the *Earth's* crust consists of a number of semirigid plates, which move relative to each other. Six major plates, associated with the continents (American, African, Antarctic, Eurasian, Indian, and Pacific), and a number of smaller ones are known. Where the plates meet each other volcanic activity and earthquakes are known to occur. When plates move away from each other (a divergent plate margin) under the sea an ocean ridge forms as material from the mantle wells up to form a new crust. This process is known as sea-floor spreading. When two plates come together (a convergent plate margin) under the sea one plate plunges under the other, forming an oceanic trench. The crust may partially melt causing a chain of volcanoes to occur in the upper plate. When two plates meet under land a mountain chain is formed (orogenesis).

platinic Containing platinum in its +4 *oxidation state*, e.g. platinum(IV) chloride, $PtCl_4$.

platinized asbestos *Asbestos* in the fibres of which a black deposit of finely divided platinum has been formed. It is used as a *catalyst*.

platinoid An *alloy* of 60% copper, 24% zinc, 14% nickel, and 2% tungsten.

platinous Containing platinum in its +2 *oxidation state*, e.g. platinum(II) chloride, $PtCl_2$.

platinum Pt. Element. R.a.m. 195.08. At. No. 78. A hard silvery-white ductile and malleable metal, r.d. 21.45, m.p. 1769°C., b.p. 3820°C., that is very resistant to both heat and acids. Its *expansivity* is very nearly equal to that of glass, which makes it useful in certain types of scientific equipment. It occurs as the metal, alloyed with osmium, iridium, and similar metals. It is used for electrical contacts, scientific apparatus, as a *catalyst* (see *platinized asbestos*), and in jewellery.

platinum chloride solution See *chloroplatinic acid*.

platinum metals A group of six *transition elements* with similar metallic properties. They are: ruthenium, rhodium, palladium, osmium, iridium, and platinum.

pleochroic Denoting certain *crystals* that have different colours, depending on the direction from which they are observed.

plumbago Black-lead, *graphite*. A natural *allotropic form* of carbon.

plumbate A *compound* formed by the action of an *alkali* on a lead *oxide* or *hydroxide*.

Pluto The outermost *planet* of the *solar system*. However, owing to its highly eccentric *orbit*, between 1979 and 1999 Pluto is nearer to the *Sun* than *Neptune*. Discovered in 1930, its mean distance from the *Sun* is 5907 million kilometres. *Sidereal period* ('year') 248.4 years. *Mass* approximately one tenth that of the *Earth*, diameter approximately 3500 kilometres. Pluto's surface temperature is probably below – 200°C.; it has one satellite provisionally called Charon.

plutonic rock See *igneous rock*.

plutonium Pu. At. No. 94. *Transuranic element*. Thirteen different *isotopes* of plutonium can be produced by suitable *nuclear reactions*. The isotope plutonium-239 is produced in *nuclear reactors* and is of considerable importance since it undergoes *nuclear fission* when bombarded by *slow neutrons*. This isotope, which has a *half-life* of 24 400 years, is also used in *nuclear weapons*, one *kilogram* having an *energy* equivalent of about 10^{14} *joules*. It is a dense silvery metal, r.d. 19.84, m.p. 641°C., b.p. 3200°C.

pneumatic Operated by, or filled with, compressed air.

pnicogens A collective term sometimes used (but not recommended) for the elements nitrogen, phosphorus, arsenic, antimony, and bismuth.

p-n-p **transistor** See *transistor*.

point-contact transistor See *transistor*.

point defect See *defect*.

point source of light A theoretical concept of a source of *light* in which all the light is emitted from a single point.

poise A unit of *viscosity* in *c.g.s. units* defined as the tangential *force* per unit area (*dynes* per sq cm) required to maintain unit difference in *velocity* (cm per second) between two parallel *planes* separated by one centimetre of *fluid*. 1 centipoise = 10^{-3} *newton* second per square *metre* (the *SI unit* of viscosity).

Poiseuille's equation The *volume V* of a *liquid* flowing through a cylindrical tube in unit time is given by the equation:

$$V = \pi p r^4/8l\eta,$$

where *p* is the *pressure* difference between two points on the *axis* of the tube at a distance *l* apart, η is the *viscosity*, and *r* is the radius of the tube. The result assumes uniform *streamline* flow, and also that the liquid in contact with the walls of the tube is at rest. Named after Jean Louis Poiseuille (1799–1869).

poison, nuclear Reactor poison. A substance that absorbs *neutrons* in a *nuclear reactor*. Poisons may be deliberately added to reduce the reactivity, or they may be *fission products*, such as xenon, some of which have to be periodically removed.

Poisson's ratio The ratio of the *lateral strain* to the *longitudinal* strain in a stretched wire. It is given by the ratio of d/D to l/L, where D = original diameter, L = original length, d = decrease in diameter, and l = increase in length. Named after Simeon Poisson (1781–1840).

polar bond An electrovalent bond. See *chemical bond*.

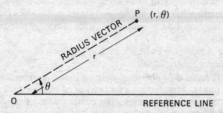

Figure 33.

polar coordinates The position of any point *P* lying in a *plane* can be completely determined by (1) its distance, *r*, from any selected point *O* in the plane, termed the *origin*, and (2) the *angle* θ that the line joining *P* and *O* (called the *radius vector*) makes with any *coplanar* reference line passing through *O*. The angle is taken as positive when measured anti-clockwise from the reference line. The polar coordinates at the point *P* are *r* and θ, denoted by (r, θ). See Fig. 33. If the *Cartesian coordinates* of a point are (x,y), then $x = r\cos\theta$ and $y = r\sin\theta$. In three dimensions, the point *P* may either be regarded as lying on the surface of a sphere (spherical polar coordinates) or on the surface of a cylinder (cylindrical polar coordinates).

polarimeter Polariscope. An apparatus for measuring the rotation of the plane of vibration of polarized light by optically active substances. See *polarization of light* and *optical activity*.

polariscope See *polarimeter*.

polarity The state of a body or system in which opposing physical properties are concentrated at one point or in one area. It usually refers to a *magnetic pole* or to *electric charge*.

polarization, angle of The *angle* of reflection from a *dielectric* medium, e.g. *glass*, at which the reflected *ray* is completely polarized, the plane of vibration being at right angles to the plane of incidence. See *Brewster's law*; *polarization of light*.

polarization, electric See *electric polarization*.

polarization, electrolytic An increase in the electrical *resistance* of an *electrolyte* due to various causes; chiefly associated with the accumulation of gaseous *molecules* on the *electrodes* at which they are liberated.

polarization of light Ordinary *light* consists of electric (E) and magnetic (H) vibrations taking place in all possible *planes* containing the *ray*, the vibrations themselves being at right angles to the direction of the light path; i.e. light is a *transverse wave* motion. For each E vibration the associated H vibration takes place in a plane at right angles to it. In plane-polarized light, the E vibrations are confined to one plane, called the plane of vibration, and hence the associated H vibrations are also confined to one plane, the plane at right angles to this, called the plane of polarization. See also *circularly* and *elliptically polarized light*.

polarization of particles Any moving *elementary particle* or *atomic nucleus* with *spin* exists in a polarization state that depends on the direction of spin relative to the particle's direction of motion. Particles spinning clockwise about their direction of motion are said to be right-polarized, while those spinning anti-clockwise are left-polarized. Particles and nuclei in stationary matter can also be polarized by aligning their spin axes parallel or *antiparallel* to the direction of a *magnetic field*.

polar molecule A *molecule* in which the configuration of *electric charge* constitutes a permanent electric *dipole*.

polarography A method of chemical *analysis* based on recording characteristic polarograms (curves representing variations of current strength with the applied voltage) for substances in solution. The compositions of solutions can be deduced from the form (characteristic 'waves') of their polarograms.

Polaroid* Trade name of thin *transparent films* that produce plane-polarized light (see *polarization of light*) on transmission. They consist of thin sheets of *cellulose nitrate* packed with ultramicroscopic doubly refracting *crystals* (see *double refraction*) with their *optic axes* parallel. The crystals produce plane-polarized light by differential absorption of the *ordinary* and *extraordinary rays*.

polaron An *excitation* in a *solid* consisting of *polar molecules* resulting from the interaction between an *electron* and its *strain field*. The presence of a polaron can be detected by irregularities in the shape of the *conduction band*.

pole, magnetic See *magnetic pole*.

pole of mirror See *mirrors, spherical*.

pole strength See *magnetic pole*.

pollen The grains containing the male *gametes* of seed plants. During pollination the pollen is transferred from the male anther of a flower to the female stigma, either of the same flower (self-pollination) or of a different flower of the same species (cross-pollination).

pollen analysis Palynology. The study of fossil pollen and spores found in *sedimentary rocks*. They provide evidence of the dominant flora and climatic conditions of the period during which the rocks were laid down.

pollution A change to the atmosphere, seas, rivers, or soil that is unwanted and results from the activities of man. Examples of pollutants are untreated sewage and oil spills in the sea, lead from *petrol* and sulphur gases (see *acid rain* and *photochemical smog*) in the atmosphere, heavy metals and nonbiodegradable insecticides (see *biodegradation*) in the soil, and industrial waste matter discharged into inland waterways. In addition, hot factory effluents discharged into rivers and lakes can cause thermal pollution, while the disposal of *nuclear waste* can also cause serious long-term pollution problems. On the global scale there are serious dangers arising from pollution of the upper atmosphere (see *greenhouse effect* and *ozone layer*). The control of pollution invariably requires international cooperation, which can only be achieved slowly.

polonium Po. A *radioactive element*. At. No. 84. The longest-lived *isotope* has a *mass number* of 209 and a *half-life* of 103 years. It is a rare metal, r.d. 9.40, m.p. 254°C., b.p. 960°C.; the isotope Po-210 occurs in some uranium ores to an extent of about 1 part in 10^{10} parts (mass of Po to mass of U).

poly- Prefix denoting many, several, numerous.

polyamide A *polymer* in which the units are linked by *amide* or thioamide groupings. See *nylon*.

polybasic An *acid* containing more than one *atom* of *acidic hydrogen* in a *molecule*.

polycarbonates *Thermoplastic resins* in which the structural units are linked through *carbonate radicals*. They usually consist of *polyesters* of *carbonic acids* and *dihydric phenols*. Their good dimensional stability and impact strength over a wide temperature range make them useful for electrical and other small components.

polychlorinated biphenyl PCB. A derivative of *biphenyl* $(C_6H_5)_2$ in which some of the hydrogen atoms on the two *benzene rings* have been substituted by chlorine atoms. These highly poisonous and *carcinogenic* compounds are used in synthetic *resins*, particularly as electrical insulators. Their increased usage has caused concern as they tend to accumulate in the *food chain*.

polychloroethene See *polyvinyl chloride*.

polychromatic radiation *Electromagnetic radiation* that consists of a mixture of *wavelengths*. Compare *monochromatic radiation*.

polycyclic Having more than one ring in a *molecule*.

polyene Any organic compound containing more than two *double bonds*.

polyester A *polymer* formed (usually) from a *polyhydric alcohol* and a *polybasic acid*. They are used in the manufacture of synthetic *resins*, *fibres*, and *plastics*.

polyethene See *polythene*.

polyethylene See *polythene*.

polyethylene terephthalate See *Terylene**.

polygon A *plane* figure bounded by straight lines. A regular polygon has equal sides and internal angles. For a regular polygon with n sides the internal angles are $180-360/n$.

polygon of forces If any number of *forces*, acting on a particle, can be represented in magnitude and direction by the sides of a *polygon* taken in order, the forces will be in *equilibrium*.

polyhedron A solid figure having *polygons* for its faces. A regular polyhedron has all its faces equal in all respects; the five possible types of regular polyhedra are: (1) *tetrahedron*, 4 triangular faces; (2) *cube*, 6 square faces; (3) *octahedron*, 8 triangular faces; (4) *dodecahedron*, 12 five-sided faces; (5) *icosahedron*, 20 triangular faces.

polyhydric Containing more than one *hydroxyl* group in the *molecule*; e.g. *ethanediol* (ethylene glycol) and *glycerol* (1,2,3-propanetriol) are polyhydric *alcohols* (polyols).

polymer A product of *polymerization*: biological polymers include *polysaccharides*, *proteins*, and *nucleic acids*. See also *atactic polymer*; *tactic polymer*.

polymerase An *enzyme* that catalyses a biological *polymerization* reaction.

polymerization Originally, the chemical union of two or more *molecules* of the same *compound* to form larger molecules, resulting in the formation of a new compound of the same *empirical formula* but of greater *relative molecular mass*. E.g. *ethanal* trimer, $(CH_3CHO)_3$, is formed by the polymerization of ethanal, CH_3CHO, and each molecule of the *polymer* is made up of three molecules of the ethanal *monomer*. The meaning of the term has been extended to cover (1) 'addition polymerization', in which the molecule of the polymer is a multiple of the monomer molecule, as in the case of ethanal trimer; (2) 'condensation polymerization', in which the monomer molecules are joined by *condensation* into a polymer molecule, which differs in empirical formula from the monomer; and (3) 'copolymerization', in which the polymer molecule is built up from two or more different kinds of monomer molecules. Many important products, such as *plastics* and textile fibres, consist of polymeric substances, either *natural* (e.g. *cellulose*), or *synthetic* (e.g. *nylon*).

polymethanal See *methanal*.

polymethyl methacrylate Poly(methyl 2-methylpropenoate), Perspex*. A colourless *transparent solid thermoplastic*, produced by the *polymerization* of methyl methacrylate (see *methacrylic acid*), which is widely used because of its optical properties in place of *glass*.

polymorphism 1. (chem.) The existence of the same substance in more than two different crystalline forms. **2.** (bio.) The existence of two (dimorphism) or more genetically distinct forms within the population of a *species*.

polynomial (math.) An expression consisting of three or more terms.

polynucleotide A chain of *nucleotides* linked together as in a *nucleic acid*. *Ribonucleic acid* consists of a single chain, while *deoxyribonucleic acid* usually consists of a double *helix* comprising two polynucleotide chains.

polypeptide A chain of two or more *amino acids* each of which is joined to its neighbours by the *peptide* linkage. Polypeptide chains may consist of up to several hundred amino acid units. *Proteins* consist of polypeptide chains cross-linked together in a variety of ways.

polyploidy Having more than twice the normal *haploid* number of *chromosomes* in the *nucleus* of a *cell*. Artificial polyploidy can be induced (e.g. by *colchicine*) and is used to produce fertile hybrids with desired characteristics.

polypropene Polypropylene. A colourless *transparent thermoplastic* material produced by the *polymerization* of *propene*. It is used where a flexible *plastic* material is required. It is similar to *polythene* but of greater strength.

polysaccharides A large class of natural *carbohydrates*. The *molecules* are derived from the *condensation* of several, frequently very many, molecules of simple *sugars* (*monosaccharides*). The class includes *cellulose* and *starch*. See also *glycan*.

polystyrene A *thermoplastic* material, produced by the *polymerization* of *styrene* (phenylethene; $C_6H_5CH:CH_2$). It is a clear glassy material possessing good electrical insulating properties. It is also used as a packing material.

polytetrafluoroethene PTFE. Teflon*. Fluon*. A *thermosetting* material produced by the *polymerization* of *tetrafluoroethene* ($CF_2:CF_2$). It is used to line saucepans and in bearings and electrical insulation because of its ability to withstand temperatures up to 400°C. and its low coefficient of friction.

polythene Polyethene, polyethylene, Alkathene*. A tough waxy *thermoplastic* material, made by the addition *polymerization* of *ethene*, C_2H_4. It is used as an insulating material and for many other purposes where a flexible, chemically resistant *plastic* material is required.

polyunsaturated See *fatty acids*.

polyurethane See *urethane resins*.

polyvalent 1. Having more than one *valence*. **2.** Having a valence of more than one. **3.** (Of a *serum*). Containing more than one type of *antibody* and therefore effective against more than one type of *microorganism*.

polyvinyl acetate PVA. Polyethenyl ethanoate. A colourless *thermoplastic* material, produced by the *polymerization* of *vinyl acetate* (ethenyl ethanoate; $CH_2:CHOOC.CH_3$). It is used in *adhesives*, inks, and lacquers for coating paper and fabric.

polyvinyl chloride PVC. Polychloroethene. A colourless *thermoplastic* material, produced by the *polymerization* of *vinyl chloride* (chloroethene; $CH_2:CHCl$) with good resistance to *water*, *acids*, *alkalis*, and *alcohols*.

polyvinylidene chloride Polydichloroethene. A white *thermoplastic* material, produced by the *polymerization* of *vinylidene chloride* (dichloroethene; $CH_2:CCl_2$). It is also used as a copolymer with *acrylonitrile* or *vinyl chloride* giving products with a wide range of flexibilities.

polywater Anomalous water. A form of water, differing in properties (*density*, *viscosity*) from normal water, that was reported in 1962. It is now accepted that these properties were due to the presence of colloidal particles derived from impurities rather than to any differences in the molecular structure of the water itself.

polyyne An *alkyne* containing two or more *triple bonds*.

population 1. The number of individuals of a *species* in a specified area. **2.** A group of individuals belonging to the same species in an ecological community.

population genetics The study of the distribution of inherited variations in a *population* of *organisms* of the same *species*.

population inversion The situation that exists in a *laser* when a large proportion of the emitting *ions* have been raised to an excited energy level by the process of optical pumping (i.e. introducing energy into the system by an external light source). This is an essential step in the process of stimulated emission. See also *maser*.

population type A classification of *stars* into two types: Population I consists of hot white young stars such as those that form the spiral arms of *spiral galaxies*; Population II consists of older stars, such as *red giants*, which are found at the centres of spiral galaxies.

porcelain A hard white material made by the firing of a *mixture* of pure kaolin (*china clay*) with *felspar* and *quartz*, or with other materials containing *silica*.

porphyrins A class of naturally occurring *pigments* derived from pyrrole. They include *chlorophyll* and the haem of *haemoglobin*. Their molecules are flat and contain a ring of 12 carbon and 4 nitrogen atoms; the latter form a small square in the centre of the molecule (compare *phthalocyanines*) and are linked to a metal atom, forming a *chelate complex*. This metal is magnesium in chlorophyll, and iron in haem.

position circle A *circle* having its centre at an observed point and a circumference that passes through the place of observation. The portion of the circumference near the place of observation approximates to a *position line* if the radius is large.

position line A line on which an observer is situated at a given time. The intersection of two position lines, determined at the same time, fixes the position of the observer.

positive (math., phys.) In any convention of signs, regarded as being counted in the plus, or positive direction, as opposed to *negative*.

positive column A luminous region in a *discharge in gases* near to the positive *electrode*.

positive electron See *positron*.

positive feedback See *feedback*.

positive magnetic pole The north-seeking pole of a *magnet*. See *magnetic pole*.

positive rays Streams of *ions* bearing positive *electric charges*. They are produced by an electric discharge in a low-pressure gas. See *discharge in gases*.

positron Positive *electron*. An *elementary particle* with the same *mass* as the electron and an *electric charge* of equal magnitude but opposite sign. Positrons are produced during several *decay* processes (see *beta decay*) and during *pair production*; they do not themselves decay spontaneously but on passing through matter they collide with negative electrons as a result of which both particles are annihilated. See *annihilation*.

positronium An unstable unit, resembling an *atom* of hydrogen, that consists of a *positron* (instead of a *proton*) and an *electron*. It decays by annihilation in less than 10^{-7} second into two or three *photons*.

potash An old name for *potassium carbonate*, *potassium hydroxide* (caustic potash), or any potassium *salt*.

potassium Kalium. K. Element. R.a.m. 39.0983. At. No. 19. A silvery-white soft highly reactive *alkali metal*, strongly resembling sodium. R.d. 0.86, m.p. 63.5°C., b.p. 770°C. Widely distributed in seawater and various *salts* (e.g. *carnallite*, *sylvine*), it is essential to life and is found in all living matter. Its salts are used as *fertilizers* and for a variety of other purposes.

potassium-argon dating A method of *dating* geological specimens based on the decay of the *radioisotope* potassium-40 to argon-40. The *half-life* of potassium-40 is about 1.3×10^9 years and an estimate of the ratio of the two *nuclides* in a specimen gives an indication of its age.

potassium bicarbonate See *potassium hydrogencarbonate*.

potassium bromide KBr. A white crystalline *salt*, m.p. 730°C., used in medicine and *photography*.

potassium carbonate Potash, pearl ash. K_2CO_3. A white very *soluble deliquescent salt*, m.p. 891°C., used in the manufacture of *glass* and soft *soap*.

potassium chlorate $KClO_3$. A white crystalline *soluble* substance, m.p. 356°C., used as an *oxidizing agent*, as a weedkiller, and in the manufacture of fireworks.

potassium chloride Potassium muriate. KCl. A white crystalline *soluble* substance, m.p. 776°C., used in medicine and as a *fertilizer*. It occurs as *sylvine* and *carnallite*.

potassium chromium sulphate Chrome alum. $K_2SO_4.Cr_2(SO_4)_3.24H_2O$. A dark purple crystalline *soluble salt*, used in *dyes*, calico printing, and *tanning*.

potassium dichromate Dichromate or bichromate of potash. $K_2Cr_2O_7$. A red crystalline *soluble salt*, m.p. 398°C., prepared from *chrome iron ore*. It is used as an *oxidizing agent*, and in the *paint* and *dye* industries.

potassium hexacyanoferrate(II) Potassium ferrocyanide. $K_4Fe(CN)_6.3H_2O$. A yellow *soluble* crystalline substance, used as a *dye* and in case-hardening.

potassium hexacyanoferrate(III) Potassium ferricyanide. $K_3Fe(CN)_6$. A red *soluble* crystalline substance, used in the manufacture of *pigments* and *paper*.

potassium hydrogencarbonate Potassium bicarbonate. $KHCO_3$. A white *soluble* substance, used in cooking and as an *antacid*.

potassium hydrogendifluoride Acid potassium fluoride. KHF_2. A *deliquescent* crystalline substance used in the electrolytic production of fluorine.

potassium hydrogentartrate Cream of tartar. $HOOC(CHOH)_2COOK$. A white crystalline powder obtained from *argol* (tartar), used in *baking powder*.

potassium hydroxide Caustic potash. KOH. A white *deliquescent solid*, m.p. 360.4°C., that dissolves in *water* to give an *alkaline* highly corrosive *solution*. It is used in medicine, in nickel-iron *accumulators*, and in the manufacture of soft *soap*.

potassium iodide KI. A white crystalline *soluble* substance, m.p. 686°C., used in *photography* and in medicine. It is also added to table salt to provide a source of iodine.

potassium nitrate Nitre, saltpetre. KNO_3. A white *soluble* crystalline *salt*, m.p. 336°C., that acts as an *oxidizing agent* when hot. It is used in medicine, for pickling meat, and in *gunpowder*.

potassium permanganate Potassium manganate(VII). $KMnO_4$. A deep purple, crystalline, *soluble salt*. Dissolved in *water* it gives a purple *solution* that acts as a powerful *oxidizing agent*. It is used as a *disinfectant* and in *volumetric analysis*.

potassium sodium tartrate See *Rochelle salt*.

potassium sulphate K_2SO_4. A white *soluble* crystalline substance, m.p. 1069°C., used in *fertilizers*, cements, and mineral waters.

potassium thiocyanate KSCN. A colourless *hygroscopic* substance, m.p. 173.2°C., used in the manufacture of *dyes* and *drugs*.

potential See *electric potential*.

potential barrier See *nuclear barrier*.

potential difference If two points have a different *electric potential* there is said to be a potential difference (p.d.) between them; if the points are joined by an electric *conductor*, an *electric current* will flow between them. Potential difference is defined as the *work* performed when a unit positive *electric charge* is moved from one of the points to the other. See also *electromotive force*, EMF. The practical unit of p.d. and EMF is the *volt*.

potential energy The *energy* that a body possesses by virtue of its position. E.g. a coiled spring, or a vehicle at the top of a hill, possesses potential energy. It is the amount of *work* done on the body in passing from a standard position in which the potential energy is considered to be zero to the position of higher energy. The potential energy of a *mass*, *m*, raised through a height, *h*, is *mgh*, where *g* is the *acceleration of free fall*.

potential series See *electromotive series*.

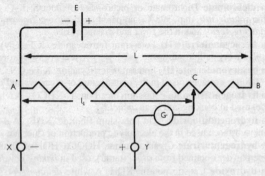

Figure 34.

potentiometer 1. An instrument for measuring *direct current EMF*, which does not draw current from the circuit containing the EMF to be measured. In its simplest form it consists of a uniform *resistance AB* (see Fig. 34) in the form of a single wire, connected to a source of EMF, *E*. A slide wire contact *C* is connected, in series with a sensitive *galvanometer G*, to one terminal of the EMF to be measured. The other terminal is connected to *A*, so that the EMFs across *XY* and *AC* are in opposition through *G*. Contact *C* is then adjusted until no current flows through the galvanometer. The required EMF is then given by El_1/L, where *L* is the total length of the resistance *AB*, and l_1 is the length *AC* for zero current through *G*. **2.** A *voltage divider*.

pound A British unit of *mass*. It was formerly defined as the mass of a platinum cylinder called the Imperial Standard Pound. The pound was redefined by statute in 1963 as 0.453 592 37 *kilogram*.

poundal A unit of *force* in the *f.p.s. system*. The force that, acting on a *mass* of 1 *pound*, will impart to it an *acceleration* of 1 foot per second per second. It is approximately 1/32 of a force of 1 pound weight.

powder metallurgy The science or practice of manufacturing small *metal* articles (such as small self-lubricating bearings) by sintering powdered metals under *heat* and *pressure* (up to 850×10^6 Pa).

power (math.) The number of times a quantity is successively multiplied by itself. Thus $2 \times 2 \times 2 \times 2$ is 2 raised to the fourth power, 2 to the fourth, denoted as 2^4, 4 being the *index* or *exponent*.

power (phys.) The rate of doing *work*. It is measured in units of work per unit time. The derived *SI unit* of power is the *watt*. See also *horsepower*.

power alcohol Industrial *ethanol* used as a *fuel*.

power factor In an electrical *circuit*, the ratio of the *power* dissipated, *P*, to the product of the *electromotive force*, *E*, and the *current*, *I*. In single-*phase* and

348

three-phase circuits the power factor is given by $\cos\phi$, where ϕ is the *phase angle* between the EMF and the current, i.e. $P = EI \cos\phi$. E and I are RMS values.

power reactor See *nuclear reactor*.

Prandtl number A dimensionless group used in calculations involving *convection* in a fluid. It is $C\eta/K\rho$, where C is the *heat capacity* per unit volume of the fluid, η is its *viscosity*, ρ its *density*, and K is its *thermal conductivity*.

praseodymium Pr. Element. R.a.m. 140.907. At. No. 59. R.d. 6.78, m.p. 935°C., b.p. 3000°C. A rare soft silvery metal occurring in *monazite* and *bastnasite*. It is used in *misch metal* and in some rare-earth *catalysts* for cracking oil. See *lanthanides*.

precessional motion A rotating body is said to precess when, as a result of an applied *couple*, the *axis* of which is at right angles to the rotation axis, the body turns about the third mutually perpendicular axis.

precession of the equinoxes A slow movement to the west of the *equinoxes* about the *ecliptic*, caused by the *precessional motion* of the *Earth*. The period of this precession is 25 800 years. The precessional motion of the Earth is a result of the gravitational attraction between its equatorial bulge and the *Sun* and the *Moon*.

precipitate (chem.) An *insoluble* substance formed in a *solution* as the result of a *chemical reaction*.

precipitation 1. (chem.) The formation of a *precipitate*. A common type of precipitation much used in chemical analysis and preparations, occurs by *double decomposition* when two *solutions* are mixed if each of the solutions contains one *radical* of an *insoluble compound*. **2.** Water deposited from the atmosphere in the form of rain, snow, hail, and dew.

precursor An intermediate substance from which another is formed in the course of a chemical process.

preon A hypothetical particle postulated as a component of *quarks* and *leptons*, which are currently regarded as the basic *elementary particles*. Such particles, if they exist, would require energies well above that available with current *accelerators*.

presbyopia Long sight. A defect of vision normally occurring after the age of 45. The subject is able to see distant objects clearly, but is unable to accommodate the eye to see near objects distinctly. It occurs as a result of loss of elasticity in the lens of the eye and is corrected by the use of *convex* spectacle *lenses*.

pressure The *force* per unit *area* in a *fluid* acting normally on a surface. 'Absolute pressure' is the pressure measured with respect to zero pressure. 'Gauge pressure' is the pressure measured by a gauge in excess of the pressure of the atmosphere. The *SI unit* of pressure is the *pascal* ($N\ m^{-2}$). The *c.g.s. system* uses the *dyne* per square centimetre ($1\ Pa = 10$ dynes cm^{-2}). Other units are the *bar* ($= 10^5$ Pa), the *atmosphere* ($= 101\ 325$ Pa), the mmHg ($= 133.322$ Pa). See also Appendix, Table 1.

pressure gauge Any device for measuring the *pressure* of a *fluid*. The simplest types are mercury-in-glass *barometers* and *manometers*. More compact are the *piston gauge*, *Bourdon gauge*, and *aneroid barometer*. Other types include the *strain gauge*, in which the extension of an element may be used to move one of the plates of a *capacitor* (capacitor gauge) or to change the electrical *resistance* of the element (resistance gauge). Vacuum gauges include *ionization gauges*, the *McLeod gauge*, *molecular gauges*, and the *Pirani gauge*.

pressurized water reactor PWR. A *nuclear reactor* in which *water* is the coolant and the *moderator*, but in which the water is maintained at a high *pressure* (usually 16×10^6 Pa) in order to prevent it *boiling*. The pressurized water leaves the core at a temperature of about 310°C. and is passed through a *heat exchanger* to generate *steam* for producing *electric power* in a conventional *turbogenerator*. Compare *boiling water reactor*.

Prévost's theory of exchanges A body emits the same *radiant energy* as it absorbs when it is in equilibrium with its surroundings. Named after Pierre Prévost (1751–1839).

primary alcohol See *alcohol*.

primary cell Voltaic cell, galvanic cell. A device, usually irreversible, for producing an *electromotive force* and delivering an *electric current* as the result of a *chemical reaction*. See *Daniell cell*, *Leclanché cell*, *Weston cell*, *mercury cell*.

primary coil Primary winding. The input coil of a *transformer* or *induction coil*. Compare *secondary coil*.

primary colours (phys.) A set of three coloured lights that when mixed in equal proportions produce white light; mixed in other proportions they can produce any colour (except black) by an *additive process*. Three coloured pigments are also called primary colours if, when mixed in equal proportions, they produce a black pigment; mixed in other proportions they can produce any colour (except white) by a *subtractive process*. There are many sets of such coloured lights and pigments, the most common is red, green, and blue.

prime number (math.) A number possessing no *factors* (i.e. divisible by no whole number, other than itself and one).

principal axis See *optical axis*.

principal focus See *mirrors*, *spherical*; *lens*.

principal plane (phys.) In a *crystal* exhibiting *double refraction*, a *plane* containing the *optic axis* and either the *ordinary ray* (principal plane of ordinary ray) or the *extraordinary ray* (principal plane of extraordinary ray).

principal point (phys.) Two points on the *optical axis* of a thick *lens* or combination lens system, such that if the object distance is measured from one and the image distance from the other, the *equations* obtained relating object-image distance, etc., are similar to those obtained for a thin lens.

principal section (phys.) A *plane* passing through the *optic axis* of a *crystal* exhibiting *double refraction* that is at right angles to one of the crystal surfaces.

principle of superposition See *Huygens principle of superposition*.

printed circuit An *electronic circuit* in which the wiring between components, and certain fixed components themselves, are printed on to an insulating board. The board is coated with copper and the portion of the metal that represents the wiring or components is photographically covered with a protective film, the rest of the metal being etched away in an *acid* bath.

prism (math.) A solid figure having two identically equal faces (bases) consisting of *polygons* in parallel *planes*; the other faces being *parallelograms* equal in number to the number of sides of one of the bases.

prism, optical A triangular *prism* made of material *transparent* to the *light* being used; e.g. *glass* for visible light, *quartz* for *ultraviolet* and near *infrared radiation*. They are used in optical instruments to deviate or disperse a ray or to turn an image upside down. See also *Nicol prism*; *Rochon prism*; *Wollaston prism*.

prismatic In the shape of a *prism* or using a prism.

prismatic optical instruments Instruments (field-glasses, etc.) in which a right-angled *prism* is used to invert the inverted image produced by the *objective*.

probability, mathematical A mathematical expresssion of the chance that a specified event will occur. If the event is certain to occur the probability is 1; if it is certain not to occur the probability is 0. Between these two extremes the probability of an event occurring is expressed as a number between 0 and 1. For example, if an event can happen in a ways and fail in b ways, and, except for the numerical difference between a and b, is as likely to happen as to fail, the mathematical probability of its happening is $a/(a + b)$ and of its failing, $b/(a + b)$. The probability that an event will occur if there are x ways it can occur in n trials is x/n. For example, the probability of throwing an even number on a dice is $3/6 = 0.5$.

process control The control of complex industrial or chemical processes by *electronic* means.

producer gas A *fuel gas* produced by the partial combustion of *coke* or *coal* in a restricted supply of air, to which *steam* may have been added. The principal constituents of the gas are *carbon monoxide* (25%–30%), nitrogen (50%–55%), and hydrogen (10%–15%). *Hydrocarbons* and *carbon dioxide* will also be present.

product (math.) The result of multiplying two or more quantities together.

production reactor See *nuclear reactor*.

progesterone $C_{21}H_{30}O_2$. A white crystalline *steroid*, m.p. 128.5°C.; a *sex hormone*, it is produced by the ovaries and is responsible for preparing the reproductive organs of mammals for pregnancy and for protecting the embryo. A fall in the *concentration* of progesterone in the blood initiates menstruation in women (see *menstrual cycle*).

progestogens *Sex hormones* that control the course of pregnancy. The most important is *progesterone*. In high doses progestogens inhibit *ovulation* and are therefore used as constituents of oral contraceptives.

program Programme. The sequence of instructions fed into a *computer* in order to enable it to carry out a process.

projectile A body that is thrown or projected. If the projectile is discharged with a *velocity v* at an *angle a* to the horizontal, the following formulae hold true if the resistance of the air is neglected (g being the *acceleration of free fall*):

Time to reach highest point of flight = $(v\sin a)/g$
Total time of flight = $(2v\sin a)/g$
Maximum height = $(v^2\sin^2 a)/2g$
Horizontal range = $(v^2\sin 2a)/g$

prokaryote Procaryote. An *organism* whose *cells* lack a true *nucleus*; the DNA is not separated from the rest of the *cytoplasm* but lies in a region called the *nucleoid*. Prokaryotic cells also lack other *membrane*-bounded *organelles*. *Bacteria* are examples.

prolactin A *hormone* that stimulates mammals to secrete milk and the ovaries to secrete *progesterone*. It is secreted by the pituitary gland.

prolate spheroid See *spheroid*.

proline A white, crystalline *amino acid*, m.p. 220°C., that occurs in most *proteins*. See Appendix, Table 5.

promethium Pm. A *radioactive element* of the *lanthanide* series. At. No. 61. The only naturally occurring isotope, Pm-147, has a *half-life* of 2.5 years, r.d. of 7.2, m.p. 1168°C., and b.p. 3300°C. It occurs as a fission product of uranium in *nuclear reactors*. The most stable *isotope*, promethium-145, has a half-life of about eighteen years.

prompt critical Capable of sustaining a *nuclear fission chain reaction* on the *prompt neutrons* alone, without contribution from *delayed neutrons*.

prompt neutrons *Neutrons* resulting from *nuclear fission* (either during the fission process or from freshly formed fission fragments) that are emitted without measurable delay, i.e. in less than a millionth of a second. See *delayed neutrons*.

proof spirit *Ethanol* containing 49.28% alcohol by *weight*, or 57.10% by *volume*, and having a *relative density* of 0.919 76 at 60°F. Formerly defined as the weakest solution of alcohol that would fire *gunpowder* when brought into contact with it and ignited.

proof spirit, degrees The number of degrees under proof is the *volume* percentage of *water* in a *solution* regarded as containing *proof spirit* and water; degrees over proof is the volume increase obtained when 100 volumes of the spirit are diluted with sufficient water to obtain proof spirit. Spirits are usually sold on the basis, '30° under proof' or '70° proof' both of which mean the same. Such spirit contains $57.1 \times 70/100 = 39.97\%$ alcohol by volume.

propanal Propionaldehyde, propyl aldehyde. CH_3CH_2CHO. A colourless *liquid aldehyde*, b.p. 48.8°C., used in the manufacture of *plastics*.

propane C_3H_8. The third *hydrocarbon* of the *alkane series*. A flammable *gas*. B.p. −42.17°C. It is used as a *fuel* in the form of bottle gas.

propanedioic acid Malonic acid. $CH_2(COOH)_2$. A white soluble *dibasic acid*, m.p. 135.6°C., used in the manufacture of *barbiturates*.

propanoic acid Propionic acid. CH_3CH_2COOH. A colourless *liquid carboxylic acid*, b.p. 141°C. It is used in the form of its calcium salt as a bread additive.

propanol Propyl alcohol. Either of two isomers (see *isomerism*). **1.** Propan-1-ol, *n*-propyl alcohol, $CH_3CH_2CH_2OH$, a colourless *liquid*, b.p. 97.2°C., used as a *solvent*. **2.** Propan-2-ol, isopropyl alcohol, $CH_3CHOHCH_3$, a colourless *liquid*, b.p. 82.4°C., used for the industrial production of *propanone* (acetone), as a *solvent*, and as an intermediate in organic *synthesis*.

propanone Acetone, dimethyl ketone. CH_3COCH_3. A colourless flammable *liquid* with a pleasant odour, b.p. 56.5°C., used as a *solvent* in making *plastics*.

propellant 1. The explosive substance used to fill cartridges, shell cases, and *solid fuel rockets*. The term is also used to include the fuel and *oxidant* of rockets when these are separate. **2.** A *gas* used in *aerosol* preparations to expel the *liquid* contents through an atomizer.

propenal Acrolein, acrylaldehyde. CH_2:$CHCHO$. A colourless flammable *liquid* with an irritating smell, b.p. 52.5°C., used to make *acrylic resins* and pharmaceutical products.

propene Propylene. CH_2:$CH.CH_3$. The second member of the *alkene* series of *hydrocarbons*. A colourless *gas*, b.p. −47°C. See also *polypropene*.

propenoate Acrylate. A *salt* or *ester* of *propenoic acid*.

propenoic acid Acrylic acid. CH_2:$CHCOOH$. A corrosive *liquid*, m.p. 13°C., b.p. 141°C., used in the manufacture of *acrylic resins*.

propenonitrile Acrylonitrile, vinyl cyanide. $CH_2:CHCN$. A colourless flammable *liquid*, b.p. 78°C., used to make *acrylic resins*, synthetic rubbers, and synthetic fibres.

proper fraction A fraction in which the *numerator* is less than the *denominator*, e.g. 3/4. In an 'improper fraction' the numerator is greater than the denominator, e.g. 4/3.

proper motion of a star The component of a *star's* motion in *space* relative to the *Sun* that is perpendicular to the line of sight.

propionaldehyde See *propanal*.

propionic acid See *propanoic acid*.

proportion (math.) An equality between two *ratios*. If $a/b = c/d$, the four quantities, a, b, c, d are in proportion.

proportional counter A *counter tube* in which the output pulse is proportional to the number of *ions* produced in the initial ionizing event (e.g. the number of ions created by the incident particle).

propulsion reactor See *nuclear reactor*.

propyl The *univalent alkyl* group C_3H_7-.

propylene See *propene*.

prostaglandins A group of compounds composed of essential fatty acids that occur in many tissues of mammals. Their effects include dilating blood vessels, and contracting the womb. They are also released at sites of inflammation resulting from tissue damage. The pain-relieving properties of such drugs as *aspirin* are due to their inhibition of prostaglandin synthesis. Synthetic prostaglandins have been used to induce labour and to procure abortions.

prostar See *stellar evolution*.

prosthetic group A non-*protein* group, often a metal *ion*, combined to a protein, e.g. the haem group in *haemoglobin* or the *nucleic acid* in *nucleoprotein*.

protactinium Pa. *Radioactive* element. At. No. 91. The most abundant natural *isotope* has a *mass number* of 231 and a *half-life* of 32 480 years. A dense metal with an r.d. 15.37, m.p. 1200°C., b.p. 4000°C. It occurs in minute quantities in uranium *ores*, as a member of the *actinium series*.

protargol A powder containing finely divided silver and *protein*; with *water*, it forms a *colloidal solution* of silver.

proteases Proteinases. A group of *enzymes* capable of breaking up *proteins* into *amino acids*, of building up amino acids into proteins, and of substituting one amino acid for another in protein *molecules*. Occurring in all living tissues, they conduct the processes of protein *metabolism* in the living *organism*. *Pepsin* and *trypsin* are examples.

proteins A class of complex nitrogenous *organic polymers* of high *relative molecular mass* (10 000 – 10 000 000), which is of great importance to all living matter. Protein *molecules* consist of hundreds or thousands of *amino acids* joined together by the *peptide* linkage into one or more interlinked *polypeptide* chains, which may be folded in a variety of different ways. Some twenty different amino acids occur in proteins and each protein molecule is likely to contain all of them arranged in a variety of sequences. It is the sequence of the different amino acids that gives individual proteins their specific properties. The particular sequence of the amino acids in proteins, which are synthesized in ribosomes in the *cytoplasms* of *cells*, is determined by the sequence of the *nucleotides* in the *nucleic acids* of the *chromosomes*, three nucleotides coding for each amino acid (see *genetic code*). Most proteins form *colloidal solutions* in water or dilute *salt* solutions, but some (notably the fibrous proteins) are *insoluble*. Proteins may be simple, i.e. yielding only amino acids on *hydrolysis*, others are 'conjugated', i.e. combined with other substances (see *prosthetic groups*). *Enzymes* are a particularly important group of proteins as they determine the *chemical reactions* that will take place in a cell, and therefore the characteristics that it will have. Proteins are also important structural molecules in living systems, but do not function (as is often thought) as major stores of energy.

protein sequencing The process of determining the sequence of *amino acids* in a protein. The usual technique is to remove the last amino acid in the chain, one by one, and identify each one by *chromatography*.

proteolytic Proteoclastic. Having the power of decomposing or *hydrolysing proteins*.

protium The hydrogen *isotope* with *mass number* of one.

protolysis A reaction involving the transfer of *protons* (*hydrogen ions*).

proton A stable *particle*, classified as a *baryon* (see also *hadron*), with *electric charge* equal in magnitude to that of the *electron* but of opposite sign, and with *mass* 1836.12 times greater than that of the electron $(1.672\ 623\ 1 \times 10^{-27}$ *kilogram*). The proton is a *hydrogen ion* (i.e. a normal hydrogen atomic *nucleus*) and is a constituent of all other atomic nuclei. See *atom, structure of*; *elementary particles*; *proton decay*.

proton decay Although there are many *elementary particles* having a lower *mass* than the *proton*, according to current theory a proton cannot *decay* into any of these as it has the lowest mass of any particle with a baryon number 1, and the baryon number is conserved in all interactions. However, some *grand unified theories* postulate that baryon numbers are not conserved and that the proton can decay into a *pion* (e.g. by $p \rightarrow e^+ + \pi^0$). The proton lifetime in these theories is thought to be about 10^{35} years and it is therefore not surprising that no such decay has been observed.

proton microscope A *microscope* that works on the principle of the *electron microscope*, but uses a beam of *protons* in place of a beam of electrons. It has a higher *resolving power* than the electron microscope and can give better contrast.

proton number See *atomic number*.

protoplasm The *matter* of which biological *cells* consist.

provitamin A substance from which a *vitamin* is formed.

Prussian blue Potassium iron(III) hexacyanoferrate(II). $KFe[Fe(CN)_6]$. A deep blue substance obtained by the action of a ferric *salt* on *potassium hexacyanoferrate(II)* (potassium ferrocyanide).

prussic acid An intensely poisonous solution of *hydrocyanic acid*, HCN.

pseudoaromatic Antiaromatic. A ring compound containing *conjugated double bonds* in the manner of an *aromatic* compound, although its properties are different to those of an aromatic compound.

pseudoscalar A *scalar quantity* that changes sign in the transition from a right-handed to a left-handed system of coordinates.

pseudovector Axial vector. A *vector* quantity defined in terms of a magnitude and the direction of an *axis*; the signs of the components are unchanged on reversing the directions of the *coordinate* axes. The vector product of two polar vectors is a pseudovector, e.g. *torque*.

psi particle J particle. A *meson* that has no charge but an anomalously long lifetime (about 10^{-20} s). The discovery of this particle in 1974 led to the extension of the *quark* model and the hypothesis that a previously unknown quark (and its antiquark) existed with a new property called *charm*. This brought the total number of quarks known at the time to four. The psi particle consists of the charmed quark plus its antiquark (i.e. c$\bar{\text{c}}$), although the particle itself has zero charm.

psychrometry The measurement of the *humidity of the atmosphere*.

PTFE See *polytetrafluoroethene*.

ptomaines A class of extremely poisonous *organic compounds* formed during the putrefaction of *proteins* of animal origin. Food poisoning, frequently misnamed ptomaine poisoning, is almost invariably due to causes other than the ptomaines.

ptyalin An *enzyme* that occurs in the *saliva* and serves to convert *starch* into *sugar*.

***p*-type conductivity** See *n-type conductivity*.

puddling process The preparation of nearly pure *wrought iron* from *cast iron* that contains a high percentage of carbon. The cast iron is heated with *haematite*, Fe_2O_3, the oxygen in which oxidizes the carbon.

pulsars *Stars* that emit *radio frequency electromagnetic radiation* in brief *pulses* at extremely regular intervals. Many such objects have been located by *radio telescopes*, a few of them have also been observed to emit pulses of light and *X-rays*. It has been suggested that pulsars are *neutron stars*, emitting pulses of radiation as they rotate.

pulsatance See *angular frequency*.

pulse A brief increase in the magnitude of a quantity whose value is usually constant (e.g. *current* or *voltage*).

pulse height analyser. An instrument incorporating an *electronic* circuit that permits only *voltage pulses* of predetermined *amplitudes* to be passed to succeeding circuits. The range of amplitudes passed through such circuits is referred to as the 'channel width' or 'window'. In a single-channel analyser the channel width is usually pre-set and the *threshold* varied to scan the amplitude spectrum of incoming pulses. In a multi-channel instrument, often called a 'kicksorter', the incoming pulses are sorted and recorded according to their amplitudes. The kicksorter is used for distinguishing between *nuclides* by sorting the characteristic 'kicks' that their *radiations* give.

pulse-jet A type of *ram-jet* in which the *combustion* process is not continuous, but is arranged to occur at intervals between which the *pressure* in the combustion chamber is allowed to build up. The German 'flying bombs' of World War II were powered by pulse-jets fitted with air intake valves that opened when the pressure resulting from the passage of the projectile through the air exceeded the pressure in the combustion chamber: each new charge being separately fired.

pulse modulation A form of *modulation* in which a series of pulses is used as the *carrier*. In the simplest form (e.g. *Morse code*) the duration of the pulse is varied. In 'pulse-amplitude modulation' the pulse height is modulated by the amplitude of the modulating *signal*. In 'pulse-code modulation (PCM)', the amplitude (or some other parameter) of the modulating signal is sampled and represented by a digital code; sample amplitudes in specified ranges are assigned discrete values, each of which is represented by a specific pattern of pulses. This enables the signal to be transmitted in a digital code, which is converted back to the analogue form at the receiver.

purine $C_5H_4N_4$. A white crystalline *organic base*, m.p. 216°C., related to *uric acid*. *Derivatives* are of great importance biologically as they occur in *adenosine triphosphate* and *nucleic acids*. *Adenine* and *guanine* are typical of such derivatives.

purple of cassius A purple *pigment*, consisting of a *mixture* of colloidal gold and tin(IV) acid. It is used for making ruby *glass*.

push-pull Denoting an *electronic circuit* in which two components are out of phase by 180°. E.g. a push-pull valve *amplifier* has two *valves* arranged so that the *control grid* input signals are 180° out of phase, the output circuits being arranged to combine the two signals so that they are in phase.

putrefaction Chemical *decomposition*, by the action of *bacteria*, of the bodies of dead animals and plants; especially the *decomposition* of *proteins* with the production of offensive substances.

putty A material composed of powdered *chalk* mixed with *linseed oil*.

putty powder Impure tin(IV) oxide, SnO_2.

PVA See *polyvinyl acetate*.

PVC See *polyvinyl chloride*.

PWR See *pressurized water reactor*.

pyknometer An apparatus for determining the *density* and *expansivity* of a *liquid*. It consists of a glass vessel graduated to hold a definite *volume* of liquid at a given *temperature*. By weighing it full of liquid at different temperatures, the variations in density, and therefore the apparent expansion, may be found.

pyramid (math.) A solid figure having a *polygon* for one of its faces (the base), the other face being *triangles* with a common *vertex*. The *volume* of a pyramid is one-third of the *product* of the area of the base and the vertical height.

pyranose A *sugar* consisting of a six-membered ring, five carbon atoms and one oxygen atom.

pyrene $C_{16}H_{10}$. A yellow crystalline *polycyclic hydrocarbon*, m.p. 149°C., found in *coal-tar*.

Pyrex* A type of *borosilicate glass* that is resistant to high *temperatures* and chemical attack; it is widely used in laboratory glassware.

pyrheliometer An instrument that measures the intensity of solar radiation at normal incidence.

pyridine C_5H_5N. A colourless *heterocyclic liquid* with an unpleasant smell. B.p. 115°C. It occurs in *bone-oil* and *coal-tar*. It is used for making *methylated spirit* unpalatable; *compounds* derived from it are used in medicine.

pyridoxine Vitamin B_6. See *vitamins*.

pyrimidine $C_4H_4N_2$. An *organic base*, m.p. 22°C., b.p. 123.5°C., consisting of a *heterocyclic* six-membered ring. *Derivatives* are of great biological importance as they occur in *nucleic acids*. *Uracil*, *thymine*, and *cytosine* are typical of such derivatives.

pyrites Natural *sulphides* of certain *metals*. Iron pyrites is FeS_2; copper pyrites (fools' gold) is $CuFeS_2$.

pyro- Prefix denoting fire, strong heat. In chemical nomenclature it denotes a substance obtained by heating; e.g. pyroboric acid, obtained by heating *boric acid*. It is also used to indicate that the *water* content of an *acid* or *salt* is intermediate between that of the *ortho-* and *meta-* compounds of the same name.

pyrocatechol See *1,2-dihydroxybenzene*.

pyroelectricity The property of certain *crystals*, e.g. *tourmaline*, of acquiring *electric charges* on opposite faces when the crystals are heated.

pyrogallol Pyrogallic acid, benzene-1,2,3-triol. $C_6H_3(OH)_3$. A white crystalline *soluble solid*, m.p. 132°C., that is a powerful *reducing agent*; *alkaline solution* rapidly absorbs oxygen. It is used in photographic *developing* and in gas analysis for the estimation of oxygen. See *Orsat apparatus*.

pyroligneous acid A watery *liquid* obtained by the *destructive distillation* of wood. It contains *ethanoic* (acetic) *acid*, CH_3COOH, *methanol*, CH_3OH, *propanone* (acetone), $(CH_3)_2CO$, and small amounts of other *organic compounds*.

pyrolusite Natural *manganese(IV) oxide* (manganese dioxide; MnO_2). A black crystalline *solid*, r.d. 4.8; it is the principal *ore* of manganese.

pyrolysis Chemical *decomposition* by the action of *heat*.

pyrometers Instruments for measuring high *temperatures*. The four main types are: (1) platinum resistance *thermometers*, which make use of the increased electrical *resistance* of platinum wire with rise in temperature; (2) thermoelectric thermometers, using the principle of the *thermocouple*; (3) optical pyrometers, in which the temperature is estimated by the intensity of the *light* emitted by the body in a narrow *wavelength* range; and (4) *radiation* pyrometers, which detect the *heat* radiation from the hot body (see *radiomicrometer*). See also *Seger cones*.

pyrophoric alloys *Alloys* that emit sparks when scraped or struck, and are therefore used as 'flints' in lighters. See *misch metal*; *Auer metal*.

pyrophosphoric acid See *phosphoric acids*.

pyrosulphuric acid See *sulphuric acids*.

pyrotechnics Fireworks.

pyroxenes A group of *minerals* consisting principally of *silicates* of magnesium, iron, and calcium.

pyrrole C_4H_5N. A colourless liquid *heterocyclic* compound, b.p. 103°C., found in coal tar.

pyruvic acid 2-oxopropanoic acid. $CH_3.CO.COOH$. A *liquid organic acid*, m.p. 13°C., of importance in the metabolic (see *metabolism*) breakdown of *glucose*. Pyruvic acid is itself broken down in the *citric-acid cycle*.

Pythagoras, theorem of In a right-angled *triangle* the square on the *hypotenuse* is equal to the sum of the squares on the other two sides. Named after the Greek mathematician (*c.* 582–500 B.C.).

Q

QCD See *quantum chromodynamics*.

QED See *quantum electrodynamics*.

Q-factor A factor associated with *resonant circuits*, defined by:
$$Q = (L/C)^{1/2}/R,$$
where R is the *resistance*, L the *inductance*, and C the *capacitance* of the circuit. In a series resonant circuit
$$Q = \omega_0 L/R = 1/\omega_0 CR,$$
where $\omega_0 = 2\pi f_r$ and f_r is the resonant frequency. In a parallel resonant circuit
$$\omega_0 = (LC)^{-1/2}(1 - 1/Q^2)^{1/2}$$

QFD Quantum flavour dynamics. See *electroweak theory*.

QSG Quasi stellar galaxy. A *quasar* that is not a radio source. Compare *QSO*; *QSS*.

QSO Quasi stellar object. See *quasar*.

QSS Quasi stellar source. A *quasar* that is also a radio source.

quadrant Quarter-circle. A *sector* of a *circle* bounded by an arc and two radii at right angles.

quadratic equation An *equation* involving the *square* or second *power* of the unknown quantity; it is satisfied by two values (known as *roots*) of the unknown quantity. Any quadratic equation may be written in the form
$$ax^2 + bx + c = 0;$$
the roots of this equation are given by the expression
$$x = [-b \pm (b^2 - 4ac)^{1/2}]/2a.$$
The sum of the roots is $-b/a$ and their products is c/a. Thus any quadratic equation may be solved by substitution of the appropriate values in the above expressions.

quadrature 1. The position of the *Moon* or an outer *planet* such that a line between it and the *Earth* makes a right angle with a line joining the Earth to the *Sun*. **2.** Two periodic quantities are said to be in quadrature when they have the same *waveform* and *frequency*, but their *phase* differs by 90°.

quadrilateral A *plane* figure bounded by four straight lines.

quadrivalent Tetravalent. Having a *valence* of four.

qualitative Dealing only with the nature, and not the amounts, of the substances under consideration.

qualitative chemical analysis The determination of the chemical nature of substances; especially the identification of substances present in a *mixture*, without attempting to assess the extent to which they are present. Compare *quantitative chemical analysis*.

quality control The application of the theory of mathematical *probability* to sampling the output of an industrial process, with the object of detecting and controlling any variations in quality.

quality of sound Timbre. Most sounds are not 'pure'; i.e. they are composed of vibrations of more than one *frequency*. A note consists of a 'fundamental', of greatest intensity and lowest *pitch*; and several *overtones*, of much lesser intensity and of frequencies that are simple multiples of that of the fundamental. The various overtones produce a characteristic quality or timbre in the note. See also *harmonics*.

quantitative Dealing with quantities as well as the nature of the substances under consideration.

quantitative chemical analysis The determination of the amounts of substances present in a *mixture*, by chemical means. Compare *qualitative chemical analysis*.

quantity of electricity The amount of *electric charge* flowing through a circuit; i.e. the product of the current and the time for which it flows. The *SI unit* is the *coulomb*.

quantized A quantity is said to be quantized if, in accordance with *quantum mechanics*, it can only have certain discrete values (each of which is called a *quantum*). Such a quantity cannot vary continuously, differences in value being separated by 'jumps'.

quantum According to the *quantum theory*, *energy* exists in discrete units, only whole numbers of which can exist: each unit is called a quantum (plural 'quanta'). The quantum of *electromagnetic radiation* is the *photon*.

quantum chromodynamics QCD. A *quantum field theory* in which the *strong interaction* is treated as the exchange of massless *gluons* between *quarks* and antiquarks. It is thus analogous to *quantum electrodynamics* with the gluon replacing the *photon* and 'colour charge' replacing *electric charge*. See *elementary particles*.

quantum electrodynamics QED. A theory used to solve problems related to the *electromagnetic interaction* in terms of *relativistic quantum mechanics*. It is a *gauge theory* in which the *electromagnetic force* can be derived by requiring that the equations describing the motions of a charged particle remain unchanged in the course of local symmetry operations. QED makes use of *Feynman diagrams* to study the scattering of *electrons* and *photons*, showing the interaction as lines representing the exchanges of *virtual* electrons and photons. Feynman diagrams using *perturbation theory* have been shown to be accurate to one part in 10^9.

quantum electronics The study of the generation or amplification of *microwave power* in *solid crystals*, in accordance with the laws of *quantum mechanics*.

quantum field theory A *gauge theory* developed from *quantum mechanics* in which particles are represented by *fields*, the normal oscillations of which are *quantized*. In *quantum electrodynamics*, for example, the *photon* is treated as the *quantum* of the *electromagnetic field*. In quantum field theories used to describe *elementary particles*, particle interactions are described by quantum fields that are relativistically invariant (these are relativistic quantum field theories).

quantum flavourdynamics QFD. See *electroweak theory*.

quantum Hall effect See *Hall effect*.

quantum mechanics The system of mechanics that, during the present century, has replaced *Newtonian mechanics* as a method of interpreting physical phenomena occurring on a very small scale (e.g. the motion of *electrons* and *nuclei* within *atoms*; see *atom, structure of*). Quantum theory originated with the discovery by Max Planck that the *heat radiation* from a black-body (see *black-body radiation*) is *quantized*, i.e. emitted in discrete *quanta* of *energy*, the magnitudes of which are given by the product of the *frequency* of the radiation and a universal constant, now known as the *Planck constant*. It was soon realized that all *electromagnetic radiations* are quantized (see *photon*) and the theory was developed by Niels Bohr so that the *spectrum* of hydrogen could by accounted for *quantitatively* (see *Bohr theory*). This early version of quantum mechanics was refined by Sommerfeld to take into account the elliptical *orbits* of electrons. Later, quantum mechanics was developed in a specialized form, known as *wave mechanics*, which is more versatile and involves fewer arbitrary assumptions than the original theory. At the same time as wave mechanics was being developed by E. Schrödinger (see *Schrödinger wave equation*), Max Born and W. Heisenberg were developing the technique of *matrix mechanics*; the two treatments have been shown to be equivalent. See also *quantum field theory*; *quantum numbers*.

quantum numbers *Integral* or half-integral numbers that specify the state of a system or its components in *quantum mechanics*. An *electron* within an *atom*, for example, is specified by four quantum numbers in the *Bohr theory*: (1) the principal quantum number, n, defining the *energy level* or *shell* in which the electron occurs; (2) the orbital or azimuthal quantum number, l, defining the shape and multiplicity of the orbit within that shell; (3) the magnetic orbital quantum number, m_l, which determines the orientation of the orbit with reference to a strong magnetic field; and (4) the magnetic *spin* quantum number, m_s, which determines the direction of spin of an electron in a magnetic field. This treatment has since been extended to replace the concept of an orbit with an *orbital*. The properties of *elementary particles* are also described by quantum numbers. See also *Pauli's exclusion principle*.

quantum statistics Statistics that deal with the distribution of a particular type of *elementary particle* among *quantized* energy states. In *Bose-Einstein statistics*, any number of particles can occupy a given quantum state (i.e. the *Pauli exclusion principle* is not obeyed). Such particles are called *bosons*. In *Fermi-Dirac statistics* only one particle can occupy each quantum state (i.e. the Pauli exclusion principle is obeyed). These particles are called *fermions*.

quantum theory The theory that grew up around Planck's introduction into *physics* of the concept of the discontinuity of *energy*. The system of *quantum mechanics* evolved from this theory during the first half of the twentieth century.

quark confinement The theory that *quarks* can never exist in isolation. In *quantum chromodynamics* (QCD) it is assumed that as quarks come closer to each other, their interactions become weaker, falling to zero as the inter-quark distance becomes zero. Conversely, as the quarks separate, their interaction strengthens, so that they can never escape from each other's influence. In some theories of the origin of the *Universe* it is assumed that at the enormous temperature of the *big bang* quarks were able to separate; the temperature at which this hypothetical event took place is known as the 'deconfinement temperature'.

quarks Originally three hypothetical particles, with corresponding antiparticles, postulated by Murray Gell-Mann to account for the composition of *hadrons*. During the 1970s particles containing two additional types of quark were found, bringing the total to five. However, the internal symmetry of the *electroweak theory* implies that quarks exist in pairs, and a sixth type, known as top, was postulated. See *elementary particles*.

quart Unit of capacity equal to one quarter of a *gallon*.

quarter-wave plate A plate (sometimes called a 'retardation plate') of doubly refracting material (see *double refraction*) cut parallel to the *optic axis* of the *crystal*, and of such a thickness that a *phase* difference of $\pi/2$ or $90°$ is introduced between the *ordinary* and *extraordinary rays* for *light* of a particular *wavelength* (usually sodium light). Plane-polarized light (see *polarization* of *light*) incident normally upon such a plate, with its plane of vibration making an angle of $45°$ with the optic axis, emerges from the plate *circularly polarized*. A quarter-wave plate is often used in the analysis of polarized light.

quartz Natural crystalline *silica*, SiO_2, which sometimes occurs in clear, colourless *crystals* (*rock crystal*); more frequently it occurs as a white, *opaque* mass. Quartz crystals exhibit the *piezoelectric effect* to a marked extent. When pure silica is melted and then cooled, it forms a *glass*, which is sometimes called 'fused quartz'. As it is no longer quartz (because it is amorphous rather than crystalline) it is best called 'vitreous silica'.

quartz clock or watch A clock or watch regulated by a *quartz crystal*, which vibrates with a definite constant *frequency* under the effect of an alternating *electric field* tuned to this *resonance* frequency of the crystal. (See *piezoelectric effect*.) Being much more accurate than a balance-wheel or pendulum-regulated clock, it is now used in all accurate clocks and watches.

quartz-iodine lamp Quartz-halogen lamp. A high-intensity electric light consisting of a high-temperature tungsten filament inside a vitreous silica (fused *quartz*) bulb containing iodine or bromine vapour. The vitreous silica is used as it withstands the high temperature.

quasar QSO, quasi stellar object. An extra-galactic source of very high *energy* from a very small region of space. The greatest part of the energy of quasars is in the *infrared*, but they also emit *X-rays*, and some are *radio sources* (these are known as QSSs, while those that are not radio sources are often called QSGs). All quasars have very large *redshifts*, which is interpreted now as a *Doppler effect*, making them extremely distant (up to 10^{10} light years away). The exact nature of quasars is not known but they are thought to be the violently active cores of galaxies, and possibly extremely massive *black holes*.

quaternary ammonium compounds *Compounds* of the general formula NR_4OH; they are theoretically derived from *ammonium hydroxide*, NH_4OH, by replacement of the hydrogen *atoms* by *organic radicals*.

quenching 1. Rapid cooling of a metal by immersion into *water* or *oil*. When *steels* are quenched they become harder, but some *non-ferrous metals* (e.g. copper) become softer on quenching. **2.** The inhibition of a continuous discharge in a *Geiger counter* by introducing a *noble gas*, usually mixed with *methane*. This enables a new ionizing event to cause a new discharge.

quicklime See *calcium oxide*, CaO.

quicksilver See *mercury*.

quiet Sun The *Sun's* condition when no *sunspots*, *solar flares*, or *solar prominences* are taking place. *Radio-frequency* emission (see *radio astronomy*) from the Sun, which has to be observed during the rare periods of the quiet Sun, has enabled temperature measurements of the various layers of the solar atmosphere to be made.

quinhydrone $C_6H_4(OH)_2.C_6H_4O_2$. An *addition* compound of *hydroquinone* and *quinone*. A green crystalline substance, m.p. 171°C., used in *photography* and as an *antioxidant*; the quinhydrone *electrode* is used in pH measurement.

quinidine $C_{20}H_{24}N_2O_2$. A colourless crystalline *alkaloid*, *isomeric* with *quinine*, m.p. 174–5°C., used in medicine.

quinine $C_{20}H_{24}O_2N_2.3H_2O$. A colourless bitter-tasting crystalline *alkaloid* that occurs in Cinchona bark, m.p. 57°C. It was used in the treatment of malaria.

quinol See *hydroquinone*.

quinoline C_9H_7N. A colourless *liquid base*, b.p. 237°C., that occurs in *coal-tar*. Used as a *solvent* and in the manufacture of *dyes*.

quinones A series of *aromatic* compounds in whose molecules two *hydrogen atoms* in the same *benzene nucleus* are replaced by *oxygen* atoms, forming *carbonyl* groups. The quinones are therefore diketones (see *ketones*). The simplest member of the series is cyclohexadiene-1,4-dione, $O:C_6H_4:O$; a yellow crystalline solid, m.p. 115.7°C., used as an oxidizing agent, in dye manufacture, and in photography.

quinquevalent Pentavalent. Having a *valence* of five.

quotient See *division*.

Q-value Nuclear energy change, nuclear heat of reaction. The net amount of *energy* released in a *nuclear reaction*; usually expressed in million *electronvolts*, *MeV*, per individual reaction.

R

racemic acid *dl*-tartaric acid, *dl*-2,3-dihydroxybutanedioic acid. The *racemic form* of *tartaric acid*.

racemic mixture Racemate. An *equimolecular mixture* of the two *optically active* forms of a substance. Such a racemic mixture is denoted by the letters *dl* (or sometimes by ±), e.g. *dl-tartaric acid*; it is optically inactive and is said to be externally compensated.

rad The former unit of absorbed *dose* of *ionizing radiation*. One rad is equal to the *energy* absorption of 100 *ergs* per gram (0.01 J kg^{-1}) of *irradiated* material. 1 rad is equivalent to 10^{-2} gray.

radar An abbreviation of RAdio Detection And Ranging. It covers any system employing *microwaves* for the purpose of locating, identifying, navigating, or guiding such moving objects as ships, aircraft, missiles, or artificial *satellites*. The system consists essentially of a generator of *electromagnetic radiation* of centimetric *wavelengths*, the output of which is *pulse* modulated (see *modulation*) at a *radio frequency* and fed to a movable *aerial* whence it is radiated as a *beam*. Distant objects that cross the path of the beam reflect the pulses back to the transmitter, which also acts as a receiver. A *cathode-ray tube* indicator displays the received signal in the correct time sequence so that the time taken for a pulse to travel to the object and back can be measured. Thus the distance of the object from the transmitter can be calculated, and its direction can be ascertained from a knowledge of the direction of the aerial. This fundamental technique has been extended so that automatic guidance and navigation can be effected by *computers* without the necessity of a display.

radial velocity See *line of sight velocity*.

radian The *SI unit* of plane *angle* defined as the angle subtended at the centre of a *circle* by an arc equal in length to the radius of the circle. 2π radians = 360°, 1 radian = 57.296°. Symbol rad.

radiance L_e. The *radiant intensity* of a source in a given direction, per unit transverse area, measured in W sr^{-1}m^{-2}.

radiant energy *Energy* that is transmitted in the form of *radiation*, particularly *electromagnetic radiation*. See *radiant flux*.

radiant flux Φ_e. The total power emitted or received by a body in the form of *radiation* (usually *electromagnetic radiation*). It is measured in *watts*. Compare *luminous flux*.

radiant intensity I_e. The *radiant flux* per unit *solid angle* emitted by a point source of *radiation*, measured in W sr^{-1}. Compare *luminous intensity*.

radiation In general, the emission of any *rays*, *wave motion*, or particles (e.g. *alpha particles*, *beta particles*, *neutrons*) from a source; it is usually applied to the emission of *electromagnetic radiation*.

radiation belts See *Van Allen radiation belts*.

radiation hazard The potential danger to health (see *radiation sickness*) resulting from exposure to *ionizing radiation* or the consumption of *radioactive* substances.

radiation law See *Planck's law of radiation*.

radiation physics The study of *radiation*, particularly the effects on various forms of matter of *ionizing radiation*.

radiation potential Resonance potential. The *energy* (expressed in *electronvolts*) necessary to transfer an *electron* from its normal position in an *atom* to some other possible position; i.e. to an *energy level* of greater energy.

radiation pressure The pressure exerted on a surface by *radiation*. As electromagnetic radiation has *mass* and *momentum*, it exerts a *force*, and therefore a *pressure*, on any surface on which it falls. In macroscopic terms this pressure is very small but on small particles it can be important. Based on the value of the *solar constant*, the pressure of the Sun's radiation on the Earth is about 4×10^{-6} pascal.

radiation sickness Illness caused by exposure to *ionizing radiation*. Initial symptoms are vomiting and diarrhoea, followed in some cases by cancer (often leukaemia).

radiation temperature The surface temperature of a *star* or other celestial body assuming that it is a *black body*. It is calculated on the basis of *Stefan's law*. If it is measured using the visible wavelengths it is called the optical temperature.

radiation units Units used to express the *activity* of a *radionuclide* and the *dose* of *ionizing radiation*. The *SI units bequerel*, *gray*, and *sievert* have replaced the older *curie*, *rad*, and *roentgen*.

radiative capture See *capture*.

radiative collision A collision between charged particles in which part of the *kinetic energy* is converted into *electromagnetic radiation*. See also *Bremsstrahlung*.

radical Radicle. **1.** (chem.). A group of *atoms*, present in a series of *compounds*, that maintains its identity through chemical changes affecting the rest of the *molecule*, but that is usually incapable of independent existence. E.g. the *ethyl* group, C_2H_5-. See also *free radical*. **2.** (math.). Relating to a *root*. The symbol $\sqrt{}$ is called the 'radical sign'.

radio The use of *electromagnetic radiation* to communicate electrical signals without wires ('wireless' transmission). In the widest sense the term incorporates sound broadcasting (including *radio telephony*), *television*, and *radar*. Transmission by radio involves a transmitter feeding a transmitting *aerial*, from which electromagnetic energy is broadcast, either as *ground waves* or *sky waves*, to a receiving aerial, which feeds a receiver (see also *transmission* of *radio waves*). The transmitter in sound broadcasting consists of a generator of a *radio frequency carrier wave* modulated (see *modulation*) in accordance with the *electric currents* provided by the amplified output of a *microphone*. The modulated carrier wave is fed to the transmitting aerial and if the receiving aerial is tuned to the *frequency* of the carrier wave (see *resonant circuit*; *tuned circuit*) it will enable the receiver selectively to amplify and demodulate the transmitted signal. *Demodulation* is achieved by *rectification* of the signal. In this way a current is produced in the output stage of the receiver, which varies in accordance with the frequency of the sound wave fed to the microphone at the transmitter. This current may then be used to operate a loudspeaker, which reproduces the original sound.

radio- See *radioactive*.

radioactive Possessing, or pertaining to, *radioactivity*. Sometimes only the prefix 'radio-' is used to describe radioactive *nuclides* or the substances containing them, e.g. radiocarbon is an abbreviation for radioactive carbon.

radioactive age The age of a *mineral, fossil*, or wooden object as estimated from its content of *radionuclides*. This method assumes that the content of radionuclides has remained unchanged except for radioactive *decay*. See also *dating*; *fission-track dating*; *potassium-argon dating*; *rubidium-strontium dating*; *radiocarbon dating*; *uranium-lead dating*.

radioactive equilibrium A state ultimately reached when a *radioactive* substance of slow *decay* (see *radioactivity*) yields a radioactive product on *disintegration*. This product may also decay to give a further radioactive substance, and so on. The amount of any of the daughter radioactive products present after equilibrium has been reached remains constant, the loss due to decay being counterbalanced by gain from the decay of the immediate parent.

radioactive series Radioactive family. A series of *radionuclides*, each except the first being the *decay* product of the previous one. The three naturally occurring series are the thorium series (starting with thorium-232), the actinium series (uranium-235), and the uranium series (uranium-238). The final member of each series, an isotope of lead, is stable. There is a fourth series, the neptunium series which starts with the artificial *isotope* plutonium-241 and ends with bismuth-209. All the isotopes in this series are short-lived. See *radioactivity*.

radioactive standard A specimen of a material containing a *radionuclide* of precisely known rate of *decay* that is used for the calibration of instruments measuring *radiation*.

radioactive tracing A method of tracing the course of an element through a biological, chemical, or mechanical system. Any two *isotopes* of an *element* are chemically identical. Thus, by introducing a small amount of a *radioisotope*, called a tracer, the course taken by the stable isotope of the same element can be followed or traced by detecting the course of the accompanying *radioisotope* by suitable means. This can be done in various ways, e.g. a *gamma camera* or a *Geiger counter*. See *labelled compound*.

radioactive waste Nuclear waste. Any waste material that contains *radioactive nuclides*. Such materials occur in the mining of radioactive *ores*, the generation of electricity by *nuclear power*, in hospitals, and in research laboratories. Nuclear wastes can be extremely dangerous and the way in which they are disposed of is strictly controlled by international agreement.

After processing to recover usable material and reducing the radioactivity of the waste, disposal is made in solid form where possible. High-level waste (e.g. spent *nuclear fuel*) has to be cooled and is therefore stored for several decades by its producer before disposal. Intermediate-level waste (e.g. filters, reactor components, etc.) is solidified and mixed with *concrete* in steel drums before being buried in deep mines or below the sea bed in concrete chambers. Low-level waste (e.g. solids or liquids contaminated by traces of *radioactivity*) is disposed of in steel drums in concrete-lined trenches in designated sites. Since 1983, by international agreement, disposal in the Atlantic and into the atmosphere have been banned.

radioactivity The property of spontaneous *disintegration* possessed by certain unstable types of atomic *nuclei* (called *radionuclides*). The disintegration is accompanied by the emission of either *alpha-* or *beta-particles* and/or *gamma rays*. The most common type of disintegration involves beta-particle emission (see *beta decay*) and occurs either: (1) when a *neutron* present in the unstable nucleus is converted into a *proton* with the emission of an *electron* and an anti-*neutrino*, or more rarely (2) when a proton is converted into a neutron with the emission of a *positron* and a neutrino. These *beta transformations* are accompanied by unit change of *atomic number* but no change in *mass number*. Alpha particles are only emitted by certain *radionuclides* of the heavier *elements* (see *alpha decay*); when this occurs the atomic number of the daughter nucleus is two less than that of the parent and its mass number is reduced by four units. Gamma-ray emission accompanies alpha or beta emission when the daughter nucleus is formed in an excited state (see *excitation*). See also *capture*.

Natural radioactivity is due to the disintegration of naturally occurring *radionuclides* (see *radioactive series*). The rate at which *radionuclides* disintegrate is not influenced by any chemical changes, any normal changes of *temperature* or *pressure*, or by the effects of *electric* or *magnetic fields*. However 'induced' or 'artificial' *radioisotopes* of most elements can be formed by bombardment with particles (e.g. neutrons) or *photons* in a *nuclear reactor* or *accelerator*.

Radiations emitted by *radionuclides* are used in the treatment of disease (see *radiotherapy*) and in *radioactive tracing*.

radio astronomy The study of heavenly bodies by the reception and analysis of the *radio-frequency electromagnetic radiation* that they emit or reflect. In general, electromagnetic radiations from extraterrestrial sources are either absorbed by the Earth's *atmosphere* or reflected away from the Earth by the *ionosphere*. The two exceptions, which allow us to experience the rest of the *Universe*, are the optical *wavelengths*, which are able to penetrate the atmosphere, and the radio wavelengths in the band 1 cm–10 *metres*, which are too long to be absorbed by the atmosphere and too short to be reflected by the ionosphere. The radiations that pass through this 'radio window' onto the Universe come from a variety of sources, ranging from objects within the *solar system* (e.g. the *Sun* and the *planet Jupiter*) to *galaxies* that are too distant to be observed by *optical telescopes*. Radio-frequency emission may be due to thermal or non-thermal causes: emission from the *quiet Sun* is of thermal origin for example, whereas the radiation from *sunspots* is of unexplained non-thermal origin. The method by which radio astronomy attempts to make sense out of the apparently incoherent radio 'noise' from the Universe, is to construct maps of the sky in terms of radio emission, at several different *frequencies*. The intensities of the sources thus located are then compared with optical observations. In this way *radio sources* and *radio galaxies* have been identified. See also *radio telescope*.

radiobiology The branch of *biology* concerned with the effects of *radiation* on living *organisms* and the behaviour of *radioactive* materials, or the use of *radioactive tracing*, in biological systems.

radiocarbon dating The estimation of the age of carbon-containing archaeological objects by measuring their content of the *radioisotope* of carbon, carbon-14. The impact of *cosmic rays* on the Earth's *atmosphere* causes a very small proportion of nitrogen *atoms* to transform into carbon-14 atoms ($^{14}_{7}N + n \rightarrow$

$^{14}_{6}C + p$). Some of these radioactive carbon atoms find their way, via *carbon dioxide* and *photosynthesis*, into living trees and other carbonaceous materials. When a tree is cut down, however, it ceases to acquire further carbon-14 atoms. Therefore by comparing the *radioactivity* of a modern piece of wood with that of a specimen of unknown age, the length of time that has elapsed since the latter ceased to be living can be estimated (provided that it is not more than about 40 000 years). This method has been calibrated by comparison with specimens of wood of known age from the tombs of the Pharaohs and has been found to be fairly reliable.

radiochemistry The study and application of chemical techniques to the purification of *radioactive* materials and the formation of *compounds* containing radioactive *elements*.

radiodiagnosis The branch of medical *radiology* concerned with the application of *X-rays* to diagnosis.

radio frequency RF. The *frequency* of *electromagnetic radiation* within the range used in *radio* i.e. 10 kilohertz to 100 000 megahertz.

radio-frequency heating Industrial *induction* or *dielectric heating*, particularly when the *frequency* of the alternating field is above about 25 *kilohertz*.

radio-frequency welding See *high-frequency welding*.

radio galaxies *Galaxies* that emit *electromagnetic radiation* of *radio frequencies* as observed by the techniques of *radio astronomy*. The exact source of this galactic radiation is not always understood, but radiation has been received from galaxies that have been observed optically to be in collision. See also *radio sources, quasars, pulsars*, and *synchrotron radiation*.

radiogenic Resulting from *radioactive decay*.

radiograph A photographic record of an image produced by short *wavelength radiation*, such as *X-rays* and *gamma rays*. See also *autoradiograph*.

radiography The formation of images on *fluorescent* screens or photographic material by short *wavelength radiation*, such as *X-rays* and *gamma rays*.

radio interferometer A type of *radio telescope* that consists of two or more separate *aerials*, each receiving *electromagnetic radiation* of *radio frequencies* from the same source, and each joined to the same receiver. The instrument works on the same principle as the optical *interferometer*, but as the *wavelengths* of the incident radiation are much greater, the distance between aerials has to be correspondingly increased. The chief advantage of radio interferometers, over single aerial *parabolic reflectors*, is that they have improved *resolving power* for detecting sources of small angular diameter. See also *radio astronomy*.

radioisotope An *isotope* of an *element* that is *radioactive*.

radiolocation The location of distant objects, such as ships or aircraft, by *radar*.

radiology The science of *X-rays* and *radioactivity*, including *radiodiagnosis* and *radiotherapy*.

radiolucent Almost *transparent* to *radiation*, especially *X-rays* and *gamma rays*, but not entirely so. An object or material that allows these radiations to pass with little or no alteration is said to be 'radiotransparent'. Objects and materials that are *opaque* to them are said to be 'radioopaque'.

radioluminescence *Fluorescence* resulting from *radioactive decay*.

radiolysis The chemical *decomposition* of substances as a result of *irradiation*. The radiation causes *ionization* and *excitation*, which promote further reactions.

radiometric dating See *dating*.

radiomicrometer An extremely sensitive instrument for measuring *heat radiations*. It consists of a *thermocouple* connected directly into a single copper loop forming the coil of a sensitive *galvanometer*.

radionuclide Radioactive nuclide. A *nuclide* that is *radioactive*.

radioopaque Opaque to *radiation*, i.e. not permitting radiation to pass through it. See *radiolucent*.

radiosonde A small balloon used to carry meteorological instruments into the *Earth's atmosphere*. Measurements of *temperature*, *pressure*, etc. are transmitted by these instruments back to Earth by *radio*.

radio source Formerly known as a 'radio star', a term that is no longer used. A discrete source of *electromagnetic radiation* of *radiofrequencies* outside the *solar system*. Such sources have been discovered by the techniques of *radio astronomy*, both within the *Galaxy* and outside it, but only a small number have been identified with *stars* that can be located with *optical telescopes*. Other sources are *supernovae* explosions and remnants, colliding *galaxies* and gas clouds, *quasars*, and *pulsars*; some sources, however, remain unexplained.

radio star See *radio source*.

radio telephony The use of *radio*, rather than wires or cables, for all or part of a *telephone* system.

radio telescope An instrument used in *radio astronomy* to pick up and analyse the *radio-frequency electromagnetic radiations* of extra-terrestrial sources. The two principal types of radio telescope are: (1) *parabolic reflectors*, which are usually steerable (a steerable dish telescope) so that they can be pointed at any part of the sky, and which reflect the incoming radiation on to a small *aerial* at the *focus* of the *paraboloid*; and (2) fixed *radio interferometers*. The latter have greater position-finding accuracy and greater ability to distinguish a small source against an intense background, while the former are more versatile owing to their mobility. See also *aperture synthesis*.

radiotherapy The branch of *radiology* concerned with the treatment of disease by means of *radiation*, particularly *X-rays* and techniques involving *radioactivity*.

radiotransparent See *radiolucent*.

radio window See *radio astronomy*.

radium Ra. Naturally occurring *radioactive element*. At. No. 88. The most stable *isotope*, radium-226, has a *half-life* of 1620 years. A very rare *metal*, chemically resembling barium; m.p. 700°C., b.p. 1500°C., r.d. 5. It occurs in *pitchblende*. See *radioactivity*.

radium emanation See *radon*.

radius See *circle*.

radius of curvature Consider any point P on a curve S lying in a *plane*. A *circle* can be drawn with centre at a unique point O on the *normal* to S at P, such that the curve and the circle are tangential at P. The radius of this circle, OP, is the radius of curvature of the curve at P. The concept may be extended to a point on a three-dimensional curved surface. In this case, an infinite number of radii of curvature exist, corresponding to the infinite number of plane curves that can form the line of intersection of the curved surface and the plane containing the normal at P. Of these curves, two are unique, one having a maximum radius of curvature at P and the other a minimum. These two are called the principal radii of curvature at P.

radius of gyration The *moment of inertia I*, of a body of *mass m* about a given *axis* can be expressed in the form $I = mk^2$, k being the radius of *gyration* about the axis.

radius vector (astr.) A line drawn from a central body (the focus) to a *planet* in any position in its *orbit*.

radius vector (math.) The position of any point P in space with respect to a given *origin O* may be completely defined by the direction and length of the line OP. This line is called the radius vector of the point P. See *polar coordinates*.

radix A number that forms the *base* of a system of numbers, *logarithms*, etc., e.g. the radix of the *binary notation* is 2.

radon Rn. Radium emanation, niton. Element. At. No. 86. The most stable *isotope*, radon-222, has a *half-life* of 3.825 days. It is a naturally occurring *radioactive* gas, produced as the immediate *decay* product of radium. Chemically it is a member of the *noble gases*.

raffinate A *refined liquid*, especially an *oil* after its *soluble* components have been removed by *solvent extraction*.

raffinose Melitose. $C_{18}H_{32}O_{16}.5H_2O$. A colourless crystalline *trisaccharide*, m.p. 80°C., that occurs in *beet sugar* but does not have a sweet flavour.

rainbow A colour effect produced by the *refraction* and internal *reflection* of sunlight in rain drops in the air; the effect is visible only when the observer has his back to the *Sun*. See *spectrum colours*.

RAM Random access memory. A *semiconductor memory* used in *computers* for recording data and retrieving stored data (compare *ROM*). The basic storage cells are microscopic *integrated circuits*, each cell storing one *bit*. A large memory can be fabricated from a rectangular array of cells, each of which is identifiable by its row and column; cells can thus be accessed rapidly in any order, i.e. cells can be randomly accessed.

r.a.m See *relative atomic mass*.

Raman effect When *monochromatic light* passes through a *transparent* medium, some of the light is scattered. If the *spectrum* of this scattered light is examined, it is found to contain, apart from light of the original *wavelength*, weaker lines differing from this by constant amounts. Such lines are called Raman lines, and they are due to the loss or gain of *energy* experienced by the *photons* of light as a result of interaction with the vibrating *molecules* of the medium through which they pass. With the use of a *laser* as the monochromatic source, Raman spectroscopy has been used to determine molecular structure and in chemical analysis. Named after Sir C. V. Raman.

ram jet Atherodyde. A simple type of aerodynamic *reaction propulsion* system in which *thrust* is obtained by the *combustion* of *fuel* in air, compressed only by the forward *speed* of the vehicle. A ram jet is also known as a 'flying drainpipe' as it consists essentially of a long duct into which fuel is fed at a controlled rate. However, the air intake and exhaust gas outlet need to be correctly designed in order to achieve maximum efficiency of the combustion process in that part of the duct that serves as a combustion chamber. The shape of the duct will depend upon whether or not the vehicle is intended to be *supersonic*. A ram jet has to be launched at high speed and cannot take off unaided from rest. See also *pulse-jet*.

Ramsden eyepiece An *eyepiece* consisting of two *planoconvex lenses* (curved surfaces inwards) of equal *focal length f*, and separated by a distance of $2f/3$. The eyepiece has low spherical *aberration*, is fairly *achromatic*, and is very useful when cross-wires or a scale are desired in the eyepiece. Named after Jesse Ramsden (1735–1820).

random access See *RAM*.

random sample A sample taken in such a way that every individual, object, or component comprising the group, set, or mass to be sampled, has an equal *probability* of forming part of the sample. In a stratified random sample, the group is divided into a number of strata, each of which is randomly sampled.

Raney nickel A black spongy form of nickel made by treating a nickel-aluminium *alloy* with *caustic soda*. It is used as a catalyst, especially for *hydrogenation* reactions.

Rankine temperature °R. The absolute *Fahrenheit scale*. Zero degrees Rankine is $-459.67°F$. and therefore $°F + 459.67 = °R$. Named after W. J. M. Rankine (1820–70).

Raoult's law When a *solute* that does not dissociate in *solution* is dissolved in a *solvent* to form a *dilute* solution, then (1) the ratio of the decrease in *vapour pressure* to the original vapour pressure is equal to N_1/N_2, N_1 and N_2 being the total numbers of *molecules* present of solute and solvent respectively; or, alternatively (2) the *elevation of the boiling point* of the solution above that of the pure solvent is proportional to N_1/N_2; or (3) the *depression of the freezing point* of the solution below that of the pure solvent is proportional to N_1/N_2. A dilute solution that obeys Raoult's law is known as an ideal solution. If the solution obeys Raoult's law over the whole range of concentrations it is said to be a perfect solution. Named after Francois Raoult (1830–1901).

rapeseed oil See *colza oil*.

rare-earth elements See *lanthanides*.

rarefaction A reduction in *pressure*. The opposite of compression.

rare gases See *noble gases*.

Raschig process An industrial process for making *chlorobenzene* and *phenol* by reacting *benzene* vapour with *hydrogen chloride* in air:
$$2C_6H_6 + 2HCl + O_2 \rightarrow 2C_6H_5Cl + 2H_2O \rightarrow 2HCl + 2C_6H_5OH.$$
The first reaction uses a copper(II) chloride catalyst and a temperature of 230°C., the conversion to phenol requires the higher temperature of 430°C. and a silicon catalyst.

raster The pattern of lines that scan the fluorescent screen of a *cathode-ray tube* in a *television* receiver.

rate constant See *rate of reaction*.

rate-determining step See *rate of reaction*.

rate equation See *rate of reaction*.

rate of reaction The speed with which a *chemical reaction* proceeds. In the reaction
$$X + 2Y = Z,$$
the rate of reaction, R, is given by the rate equation:
$$R = K [X] [Y]^2,$$
where K is the rate constant and [X] [Y] are the *concentrations* or *activities* of X and Y. See also *order*. The form of the rate equation depends on the mechanism

of the reaction. If there are a number of steps in the reaction, the overall rate will be determined by the slowest reaction, which is known as the 'rate-determining step'.

ratio The numerical relation one quantity bears to another of the same kind. E.g. 6 tons and 4 tons, and 30 and 20, are both in the ratio of 3:2.

rationalized units Units of measurement, such as *SI units*, in which the definitions conform logically to the geometry of system. Definitions involving circular symmetry contain the factor 2π; those involving spherical symmetry contain 4π.

rational number (math.). A whole number, or a number that can be expressed as the *ratio* of two whole numbers. Compare *irrational number*.

ray The *rectilinear* path along which any *radiation*, e.g. *light*, travels in any direction from a point in the source of the radiation. It is loosely used to denote radiation of any kind.

Rayleigh scattering See *scattering*.

rayon Formerly called 'artificial silk', rayon is now restricted to two types of man-made *cellulose* fibres: (1) *viscose* rayon, made by forcing a solution of viscose through fine holes into a *solution* that decomposes the viscose to give threads of cellulose, and (2) *cellulose acetate* rayon, made by forcing a solution of cellulose acetate (*cellulose ethanoate*) through fine holes into warm air and allowing the *solvent* to *evaporate*, thus leaving threads of cellulose acetate.

RDX* See *cyclonite*.

reactance X. A property of *alternating current* circuits that together with the *resistance*, R, makes up the *impedance* Z, according to the relation,
$$Z = (R^2 + X^2)^{1/2}.$$
If the circuit comprises the resistance, an *inductance L*, and *capacitance C* all in series, the reactance is given by:
$$X = \omega L - 1/\omega C$$
where ω is the angular frequency ($\omega = 2\pi f$, f being the *frequency* of the *alternating current*).

reactant A substance that takes part in a *chemical reaction*.

reaction A *force* that is equal in magnitude but opposite in direction to some other force (see *Newton's laws of motion*).

reaction, chemical See *chemical reaction*.

reaction propulsion Jet propulsion. A form of aerodynamic propulsion in which a high-speed stream of *gas* (usually produced by *combustion*) reacts upon the vehicle in which it was produced in accordance with *Newton's* (third) *law of motion*, so that the vehicle is propelled through the medium in which it is travelling. The lower the *density* of the medium, the higher the *efficiency* of the propulsion. Reaction propulsion is the only known method of propulsion through *space* where there is no supporting medium. It is also used in the Earth's atmosphere in the form of a *rocket*, *jet engine*, or *ram jet*. See also *ion engine*.

reactive (chem.) Readily entering into *chemical reactions*; chemically active.

reactive dyes *Dyes* that react chemically with the substances being dyed, to form covalent bonds (see *chemical bonds*) with the atoms of the substrate. They are used for cellulose fibres.

reactor (chem.) Any vessel in which a *chemical reaction* (especially industrial) is conducted.

reactor (phys.) **1.** A device for introducing reactance into an electrical circuit (e.g. a *capacitor*). **2.** See *nuclear reactor*.

reagent A chemical substance used to produce a *chemical reaction*.

realgar Natural red *tetrarsenic tetrasulphide* (arsenic(II) sulphide), As_4S_4.

real gas Any gas that does not have the properties of a *perfect gas*. It has molecules with a finite volume and intermolecular forces. See *equation of state*.

real-time working A method of operating a *computer* as part of a larger system, in which information from the computer output is available at the time it is required by the rest of the system.

Réaumur scale A *temperature* scale in which the *melting point* of *ice* is taken as $0°R$. and the *boiling point* of *water* as $80°R$. Named after Rene Antoine Réaumur (1683–1757).

receptor 1. Membrane receptor. A specific *protein*, usually on the *cell* surface, that binds a signalling molecular (e.g. a *hormone*) thus changing in shape (*allosterism*) and thereby altering some process within the cell. For example, receptor molecules at a *synapse* when combined with *neurotransmitter* molecules alter the permeability of the *membrane* of the receiving cell. **2.** Sense cell. A *cell* (or part of a cell) that is specialized to respond to a stimulus from the external or internal environment of an *organism* and to convert (transduce) it to an *impulse* in a *neurone*.

recessive Denoting the *allele* that does not function when two different alleles of a *gene* are present in a *cell*. Compare *dominant*.

reciprocal of a quantity 1 divided by the quantity; e.g. the reciprocal of 5 is 1/5.

reciprocal ohm See *mho* and *siemens*.

reciprocal proportions, law of See *chemical combination, laws of*.

recoil electron See *Compton effect*.

recombinant DNA *Deoxyribonucleic acid* (DNA) containing *genes* from different sources that have been combined by *genetic engineering*.

recombination 1. The formation of a neutral *atom* from a positive *ion* and a negative ion or an *electron*. In a recombination process the neutral atom may be in an excited state (see *excitation*), from which it can *decay* with the emission of *electromagnetic radiation*. **2.** The joining together of an electron and a *hole* in a *semiconductor*. The rate at which this occurs is the 'recombination rate'.

rectangle A *quadrilateral* with right angles between all four sides.

rectification (chem.) The purification of a *liquid* by *distillation*.

rectification (math.) The process of determining the length of a curve.

rectification (phys.) The conversion of an *alternating* into a *direct current*. See *rectifier*.

rectified spirit *Ethanol*, usually obtained by *fermentation* on an industrial scale, and purified by *fractional distillation* to give an ethanol/water mixture containing 95.6% ethanol.

rectifier (phys.) A device for transforming an *alternating current* into a *direct current*; it consists of an arrangement that presents a much higher *resistance* to an *electric current* flowing in one direction than in the other. Rectification may be *half-wave* or *full-wave*. See *rectifying valve*; *crystal rectifier*; *barrier-layer rectifier*; *junction rectifier*; *semiconductor*.

rectifying valve The *thermionic valve* used for *rectification* is the *diode*. The valve will pass current only when the *anode* is at a positive *potential* with respect to the cathode. Hence if an alternating potential is applied to a circuit containing such a valve, a *direct current* will flow through the circuit. For most purposes rectifying valves have now been replaced by *semiconductor diodes*.

rectilinear In a straight line; consisting of straight lines.

rectilinear propagation of light To a first approximation *light* travels in straight lines. This is evident from the formation of *shadows* and other everyday experience; see, however, *diffraction*.

r.d See *relative density*.

red blood cells See *erythrocytes*.

red giant A type of *star*; see *stellar evolution*.

red lead Minium. Dilead(II) lead(IV) oxide. Pb_3O_4. A bright scarlet powder, used as a *pigment*, in *glass* manufacture, and as an *oxidizing agent*. See *lead oxides*.

redox exchanger Electron exchanger. A substance, usually a polymer, that can "exchange" (i.e. transfer) *electrons*, thereby effecting *redox reactions*, when in contact with reacting *ions* or *molecules*. Redox exchangers may also act as ion exchangers. See *ion exchange*.

redox reaction Oxidation-reduction reaction. A *chemical reaction* in which an *oxidizing agent* is reduced and a *reducing agent* is oxidized, thus involving the transfer of *electrons* from one *atom*, *ion*, or *molecule* to another. For example, in
$$4K + O_2 \rightarrow 2K_2O,$$
the potassium atoms lose electrons to become ions (K^+) and are oxidized by the oxygen atoms, which gain the electrons and are thereby reduced. This exchange of electrons, involving ionic bonds (see *chemical bonds*), can occur in the absence of oxygen and still be regarded as a redox reaction, e.g.
$$2K + Br_2 \rightarrow 2KBr;$$
in this reaction the potassium is also said to be oxidized because it loses electrons to the bromine, which is also said to be reduced. This oxidation-reduction process involving ions also occurs at the electrodes of *cells*. The 'redox potential', is the *potential* required in a cell to produce *oxidation* at the *anode* and *reduction* at the *cathode*. This potential is measured relative to a standard *hydrogen electrode*, which is taken as zero. The oxidation-reduction concept can also be extended to reactions involving *covalent compounds* (see *oxidation number*). Most reactions that yield or absorb energy in living systems are redox reactions, catalyzed by an enzyme called an *oxidoreductase*.

red shift 1. A *Doppler effect* in which the displacement of spectral lines towards the red end of the *spectrum* observed in the *light* from certain *galaxies* is taken to mean that they are receding into *space*, as a consequence of the *expansion of the Universe*. **2.** A gravitational effect, often called the Einstein shift, in which a similar displacement of spectral lines towards the red is caused by a gravitational field, as predicted by Einstein.

reduced equation An *equation of state* of a *gas* in which the temperature, pressure, and volume are replaced by their reduced values. See *reduced temperature*, *pressure*, and *volume*.

reduced temperature, pressure, and volume Ratios of the *temperature*, the *pressure*, and the *volume* to the *critical temperature*, *critical pressure*, and *critical volume* respectively.

reducing agent A substance that removes oxygen from, or adds hydrogen to, another substance: in the more general sense, one that donates *electrons*. See *reduction*.

reductase An *enzyme* that promotes a *reduction* reaction. See *oxidoreductase*.

reduction The removal of oxygen from a substance, or the addition of hydrogen to it. The term is also used more generally to include any reaction in which an *atom* gains *electrons*. See *redox reaction*.

redundancy A term used in *information theory*: the amount by which the *ratio* of the information rate to its hypothetical maximum value falls below unity; usually expressed as a percentage.

re-entry The position, time, or act of re-entering the *Earth's* atmosphere after a journey into *space*. The 'angle of re-entry' is critical because of the enormous quantity of *heat* generated by a *spacecraft* as it enters the atmosphere. This heat is generated by *friction* between the *atoms* and *molecules* of the atmosphere and the great speed of the moving spacecraft; it is normally absorbed by the *heat shield*. Too sharp an angle of re-entry would cause the spacecraft to burn up, too oblique an angle would cause the spacecraft to bounce off the atmosphere.

refine Purify: remove the impurities from (*sugar*, *metals*, *oil*, etc.).

refinery gas A mixture of gases produced during the *fractional distillation* of *petroleum* in an oil refinery. It consists of a mixture of *methane*, *ethane*, *butane*, and *propane*. It is used as a *fuel* and in the manufacture of a variety of *petrochemicals*.

reflectance ρ. A measure of the extent to which a surface is capable of reflecting *radiation*, defined as the *ratio* of the reflected *radiant flux* or *luminous flux* to the intensity of the incident flux.

reflecting telescope Reflector. See *telescope*.

reflection, angle of The angle between a *ray* of *light* reflected from a surface, and the *normal* to the surface at that point.

reflection of light Certain surfaces have the property of reflecting or returning *rays* of *light* that fall upon them, according to definite laws (see *reflectance*; *reflection of light, laws of*).

reflection of light, laws of 1. The incident *ray*, the reflected ray, and the *normal* to the reflecting surface at the point of incidence lie in the same *plane*. **2.** The *angle* between the incident ray and the normal (i.e. the angle of *incidence*) is equal to the angle between the reflected ray and the normal (i.e. the angle of *reflection*).

reflector 1. Any surface that reflects *radiation*, particularly *electromagnetic radiation* (See also *parabolic reflector*). **2.** A reflecting telescope. See *telescope*. **3.** A layer of material (which may contain *moderator*) surrounding the core of a *nuclear reactor* that reflects back into the core some of the *neutrons* that would otherwise escape.

reflex angle An *angle* greater than 180° and less than 360°.

reflex camera A *camera* that allows the photographer to view and focus the exact scene he is photographing. In a single-lens reflex (SLR), the light enters through the main camera lens and falls on the film when the viewfinder mirror is retracted. In a twin-lens reflex (TLR), the light enters through a separate lens and is deflected onto a viewfinder screen.

reflux condenser A *condenser* in which the *vapour* over a *boiling liquid* is condensed to a liquid, which flows back into the vessel, so preventing its contents from boiling dry.

reforming A catalytic reaction in which straight-chain *alkanes* are converted into branched-chain compounds or *aromatics*. In some cases it involves *catalytic cracking*. It is used to produce *petrol* from crude oil. See also *steam reforming*.

refracting telescope Refractor. See *telescope*.

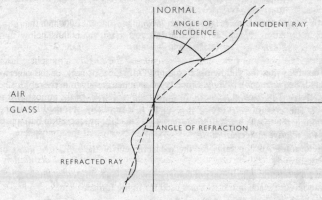

Figure 35.

refraction, angle of The angle between the refracted *ray* and the *normal* to the surface at the point of *refraction*. See Fig. 35.

refraction, laws of 1. The incident *ray*, the refracted ray, and the *normal* to the surface of separation of the two media at the point of incidence lie in the same *plane*. **2.** Snell's law. The radio of the sine of the *angle* of *incidence* to the sine of the angle of *refraction* is a constant for any pair of media. See *refractive index*. Named after Willebrord Snell (1591 – 1626).

refraction correction The small correction that has to be made to the observed *altitude* of a heavenly body due to the *refraction* of the *light* it emits or reflects by the Earth's *atmosphere*. All bodies appear to be slightly higher than they actually are.

refraction of light When a *ray* of *light* travels obliquely from one medium to another, it is bent or refracted at the surface separating the two media. The refraction occurs because light travels at slightly different *velocities* in different media; thus at the interface between media there is a slight change of *wavelength*. See Fig. 35. The ray before refraction is called the incident ray; on being refracted it becomes the refracted ray. A line perpendicular to the refracting medium at the point where the incident ray enters it is the *normal*. *Glass*, *water*, etc., cause the incident ray to be turned towards the normal when the ray enters from a medium less optically dense, such as air. Similar considerations apply to *wave motions* other than light.

refractive index of a medium *n.* The ratio of the sine (see *trigonometrical ratios*) of the angle of *incidence* to the sine of the angle of *refraction* when *light* is refracted from a *vacuum* (or, to a very close approximation, from air) into the medium. This is equivalent to the ratio of the *speed* of *light* in free space to that in the medium. As the refractive index varies with *wavelength*, the wavelength

	Refractive indices
Diamond	2.4173
Glass	1.5–1.7
Quartz (fused)	1.458
Ethanol (at 25°C.)	1.359
Water (at 25°C.)	1.332
Carbon Dioxide (at 0°C. and 760 mm)	1.000450
Air (at 0°C. and 760 mm)	1.000293
Oxygen (at 0°C. and 760 mm)	1.000272

is taken as that for yellow light (sodium D-line, 589.3 nm), unless otherwise stated. See *refraction of light*. Some typical values are given in the table.

refractivity If the *refractive index* of a medium is n, its refractivity is defined as $n-1$. The 'specific refractivity' is given by $(n-1)/\rho$ where ρ is the *density* of the medium; the 'molar refractivity' is defined as the specific refractivity multiplied by the *relative molecular mass*.

refractometer An apparatus for the measurement of the *refractive index* of a substance.

refractory (chem.) A material not damaged by heating to high *temperatures*. Such materials are made into bricks and used for lining furnaces, etc.

refrigerant A *fluid* used in the *refrigerating cycle* of a refrigerator, usually consisting of a *liquid* that will vaporize at a low *temperature* (e.g. *freon* or *ammonia*).

refrigerating cycle The cycle of operations that takes place in a refrigerator. In the vapour-compression cycle, the volatile *refrigerant* absorbs *heat* from the cold chamber and its contents, which causes it to vaporize; it is then pumped to a compressor after which it passes to a condenser, where it gives up heat and condenses back to a *liquid*; after passing through an expansion valve, it again passes to the cold chamber, thus constituting a continuous cycle. In the vapour-absorption cycle there is no pump, energy being supplied as heat. This type of refrigerator may therefore be used where there is no electrical supply. The refrigerant, usually an aqueous solution of ammonia, is moved round the circuit by a stream of pressurized hydrogen. In a generator, heat is supplied to the liquid aqueous solution causing it to vaporize. In the separator to which it rises, the ammonia separates from the water, passing to the condenser where it liquefies, giving off its latent heat. The liquid ammonia then mixes with hydrogen, which carries it through the evaporator in the cold chamber. The hydrogen and ammonia then pass to an absorber, where the water from the separator dissolves it before it passes to the generator again.

regelation of ice The *melting point* of *ice* is lowered by increased *pressure*; therefore ice near its melting point is melted by sufficient pressure, and solidification or regelation takes place again when the pressure is removed.

regenerative braking A method of braking a vehicle driven by one or more *electric motors*, in which a motor becomes a *generator*, the output of which is returned to the power supply (e.g. batteries). This method is used with electric cars to conserve *energy*.

regenerator A *heat exchanger* usually consisting of a chamber filled with bricks arranged in checkerwork. The exhaust gases from a furnace, and the cold air to be used in the *combustion*, are passed alternately through the chamber for specified periods. *Heat* from the exhaust gases raises the temperature of the brickwork and is thus transferred to the cold air, increasing the efficiency of the combustion process.

relative aperture See *aperture*.

relative atomic mass Atomic weight. The ratio of the average mass per *atom* of a specified isotopic composition of an *element* to 1/12 of the mass of a carbon-12 atom. The natural isotopic composition is assumed unless otherwise stated. The relative atomic masses of the elements are given in the appendix, Table 3.

THE RELATIVE DENSITY OF SOME COMMON MATERIALS

Material	r.d.	Material	r.d.
Cork	0.24	Aluminium	2.7
Pine	0.45	Diamond	3.5
Oak	0.80	Titanium	4.5
Water	1.00	Iron (cast)	6.9–7.5
Brick	1.6	Steel	7.6–7.8
Earth	1.9–2.1	Brass (cast)	8.1
Cement (set)	2.2	Lead	11.3
Granite	2.6	Mercury	13.6
Marble	2.7	Gold	19.3

relative density (r.d.) Specific gravity. The ratio of the density of a solid or liquid at a specified temperature (often 20°C.) to the density of *water* at the temperature of its maximum density (4°C.). It is a pure number, but is numerically equal to the density in grams per cubic centimetre. The density in *SI units* ($kg\ m^{-3}$) is 1000 times greater than the relative density. If the r.d. of a substance is less than 1 it will float on water, if it is greater than 1 it will sink. The r.d. of gases is usually expressed with reference to air, both gases being at *s.t.p.* The table gives the relative densities of some common materials. The term has now replaced the older term, *specific gravity* (see *specific*).

relative humidity The hygrometric state of the atmosphere can be defined either as: (1) the ratio of the *pressure* of the *water vapour* actually present in the atmosphere to the pressure of the vapour that would be present if the vapour were *saturated* at the same *temperature*; or (2) the ratio of the *mass* of water vapour per unit *volume* of the air to the mass of water vapour per unit volume of saturated air at the same temperature. The numerical difference between the two is very small and can normally be neglected. The relative humidity is usually expressed as a percentage. Its value may be determined from a knowledge of the *dew-point*, since the *saturated vapour pressure* at the dew-point is equal to the aqueous vapour pressure at the temperature of the experiment. The result is then obtained by reference to tables, which give the saturated vapour pressure at different temperatures.

relative molecular mass Molecular weight. The ratio of the average *mass* per *molecule* of a specified isotopic composition of a substance to 1/12 the mass of a carbon-12 *atom*. It is equal to the sum of the *relative atomic masses* of the atoms that make up the molecule.

relative permeability See *magnetic permeability*.

relative permittivity See *permittivity*.

relativistic mass The *mass* of a body that is travelling at a speed comparable to the *speed* of *light*. The relativistic mass, *m*, of a body travelling at a speed, *v*, is given by:

$$m = m_0(1 - v^2/c^2)^{-1/2}$$

where m_0 is the *rest mass* and *c* is the speed of light.

relativistic mechanics The form of *mechanics* that replaces *Newtonian mechanics* for bodies travelling at *relativistic speeds*.

relativistic particle A particle that has a speed comparable to the *speed* of *light*; i.e. a particle with a *relativistic mass* substantially in excess of its *rest mass*.

relativistic quantum field theories See *quantum field theory*.

relativistic speed A *speed*, approaching the *speed of light*, at which the effect of the theory of *relativity* is significant.

relativity, theory of A theory, formulated by Einstein, that recognizes the impossibility of determining absolute motion and leads to the concept of a four-dimensional *space-time continuum*. See also *Minkowski's geometry*. The special theory, which is limited to the description of events as they appear to observers in a state of uniform motion relative to one another, is developed from two axioms: (1) the laws of natural phenomena are the same for all observers, and (2) the *speed of light* is the same for all observers irrespective of their own speed. The more important consequences of this theory are (*a*) the *mass* of a body is a function of its speed (see *relativistic mass*); (*b*) the *mass-energy equation*, $E = mc^2$, where *c* is the speed of light; (*c*) the *Fitzgerald-Lorentz contraction* appears as a natural consequence of the theory; (*d*) time has no absolute value (see *time dilation*). The general theory, applicable to observers not in uniform relative motion, leads to a novel concept of *gravitation*. In this theory the presence of *matter* in *space* causes space to 'curve' in such a manner that the *gravitational field* is set up. Thus gravitation becomes a property of space itself. The validity of the theory of relativity has been amply confirmed in modern *physics*.

relaxation oscillator See *oscillator*.

relay, electrical A device by which the *electric current* flowing in one circuit can open or close a second circuit and thus control the switching on and off of a current in the second circuit. Electrical relays may be mechanical switches operated by *electromagnets*, or they may be *electronic* switches based upon such solid-state devices as the *thyristor*, which have replaced the older *thyratron* valve.

reluctance Magnetic resistance. The ratio of the *magnetomotive force* acting in a *magnetic circuit* to the *magnetic flux*. It is measured in henries^{-1}.

reluctivity The *reciprocal* of *magnetic permeability*.

rem Roentgen equivalent man. The unit dose of *ionizing radiation* that gives the same biological effect as that due to one *roentgen* of *X-rays*. This has now been replaced by the *sievert*.

remanence Retentivity. The residual *magnetic flux density* of a *ferromagnetic* *substance* subjected to a *hysteresis cycle* when the magnetizing field is reduced to zero.

renewable energy sources Alternative energy sources. Energy sources that do not rely on *fuels* of which there are only finite stocks. The nonrenewable sources are *fossil fuels* and *nuclear fuels* (especially those used in *nuclear fission*). The most widely used renewable source is *hydroelectric power*, which currently supplies some 6.6% of the world's energy needs. Other renewable sources are *biomass energy*, *solar energy*, *tidal energy*, *wave energy*, and *wind energy*.

Most renewable energy sources require a high capital investment but have low running costs; most avoid *pollution* and adding to the dangers of the *greenhouse effect* (biomass energy does not avoid this danger).

rennet An extract of the fourth stomach of the calf, containing *rennin*.

rennin An *enzyme* having the power of coagulating the *protein* in milk.

replication The process in which DNA *molecules* (see *deoxyribonucleic acid*) copy themselves and so reproduce. The two *polynucleotide* strands separate, *nucleotides* (as triphosphates) assemble along each strand by *base pairing* and a DNA *polymerase* joins them together so that there are then two copies, each containing one original strand and each identical to the original double strand.

repressor A specific *protein* that binds to the operator region of an *operon*, thereby blocking *transcription* of the adjacent *genes*.

resins Natural resins are *amorphous organic compounds* secreted by certain plants and insects; they are usually *insoluble* in *water* but *soluble* in various *organic solvents*. Typical natural resins are *rosin* and *shellac*. *Synthetic* resins were originally described as a group of synthetic substances whose properties resembled natural resins. The term is now applied more generally to any synthetic *plastic* material produced by *polymerization*, although chemically modified natural *polymers*, such as those based on *cellulose* or *casein*, are not usually classed as synthetic resins.

resistance, electrical *R*. The *potential difference* between the ends of a conductor divided by the *electrical current* flowing in the conductor. See *Ohm's law*. All materials except *superconductors* resist the flow of an electric current, using a proportion of the *energy* to increase the *temperature* of the material. The extent to which a conductor resists the flow of a given current depends upon its physical dimensions, the nature of the material of which it is made, its *temperature*, and in some cases the extent to which it is illuminated. See *photoconductive effect*. The derived *SI unit* of resistance is the *ohm*.

resistance strain gauge See *strain gauge, electrical*.

resistance thermometer The electrical *resistance* of a *conductor* varies with *temperature*, normally increasing with rise in temperature, according to the relationship:

$$R = R_0(1 + at + bt^2)$$

where R is its resistance at temperature t, and R_0 is its resistance at 0°C. (or some other reference temperature); a and b are constants. This forms the basis of a convenient and accurate *thermometer*, in which the temperature is deduced from the measurement of the resistance of a spiral of a metal (usually platinum) in the form of a wire. See also *thermistor*.

resistivity Specific resistance. A constant for any material equal to the *reciprocal* of its *conductivity*. The resistivity, ρ equals RA/l where R is the resistance of a uniform *conductor* of length *l* and cross-sectional area *A*. It is usually expressed in *ohm metres*.

resistor A device used in *electronic* circuits primarily for its *resistance*. The most common types are either 'wire-wound', or made of finely ground carbon particles mixed with a *ceramic* binder.

resolution (chem.) The separation of a *racemic mixture* into its optically active components.

resolution of vectors The divison of *vectors* into components that act in specified directions, usually perpendicular to each other.

resolving power The ability of an optical system (e.g. *microscope*, *telescope*, the eye, etc.) to produce separate images of objects very close together.

resonance (chem.) Quantum-chemical resonance, mesomerism. The description of the structure of a molecule in terms of definite *valence* states of its atoms, and integral numbers of *chemical bonds* between the atoms, gives an oversimplified picture of the actual state of the molecule, whose characteristics, e.g. *electron-density* distribution, may be inconsistent with any classical formula. The resonance or valence-bond method of describing approximately the actual structure of a compound uses a number of classical structures ('canonical forms'), in terms of which the actual structure (the 'resonance hybrid') is described. See *benzene ring*.

resonance (phys.) 1. If, to a system capable of oscillation, a small periodic *force* is applied, the system is in general set into forced oscillations of small *amplitude*. As the *frequency*, *f*, of the exciting force approaches the *natural frequency* of the system, f_0, the amplitude of the oscillations builds up, becoming a maximum when $f = f_0$. The system is then said to be in resonance with the exciting force, or simply in resonance. 2. An extremely short-lived (10^{-24} second) *hadron*, which is regarded as an excited state of a more stable *elementary particle*.

resonance, nuclear Resonance is said to occur in *nuclear reactions* if the *energy* of an incident particle or *photon* is equal, or near, to the value of an appropriate *energy level* of the compound *nucleus*. Thus a resonance *neutron* is one whose energy corresponds to a particular energy level of a nucleus that will readily absorb it.

resonance neutron see *resonance, nuclear*.

resonant cavity A space enclosed by electrically conducting surfaces, in which electromagnetic *energy* may be stored or excited. The *frequency* of the oscillations (resonant frequency) within a resonant cavity will depend upon its physical dimensions.

resonant circuit An electronic circuit containing *resistance, inductance,* and *capacitance*. If these elements are *in series* (a series resonant circuit), resonance occurs when the *impedance* is a minimum, i.e. the impedance Z is equal to

$$R + i[\omega L - 1/\omega C]$$

and at resonance $\omega L = 1/\omega C$. Thus, at resonance the circuit has only resistance ($\omega = 2\pi f$, where *f* is the frequency; *R* is resistance, *L* is inductance, and *C* is capacitance).

In a parallel resonant circuit, the inductance and capacitance are *in parallel* and resonance occurs at maximum impedance,

$$\text{i.e. when } R^2 + \omega^2 L^2 = L/C,$$

which is also often approximately when $\omega L = 1/\omega C$. In *radio*, resonant circuits are used to generate *radio-frequency* oscillations in transmitters and to selectively detect them in receivers (see *tuned circuit*). See also *Q-factor*.

resorcinol Benzene-1,3-diol. $C_6H_4(OH)_2$. A solid *dihydric phenol*, m.p. 110°C. It is used in tanning and as an *intermediate* in the manufacture of *resins, drugs,* and other products.

respiration *Aerobic* respiration is the process by which living *organisms*, or their components, take oxygen from the *atmosphere* to oxidize *organic compounds* to obtain *energy* (see also *mitochondria*; *electron carrier system*). *Anaerobic* respiration is the process by which organisms, or their components, obtain energy when they do not have access to free oxygen; anaerobic respiration of *sugars* yields *lactate* (e.g. in muscle) or *ethanol* (e.g. in *yeasts*). Many organisms can respire anaerobically for a short time only but certain *bacteria* depend entirely on anaerobic respiration.

respiratory chain See *electron transport chain*.

respiratory pigment A substance formed in *blood cells* or *blood plasma* that is capable of combining loosely and reversibly with oxygen, e.g. *haemoglobin*. See also *cytochrome*.

respiratory quotient RQ. The *volume* of *carbon dioxide* expired by an *organism* or tissue divided by the volume of oxygen consumed by it over the same period.

rest energy The equivalent of the *rest mass* of a body expressed in *energy* units, i.e. m_0c^2, where m_0 is the rest mass and c is the *speed of light*.

restitution, coefficient of *e*. A measure of the *elasticity* of bodies upon impact. For two bodies of a given material colliding, *e* is equal to the ratio of the relative *velocity* of the bodies along their line of centres immediately after impact to their relative velocity before impact.

rest mass The *mass* of a body when at rest relative to the observer. The mass of a body varies with its *speed* (see *relativity, theory of*), a result of great importance when speeds approaching those of *light* are considered, e.g. in *nuclear physics*. See *relativistic mass*.

resultant (phys.) A single *vector* representing a *force* or *velocity* that produces the same effect as the two or more forces or velocities acting together.

retardation (phys.) Deceleration. The rate of decrease of *velocity* or *speed*.

retardation plate Either a *half-wave plate* or a *quarter-wave plate*.

retentivity See *remanence*.

retina The several cell layers at the back of the inside of the eye that contains the *rods* and *cones*, which respond to light by sending *impulses* to the optic nerve. These impulses are sent by the optic nerve to the brain, where the visual image is formed.

retinol See *vitamin A*.

retort (chem.) **1.** A *glass* vessel consisting of a large bulb with a long neck narrowing somewhat towards the end. **2.** In industrial processes, any vessel in which a *chemical reaction* or process takes place, especially *distillation*. **3.** In the canning industry, a large *autoclave* for heating sealed cans by *superheated steam* under pressure.

retort carbon See *gas carbon*.

retrograde motion See *direct motion*.

retrorocket A small *rocket*, forming part of a larger one, that produces *thrust* in the opposite direction to that of the main rocket with the object of decelerating it; e.g. to enable a lunar *module* to make a 'soft' landing on the *Moon*.

retrovirus A *virus* containing *ribonucleic acid* (RNA) in which a *reverse transcriptase* converts the RNA into *deoxyribonucleic acid* (DNA), enabling the virus to become integrated with the DNA of its host cell. Retroviruses can cause cancer in animals.

reverberation time The time taken for a *sound* to fall by 60 *decibels* in an auditorium, i.e. to the limit of audibility from a value 10^6 times this.

reverberatory furnace A furnace designed for operations in which it is not desirable to mix the material with the *fuel*; the roof is heated by *flames*, and the *heat* is radiated down on to the material off the roof.

reverse osmosis A method of *desalination* in which brine and pure *water* are separated by a *semipermeable membrane*. The pressure on the brine side is raised to some 25 *atmospheres*, which causes water from the brine to pass through the membrane into the pure water. The high pressure makes the process difficult to apply on a large scale.

reverse transcriptase An *enzyme* that occurs in *retroviruses* and has the ability to *catalyse* the conversion of *ribonucleic acid* (RNA) into *deoxyribonucleic acid* (DNA). This enables the viral *genome* to be inserted into the DNA of the host cell and to be replicated by the host. Reverse transcriptase is used in *genetic engineering* to produce *complementary DNA* from messenger RNA.

reversible process (in *thermodynamics*) A hypothetical process that can be performed in the reverse direction, the whole series of changes constituting the process being exactly reversed. A reversible process can take place only in infinitesimal steps about equilibrium states of the system. In practice, all real processes are irreversible.

reversible reaction A *chemical reaction* that may be made, under suitable conditions, to proceed in either direction. See *chemical equilibrium*.

Reynolds number (*Re*). A dimensionless quantity applied to a fluid flow. For a *liquid* flowing through a tube,

$$(Re) = u\rho l/\eta,$$

where u = *speed* of flow, ρ = *density* of the liquid, l = the diameter of the tube, and η = the coefficient of *viscosity* of the liquid. At low speeds the flow of the liquid is *streamline*. At a certain value of (*Re*), corresponding to a critical speed u_c, the flow becomes *turbulent*; in a smooth straight pipe of uniform bore, if (*Re*) exceeds 3000 the flow becomes turbulent. The concept does not only apply to flow through pipes, it is also used in *aerodynamics*. Named after Osborne Reynolds (1842–1912).

RF See *radio frequency*.

rhe A unit of *fluidity*. The *reciprocal* of the *poise*.

rhenium Re. Element. R.a.m. 186.207. At. No. 75. A hard heavy grey *metal*, r.d. 21.023, m.p. 3180°C., b.p. 5600°C. It is used in *thermocouples* and as a *catalyst*.

rheology The study of the deformation and flow of *matter*.

rheopexy The acceleration of a thixotropic (see *thixotropy*) increase of *viscosity* by gentle stirring.

rheostan An *alloy* of 52% copper, 25% nickel, 18% zinc, and 5% iron that is used for electrical resistance wire.

rheostat A variable electrical *resistor*. In a simple wire-wound rheostat a sliding contact moves along a coil of wire.

rhesus factor Rh factor. A group of *antigens* in the red *blood cells* of some humans (said to be Rh positive) but absent in some individuals (Rh negative). If a Rh negative mother conceives a Rh positive foetus, anti-Rh *antibodies* may form on her blood cells, which could cause anaemia in subsequent Rh positive foetuses.

rhodium Rh. Element. R.a.m. 102.9055. At. No. 45. A silvery-white hard *metal*, r.d. 12.4, m.p. 1963°C., b.p. 3700°C. It occurs with and resembles platinum. It is used in *alloys*, *catalysts*, and *thermocouples*.

rhodopsin Visual purple. A complex *organic compound* formed in the *rods* of the *retina* of the eye. It makes the eye more sensitive in very dim light; lack of it causes night blindness. The *aldehyde* derivative of *vitamin* A, retinal, is a constituent of rhodopsin.

rhombus A *quadrilateral* having all its sides equal.

r_H scale A scale of hydrogen pressures that gives a measure of the strength of a *reducing agent*. The r_H value is defined as $\log_{10} 1/[\text{H}]$, where [H] is the hydrogen pressure that would produce the same electrode potential as that of a given *redox reaction* at the same *pH value*.

ribbon microphone A *microphone* in which a thin aluminium ribbon is loosely fixed in a strong *magnetic field* parallel to the plane of the ribbon. An EMF is generated in the ribbon as it is made to oscillate by the incident sound waves. If the *resonant frequency* of the ribbon is lower than the frequency of the sound, the EMF is independent of the frequency. A ribbon microphone therefore has strong directional characteristics and a low *noise* level.

riboflavin Lactoflavin. Vitamin B_2. See *vitamins*.

ribonuclease RNAase. An *enzyme* that *catalyzes* the *hydrolysis* of *ribonucleic acid*.

ribonucleic acid RNA. Long thread-like *molecules* consisting of single *polynucleotide* chains. The *sugar* of the *nucleotides* is *ribose*, and the four nitrogenous bases that occur in them are the same as those found in *deoxyribonucleic* acid, except that *uracil* replaces *thymine*. RNA is the chief constituent, together with *protein*, of many types of *virus*, and it appears to be responsible for the self-replication of the virus. 'Messenger' RNA (mRNA) transmits the coded information contained by the *chromosomes* of the *nucleus* of a *cell* to the *protein*-making *ribosomes* of the *cytoplasm*. 'Ribosomal' RNA (rRNA) combined with proteins forms the ribosomes. 'Transfer' or soluble RNA (tRNA) transfers the activated *amino acids* on to the mRNA (see *translation*).

ribose $C_5H_{10}O_5$. A *pentose*, m.p. 95°C.; the *dextrorotatory* form is of great biological importance as it occurs in the *nucleotides* of *ribonucleic acid*.

ribosomes Small granules (about 10^{-8} *metre* in diameter) that occur in the *cytoplasm* of *cells* and are the sites of *protein* synthesis. They are composed of ribosomal RNA (see *ribonucleic acid*) and proteins and they hold together all the component molecules required to *synthesize* a *polypeptide* (see *translation*).

Richardson equation Richardson-Dushman equation. The equation that relates the number of *electrons* emitted by a heated metal surface to the *thermodynamic temperature*, T, in *thermionic emission*. If the emitted current density is given by j, then

$$j = AT^2 \exp(-b/T),$$

where A is a constant related to the surface properties of the metal and b is a constant equal to W/k; W is the *work function* of the metal and k is the *Boltzmann constant*.

Richter scale A scale used to compare the strength of earthquakes. It is a logarithmic scale ranging from 0 to 10, the strength of the earthquake being based on the ratio of the logarithm of the amplitude of the ground movement to the period of the dominant wave. On this scale damage to buildings can occur for values greater than 6. Named after C. F. Richter (1900–85).

ricinoleic acid $C_{17}H_{32}OHCOOH$. A yellow *liquid*, b.p. 227°C., that occurs in *castor oil* and is used in the manufacture of *soap*.

rigidity modulus See *shear stress*.

ring compound (chem.) A chemical *compound* in the *molecule* of which some or all of the *atoms* are linked in a closed ring. See *carbocyclic compounds*; *heterocyclic compounds*.

Ringer's fluid *Physiological saline* containing sodium, potassium, and calcium *chlorides*; it is widely used for sustaining *cells* or tissues during *in vitro* biochemical experiments. Named after Sydney Ringer (1835–1910).

RMS value See *root-mean-square value of a variable*.

RNA See *ribonucleic acid*.

Rochelle salt Potassium sodium tartrate, potassium sodium 2,3-dihydroxybutane-dioate. $COOK.(CHOH)_2.COONa.4H_2O$. A white crystalline soluble salt, m.p. 70–80°C., that has *piezoelectric* properties. It is used for piezoelectric crystals and in *baking powder*.

Rochon prism A device used for obtaining plane-polarized light (see *polarization of light*) and in other related problems. It consists of two prisms made of *quartz*, one cut parallel to the *optic axis* and the other perpendicular to it. It is used for work with *ultraviolet radiation*.

rock In the scientific sense, a rock is any distinct material present in the Earth's crust but, in distinction from a *mineral*, it need not have a definite chemical composition and may consist of more than one mineral. A rock need not necessarily be hard or stone-like; e.g. *clays* are regarded as rock materials. Rocks are classified as *igneous rocks*, *metamorphic rocks*, or *sedimentary rocks*.

rock crystal A pure natural crystalline form of *silica*, SiO_2. See *quartz*.

rocket A projectile driven by *reaction propulsion* that contains its own *propellants*. A rocket is therefore independent of the Earth's *atmosphere* both with respect to *thrust* and *oxidant* and provides the only known practicable means of propulsion in *space*. 'Chemical' rockets may be powered by either *solid* or *liquid fuels* that burn in oxygen, while 'nuclear' rockets would be powered by a propulsion reactor (see *nuclear reactor*). 'Multistage' or 'step' rockets are rockets built up of several separate sections, each stage being jettisoned when it has burnt out. The 'booster', or first stage, of a space rocket accelerates the projectile up to the thinner regions of the atmosphere, when subsequent stages take over the propulsion. Thus the necessarily high *escape velocity* is not achieved in denser parts of the atmosphere (which would introduce *friction* heating problems), moreover as each stage is jettisoned the projectile becomes subtantially lighter, and higher velocities can be achieved with less thrust (see *specific impulse*). Deceleration of rockets is obtained by the use of *retrorockets*.

'Rocket motors' are also used on certain types of aircraft for take-off, or when a high thrust is required for a short period. See also *ion engine*.

rock salt Halite. Natural crystalline *sodium chloride*, NaCl.

rod A photosensitive cell in the *retina* of vertebrate eyes. They are essential for vision in dim light and occur in the margins of the retina but not in the fovea. Rods contain the pigment *rhodopsin*. Compare *cone*. See *scotopic* vision.

Rodinal* A photographic *developer* consisting of an *alkaline solution* of 4-aminophenol, $NH_2C_6H_4OH$, with sodium bisulphite, $NaHSO_3$.

roentgen An obsolete unit used to measure a *dose* of *X*- or *gamma-radiation* that will produce *ions* carrying 2.58×10^{-4} *coulomb* of electric charge of either sign in 1 kg of dry air. Named after Wilhelm Konrad Roentgen (1845–1923).

Roentgen rays See *X-rays*.

rolling friction See *friction, coefficients of.*

ROM Read only memory. A *semiconductor memory* used in *computers* for storing information that does not require to be updated or changed. It is a form of *RAM*, but the memory can only be read, the contents of the storage elements being fixed during manufacture.

rongalite A *compound* of sodium sulphoxylate and *methanal*, $NaHSO_2.HCHO$. It is used as a *reducing agent* in dyeing.

root (math.) **1.** One of the equal *factors* of a number or quantity. The square root, $\sqrt{}$ or $\sqrt[2]{}$, is one of two equal factors; e.g. $9 = 3 \times 3$ or -3×-3; hence $\sqrt{9} = \pm 3$. Similarly the *cube* or third *root* is denoted by $\sqrt[3]{}$, etc. It may also be denoted by a fractional *index*; thus $\sqrt{x} = x^{1/2}$. **2.** The root of an *equation* is a value of the unknown quantity that satisfies the equation.

root-mean-square value of an alternating quantity If y is a periodic *function* of t, of period T, the root-mean-square (RMS) value of y is the square *root* of the mean of the square of y taken over a period. The RMS value I of an *alternating current* is important since it determines the *heat* generated (RI^2) in a *resistance* R (see *electric current, heating effect of*). All ordinary AC measuring instruments give RMS values of current, etc. If the alternating quantity can be represented by a pure *sine wave*, the RMS value of the quantity A is related to the maximum value a of the quantity (i.e. *amplitude*) by the expression $A = a/\sqrt{2}$. The RMS value of a current is also known as the 'effective value of the current'. Similarly, the RMS value of an alternating *EMF* is known as the 'effective EMF'.

root-mean-square value of a variable RMS. The square root of the average of the squares of a number of values, given by:

$$RMS = \frac{\sqrt{(\text{sum of squares of the individual values of the variable})}}{(\text{total number of values})}$$

See also *root-mean-square value of an alternating quantity*.

Rose's metal An *alloy* of 50% bismuth, 25% lead, and 25% tin; m.p. 94°C.

rosin Colophony. A yellowish *amorphous resin* obtained as a residue from the *distillation* of *turpentine*. R.d. 1.08, m.p. 120°–150°C. It is used in varnishes, *soaps*, and *soldering* fluxes. See also *ester gum*.

Rotameter* A device for measuring the rate of flow of *fluids*; it consists of a small float that is suspended by the fluid in a vertical calibrated tube. The weight of the float gives a measure of the rate of flow.

rotary converter An *alternating-current electric motor* mechanically coupled to a *direct-current generator*. It is used for converting an AC supply into DC.

rotary dispersion See *optical activity*.

rotation The angular displacement of a body or line is the angle, θ (in radians), through which the body or line is rotated about a specified axis in a specified direction. The angular velocity, ω, is given by $d\theta/dt$, where t is the time. The angular acceleration, α, is given by:

$$\alpha = d\omega/dt = d^2\theta/dt^2.$$

The torque, T, causing an angular acceleration is given by $T = I\alpha$, where I is the *moment of inertia* of the body.

rotor The rotating part of a *turbine*, *electric motor*, or *generator*. Compare *stator*.

RQ See *respiratory quotient*.

rubber An *elastomer* obtained from the *latex* of the *Hevea brasiliensis* tree. Raw natural rubber consists mainly of the *cis*-form of poly*isoprene*, $(CH_2.CH:C(CH_3):CH_2)_n$, a *hydrocarbon polymer*, with a *relative molecular mass* of about 300 000. Nearly all rubber articles are made by 'compounding' raw rubber, i.e. mixing it with other ingredients and then vulcanizing it in moulds by heating with sulphur and *accelerators*.

rubber, synthetic A class of synthetic *elastomers* made from *polymers* or copolymers (see *polymerization*) of simple *molecules*. See *butyl rubber*, *neoprene*, *nitrile rubber*, *styrene-butadiene rubber* (SBR), *silicone rubber*, and *stereoregular rubbers*.

rubidium Rb. Element. R.a.m. 85.4678. At. No. 37. A soft extremely *reactive* white *metal* resembling sodium. R.d. 1.53, m.p. 38.9°C., b.p. 705°C. It occurs in a few rare *minerals*. See also *rubidium-strontium dating*.

rubidium-strontium dating A method of *dating* some rocks, used for specimens over 10^9 years old. It is based on the *decay* of the *radioisotope* rubidium-87 (*half-life* 5×10^{10} years) to yield strontium-87. An estimate of the sample's age is given by the ratio of the two *isotopes*.

ruby A red form of *corundum*, Al_2O_3, that owes its colour to traces of chromium. It is used in *lasers* and as a gem stone.

rules of Fajans Rules that describe the conditions determining whether an electrovalent or a covalent bond (see *chemical bond*) will be formed between *atoms*. Fajans' rules state that an electrovalent bond will be replaced by a covalent bond if: (1) the charge on either of the *ions* resulting from an electrovalent donation of *electrons* is large (i.e. if more than 1 or 2 electrons are donated); or (2) the volume of the *cation* is small or that of the *anion* is large. Named after Kasimir Fajans (1887–1975).

rust (chem.) An *hydrated oxide* of iron, mainly $Fe_2O_3.H_2O$, formed on the surface of iron when it is exposed to moisture and air. The process is *electrochemical* with different parts of the iron surface acting as the *anode* and *cathode* of an *electrolytic cell*. The reactions taking place are:

$$
\begin{array}{ll}
\text{Fe (anode)} & \tfrac{1}{2}O_2 + H_2O \text{ (cathode)} \\
\downarrow & \downarrow \\
Fe^{2+} \quad + \quad 2OH^- \rightarrow Fe(OH)_2 \xrightarrow{O_2} Fe_2O_3 &
\end{array}
$$

rust (bio.) A fungal disease of plants.

ruthenium Ru. Element. R.a.m. 101.07. At. No. 44. A hard brittle *metal*, r.d.

12.36, m.p. 2330°C., b.p. 4100°C. It occurs together with platinum. It is used in *alloys* and as a *catalyst*.

rutherfordium A name proposed for the element with *atomic number* 104. See *transactinides*.

rutile A crystalline form of natural *titanium dioxide*, TiO_2.

Rydberg constant A constant relating to those atomic *spectra* that are similar to the hydrogen *atom* spectrum (see *Balmer series*). The Rydberg constant for hydrogen is $1.096\ 77 \times 10^7$ m^{-1}. The general Rydberg formula is:

$$1/\lambda = R(1/n^2 - 1/m^2),$$

where R is the Rydberg constant and n and m are positive *integers*. The quantity R_ihc, where h is the *Planck constant* and c is the *speed of light*, is sometimes treated as a unit of *energy* called the rydberg, symbol Ry, such that Ry = $2.179\ 72 \times 10^{-18}$ *joule*. R_i is defined as

$$m_e e^4/8\varepsilon_0^2 h^3 c,$$

where m_e is the *mass* of an *electron* and e its charge; ε_0 is the *electric constant*. Named after J. R. Rydberg (1854–1919).

S

saccharide A simple *sugar*; a *monosaccharide*.

saccharimeter An apparatus for determining the *concentration* of a *sugar solution* by measuring the *angle* of rotation of the plane of vibration of polarized light passing through a tube containing the solution. See *optical activity*; *polarization of light*.

saccharin $C_6H_4SO_2CONH$. A white crystalline sparingly *soluble solid*; m.p. 227°C. When pure, it has about 550 times the sweetening power of *sugar*, but has no food value, and may have harmful effects if used to excess. It is manufactured from *toluene*, $C_6H_5CH_3$; it is also used in the form of a sodium *salt* called 'saccharin sodium', $C_6H_4COSO_2NNa.2H_2O$.

saccharometer A type of *hydrometer* used for finding the *concentration* of *sugar solutions* by determining their *density*; it is usually graduated to read the percentage of sugar direct.

saccharose See *sucrose*.

sacrificial protection The protection of ferrous metals against *rusting* by the use of a more reactive metal. For example, if *galvanized iron* is scratched so that the iron is exposed, it does not rust as the zinc *ions* go into solution before the iron ions.

safety glass Glass that has been treated so that it does not break or splinter on impact. Toughened glass is heat treated by heating nearly to the softening point, followed by controlled cooling. Laminated glass has a thin layer of transparent plastic sheet between two sheets of glass. Wire glass has a wire mesh built into it.

safety lamp Davy lamp. An oil-lamp that will not ignite flammable *gases*, e.g. *methane* (fire-damp). It has a cylinder of wire gauze acting as a chimney; the *heat* of the *flame* is conducted away by the gauze, and while fire-damp will burn inside the gauze, the *temperature* of the gauze does not rise sufficiently to ignite the gas outside.

safrole $CH_2:CHCH_2C_6H_3O_2CH_2$. A yellowish crystalline substance, m.p. 11.2°C., b.p. 234.5°C., used in the manufacture of perfumes, flavours, and *soaps*.

sal ammoniac See *ammonium chloride*, NH_4Cl.

salicin $CH_2OHC_6H_4OC_6H_{11}O_5$. A colourless *soluble glucoside*, m.p. 200°C., used as an *antipyretic* and *analgesic*.

salicylate A *salt* or *ester* of *salicylic acid*.

salicylic acid 2-hydroxybenzoic acid. $OH.C_6H_4COOH$. A white crystalline solid, m.p. 159°C. It is used as an antiseptic and in the form of a derivative as *aspirin*.

saline Containing *salt*, especially the salts of *alkali metals* and magnesium. A 'saline solution' is a solution of salts in *water*, especially one that is *isotonic* with body fluids. See also *physiological saline*.

salinometer A type of *hydrometer* used for determining the *concentration* of *salt* *solutions* by measuring their *density*. Other types measure the electrical *conductivity* of the solution.

saliva An *alkaline aqueous liquid* secreted by the salivary glands in response to chewing or the thought, sight, or smell of food. It contains the *glycoprotein* mucin to lubricate the food's passage through the oesophagus and the *enzymes* *amylase* and *maltase* to start the digestion of *starch*.

salt (chem.) A chemical *compound* formed when the hydrogen of an *acid* has been replaced by a *metal* or other positive *ion*. A salt is produced, together with *water*, when an acid reacts with a *base*. Salts are named according to the acid and the metal from which the salt is derived; thus *sodium sulphate* is a salt derived from sodium and *sulphuric acid*.

salt, common See *sodium chloride,* NaCl.

salt bridge A tube of *potassium chloride* in the form of a *gel*, used to connect two *half cells* without mixing the *electrolytes*.

saltcake See *sodium sulphate*, $Na_2SO_4.10H_2O$.

salt effect See *salting-out*.

salting-out *Precipitation* of a dissolved substance by addition of another (usually a *salt*) that lowers its *solubility*; e.g. *soaps* can be salted-out by common salt (*sodium chloride*) from solutions in water.

saltpetre Nitre. See *potassium nitrate*.

salts of lemon Potassium quadroxalate, $KH_3C_4O_8.2H_2O$. A white *soluble* poisonous crystalline *salt*, used for removing ink-stains.

sal volatile Commercial 'ammonium carbonate', actually consisting of a mixture of ammonium hydrogencarbonate, NH_4HCO_3, ammonium carbamate, $NH_4O.CO.NH_2$, and ammonium carbonate, $(NH_4)_2CO_3$.

samarium Sm. Element. R.a.m. 150.36. At. No. 62. R.d. 7.536, m.p. 1050°C., b.p. 1600°C. A soft silvery *metal* used in *nuclear reactors* as a *neutron* absorber, in *ferromagnetic alloys* (e.g. $SmCo_5$), and, in the form of its *oxide*, in some special optical glasses. See *lanthanides*.

sample See *random sample*.

sand Hard, granular powder (usually with a particle diameter of less than 2 mm), generally composed of granules of impure *silica*, SiO_2. See also *oil sand*.

sandstone A *sedimentary rock* formed from *sand* or *quartz* particles cemented together with *clay*, *calcium carbonate*, and iron oxide.

sandwich compound A *complex* in which an atom of a *transition element* is sandwiched between two substantially planar *hydrocarbon* molecules or groups containing *pi bonds* (see *orbital*). Typical examples are *ferrocene* and its analogues.

saponification The *hydrolysis* of an *ester*; the term is often confined to the hydrolysis of an ester using an *alkali*, thus forming a *salt* (a *soap* in the case of some of the higher *fatty acids*) and the free *alcohol*.

saponification number One of the characteristics of a *fat or oil*; the number of *milligrams* of *potassium hydroxide* required for the complete *saponification* of one gram of the fat or oil.

saponins *Glucosides*, derived from plants, that form a lather with water. They are used as foaming agents and *detergents*.

sapphire A natural crystalline form of blue, *transparent corundum* (*alumina*, Al_2O_3); the colour being due to traces of cobalt or other *metals*.

satellites Bodies that rotate in *orbits* round other bodies of greater *mass* under the influence of their mutual *gravitational field*. Particularly bodies, or moons, that rotate around *planets*. E.g. the *Moon* is a satellite of the *Earth*. See also *satellites, artificial*.

satellites, artificial In 1957 the first man-made artificial *satellite* was launched by the USSR into *orbit* around the *Earth*. This, and subsequent Soviet and American artificial satellites, have been used to obtain, and *radio* back to Earth, information concerning conditions prevailing in the *upper atmosphere* and the *ionosphere*. Valuable information has also been obtained relating to *cosmic rays*, the *density* of *matter* and the frequency of *meteors* in *space*, the shape and *magnetic fields* of the Earth, and the nature of solar *radiations*. As a result of the earlier American satellites the *Van Allen radiation belts* were discovered.

'Communication' satellites are artificial Earth satellites used for relaying radio, television, and telephone signals around the curved surface of the Earth. 'Passive' satellites merely reflect the transmissions from their surfaces, while 'active' satellites are equipped to receive and retransmit signals. See also *synchronous orbit*.

saturated compound (chem.) A *compound* that does not form *addition compounds*; a compound the *molecule* of which contains no double or multiple *chemical bonds* between the *atoms*. Compare *unsaturated compound*.

saturated solution A *solution* that can exist in *equilibrium* with excess of *solute*, i.e. a solution that contains the maximum amount of solute that will remain in solution at a given temperature. The saturation *concentration* is a function of the *temperature*. See also *supersaturation*.

saturated vapour A *vapour* that can exist in *equilibrium* with its *liquid*.

saturated vapour pressure SVP. The *pressure* exerted by a *saturated vapour*. This pressure is a function of the *temperature*.

saturation 1. The characteristic of a *colour* that is determined by the degree to which it departs from white and approaches a pure spectral colour. **2.** The state of a *ferromagnetic substance* when all its *magnetic domains* are orientated in the direction of an external field and it cannot be magnetized more strongly.

Saturn (astr.) A *planet*, with at least 22 small *satellites*, that is surrounded by characteristic rings (see *Saturn's rings*). Its *orbit* lies between those of *Jupiter* and *Uranus*. Mean distance from the *Sun*, 1427.01 million kilometres. *Sidereal period* ('year'), 29.46 years. Mass, approximately 95.14 times that of the *Earth*, diameter 120 800 kilometres. Surface temperature, about $-150°C$. Voyagers I (1980) and II (1981) carried out studies of the planet.

Saturn's rings A system of concentric rings, probably composed of ice particles or the remains of a broken-up *satellite*, which are seen round the *planet Saturn*.

sawtooth waveform A waveform in which the shape resembles the teeth of a saw. The variable builds slowly and linearly up to a maximum value and then falls perpendicularly to zero in each cycle. Sawtooth generators are used for many purposes, e.g. to provide a time base for an electronic circuit. Sawtooth waveforms are produced by relaxation *oscillators*.

SBR See *styrene-butadiene rubber*.

scalar quantity Any quantity that is sufficiently defined when the magnitude is given in appropriate units. Compare *vector*.

scalene (Of a *triangle*) having three unequal angles and sides. (Of a *cone*) having its *axis* inclined to its *base*.

scaler Scaling circuit. An *electronic* device or circuit that produces an output *pulse* when a prescribed number of input pulses has been received. If the prescribed number is two (or ten) the circuit is referred to as a binary (or decade) scaling circuit or scaler.

scandium Sc. Element. R.a.m. 44.95591. At. No. 21. R.d. 2.99, m.p. 1500°C., b.p. 2800°C. A rare *metal* that occurs in small quantities as the oxide Sc_2O_3. Only scandium-45, of its ten isotopes, is not radioactive.

scanning The repeated and controlled traversing of: (1) a *mosaic* in a television *camera*, or a screen in a *cathode-ray tube*, with an *electron* beam; (2) an airspace with a *radar aerial*; (3) the sky with a steerable dish *radio telescope*; or more generally (4) any area or volume with a moving detector in order to measure some quantity or detect some object.

scanning electron microscope SEM. See *electron microscope*.

scanning tunnelling microscope STM. A *microscope* in which *electrons* from a fine probe tunnel (see *tunnel effect*) between the probe and the surface of a conducting sample to be examined, produce an electrical signal. The probe is made to scan the surface and is raised and lowered to maintain a constant signal strength. With the aid of a computer, the signal can be used to generate a contoured map of the surface, in which individual atoms can be resolved. See also *atomic force microscope*.

scattering The deflection of any *radiation* as a result of its interaction with *matter*. E.g. the change in direction of a particle or *photon* on interacting with a *nucleus* or *electron*. If the scattered particle or photon loses *energy* by causing *excitation* of the struck nucleus the scattering is said to be 'inelastic' (see *Raman effect*; *Compton effect*; *Tyndall effect*); if energy is not lost in this way the scattering is 'elastic' (even though some energy may be lost because of the recoil of the scattering particle). This is known as Rayleigh scattering; in this case there may be a change in the *phase* of the radiation but there is no *frequency* change. See also *Coulomb scattering*; *scattering of light*; *Thomson scattering*.

scattering of light When a *beam* of *light* traverses a material medium, scattering of the beam takes place. Two types of scattering occur: (1) by random *reflection*; i.e. small particles suspended in the medium act as tiny *mirrors* and, being randomly orientated with respect to the beam, produce random reflections. This type occurs when the size of the particles is large in comparison with the *wavelength* of the light: (2) by *diffraction*; this occurs when particles that are small compared with the wavelength of the light are present in the medium. Owing to diffraction phenomena, the particles act as centres of radiation and each particle scatters the light in all directions. In this type, the degree of scattering is proportional to the inverse fourth *power* of the wavelength of the light. Thus, blue light is scattered to a greater extent than red. The blue colour of the sky is due to scattering by the actual *molecules* of the *atmosphere*.

Scheele's green A bright green *precipitate*, probably consisting of copper(II) arsenite, $Cu_3(AsO_8)_2.2H_2O$. It is used as a *pigment* and *insecticide*. Named after Karl Wilhelm Scheele (1742–86).

scheelite A naturally occurring *ore* of tungsten, $CaWO_4$.

Schiff's reagent A reagent used to test for *aldehydes* and *ketones*. It consists of the dye *magenta*, which has been decolorized with *sulphur dioxide* or *sulphurous acid*. Aliphatic or aromatic aldehydes oxidize the reduced form of the dye back to its original colour and aliphatic ketones restore the colour slowly. Aromatic ketones have no effect. Named after Hugo Schiff (1834–1915).

Schlieren photography In a fast moving *fluid* in which there is *turbulent flow*, streaks (German, 'Schliere') become visible because they have a different *density* and *refractive index* from the bulk of the fluid. These streaks can be photographed using *spark photography*, or other high speed photographic methods.

SCHMIDT TELESCOPE

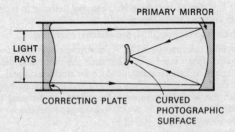

Figure 36.

Schmidt telescope (camera) A type of astronomical reflecting *telescope* consisting of a primary spherical *mirror* with a correcting plate at, or near, its *centre of curvature*. This plate corrects for *aberration*, *coma*, and *astigmatism*, enabling a wide area of the sky to be photographed with good *definition*. The instrument is not used visually but *images* are photographed on a curved surface. See Fig. 36. Named after Bernhard Schmidt (1879–1935).

Schottky defect See *vacancy*.

Schrödinger wave equation The *wave equation* used in *wave mechanics* to describe the behaviour of a particle in a field of force. It is based on de Broglie's concept that every moving particle is associated with a wave of wavelength h/mv (where h is the *Planck constant* and m and v are the mass and velocity of the particle). In three dimensions the equation has the form:

$$\nabla^2 \psi - 4\pi m/ih(\partial \psi/\partial t) - 8\pi^2 mU\psi/h^2 = 0$$

where ∇^2 is the *Laplace operator*, ψ is the *wave function* of the coordinates and time, and U is the potential energy of the particle. See also *eigenfunction*. Named after Erwin Schrödinger (1887–1961).

Schwartzchild radius See *black hole*.

Schweitzer's reagent A deep blue solution of a copper *ammine* in copper(II) hydroxide, $Cu(OH)_2$. A *solvent* for *cellulose*; it was formerly used for this purpose in the obsolete *cuprammonium* process for *rayon* manufacture.

science The study of the physical universe and its contents by means of reproducible observations, measurements, and experiments to establish, verify, or modify general laws to explain its nature and behaviour.

scintillation counter A device in which *light* flashes, produced by a scintillator (see *phosphor*) when exposed to *ionizing radiation*, are converted into electrical *pulses* by a *photomultiplier*, thus enabling the number of ionizing events to be counted. Scintillation counters may also be used to determine the energies of the ionizing events.

scintillation spectrometer A device for determining the *energy* distribution of a given *radiation*. It consists of a *scintillation counter* that incorporates a *pulse height analyser*.

scintillator See *phosphor*.

sclerometer An instrument for measuring the hardness of a material, usually by measuring the pressure required to scratch it, or by measuring the height to which a standard ball will rebound from it when dropped from a fixed height. See *Mohs scale of hardness*; *Brinell test*.

scleroprotein A class of complex, *insoluble*, fibrous *proteins*, (e.g. *keratin*, *collagen*, *elastin*) that occur in the surface coatings of animals and form the framework binding *cells* together in animal tissues.

-scope Suffix applied to names of instruments for observing or watching, usually as distinct from measuring. E.g. *telescope*.

scopolamine See *hyoscine*.

scotopic vision Vision in which the *rods* in the eye are the principal receptors. This type of vision occurs when the level of light is low and colours cannot be distinguished. Compare *photopic vision*.

SCR See *silicon-controlled rectifier*.

screen grid A grid placed between the *anode* and *control grid* of a *thermionic valve*, usually held at a fixed positive *potential*.

scruple 1/24 ounce Troy. See *Troy weight*.

S-drops See *strange matter*.

sea-floor spreading See *plate tectonics*.

sea-water The approximate composition (not including inland seas such as the Dead Sea) is water, 96.4%; common *salt*, NaCl, 2.8%; *magnesium chloride*, $MgCl_2$, 0.4%; *magnesium sulphate*, $MgSO_4$, 0.2%; *calcium sulphate*, $CaSO_4$, and *potassium chloride*, KCl, 0.1% each.

sebacic acid Decanedioic acid. $HOOC(CH_2)_8COOH$. A *dibasic* crystalline *fatty acid*, m.p. 134.5°C., used in the manufacture of *plasticizers* and *resins*.

secant 1. A straight line cutting a *circle* or other curve. 2. See *trigonometrical ratios*.

second 1. The *SI unit* of time defined as the duration of 9 192 631 770 *periods* of the *radiation* corresponding to the transition between two hyperfine levels of the *ground state* of the caesium–133 atom. Symbol s. 2. A measure of *angle*: 1/60 of a *minute*.

secondary cell See *accumulator*.

secondary coil Secondary winding. The output coil of a transformer or induction coil. Compare *primary coil*.

secondary colour A colour, e.g. orange, obtained by mixing two *primary colours*.

secondary emission of electrons When a primary *beam* of rapidly moving *elec-*

trons strikes a *metal* surface, secondary electrons are emitted from the surface. The effect is of importance in the *thermionic valve*, the *photomultiplier*, etc. In the thermionic valve, the emission occurs when the electrons strike the *anode*, and may be suppressed or controlled in multi-electrode tubes (*tetrode, pentode*) by various grids called the *suppressor* and *screen grids*. Secondary emission may also occur when such a surface is bombarded with positive *ions*, as in a *discharge in gases*.

second derivative (math.). The *derivative* of a derivative, written d^2x/dy^2. E.g. *acceleration* (a) is the second derivative of *displacement* (s) with respect to time (t), or the first derivative of *speed* (v) with respect to time, i.e.,

$$a = d^2s/dt^2 = dv/dt.$$

secretion The process in which substances within a living *cell* are passed out of the cell, usually by a *membrane*-bounded vesicle fusing with the *plasma membrane*.

sector See *circle*.

secular variation of magnetic declination If the *Earth's* magnetic North Pole is considered to rotate round the geographical North Pole, completing a cycle in about 930 years, a representation of a steady variation of *magnetic declination*, known as the secular variation, will be seen. Thus, the magnetic declination in London is at present westerly, and decreasing until it is due to become zero at the beginning of the twenty-second century.

sedative A *drug* that reduces nervousness and excitement.

sedimentary rock Rock formed by the accumulation and consolidation of sediments. Chemical sedimentary rocks are formed from chemical precipitates, these include *coal* and most *limestone*. Clastic sedimentary rocks are formed from existing rocks that have broken down into small particles as a result of attrition, etc., and been transported and redeposited elsewhere; these include *sandstone* and *clay*. Compare *igneous rock*; *metamorphic rock*.

sedimentation The process of separating an *insoluble solid* from a *liquid* in which it is suspended by allowing it to fall to the bottom of the containing vessel, with or without agitation or *centrifuging*.

Seebeck effect If two wires of different *metals* or two *semiconductors* are joined at their ends to form a *circuit* and the two junctions are maintained at different *temperatures*, an *electric current* flows round the circuit. Compare *Peltier effect*. Named after T. J. Seebeck (1770–1831).

seeding Impfing. The addition of fine particles to a *solution* to induce crystallization. Each particle (often a tiny crystal of the *solute*) acts as a *nucleus* upon which the new crystal grows.

Seger cone Pyrometric cone. A device for estimating the approximate *temperature* of a furnace; cones are made of materials softening at definite temperatures and when inserted in the furnace give an indication of temperature as they soften (i.e. when the vertex of the cone is seen to droop). Named after Hermann Seger (d. 1893).

segment See *circle* and *sphere*.

seismograph An instrument for recording earthquake shocks, volcanic activity, or nuclear explosions. See also *Richter scale*.

seismology The scientific study of earthquakes and the phenomena associated with them.

selenate A *salt* or *ester* of *selenic acid*.

selenic acid H_2SeO_4. A strongly corrosive crystalline *acid*, m.p. 58°C., with properties resembling those of *sulphuric acid*.

selenide A *binary compound* of *selenium*. Selenides of non-metals are *covalent compounds*.

selenium Se. Element. R.a.m. 78.96. At. No. 34. It is a non-metal resembling sulphur in its chemical properties. R.d. 4.81, m.p. 220°C., b.p. 685°C. It exists in several *allotropic forms*. The so-called 'metallic' selenium, a silvery-grey crystalline *solid*, is a *semiconductor* that varies in electrical *resistance* on exposure to *light* and is used in *photoelectric cells* (see *selenium cell*). Selenium occurs as *selenides* of metals, together with their *sulphides*; it is used in the manufacture of *rubber* and of ruby *glass*.

selenium cell 1. A *photoelectric cell* consisting of a layer of selenium covered by a thin transparent layer of gold. Light falling on the cell produces a voltage by the *photovoltaic effect*. **2.** A photoelectric cell in which a selenium element changes its *resistance* on exposure to light. An external *EMF* is applied to the element and the current produced is a measure of the intensity of the light. Thus this type of cell relies on the *photoconductive effect*.

selenium rectifier A *rectifier* that consists of alternate layers of iron and selenium in contact.

selenology The scientific study of the *Moon*, its nature, origin, and movements. Now that samples of the Moon's surface are available for study on Earth, selenology has become a branch of *chemistry* as well as *astronomy*.

self-absorption The decrease in the *radiation* from a *radioactive* material caused by the absorption of a part of the radiation by the material itself.

self-exciting (Of a *generator*) having *magnets* that are excited by *current* drawn from the output of the generator.

self-inductance The coefficient of *self-induction*.

self-induction The *magnetic field* associated with an *electric current* cuts the *conductor* carrying the current. When the current changes, so does the magnetic field, resulting in an induced *EMF* (see *induction, electromagnetic*). This phenomenon is called self-induction. The induced EMF (E) is proportional to the rate of change of the current (in the absence of any *ferromagnetic* material), the constant of proportionality being called the the self-inductance. Thus, $E = -L.dI/dt$ where I is the instantaneous current and L is the self-inductance. The minus sign indicates that the EMF opposes the change in the current. The magnitude of the self-inductance is a function only of the geometry of the electrical circuit and can be calculated in a few simple cases. The derived *SI unit* of inductance is the *henry*.

SEM Scanning electron microscope. See *electron microscope*.

semicarbazone A *compound* containing the unsaturated group =C:N.NH-CO.NH$_2$. They are used to identify *aldehydes* and *ketones* in *chemical analysis*.

semiconductor An electrical *conductor* whose *resistance* decreases with rising *temperature* and the presence of impurities, in contrast to normal metallic conductors for which the reverse is true. Semiconductors, which may be *elements* or *compounds*, include germanium, silicon, selenium, and lead-telluride. In general, semiconductors consist of *covalent crystals*, 'ideal' examples of which at the *absolute zero* of temperature would pass no *electric current* as all the *valence electrons* would be held by the covalent bonds. At normal temperatures, however, some of the electrons have sufficient thermal *energy* to break free from the bonds leaving *holes*. Electrons liberated in this way will have random thermal motions, but in an imposed *electric field* there will be a net drift against the field resulting in so called *n-type conductivity*. The behaviour of the holes is more complex, but they may be regarded as positive charges free to move about the crystal giving rise to *p-type conductivity*. The total current passed by such an *intrinsic semiconductor* is therefore the sum of the electron current and the hole current in the direction of the field. A rise in temperature will create more *carriers*, due to more bonds being broken by thermal energy, and thus lower resistance. The foregoing refers to 'ideal' crystals, but real crystals will have inherent *defects*, *dislocations*, and impurities that will produce additional carriers (see *extrinsic semiconductor*). In practical semiconductors impurities are added in controlled quantities during crystal growth, the number of *valence electrons* of the impurity *atoms* determining whether the *majority carriers* will be *p*- or *n*-type. A *p-n semiconductor junction* is formed when there is a change along the length of a crystal from one type of impurity to the other. At a *p-n* junction an internal electric field is created between the charged impurity *ions* of the two types. This field is sufficient to prevent the drift of electrons from the *n*-side to the *p*-side of the junction, and the drift of holes in the opposite direction. If an external positive *voltage* is applied to the *p*-side and a negative voltage to the *n*-side, the internal field can be overcome and a substantial current will flow as a result of the tendency of the majority carriers on each side to migrate to the other side: the magnitude of the current will depend upon the applied voltage. Reversing the voltage increases the effect of the internal field and the only current to flow will be the small number of *minority carrier* electrons on the *p*-side carried over to the *n*-side; similarly minority carrier holes will be carried from the *n*- to the *p*- regions. The reverse current is therefore small and does not depend upon the applied voltage. The *p-n* junction is thus a very efficient *rectifier* and is widely used for this purpose (see *semiconductor diode*); it is also the basis of the *transistor*, which has replaced the earlier thermionic valves. See also *energy band*.

semiconductor diode A *semiconductor* device, either based on a *semiconductor junction* or on point contact, with two *electrodes*. It is used for rectification.

semiconductor junction A plane that separates two layers of a *semiconductor* each of which have different electrical characteristics. For example, a *p-n* junction separates the *p*-region (in which *holes* are the *majority carriers*) from the *n*-region (in which *electrons* are the majority carriers).

semiconductor memory Solid-state memory. A form of *computer memory* consisting of one or more *integrated circuits* comprising an array of thousands of microscopic *electronic* elements each of which can store one *bit*. Such a memory provides very rapid random access (see *RAM*).

semimetal See *metalloid*.

semipermeable membrane A membrane allowing the passage of pure *solvent molecules* but not those of the dissolved substance. They often consist of a film of cellulose supported on a wire gauze. A semipermeable membrane is used as a partition between *solution* and solvent in osmotic measurements (see *osmotic pressure*) and in *dialysis*.

semipolar bond Coordinate bond, dative bond. A *chemical bond* in which two *electrons* are donated by one *atom* (usually nitrogen or oxygen) to another atom, which requires both of them to complete its *octet*. This is equivalent to one electrovalent bond and one covalent bond and is therefore called a semipolar bond.

Semtex Tradename for a nitrogen-based plastic *explosive* that is odourless and safe to handle. It is therefore widely used by terrorists.

senescence The process of ageing. It is characterized by a deterioration in the function of cells, as a result of a decline in their ability to dispose of waste materials and to replace essential components. These changes cause a decline in both mental and physical ability in animals.

sensitization (phot.) Photographic *silver bromide* emulsions are sensitive only to short-wave visible light (violet and blue), so that light of longer wavelength (e.g. red, green) is not registered. Emulsions for correct rendering of relative intensities of light of different *colours* (*panchromatic*) in colour photography, can be rendered sensitive to radiation in particular wavelength ranges by the use of certain dyes, known as sensitizers, which absorb radiation in these ranges (including *infrared*) and are able to utilize the energy absorbed in the breakdown of the silver bromide. *Cyanine dyes* are particularly useful for this purpose.

sensitometer An instrument for measuring the sensitivity of a photographic plate or film (see *photography*).

separation energy The *energy* required to remove a particle (a *proton* or a *neutron*) from a particular atomic *nucleus*.

septivalent Heptavalent. Having a *valence* of seven.

sequestering agent See *chelation*; *sequestration*.

sequestration The process of 'locking-up' metal *ions* in *coordination compounds* to make them ineffective. The sequestering agents used for this purpose are usually chelating agents. See *chelation*.

series (math.) A sequence of numbers or mathematical expressions such that the nth term may be written down in general form, and any particular term (say, the rth) may be obtained by substituting r for n; e.g. x^n is the general term of the series $1, x, x^2, x^3 \ldots x^n$. See also *convergent series*; *divergent series*.

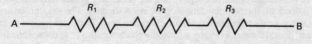

Figure 37.

series, resistances in If a number of *conductors* of electricity are connected in series, i.e. one after the other, so that the current flows through each in turn, the total *resistance* is the sum of the separate resistances of the conductors. See Fig. 37. Compare *parallel, conductors in*. For *capacitors* in series, the total capacitance C is given by $1/C_1 + 1/C_2 + 1/C_3 \ldots$

serine A white crystalline *amino acid*, m.p. 246°C., that occurs in many *proteins*. See Appendix, Table 5.

Serpek process A process for the *fixation of atmospheric nitrogen*. Aluminium is made to react with nitrogen to form aluminium nitride, which is then decomposed by *steam* to give *ammonia*.

serpentine A group of magnesium silicate *minerals* with the general formula $Mg_3Si_2O_5(OH)_4$. It occurs in two main forms, the fibrous source of *asbestos*, called chrysotile, or the mottled green and white antigorite, from which it gets its name.

serum The *liquid* that remains after the clotting and removal of *blood cells* and *fibrin* from the *blood*; any similar body liquid.

servomechanism A mechanism that converts a small low-powered mechanical motion into a mechanical motion requiring considerably greater power. The output power is always proportional to the input power, and the system may include a negative *feedback* device (usually *electronic*).

sesame oil A yellow oil obtained from sesame seeds, m.p. $-6°C$., r.d. 0.919, used in the manufacture of *margarine* and cosmetics.

sesqui- Prefix denoting that the elements in a chemical *compound* are present in the ratio 2:3.

sets A set is a group of objects or elements that have at least one common characteristic. If these objects or elements are represented by m_1, m_2, m_3, etc., then $\{m_1, m_2, m_3 ...\} = M$ is the way of writing that m_1, m_2, m_3, etc. belong to the set M.

$m_1 \in M$ means that m_1 is a member of set M. If some of the objects or elements m_1, m_2, m_3, etc. can be classified into a subset A, and some others into subset B, then $A \subset M$ (read as subset A is contained in set M) and $B \subset M$ (read as subset B is contained in set M). If, for example, m_2 belongs to both subsets A and B, then $m_2 \in A \cap B$, means that m_2 is a member of subsets A and B, or m_2 belongs to the intersection of subsets A and B. The mathematical theory dealing with relationships between sets is known as 'set theory' and the diagrams used to illustrate them are known as Venn diagrams (after John Venn; 1834–1923).

sex chromosome A *chromosome* that determines the sex of an individual. In most animals there are two kinds: the X-chromosome, which is similar in size to other chromosomes, and the Y-chromosome, which is smaller. In most animals, including man, two X-chromosomes give a female, whereas an X- and a Y-chromosome yield a male. The sex chromosomes carry the *genes* controlling the development of the sex organs and the secondary sexual characteristics of the individual, through the action of the *sex hormones*. The sex chromosomes also carry genes unrelated to sexual characteristics. These are known as 'sex-linked' genes; there is a tendency for some characteristics (e.g. *colour blindness*, haemophilia, etc.) controlled by sex-linked genes to occur more frequently in men than in women. This is because they occur on the X-chromosome; thus, in women, if one X-chromosome carries an abnormal *allele* its effect is likely to be masked by the normal allele on the other X-chromosome. As men have only one X-chromosome, an abnormal allele on this chromosome will not be masked.

sex hormone A steroid hormone that controls the development of the reproductive organs and secondary sexual characteristics (the features of a mature animal that are significant in sexual behaviour although not directly involved in copulation). The most important are the *androgens* and the *oestrogens*.

sexivalent Hexivalent. Having a *valence* of six.

sex-linked gene See *sex chromosome*.

sextant An instrument for determining the *angle* between two objects (e.g. the horizon and a *star*). It is commonly employed for determining the radius of a *position circle*.

Seyfert galaxies Spiral *galaxies* with exceptionally bright central regions and strong emissions of *infrared*, *ultraviolet*, and *X-rays* in addition to *light*. It is thought the central activity could be a result of a *black hole*. Named after C. K. Seyfert (1911–60).

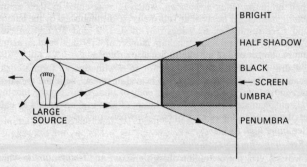

BRIGHT

HALF SHADOW

BLACK

← SCREEN

UMBRA

LARGE SOURCE

PENUMBRA

Figure 38.

shadow A dark patch formed by a body that obstructs *rays* of *light*. A shadow cast by an object in front of a *point source* of light is a sharply defined area; a source of light of appreciable size produces two distinct regions, the *umbra* or full shadow, and the *penumbra* of half-shadow. See Fig. 38.

shadow bands A series of wavy *shadow* bands that fall across the *Earth* just before and after *totality* in a solar *eclipse*. It is due to differences in *density* of the *atmosphere*.

shale A consolidated laminated *clay* rock. *Slate* consists of metamorphosed (see *metamorphic rock*) shale. See also *oil shale*.

shear modulus See *shear stress*.

shear stress A combination of four forces, acting (in the simplest case) over the surfaces of a block, producing two equal and opposite *couples*. The shear strain is the resulting change in shape, usually expressed as the angle, in circular measure, by which the block is deformed. The shear modulus (or rigidity modulus) is the ratio of the shear stress to the shear strain.

shellac A yellowish natural *resin* secreted by the lac insect (*Laccifer lacca*), which is parasitic on certain trees native to India and Thailand. It consists of several *polyhydroxy organic acids* (predominantly aleuritic acid, $C_{16}H_{32}O_5$, and shellolic acid, $C_{15}H_{20}O_6$) together with 3%–5% of *wax*. Shellac produces smooth, durable *films* from alcoholic *solutions* and *alkaline* dispersions, which adhere to a variety of surfaces: it is used in varnishes, polishes, leather dressings, and sealing wax. Owing to its electrical insulation properties it is used in Micanite* (see *mica*).

shells, electron According to the interpretation of *quantum mechanics*, the *electrons* contained within an *atom* circle round the *nucleus* in *orbits* at various distances from the nucleus. These orbital electrons may be visualized as forming a series of concentric shells: electrons in the same shell have the same principal *quantum number*, *n*. The shells are designated by the letters K-P (equivalent to values of *n* from 1–6) in order of increasing distance from the nucleus. The number of electrons in each shell is restricted (see *Pauli's exclusion principle*), but each shell is capable of containing $2n^2$ electrons. Table 7 in the Appendix gives the electronic configuration of the commoner *elements*. Within each shell, electrons are further classified into sub-shells (or *energy* sub-levels) according to their orbital *angular momentum*, which is represented by their azimuthal quantum number, *l*. The separate sub-shells are distinguished by the letters s, p, d, and f (corresponding to values of *l* of 0, 1, 2, and 3). E.g. an electron designated 4f, has a principal quantum number of 4 (N shell) and an azimuthal quantum number of 3 (f sub-shell). See also *orbital*.

sherardizing A method of plating iron or *steel* with zinc, to form a *corrosion* resistant coating. The iron or steel is heated in contact with zinc powder to a *temperature* slightly below the *melting point* of zinc. At this temperature the two metals amalgamate forming internal layers of zinc-iron *alloys* and an external layer of pure zinc. Named after Sherard Cowper-Coles (died 1935).

SHF See *super high frequency*.

SHM See *simple harmonic motion*.

shock wave A very narrow region of high *pressure* and *temperature* in which air flow changes from *subsonic* to *supersonic*. See also *sonic boom*.

shooting star See *meteor*.

short circuit If a *potential difference* exists between two points A and B (e.g. the terminals of an electrical supply), a system of *conductors* connecting A and B constitutes a *circuit*. If now A and B are placed in contact, or joined by a conductor of much lower *resistance* than the rest of the circuit, most of the current will flow direct between A and B, which are then said to be short-circuited or 'shorted'.

short sight See *myopia*.

shower The production by one high-*energy* particle, originating from *cosmic rays* or *accelerators*, of several fast particles. 'Cascade' showers (or soft showers) consist of *electrons*, *positrons*, or *photons* formed by successive *pair productions* or *radiative collisions*. 'Penetrating' showers contain *nucleons* and *muons* capable of penetrating up to about 20 cm of lead. 'Auger' showers (or extensive showers) extend over areas of up to 1000 square metres.

shunt, electrical A device for reducing the amount of *electric current* flowing through a piece of apparatus, such as a *galvanometer*. It consists of a *conductor* connected in *parallel* with the apparatus.

sial Rocks that form the Earth's crust below the continents. They are rich in *si*lica and *al*uminium, hence the name. Compare *sima*.

sideband The band of *frequencies* lying on either side of a *modulated carrier wave*; the width of each sideband is equal to the highest modulating frequency.

side chain (chem.) An *aliphatic* group attached to a *straight chain* or to a *benzene ring* or other cyclic group in the *molecule* of an *organic compound*. E.g. in *methylbenzene* (toluene), $C_6H_5.CH_3$, the *methyl group*, CH_3, is a side chain attached to a benzene ring.

sidereal day The period of a complete rotation of the *Earth* upon its *axis*, with respect to the *fixed stars*. It is 4.09 minutes shorter than a mean *solar day*.

sidereal period of a planet The 'year' of a *planet*. The actual period of its revolution round the *Sun*. See Appendix, Table 4.

sidereal year See *year*.

siderite 1. Natural iron(II) carbonate, $FeCO_3$. An important *ore* of iron. **2.** A *meteorite* consisting of *metals* (principally iron) and metallic *compounds*.

siemens The *SI unit* of electric *conductance* defined as the conductance of a circuit or element that has a resistance of 1 *ohm*. The unit was formerly called the reciprocal ohm or *mho*. Symbol S. Named after Sir William Siemens (1823 – 83).

Siemens-Martin process See *open-hearth process*.

sievert The *SI unit* of *dose* equivalent. It is the dose equivalent when the absorbed dose of *ionizing radiation* multiplied by a stipulated dimensionless factor gives 1 joule per kilogram. The dimensionless factors are stipulated by the International Commission on Radiological Protection. The former unit of dose equivalent, the rem, is equal to 10^{-2} sievert. Symbol Sv.

sigma bond σ bond. See *orbital*.

sigma particle Σ-particle. An *elementary particle* classified as a *hyperon*. It exists in three charged states: positive, negative, and neutral. See Appendix, Table 6.

sigma pile An assembly consisting of a *neutron* source and a *moderator*, without any *fissile* material, which is used to study the properties of moderators.

sign, algebraical The plus or minus sign, + or –, indicating opposite senses or directions; thus +5 is numerically equal, but opposite in sign, to –5.

signal The means by which *information* is conveyed through an electronic system. It consists of a varying parameter, such as *voltage* or *current*. A 'signal generator' is the device or circuit that produces a continuous-wave signal, usually supplying a voltage with varying *amplitude*, *frequency* or *waveform*.

signal-to-noise ratio The ratio of the information-carrying *signal* strength to the *noise*, often measured in *decibels*, in an electronic *circuit* or the output from an electronic device.

significant figures The number of digits in a number, as a measure of its accuracy. For example, 7631 is accurate to four significant figures; 0.0063 is accurate only to two significant figures as the zeros only indicate the order of magnitude. 603, however, has three figures as this zero is expressing a magnitude.

silage A stored form of cattle-fodder produced by a limited *fermentation* of green fodder pressed down and stored in a pit. *Lactic acid* is formed during the process.

silane Silicane, silicon hydride. SiH_4. A colourless *gas*, m.p. $-185°C.$, b.p. $-112°C$. It is used as a dopant for *semiconductors*, and is the first member of the *silanes*.

silanes A class of silicon *hydrides* of the general formula Si_nH_{2n+2}, forming a *homologous series* analogous to the *alkanes*. The first member of the series is *silane*, SiH_4, the second is disilane, Si_2H_6, etc. They are less stable than the alkanes. There are no analogues to the *alkenes* or *alkynes*.

silica Silicon(IV) oxide, silicon dioxide. SiO_2. A hard *insoluble* white or colourless *solid* with a high *melting point* (1610 – 1713°C.). It is very abundant in nature in the forms of *quartz*, *rock-crystal*, *flint*, and as *silicates* in rocks. It is used in the form of a white powder in the manufacture of *glass*, *ceramics*, and *abrasives*. See also *silica gel*.

silica gel A form of *silica*, SiO_2, with a highly porous structure capable of adsorbing (see *adsorption*) 40% of its *weight* of *water* from a *saturated vapour*. It is used in *gas* drying and as a *catalyst* support.

silicane See *silane*.

silicates A vast range of compounds, *salts* of or derived from *silicic acids*, that may be conveniently regarded as compounds of *silica* with various metal *oxides*. Most of the *Earth's* crust is composed of the silicates of calcium, aluminium, magnesium, and other metals. Various glasses, ceramics, and cements consist largely of silicates. See also *aluminosilicates*.

silicic acids Various *hydrated* forms of *silica*, obtained in colloidal or *gel* form by the action of acids on soluble *silicates* in solution. E.g. metasilicic acid, H_2SiO_3 ($SiO_2.H_2O$), and orthosilicic acid, H_3SiO_4 ($SiO_2.2H_2O$), giving rise to the meta- and orthosilicates.

silicol process The manufacture of hydrogen by the action of *sodium hydroxide* (caustic soda, NaOH) *solution* on silicon.

silicon Si. Element. R.a.m. 28.0855. At. No. 14. A non-metal similar to carbon in its chemical properties. It occurs in two *allotropic forms*: a brown *amorphous* powder and dark grey *crystals*; r.d. 2.33, m.p. 1410°C., b.p. 3200°C. It is the second most abundant element in the *Earth's crust*, occurring in sand and rocks as *silica* and as *silicates*. The element is obtained by reducing silica with carbon in an electric furnace. The pure element is used in *semiconductors*; it is also used in *alloys* and in the form of *silicates* in *glass. Silicones* are also widely used.

silicon carbide Carborundum. SiC. A hard black *insoluble* substance, m.p. 2700°C., used as an *abrasive refractory*, and in *resistors* required to withstand high *temperatures*. It is made by heating *silica* with carbon in an *electric furnace*.

silicon chip See *chip*.

silicon-controlled rectifier SCR. A solid-state electronic device, consisting of three *p-n* junctions, in which the forward anode-cathode current is controlled by a third *electrode*, called the *gate*. It functions as two junction *transistors* with two common electrodes. The common form is the *thyristor*, which has taken over many of the uses of the *thyratron*.

silicone rubbers Rubber-like *polymers* of various *organosilicon compounds*, such as *siloxanes* (in particular, dimethylsiloxane, $(CH_3)_2SiO$), having valuable characteristics, such as high stability over wide ranges of temperature, outstanding water repellence, high resistance to chemical action, good electrical properties, etc.

silicones A term originally applied to *compounds* of the general formula R_2SiO, where R stands for *hydrocarbon radicals*. They are now defined as polymeric (see *polymerization*) *organic siloxanes* of the general type $(R_2SiO)_n$. They are used as lubricants, for water-repellent finishes, high-*temperature* resisting *resins*, and lacquers.

silicon hydride See *silane*.

silicon tetrachloride $SiCl_4$. A colourless fuming *liquid*, b.p. 57.57°C., used in making silicon *compounds* and smokescreens.

silk A thread-like substance produced by the silkworm. It consists mainly of the *proteins* sericin and fibroin.

siloxanes A group of *compounds* with the general *formula* R_2SiO, where R stands for an *organic* group or hydrogen. See also *silicones*.

silver Ag. Element. R.a.m. 107.87. At. No. 47. A white, rather soft, extremely malleable metal; r.d. 10.5, m.p. 961.93°C., b.p. 2170°C. It occurs as the metal, and as *argentite* or silver glance, Ag_2S; *horn silver*, AgCl; and other *compounds*. It is extracted by alloying with lead and then separating the lead by *cupellation* and other methods. It is used in coinage and jewellery; *compounds* are used in *photography*.

silver bromide AgBr. A pale yellow *insoluble salt*, m.p. 432°C., used in *photography*.

silver chloride AgCl. A white *insoluble salt*, m.p. 455°C., that occurs naturally as *horn silver* (cerargyrite) and is used in *photography* and *antiseptics*.

silver glance See *argentite*.

silver iodide AgI. A yellow *insoluble salt*, m.p. 558°C., that occurs naturally as iodyrite and is used in *photography*, medicine, and in seeding clouds to produce artificial rain.

silver nitrate Lunar caustic. $AgNO_3$. A white *soluble* crystalline *salt*, m.p. 212°C. It is used in marking-inks, medicine, and chemical analysis.

silver plating The process of depositing a layer of silver on the surface of metal articles, usually by electrolytic methods. See *electroplating*.

sima Rocks that form the *Earth's* crust below the oceans. They are rich in *si*lica and *ma*gnesium, hence the name. Sima is denser and more plastic than *sial*.

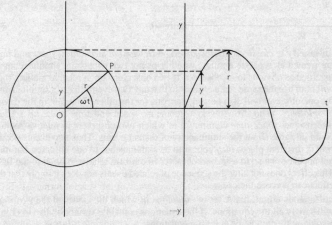

Figure 39.

simple harmonic motion SHM. A point is said to move in simple harmonic motion when it oscillates along a line about a central point, O, so that its *acceleration* towards O is always proportional to its distance from O. Thus, if a point P moves in a *circle*, centre O and radius *r*, with a constant *angular velocity* ω, the projection of P on any diameter will move in SHM. If the distance from

403

O of the projection of P on a vertical diameter is y, at time t, then a *graph* of y against t will give a curve of *amplitude r* and *equation* $y = r \sin\omega t$. (See Fig. 39.) An oscillator executing SHM will propagate a *sine wave* through a medium, having the characteristics of the curve in the illustration. The equation of this curve may be rewritten in the more general form:

$$y = r \sin 2\pi \, (t/T - x/\lambda)$$

where T is the *period* of the wave, λ its *wavelength* and x the distance it has travelled from O in time t.

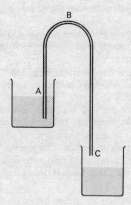

Figure 40.

simultaneity In *classical physics*, based on *Newtonian mechanics*, *space* and *time* are treated as separate entities; whether or not two events are simultaneous is regarded as obvious to an observer. If two observers are in relative motion, they will both use the same scale of time so that any two events that are simultaneous for one observer will also be simultaneous for the other. However, in the theory of *relativity*, using Minkowski's geometry, space and time make up a four-dimensional *space-time* continuum in which two observers in relative motion could disagree about the simultaneity of distant events. Thus two events occurring in different places may appear to be simultaneous to one observer, but may not appear to be so to a second observer in uniform relative motion to the first. This effect does not alter the sequence of related events nor does it imply that the effect can precede the cause.

simultaneous equations A set of *equations* in which the values of the *variables* will satisfy all the equations; if the equations contain n variables, then to obtain a *solution* there must be at least n equations.

sine See *trigonometrical ratios*.

sine rule In any triangle:

$$a/\sin A = b/\sin B = c/\sin C,$$

where a, b, and c are the sides opposite the angles A, B, and C.

sine wave Sinusoidal wave. A wave that has an equation in which one variable is proportional to the *sine* of the other. See *simple harmonic motion*.

sintering Compressing *metal* particles into a coherent *solid* body. The process is carried out under *heat*, but at a *temperature* below the *melting point* of the metal. Certain non-metals, such as *ceramics* and *glass*, may also be sintered.

sinusoidal Having the characteristics of a *sine wave*. See *simple harmonic motion*.

siphon A bent tube used for transferring *liquid* from one level to a lower level via a third level higher than either. If the shorter arm of an inverted U-tube filled with liquid is immersed below the liquid surface in A (see Fig. 40), liquid will flow from A to C through the tube. The siphon depends for its action on the fact that the *pressure* at A tending to force the liquid up the tube is $P - P_{ab}$ and the pressure acting upwards on the liquid at C is $P - P_{ac}$, where P = external atmospheric pressure, and P_{ab} and P_{ac} are the pressures due to the *weights* of the liquid columns AB and AC respectively. Hence flow from A to B will occur provided BC is greater than AB.

SI units Système International d'Unités. An internationally agreed *coherent* system of *units*, derived from the *m.k.s. system*, now in use for all scientific purposes and thereby replacing the *c.g.s. system* and the *f.p.s. system*. The seven base units are: the *metre* (symbol m), *kilogram* (kg), *second* (s), *ampere* (A), *kelvin* (K), *mole* (mol), and *candela* (cd). The *radian* (rad) and *steradian* (sr) were originally treated as supplementary units but are now regarded as dimensionless derived units. Derived units having special names and symbols are the *hertz* (Hz), *newton* (N), *joule* (J), *watt* (W), *coulomb* (C), *volt* (V), *farad* (F), *ohm* (Ω), *weber* (Wb), *tesla* (T), *henry* (H), *lumen* (lm), *lux* (lx), *pascal* (Pa), *siemens* (S), *becquerel* (Bq), *sievert* (Sv), and *gray* (Gy). Decimal multiples are given in the table (where possible a prefix representing 10 raised to a *power* that is a multiple of three should be used). See Appendix, Table 1.

Factor	Name of Prefix	Symbol	Factor	Name of Prefix	Symbol
10	deca-	da	10^{-1}	deci-	d
10^2	hecto-	h	10^{-2}	centi-	c
10^3	kilo-	k	10^{-3}	milli-	m
10^6	mega-	M	10^{-6}	micro-	μ
10^9	giga-	G	10^{-9}	nano-	n
10^{12}	tera-	T	10^{-12}	pico-	p
10^{15}	peta-	P	10^{-15}	femto-	f
10^{18}	exa-	E	10^{-18}	atto-	a
10^{21}	zetta-	Z	10^{-21}	zepto-	z
10^{24}	yotta-	Y	10^{-24}	yocto-	y

skatole C_9H_9N. A white *soluble* crystalline substance, m.p. $265°C.$, with a strong odour, used in the manufacture of perfumes.

skip distance There is a minimum angle of *incidence* at the *ionosphere* below which a *sky wave* of a given *frequency* is not reflected, but is transmitted through to outer *space*. Consequently there is a region surrounding a *radio* transmitter within which no sky wave can be received. The minimum distance at which reception of the sky wave is possible is called the skip distance.

sky wave Ionospheric wave. A *radio* wave may travel from transmitting *aerial* to receiving aerial by one of two paths: either directly parallel to the ground (see *ground wave*), or by reflection from the *ionosphere*. In the latter case it is called a sky wave or ionospheric wave.

slag Non-metallic material obtained during the *smelting* of metallic *ores*; it is generally formed as a molten mass floating on the molten *metal*.

slaked lime See *calcium hydroxide*.

slaking The addition of *water*.

slate A natural form of aluminium silicate formed from *clay* hardened by *pressure*.

slide rule A mathematical instrument formerly used for rapid calculations; it consists of a grooved ruler with a scale, with another similarly marked ruler sliding inside the groove. Multiplication and division are carried out by adding or subtracting lengths on the two rulers, the divisions on which are in a logarithmic scale. Slide rules have been replaced by pocket calculators.

slip ring A copper ring on the *armature* of an electrical machine that is connected to one or more windings in that machine; it enables contact to be made to the windings from external circuits, by means of carbon brushes that make contact with the surface of the ring.

slow neutron A *neutron* whose *kinetic energy* does not exceed about 10 *electronvolts*.

SLR Single lens reflex. See *reflex camera*.

slug A *unit* of *mass* in the *f.p.s. system* defined as the mass that will acquire an *acceleration* of 1 ft/sec^2 when acted upon by a *force* of 1 lb. 1 slug is equal to 32.174 lbs.

slurry A thin paste consisting of a suspension of a solid in a liquid.

smelting The extraction of a *metal* from its *ores* by a process involving *heat*. Generally the process is one of chemical *reduction* of the *oxide* of the metal with carbon in a suitable furnace.

smectic crystals *Liquid crystals* in which the molecules are arranged in layers with their axes parallel and perpendicular to the plane of the layers. See also *cholesteric crystals*; *nematic crystals*.

smog A dark, thick, dust- and soot-laden, sulphurous *fog* that, under certain meteorological conditions, pollutes the atmosphere of some industrial cities and the lungs of their inhabitants. See also *photochemical smog*.

smoke A suspension of fine particles of a *solid* in a *gas*; smoke from *coal* consists mainly of fine particles of carbon.

Snell's law See *refraction, laws of*.

SNG Synthetic natural gas. A mixture of gases, including *hydrocarbons*, hydrogen, and *carbon monoxide*, produced from *coal* and *petroleum* for use as a *fuel*.

soap A mixture of the sodium *salts* of *stearic acid*, $C_{17}H_{35}COOH$, *palmitic acid*, $C_{15}H_{31}COOH$, and *oleic acid*, $C_{17}H_{33}COOH$; or of the potassium salts of these acids ('soft soap'). Soaps are made by the action of *sodium* or *potassium hydroxide* on *fats*, the process of *hydrolysis* or *saponification* giving the soap, with *glycerol* as a by-product. The term soap is also applied to *fatty acid* salts of *metals* other than sodium or potassium, although such compounds are unlike the ordinary soaps.

sociobiology The study of social behaviour in animals (including man). It is believed that many aspects of social behaviour, such as aggression, and separate roles for the sexes, have evolved in much the same way as physical characteristics, i.e. by *natural selection*.

soda Any of various sodium *compounds*; *washing soda*, *sodium carbonate*, $Na_2CO_3.10H_2O$; *baking soda*, *sodium hydrogencarbonate*, $NaHCO_3$; *caustic soda*, *sodium hydroxide*, $NaOH$.

soda ash The common name for *anhydrous sodium carbonate*, Na_2CO_3.

soda-lime A *solid mixture* of *sodium hydroxide*, $NaOH$, and *calcium hydroxide*, $Ca(OH)_2$, made by slaking quicklime (see *calcium oxide*) with a *solution* of *sodium hydroxide* and drying by *heat*.

soda nitre *Caliche*. Impure natural *sodium nitrate*.

soda water *Water* containing *carbon dioxide*, CO_2, under *pressure*; releasing the pressure lowers the *solubility* of the gas, and thus causes *effervescence*.

sodium Na. (Natrium.) Element. R.a.m. 22.98977. At. No. 11. A soft silvery-white *metal*, r.d. 0.966, m.p. 97.9°C., b.p. 900°C. It is very reactive, tarnishing rapidly in air. It reacts violently with water, forming *sodium hydroxide* and hydrogen gas. *Compounds* are very abundant and widely distributed; the commonest is *sodium chloride*, $NaCl$ (common salt). The metal is used in the preparation of *organic compounds* and as a *coolant* in some types of *nuclear reactor*.

sodium azide NaN_3. A colourless crystalline substance, used in the manufacture of *explosives*. It decomposes on heating.

sodium benzenecarboxylate Sodium benzoate. C_6H_5COONa. A white *soluble* powder, used as a food preservative.

sodium bicarbonate $NaHCO_3$. See *sodium hydrogencarbonate*.

sodium carbonate Washing soda. $Na_2CO_3.10H_2O$. A white crystalline *soluble salt*, m.p. 850°C. It is used in the household, in the manufacture of *glass*, *soap*, *paper*, and for *bleaching*.

sodium chlorates 1. Sodium chlorate(V). $NaClO_3$. A colourless soluble crystalline substance, m.p. 248°C., used in the manufacture of *explosives*, as a weed-killer, *oxidizing agent*, *mordant*, and *antiseptic*. **2.** Sodium chlorate(I). Sodium hypochlorite. $NaOCl$. A white unstable crystalline solid, m.p. 18°C., usually kept in aqueous solution. It is used in bleaching paper and textiles and as an oxidizing agent, antiseptic, and fungicide.

sodium chloride Common salt, salt. $NaCl$. A white crystalline *soluble salt*, m.p. 801°C. It occurs extensively in *sea water* and as halite. It has many uses in addition to its common use for seasoning and preserving food.

sodium cyanide $NaCN$. A white *soluble deliquescent* highly poisonous substance, m.p. 563.7°C., used in *electroplating*, case-hardening, and fumigation.

sodium cyclamate $C_6H_{11}NHSO_3Na$. A white crystalline *soluble* powder, formerly used as a sweetening agent in soft drinks and for diabetics, but now banned from such use owing to possible side-effects.

sodium dichromate (bichromate) $Na_2Cr_2O_7.2H_2O$. An orange *soluble* crystalline substance, m.p. 356.7°C. (after losing its *water of crystallization* at 100°C.), used as a *mordant*, *corrosion* inhibitor, *oxidizing agent*, and in *electroplating*.

sodium ethoxide Sodium ethylate. C_2H_5ONa. A white *hygroscopic* substance, used in organic synthesis.

sodium fluoride NaF. A colourless crystalline substance, m.p. 988°C., used in the *fluoridation* of *water* and as an *insecticide*.

sodium hydrogencarbonate Sodium bicarbonate. $NaHCO_3$. A white *soluble* powder, used in *baking powder*, *fire extinguishers*, and in medicine as an *antacid*.

sodium hydrogenglutamate Monosodium glutamate. MSG. $HOOC(CH_2)_2$-$CH(NH_2)COONa$. A white *soluble* crystalline substance, used to intensify the flavour of foods.

sodium hydroxide Caustic soda. NaOH. A white *deliquescent solid*, m.p. 318.4°C., that dissolves in *water* to give an *alkaline solution*. It is used in the manufacture of *soap*, *rayon*, and other chemicals.

sodium hypochlorite See *sodium chlorates*.

sodium nitrate $NaNO_3$. A white *soluble* crystalline *salt*, m.p. 306.8°C., that occurs naturally as *Chile saltpetre*. It is used as a *fertilizer* and in the manufacture of *nitric acid* and *explosives*.

sodium perborate $NaBO_3.4H_2O$. A white *soluble* crystalline substance, m.p. 63°C., used in *bleaching* and as a disinfectant.

sodium peroxide Na_2O_2. A yellow powder, formed when sodium *metal* burns in air. It reacts with *water* to give *sodium hydroxide* and oxygen *gas*. It is used in *bleaching* and as an *oxidizing agent*.

sodium phosphate 1. Sodium dihydrogenphosphate(V) (sodium dihydrogen orthophosphate), NaH_2PO_4, a white *soluble* crystalline substance used in *electroplating* and dyeing. **2.** Disodium hydrogenphosphate(V) (sodium hydrogen orthophosphate), Na_2HPO_4, a white soluble crystalline substance used in *dyes*, *fertilizers*, *detergents*, *baking powder*, and medicine. **3.** Trisodium phosphate(V) (sodium orthophosphate), $Na_3PO_4.12H_2O$, a colourless soluble crystalline substance used in detergents, and in the manufacture of *paper* and water softeners.

sodium pump The biological process by means of which sodium *ions* move across a *cell membrane*. It is energized by *adenosine triphosphate*.

sodium silicate Na_2SiO_3. A white *soluble* crystalline *salt*, used in the household as 'water-glass', in fireproofing textiles, and in the manufacture of *paper* and *cement*.

sodium sulphate Glauber's salt. Saltcake. $Na_2SO_4.10H_2O$. A white *soluble* crystalline *salt*. It occurs naturally as *threnardite*. It is used in the manufacture of *soap*, *detergents*, and *dyes*.

sodium sulphide Na_2S. An orange *soluble deliquescent* substance, m.p. 1180°C., used in the manufacture of *soaps* and *dyes*.

sodium sulphite Na_2SO_3. A white crystalline *soluble* powder, used as a food preservative, in *bleaching*, and in *photography*.

sodium tetraborate See *disodium tetraborate*.

sodium thiocyanate NaSCN. A colourless *deliquescent* crystalline substance, m.p. 287°C., used in medicine.

sodium thiosulphate Sodium hyposulphite, *hypo*. $Na_2S_2O_3.5H_2O$. A white crystalline very *soluble salt*, used in *photography* for fixing.

sodium-vapour lamp A luminous *discharge* obtained by passing an *electric current* between two *electrodes* in a tube containing sodium *vapour* at low *pressure*. It is used in street lighting as the characteristic yellow light is less absorbed by fog and mist than *white light*. The pink light emitted during starting up is caused by the small percentage of neon added to warm the lamp up sufficiently for the sodium to vaporize.

soft iron Iron containing little carbon, as distinct from *steel*; iron that does not retain *magnetism* permanently, but loses most of it when the magnetizing field is removed.

soft radiation *Ionizing radiation* (usually X-rays) of relatively long *wavelength* and low penetrating power, as opposed to 'hard' radiation, which is of shorter wavelength and high penetrating power.

soft soap See *soap*.

software The *programs* used in a *computer*; the 'hardware' is the actual equipment of the computer itself. Software is of two basic kinds: the 'systems software' is supplied by the manufacturer and contains the programs required to make the system function (including the operating system); 'applications software', which may or may not be supplied by the manufacturer, makes the system perform the specific tasks required of it.

soft water *Water* that forms an immediate lather with *soap*. See *hard water*.

soil The layer of *inorganic* weathered unconsolidated *rock* material, *organic* matter, and *water* that covers most of the *Earth*. It is the material in which plants grow. Soils vary enormously in their chemical composition. The inorganic portion of a soil is composed of *silicates* of various *metals*, mainly of aluminium, but also of iron, calcium, magnesium, etc., free *silica* (sand) and other inorganic matter, depending on the source. Organic matter in the soil is mainly derived from decomposed plants; much of it is in the form of a black sticky substance known as *humus*.

sol See *colloidal solution*.

solar cell (battery) An electric *cell* that converts *energy* from the *Sun* into *electrical energy*. It usually consists of a *semiconductor* device sensitive to the *photovoltaic effect*; e.g. a *p-n semiconductor junction* in a *crystal* of silicon. They are used in artificial *satellites* and *space probes* to power *electronic* equipment. Individual solar cells are unable to deliver much more than ½W at about ½V. They therefore need to be used in large panels. See also *solar energy*.

solar constant The *energy* that would (in the absence of the *atmosphere*) be received per second by an area of 1 sq metre placed at the mean distance of the *Earth* from the *Sun* and at right angles to the incident *radiation*; its value is approximately 1400 J s^{-1} m^{-2} (2 cals min^{-1} cm^{-2}).

solar day The variable interval between two successive returns of the *Sun* to the *meridian*. The mean solar day is the average value of this. See *time measurement*.

solar energy Energy from the Sun. Life on Earth relies almost entirely on solar energy. It provides the energy needed for plant growth by *photosynthesis* and animals obtain their energy from plants and other animals. *Fossil fuels* also depend ultimately on photosynthesis. *Hydroelectric power*, *wind energy*, and *wave energy* all depend on the Sun's energy through its influence on the weather.

As a *renewable energy source*, the Sun is a prolific potential reservoir of energy: the amount of energy falling on the Earth from the Sun is given by the *solar constant*. If all this energy could be harnessed, every inhabitant of the Earth could burn 12 000 2kW heaters continuously. But, in fact, very little direct use has been made of solar energy. Broadly, there are two ways of using solar energy directly. The thermal methods involve absorbing the Sun's radiation on a metal plate and using the absorbed heat to raise the temperature of a fluid. This is the principle of the domestic and industrial *solar heater*. Non-thermal methods use devices, such as *solar cells*, to produce electricity from sunlight. This is the method used in spacecraft, satellites, etc. In order for solar cells to be useful as a source of energy on Earth (domestically or industrially) their price would have to drop substantially to make them competitive with other energy sources.

solar flares Short high *temperature* outbursts seen as bright areas in the *chromosphere* of the Sun. Jets of particles (known as the *solar wind*) and strong *radio frequency electromagnetic radiations* (see *radio astronomy*) are emitted during solar flares. Solar flares are associated with *sunspots* and usually cause magnetic and *radio* disturbances on *Earth*.

solar heater A domestic or industrial heater that makes direct use of *solar energy*. The simplest form consists of a collector through which a *fluid* is pumped. The circuit also contains some form of heat storage tank and an alternative energy source to provide energy when the sun is not shining. The collector usually consists of a black surface through which water is piped, the black surface being enclosed behind glass sheets to make use of the *greenhouse effect*. Solar heating systems can be adapted to provide summer air-conditioning with winter heating. More sophisticated solar heaters use reflectors to focus the Sun's rays.

solar neutrinos *Neutrinos* produced in the *thermonuclear reactions* within the *Sun*. An experiment to detect the particles arriving at the Earth has been underway since the late 1960s but has found only a third of the number predicted by theory. The mystery of the 'missing solar neutrinos' has produced many hypotheses but remains unexplained.

solar parallax The *angle* subtended by the mean equatorial radius of the *Earth* at a distance of one *astronomical unit*.

solar prominences Large eruptions of luminous *gas* that rise several thousands of kilometres above the *Sun's chromosphere*.

solar system The system of nine planets – *Mercury*, *Venus*, the *Earth*, *Mars*, *Jupiter*, *Saturn*, *Uranus*, *Neptune* and *Pluto* – together with their *satellites* and the belt of *asteroids* revolving in elliptical *orbits* round the *Sun*. The orbits are nearly circular, and lie very nearly in the same *plane*. See Appendix, Table 4.

solar wind Streams of electrically charged particles (*protons* and *electrons*) emitted by the *Sun*, predominantly during *solar flares* and sunspot activity. Some of these particles become trapped in the Earth's magnetic field (see *magnetism, terrestrial*) forming the outer *Van Allen radiation belt*, but some penetrate to the *upper atmosphere* where they congregate in narrow zones in the region of the Earth's *magnetic poles* producing auroral displays (see *aurora borealis*).

solar year See *year*.

solder An *alloy* for joining *metals*. Soft solders are alloys of tin and lead in varying proportions, they melt in the range 200–300°C.; brazing solders are usually composed of copper and zinc, with melting points over 800°C.

solenoid A coil of wire wound uniformly on a cylindrical former, having a length that is large compared with its radius. When a current I is passed through the solenoid, a uniform *magnetic field* H is produced inside the coil parallel to its *axis*. If I is in *amperes* and n is the number of turns per *metre*, $H = nI$ amperes per metre.

solid (math.) A three-dimensional figure, having length, breadth, and thickness; a figure occupying *space* or having a measurable *volume*.

solid angle Ω. The ratio of the area of the surface of the portion of a *sphere* enclosed by the conical surface forming the angle, to the square of the radius of the sphere. See *steradian*.

solidifying point The constant *temperature* at which a *liquid* solidifies under a given *pressure*, usually the standard *atmosphere*.

solid solution A crystalline *homogeneous mixture* of two or more substances. E.g. some *alloys* are solid solutions of the *metals* in each other, the process of *solution* having taken place in the molten state.

solid state The *physical state of matter* in which the constituent *molecules*, *atoms*, or *ions* have no *translatory motion* although they vibrate about the fixed positions that they occupy in a *crystal lattice*. A solid is said to possess *cohesion*, remaining the same shape unless changed by external *forces*. Certain solids are not crystalline, they are then said to be *amorphous*. A crystalline solid has a definite *melting point* at which it becomes a *liquid*; amorphous solids have no precise melting point, but when heated become increasingly pliable until they assume the properties usually associated with liquids, they may therefore be thought of as 'supercooled' liquids.

solid-state memory See *semiconductor memory*.

solid-state physics Condensed-matter physics. The branch of *physics* that deals with the nature and properties of *matter* in the *solid state*. The term is often used to refer especially to the study of the properties of *semiconductors* and 'solid-state devices', i.e. *electronic* devices consisting entirely of solids, without moving parts, *gases*, or heated *filaments*, e.g. semiconductors, *transistors*, *integrated circuits*, etc.

solstice The points at which the *Sun* reaches its greatest *declination* North or South. The points are situated upon the *ecliptic* half-way between the *equinoxes*; the times at which the Sun reaches these points are 21 June and 21 December, the summer and winter solstices, respectively.

solubility The extent to which a *solute* will dissolve in a *solvent* to give a *saturated solution*. It is usually expressed in kilograms per cubic metre or moles per kilogram of solvent at a specified *temperature*. See also *concentration*.

solubility product The product of the *concentrations* of the *ions* of a dissolved *electrolyte* when in *equilibrium* with undissolved substance. For sparingly *soluble* electrolytes, the solubility product is a constant for a given substance at a given *temperature*. When the solubility product for a given compound is exceeded in a *solution*, some of it is precipitated until the product of the ionic concentrations falls to the constant value.

soluble Capable of being dissolved (usually in *water*).

solute A substance that is dissolved in a *solvent* to form a *solution*.

solution (chem.) A *homogeneous* molecular *mixture* of two or more substances of dissimilar molecular structure; the word is usually applied to solutions of *solids* in *liquids*. Other types of solutions include *gases* in liquids, the *solubility* of gases decreasing with rise in *temperature*; gases in solids; liquids in liquids; and solids in solids (e.g. some *alloys*), see *solid solutions*.

solution (math.) A set of values that when substituted for the variables in an *equation* give a true statement.

solvation The combination of *solvent molecules* with *molecules* or *ions* of the *solute*. The *compound* so formed is called a 'solvate'. If the solvent is water, the process is called 'hydration'.

411

Solvay process Ammonia-soda process. An industrial preparation of *sodium carbonate* or *washing-soda*, $Na_2CO_3.10H_2O$, from common *salt*, NaCl, and *calcium carbonate*, $CaCO_3$. By the action of *ammonia*, NH_3, and *carbon dioxide* (obtained by heating $CaCO_3$) on a salt *solution*, the less soluble *sodium hydrogencarbonate*, $NaHCO_3$, is precipitated. The action of *heat* on this compound gives the required sodium carbonate, while the ammonia is recovered from solution by the action of the *lime* (*calcium oxide*). Named after Ernest Solvay (1838–1922).

solvent A substance (usually *liquid*) having the power of dissolving other substances in it; that component of a *solution* that has the same *physical state* as the solution itself. E.g. in a solution of *sugar* in *water*, water is the solvent, while sugar is the *solute*. 'Polar solvents', such as *water* have *dipole moments* and can therefore dissolve *ionic compounds* or *covalent compounds* that can ionize. 'Nonpolar solvents', such as *benzene*, do not have permanent dipole moments and do not therefore dissolve ionic compounds, but they do dissolve nonpolar covalent compounds. 'Amphiprotic solvents', such as water, ionize themselves and can therefore act as both *proton* donors and acceptors (e.g. $2H_2O \rightarrow H_3O^+ + OH^-$); 'aprotic solvents' neither accept nor donate protons.

solvent extraction See *extraction*.

somatic 1. Pertaining to the body. Somatic *cells* are the cells of which the body of an *organism* is constructed, as opposed to the reproductive or germ cells. See also *mutation*. **2.** The body as opposed to the mind: e.g. psychosomatic medicine is the study of the influence of psychological factors upon physiological illness.

sonar SOund NAvigation Ranging. An apparatus for locating submerged objects by transmitting a high-frequency *sound* wave and collecting the reflected wave. The time for the wave to travel to the object and return gives an indication of the depth.

sonde A small system carried by a *rocket*, *balloon*, or *satellite*, used to make meteorological or astronomical measurements. See also *radiosonde*.

sonic boom The loud noise created by the *shock wave* set up by an aircraft or missile travelling at *supersonic speeds*. A subsonic aircraft produces pressure waves ahead of itself, which travel at the speed of *sound*, and 'clear a path' for the oncoming aircraft. In supersonic flight the aircraft overtakes the pressure waves so that a shock wave *cone* is created with the nose of the aircraft at its *vertex*. In level flight the intersection of the shock wave cone with the ground produces a *hyperbola*, at all points along which the sonic boom is simultaneously experienced; subsequently the boom will be experienced at all points within the hyperbola's path over the ground.

sorbent Any agent used for *sorption*.

sorbite 1. The constituent of *steel* produced when *martensite* is *tempered* above 450°C., consisting of *ferrite* and finely divided *cementite*. **2.** Sorbitic pearlite. The constituent of steel produced by the decomposition of *austenite* when cooled at a slower rate than will yield *troostite* and a faster rate than will yield *pearlite*.

sorbitol $CH_2OH(CHOH)_4CH_2OH$. A white crystalline sweet *soluble polyhydric alcohol*, m.p. 110°C. (for the *dextrorotatory compound*), obtained from *dextrose*; it is used as a *sugar* substitute and in the manufacture of synthetic *resins*.

Sorel's cement See *magnesium chloride*.

sorption *Adsorption* (a surface process) or *absorption* (a volume process). The term is often used when the mechanism of a particular process is not known or is not specified.

sorption pump A *vacuum pump* that removes gas from an enclosed vessel by *adsorption* on a solid at low temperature. It can be used to obtain a vacuum down to 10^{-6} Pa for small gas volumes.

sound A physiological sensation received by the ear. It is caused by a vibrating source with a *frequency* in the range 20–20 000 *hertz* and is transmitted as a longitudinal pressure *wave motion* (see *longitudinal waves*) through an *elastic* material medium such as air. See also *ultrasonics; infrasound.*

sound, speed of *c*. The *speed* of propagation of *sound* waves (see *wave motion*). This speed is a function of the *temperature* and of the nature of the propagating medium. In *gases* it is independent of the *pressure*. In air at 0°C. it is 332 metres per second or approximately 760 miles per hour.

In general, the speed of sound through a medium of *density* ρ and *elastic modulus E* is given by:
$$c = (E/\rho)^{1/2}.$$
For a gas, $E = \gamma p$, where γ is the ratio of the principal *specific heat capacities* of the gas and *p* is its pressure.

sound intensity *I*. The rate of flow of sound energy through unit area perpendicular to the direction of the flow.
$$I = p^2/\rho c,$$
where *p* is the root-mean-square sound pressure, *c* the speed of sound in the medium, whose density is ρ. In SI units it is measured in watts per square metre.

source The *electrode* in a *field-effect transistor* from which electrons or *holes* enter the inter-electrode space.

Soxhlet extraction apparatus A device for extracting the *soluble* portion of any substance by a continuous circulation of the boiling *solvent* through it.

space That part of the boundless four-dimensional *continuum* in which *matter* is physically (rather than temporally) extended. (See *space-time*.) More colloquially, space (or 'outer' space) is that part of the *Universe* that lies beyond the Earth's *atmosphere*, in which the density of matter is very low.

spacecraft A vehicle capable of travelling in *space*.

space probe A *rocket*-propelled missile that has sufficient *speed* to escape from the Earth's *atmosphere*. Space probes are used for making measurements of conditions within the *solar system* that cannot be made by terrestrial observation. The measurements are made by miniaturized *electronic* equipment within the probe, the results of which are signalled back to Earth by *radio*. The radio and other equipment is often powered by *solar cells*. A Moon-probe, or Lunar-probe, is one intended to study the *Moon* and its environment.

space-reflection symmetry See *parity*.

space-time The development of the theory of *relativity* has led to the disappearance of a clear-cut distinction between a three-dimensional *space* and an independent *time*; in this view, space and time are considered as being welded together in a four-dimensional space-time *continuum*. See also *Minkowski's geometry*. This has important repercussions on the concept of *simultaneity*.

spallation A *nuclear reaction* in which a high-energy incident particle, or *photon*, causes several particles or fragments to be emitted by the target *nucleus*. The *mass number* and *atomic number* of the target nucleus may thus be reduced by several units.

spark See *electric spark*.

spark chamber A device for detecting radiation or *elementary particles*. A 'spark counter' consists of a pair of electrodes with a high potential difference between them, placed close together in a gas, such as neon. If a particle passes between anode and cathode it causes a spark and, at the moment of discharge, a measurable drop in the anode voltage. The passage of particles through the device is recorded photographically or electronically. The 'spark chamber' usually has many pairs of electrodes and is often filled with neon at atmospheric pressure. The plates are connected alternately to the positive and negative terminals of a source of high *p.d.* (10 kV or more). A charged particle creates *ion* pairs, which create sparks between the plates, enabling photographs of the particle's tracks to be obtained.

spark coil See *induction coil*.

sparking-plug A device for providing an *electric spark* for exploding the *mixture* of air and *petrol vapour* in the cylinder of the *internal-combustion engine*.

sparking potential Sparking voltage. The difference in *potential* (i.e. the *voltage*) required for an *electric spark* to pass across a given gap. See *Paschen's law*.

spark photography Flash photography. *Photography* in which the source of *light* is an *electric spark*, usually of predetermined duration. Photographs are taken in the dark (or low light) and the *camera lens* is left open, thus the exposure time can be made very short and rapidly moving objects can be photographed.

sparteine $C_{15}H_{26}N_2$. A bitter colourless *alkaloid*, m.p. 30°C., used in medicine.

species A group (*taxon*) of *organisms* that are so similar that they can breed together. Different species cannot interbreed.

special theory of relativity See *relativity, theory of*.

specific When the adjective 'specific' is used before the name of an extensive physical quantity, it implies 'divided by *mass*'. E.g. *specific heat capacity* is *heat capacity* per unit mass. When the extensive quantity is denoted by a capital letter (e.g. *V* for *volume*), the specific quantity is usually denoted by the corresponding small letter ($v = V/m$ for *specific volume*). In some older physical quantities the word has had other meanings (e.g. *specific resistance*), but such uses are now deprecated. See *specific gravity*.

specific activity *a*. The *activity* per unit *mass* of a pure *radioisotope*; or the activity of a radioisotope in a material per unit mass of that material. It is expressed in *disintegrations* per second per kg.

specific charge The *electric charge* to *mass* ratio of an *elementary particle*.

specific gravity The former term for the ratio of the density of a substance to that of water. As the word *specific* now has a different usage, the term *relative density* is now used for this concept.

specific heat capacity Specific heat. *c*. *Heat capacity* divided by *mass*. The quantity of *heat* required to raise the *temperature* of unit mass of a substance by one degree. It is expressed in *joules* per kg per *kelvin* (*SI units*), *calories* per gram per °C. (*c.g.s. units*), or *British thermal units* per lb per °F (*f.p.s. units*). For *liquids* and solids *c* is measured at constant pressure.

The two most important specific heat capacities of a *gas* are (1) that measured at constant pressure, c_p, and (2) that measured at constant volume, c_v. c_p is greater than c_v because when a gas is heated at constant pressure it has to do work against the surroundings in expanding. The ratio c_p/c_v, usually denoted by γ(gamma), is about 1.66 for *monatomic* gases and about 1.4 for *diatomic* gases; for polyatomic gases it is still greater than 1. The value of gamma therefore gives some indication of the number of atoms in the molecules of a gas.

specific impulse A term used in connection with *rockets*. The ratio of the *thrust* produced to the rate of *fuel* consumption.

specific latent heat See *latent heat*.

specific resistance See *resistivity*.

specific surface The total surface area per unit mass of a given substance, e.g. a powder or a porous material. It is usually expressed in m^2 kg^{-1} or square centimetres per gram. It represents the actual surface area available for processes, such as *adsorption*, and may be very large for fine powders and highly porous substances.

specific volume The *volume*, at a specified *temperature* and *pressure*, occupied by unit mass (usually 1 kg) of a substance. The *reciprocal* of the *density*.

spectral 1. Relating to a *spectrum*. **2.** Relating to a specific *frequency* or *wavelength*.

spectral lines See *line spectrum*.

spectral series The emission *spectrum* of the *atoms* or *ions* of a gas may be analysed into one or more groups of *frequencies* (or *wavelengths*), the frequencies in each group forming a series. For example, the spectrum of the hydrogen atom possesses series given by the expression:
$$f = k(1/n_0^2 - 1/n^2),$$
where f is the frequency of the spectral *lines* and k is a constant. For the different series, n_0 takes the values 1, 2, 3, 4, etc. For any one value of n_0, n may have all *integral* values from $n_0 + 1$ upwards, the expression then giving the frequencies of all the lines in that particular series. See also *Balmer series* and *Rydberg constant*.

spectral types Spectral classes. The classification of *stars* based on the *spectrum* of the *light* they emit. The system now used is the Harvard classification, which comprises seven types of star:

O the hottest blue stars (helium lines dominant)
B hot blue stars (neutral helium dominant)
A blue white stars (hydrogen lines dominant)
F white stars (metallic lines)
G yellow stars (calcium lines dominant)
K orange stars (some molecular bands)
M coolest red stars (molecular bands dominant).

spectrograph 1. An instrument by which *spectra* may be photographed. **2.** A photograph taken by means of such an instrument. See *spectrographic analysis*.

spectrographic analysis An investigation of the chemical nature of a substance by the examination of its *spectrum*, using the fact that the position of emission and absorption *lines* and *bands* in the spectrum of a substance is characteristic of it. See *atomic absorption spectroscopy*; *atomic emission spectroscopy*.

spectroheliograph An instrument used to photograph the *Sun* with *light* of a particular *wavelength*.

spectrometer 1. A type of *spectroscope* so calibrated that it is suitable for the

415

precise measurements of *refractive indices*. **2.** An instrument for measuring the *energy* distribution of a particular type of *radiation*, e.g. a *scintillation spectrometer*.

spectrophotometer A *photometer* for comparing two *light* radiations *wavelength* by wavelength.

spectroscope An instrument for *spectrographic analysis* or the observation of *spectra*. The simplest type is the prism spectroscope. This consists of a *collimator*, which collects the light from the source and throws it onto the face of a glass *prism*. The *spectrum* so formed, after refraction by the prism, is viewed through a *telescope*. The angle between the collimator and the telescope can be varied. For visible light a *diffraction grating* can be used in place of the prism. See also *spectrograph*; *spectrometer*; *spectrophotometer*.

spectroscopic binary A *binary star* system that cannot be seen as two stars by a *telescope*, but which show a *Doppler effect* in their *line spectrum* as these stars revolve about each other. See *visual binary*.

spectroscopy The study of *matter* and *energy* by the use of a *spectroscope*. See *spectrographic analysis*.

spectrum 1. The result obtained when *electromagnetic radiations* are resolved into their constituent *wavelengths* or *frequencies*. In the visible region (i.e. *light* waves) a well-known example is provided by the coloured bands produced when white light is passed through a *prism* or *diffraction grating*. (See *spectrum colours*). Spectra formed from bodies emitting radiations, which may be *infrared*, light, *ultraviolet*, or *X-rays*, are termed *emission spectra*. When radiation is passed through a semitransparent medium, selective absorption of certain wavelengths or bands of wavelengths takes place; the spectrum of the transmitted radiation is called an *absorption spectrum*. A continuous spectrum is one in which all wavelengths, between certain limits, are present. A *line spectrum* is one in which only certain wavelengths or 'lines' appear. The emission and absorption spectra of a substance are fundamental characteristics of it and are often used as a means of identification. Such spectra arise as a result of transitions between different *stationary states* of the *atoms* or *molecules* of the substance, electromagnetic waves being emitted or absorbed simultaneously with the transition. The frequency f of the emitted or absorbed radiation is given by $E_1 - E_2 = hf$, where E_1 and E_2 are the *energies* of the first and second states respectively between which the transition takes place, and h is the *Planck constant*. When E_1 is greater than E_2, electromagnetic waves are emitted; in the converse case, they are absorbed. **2.** Any distribution of energies, velocities, etc., associated with a system (e.g. see *mass spectroscopy*).

spectrum colours The *colours* visible in the continuous *spectrum* of *white light*. These colours, their *wavelengths* and *frequencies* are given in the table. These are the colours seen in a *rainbow*.

specular reflection Perfect or regular reflection of *electromagnetic radiation*, e.g. *light*. It occurs whenever the reflecting surface is flat to approximately 1/8 of a *wavelength* of the radiation incident upon it.

speculum A reflecting *mirror*, especially a metallic mirror (see *speculum metal*) used in a reflecting *telescope*. Most telescopes now have glass mirrors.

speculum metal An *alloy* of 2/3 copper and 1/3 tin; formerly used for *mirrors* and *reflectors*. See *speculum*.

Colour of Light	Wavelength/ 10^{-7} metres	Frequency/ 10^{14} hertz
Red	6.470–7.000	4.634–4.284
Orange	5.850–6.470	5.125–4.634
Yellow	5.750–5.850	5.215–5.125
Green	4.912–5.750	6.104–5.215
Blue	4.240–4.912	7.115–6.104
Violet	4.000–4.240	7.495–7.115

speed The ratio of the distance covered to the time taken by a moving body. Speed is a *scalar* quantity, *velocity*, a vector quantity has a specified direction. See also *light, speed of; sound, speed of.*

spelter Commercial zinc, about 97% pure, containing lead and other impurities.

spermaceti A white, waxy *solid* consisting mainly of cetyl palmitate, $C_{15}H_{31}COOC_{16}H_{33}$. M.p. 40°–50°C. It is obtained from the head of the sperm whale and is used in the manufacture of *soaps* and cosmetics.

spermatocyte A *gametocyte* within an animal testis that undergoes *meiosis* to form spermatids, which change into *spermatozoa*.

spermatozoon Sperm. A *gamete* of male origin, four of which are derived by *meiosis* from a single *spermatocyte*. Some spermatozoon are produced by *mitosis*. The plural of 'spermatozoon' is 'spermatozoa'.

sphalerite See *zinc blende*.

sphere (math.) A *solid* figure generated by the revolution of a semicircle about a diameter as *axis*. The flat surface of a section cut by a *plane* passing through the centre is a *great circle*; the surface of a section cut off by any other plane is a small circle. The solid cut off by a plane of a great circle is a hemisphere; that cut off by a small circle is a segment. The *volume* of a sphere having radius $r = 4\pi r^3/3$; surface area $= 4\pi r^2$. In *Cartesian coordinates* the equation of a sphere with its centre at the *origin* is $r^2 = x^2 + y^2 + z^2$.

spherical aberration See *aberration, spherical*.

spherical coordinates Three-dimensional *polar coordinates*. A point in *space* is defined by the length of its *radius vector*, the *angle* this vector makes with an axis, and the angle this vector makes with a perpendicular *plane*.

spherical mirror See *mirrors, spherical*.

spherical triangle A *triangle* drawn on a spherical surface, bounded by the arcs of three *great circles*. The properties of such triangles differ from those of *plane* triangles; calculations relating to them form the purpose of spherical trigonometry.

spherical trigonometry *Trigonometry* that deals with *spherical triangles*.

spheroid A *solid* figure generated by an *ellipse* rotating about is minor *axis* (oblate spheroid, a 'flattened sphere') or about its major axis (prolate spheroid, an 'elongated sphere').

spherometer An instrument for the accurate measurement of small thicknesses, or curvature of spherical surfaces. It consists of a tripod which rests on the spherical surface, with a central screw, the height of which is controlled by a *micrometer*. The reading on the micrometer scale enables the radius of the sphere to be calculated.

spiegel Spiegeleisen. An *alloy* of iron, manganese, and carbon, used in the manufacture of *steel* by the *Bessemer process*.

spin Intrinsic angular momentum. A term of special significance in *particle physics*. Sub-atomic particles (*electrons*, *neutrons*, *nuclei*, *mesons*, etc.) may possess, in addition to other forms of *energy*, such as energy of translation, energy due to the spinning of the particle about an *axis* within itself. Quantum considerations limit the magnitude of the spin *angular momentum* of *orbital electrons* to two values, given by $sh/2\pi$ (where h is the *Planck constant*). For an electron, the spin *quantum number*, s, can have the values $\pm\frac{1}{2}$. The plus and minus signs indicate that the spin can be clockwise or anti-clockwise. For all *baryons* and *leptons* s is half integral ($\frac{1}{2}$, $1\frac{1}{2}$), but for *mesons* and *photons* it is integral (0, 1, 2). Spinning particles have an intrinsic *magnetic moment*; in a *magnetic field* the spin of a particle lines up at an angle to the direction of the field, precessing around this direction. Here, too, there is quantization, with the angular momentum *vector* having allowed directions in which the component is m_s ($h/2\pi$), where m_s is the magnetic spin *quantum number*. See Appendix, Table 6.

spinels A group of *minerals* having the general composition $MO.R_2O_3$, M being a *bivalent metal* (magnesium, iron(II) ion, manganese, zinc) and R a *tervalent* metal (aluminium, chromium, iron(III) ion). See *ferrites*.

spin glass An alloy of a nonmagnetic metal, such as gold or copper, with a small proportion (up to 10%) of a magnetic metal, such as iron or magnesium, distributed randomly in the crystal *lattice*. Spin glasses have a complicated magnetic behaviour.

spiral 1. One of several plane curves in which a point winds about a fixed central point at ever increasing distances from it. **2.** A *helix*.

spiral galaxies Spiral nebulae. *Galaxies* in which the *stars*, dust, and *gas* clouds are concentrated in the arms of a spiral. Spiral galaxies are believed to have evolved from 'elliptical' galaxies. The *Galaxy* to which the *solar system* belongs is also spiral in form.

spirans Spiro compounds. *Compounds* whose *molecules* contain two rings sharing a common atom.

spirillum A spiral-shaped *bacterium*.

spirits of salt A *solution* of *hydrochloric acid*, made by adding common *salt* to *sulphuric acid*.

spirits of wine See *ethanol*.

spiro compounds See *spirans*.

spontaneous combustion The *combustion* of a substance of low *ignition point*, which results from the *heat* produced within the substance by slow *oxidation*.

spontaneous generation See *abiogenesis*.

spore A microscopic thick-walled dormant reproductive stage of some *organisms* (e.g. *fungi*, protozoa, ferns). Spores can withstand adverse conditions, such as drying or lack of food, and can be carried long distances to colonize new habitats or new hosts.

spring balance A device for obtaining an approximate value of the *weight* of a body. It consists of a helical spring, which is stretched by the weight of the body, the extension of the spring being read off a scale. According to *Hooke's law* the extension is proportional to the force producing it, if the spring is not over-stretched.

sputtering A process for depositing a thin uniform film of a *metal* on to a surface. A disc of the metal to be 'sputtered' is made the *cathode* of a low-pressure discharge system (see *discharge in gases*). The material to be coated is placed between cathode and *anode*, the whole arrangement being enclosed and evacuated to a *pressure* of between 1 and 0.01 mm. A discharge is set up by applying a *voltage* (1000–20 000 volts) between anode and cathode. Metallic *atoms* are ejected from the cathode and are deposited on the surface to be coated.

sputter-ion pump See *ion pump*.

square 1. A *quadrilateral* having all its sides equal and all its *angles* right angles. **2.** The square of a quantity is that quantity raised to the second *power*, i.e. multiplied by itself.

square root See *root*.

square wave A *wave motion* that alternates between two fixed values for equal lengths of time, the time of transition between the two values being negligible compared to the duration of each fixed value. It is produced by a relaxation *oscillator*.

squaring the circle The problem of constructing a *square* exactly equal in area to a given *circle*. The exact area of a circle cannot be determined, except in terms of π, which cannot be expressed as an exact fraction or decimal, although any required degree of approximation can be obtained. The problem, therefore, is impossible of solution.

squid *S*uperconducting *qu*antum *i*nterference *d*evice. A superconducting device used to measure small *magnetic fields*, *voltages*, or *currents* using Josephson junctions (see *Josephson effects*).

stabilization (chem.) The prevention of chemical *decomposition* of a substance by the addition of a 'stablizer' or 'negative *catalyst*'.

stable (chem.) Not readily decomposed.

stable equilibrium (phys.) A body at rest is in stable *equilibrium* if, when slightly displaced, it tends to return to its original position of equilibrium. If the displacement tends to increase, the body is said to be in unstable equilibrium. Positions of stable equilibrium are positions of minimum *potential energy*; those of unstable equilibrium are of maximum potential energy.

stainless steel A class of chromium *steels* containing 70%–90% iron, 12%–20% chromium, 0.1%–0.8% carbon. As it neither rusts nor stains it has many industrial and domestic uses, especially as the alloy 18-8 (18% Cr, 8% Ni, and 0.08% C).

stalactite The downward growth of *calcium carbonate*, $CaCO_3$, formed on the roof of a cave by the trickling of *water* containing calcium *compounds*. See also *stalagmite*.

stalagmite The upward growth from the floor of a cave of *calcium carbonate*, $CaCO_3$; it is of the same nature and origin as a *stalactite*.

stalagmometry The measurement of *surface tension* by determining the *mass* (or *volume*) of a drop of the *liquid* hanging from the end of a tube.

Stalloy* Tradename for a *steel* containing 3.5% silicon, having low *energy* losses due to *hysteresis*. It is used in portions of electrical apparatus that are subjected to alternating *magnetic fields*.

standard atmosphere See *atmosphere*.

standard cell A specially prepared *primary cell*, e.g. the *Weston cell*, characterized by a high constancy of *EMF* over long periods of time. The EMF is a function of the *temperature*, and in the Weston cell it decreases by about 1 part in 10^5 per 1°C. rise. See also *Clark cell*.

standard deviation A measure, used in *statistics*, of the scatter of a series of numbers or measurements about their *mean* value. It is defined as the *square root* of the average value of the *squares* of the deviations from their mean value.

standard electrode See *hydrogen electrode*; *calomel electrode*.

standard electrode potential See *electrode potential*.

standard form A form in which large and small numbers are written so that only one digit appears before the decimal point, and the magnitude is indicated by multiplying by 10 raised to the appropriate power (positive for large numbers and negative for small numbers). For example, 106 452 would be written in the standard form as $1.064\ 52 \times 10^5$; similarly 0.000 106 452 would be written $1.064\ 52 \times 10^{-4}$.

standard model The theory of *elementary particles* and their *fundamental interactions*, which consists of the combination of *electroweak theory* with *quantum chromodynamics*.

standard temperature and pressure See *s.t.p.*

standing wave Stationary wave. A wave produced by the simultaneous transmission of two similar *wave motions* in opposite directions. The profile of the wave is stationary and does not move through the medium, as it does in a 'travelling wave', at the speed of propagation. See also *antinode*; *node*.

stand oil A *drying oil* that has been thickened by heating in an inert atmosphere (without the addition of driers). The thickening is due to *polymerization* of some of the constituents.

stannate A *compound* formed when *tin(IV) oxide* reacts with an *alkali*.

stannic Containing tin in its +4 *oxidation state*, e.g. stannic chloride, tin(IV) chloride, $SnCl_4$.

stannic acid See *tin(IV) oxide*.

stannous Containing tin in its +2 *oxidation state*, e.g. stannous chloride, tin(II) chloride, $SnCl_2$.

stannum See *tin*.

starch Amylum. *Polysaccharides* consisting of chains of *glucose* units arranged in one of two forms: *amylose* and *amylopectin*. Most *natural* starches are mixtures of these two forms (e.g. potato and cereal starches are 20%–30% amylose and 70%–80% amylopectin). Starch is a white tasteless *insoluble* powder that on *hydrolysis* (by boiling with *dilute acids*, or by reacting with *amylases*) gives first *dextrin* and finally glucose. Starch is stored by plants in the form of granules and occurs in most seeds.

starch gum See *dextrin*.

stars Heavenly bodies of a similar nature to the *Sun*, i.e. intensely hot, glowing masses that produce their *energy* by *thermonuclear reactions*. The nearest star to the Sun is over 4 *light-years* away; the other *fixed stars* visible to the naked eye are all members of the *Galaxy* and many of them are members of *binary star* systems. The stars are not uniformly distributed throughout the *Universe*, being grouped into enormous clusters called *galaxies*. The nearest galaxy to ours is some 16×10^5 *light-years* away. See also *stellar evolution*.

Stassano furnace See *electric-arc furnace*.

Stassfurt deposits Natural deposits of several *inorganic salts*. The deposit consists of several strata, of a total estimated thickness of 800 *metres*. They are a source of potassium and sodium *compounds* in the form of *carnallite*; also of magnesium bromide, $MgBr_2.6H_2O$, and *rock-salt*.

stat- Prefix attached to the name of electrical *units* to indicate the corresponding *electrostatic unit* (e.g. *statcoulomb*).

statcoulomb The *electrostatic unit* of *electric charge* in the *c.g.s. system*. It is equal to 3.3356×10^{-10} *coulomb*.

states of matter See *physical states of matter*.

static electricity See *electricity, static*.

statics A branch of *mechanics*; the mathematical and physical study of the behaviour of *matter* under the action of *forces*, dealing with cases where no *acceleration* is produced. Compare *dynamics*.

stationary orbit See *synchronous orbit*.

stationary phase See *chromatography*.

stationary states A term used in *quantum mechanics*. If only certain energy values or *energy levels* for the total energy of a system are permissible, the energy is said to be *quantized*. These levels are characteristic of the state of the system; such states are called stationary states. A transition from one stationary state to another can only occur with the emission or absorption of energy in the form of *photons*; i.e. *electromagnetic radiation* is emitted or absorbed. See also *internal conversion*.

stationary wave See *standing wave*.

statistical mechanics The study of the properties of large assemblies of particles or components in terms of *statistics*. E.g. the *kinetic theory of gases* treats the *molecules* of a *gas* in terms of statistical mechanics. In *classical physics*, statistical mechanics assumed that every particle has an exact position in space and an exact *momentum* at any instant. With the advent of *quantum theory*, and particularly the Heisenberg *uncertainty principle*, the precise position and momentum could not be given for a particle. This led to the development of *quantum statistics*.

statistics The collection and study of numerical facts or data and their interpretation in mathematical terms, with special reference to the theory of *probability*. See also *random sample*.

stator The fixed part of any *electric motor* or *generator* that contains the stationary magnetic circuits. Compare *rotor*.

steady-state theory A theory in *cosmology* that postulates that the *Universe* has always existed in a steady state, that the *expansion of the Universe* is compensated by the continuous creation of *matter*, which is viewed as a property of *space*, and that despite local evolutionary processes, the Universe as a whole is not evolving. The rate at which matter would have to be spontaneously created to compensate for the Universe's expansion (about 10^{-43} kg m^{-3} s^{-1}) is far too low to be measurable and therefore evidence to support this theory has to be sought in other directions. If it could be established that the *density* of matter throughout the Universe does not vary with distance or time, this would support the steady-state theory rather than its main competitor the *big-bang theory*. *Radio astronomy* has been used to assess the density of matter at the most distant parts of the observable Universe in order to decide between these two theories. On the present evidence this theory has been discredited in favour of the big-bang theory.

steam *Water*, H_2O, in the gaseous state; water above its *boiling point*. An invisible *gas*; the white clouds that are frequently termed 'steam' consist of droplets of *liquid* water formed by the *condensation* of steam. See also *superheated steam*.

steam engine A machine utilizing *steam* power; either a steam turbine (see *turbine*) or a reciprocating steam engine, consisting essentially of a cylinder in which a piston is moved backwards and forwards by the expansion of steam under pressure.

steam point The *temperature* at which the maximum *vapour pressure* of *water* is equal to standard atmospheric pressure (see *atmosphere*), i.e. the normal *boiling point*. In the *Celsius temperature* scale the steam point is given the value of $100°C$.

steam reforming The process of converting *methane* from *natural gas* into *carbon monoxide* and hydrogen as starting materials for organic synthesis, i.e.
$$CH_4 + H_2O \rightarrow 3H_2 + CO.$$
In this reaction the steam is heated to about $900°C$. and the reaction takes place over a nickel *catalyst*.

steam turbine See *turbine*.

stearate A *salt* or *ester* of stearic acid (*octadecanoic acid*).

stearic acid See *octadecanoic acid*.

stearin 1. See *tristearin*. **2.** See *stearine*.

stearine Stearin. A hard white waxy *solid* consisting mainly of *stearic* and *palmitic acids*. It is made by the *saponification* of natural *fats*.

stearoyl The univalent group $CH_3(CH_2)_{16}CO-$ (from stearic acid).

steel Iron containing from 0.1% to 1.5% carbon in the form of *cementite* (iron carbide Fe_3C). The properties of different steels vary according to the percentage of carbon and of *metals* other than iron present, and also according to the method of preparation. Steel is prepared by the *basic-oxygen*, *open-hearth,* and *Bessemer processes* and in *electric-arc furnaces*. See also *stainless steel*.

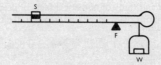

Figure 41.

steelyard A weighing-machine consisting of a long rigid bar, with a pan or hook at one end for taking the load to be weighed. The rod is pivoted about a fixed point or fulcrum near the *centre of gravity*, which is fairly near the end with the pan or hook. The other portion of the bar is graduated, and a movable weight slides along this, the weight balanced by it being proportional to its distance from the centre of gravity. See Fig. 41.

steerable dish See *radio telescope*.

Stefan's law The total *energy* emitted per unit time from unit area of a black body is proportional to the fourth power of its *thermodynamic temperature* (see *black-body radiation*). The constant of proportionality, Stefan's constant, = $5.670\,51 \times 10^{-8}$ W m^{-2} K^{-4}. Stefan's constant, σ is given by:
$$2\pi^5 k^4/15h^3 c^2,$$
where k is the *Boltzmann constant*, h is the *Planck constant*, and c is the *speed of light*. It is also known as the Stefan-Boltzmann law and constant. Named after Josef Stefan (1853–93).

stellar evolution According to current views *stars* evolve during the course of their history. It is thought that they are born from a *condensation* of *gas* (mostly hydrogen) called a *prostar*, which is compressed as a result of the *gravitational field* between the constituents. The compression is so great in the interior of the gas that *thermonuclear reactions* occur during which hydrogen is converted to helium (and possibly heavier *elements*) with the evolution of *energy*. On a *Hertzsprung-Russell diagram* the stars remain on the 'main sequence' until they have consumed some 10% of their hydrogen. They then become *red giants* and consume their hydrogen at increased rates so that eventually they contract and become *white dwarfs* (see also *gravitational collapse*). See also *novae* and *supernovae*; *neutron star*; *black hole*.

Stellite* An *alloy* of cobalt (35%–80%), chromium (15%–40%), tungsten (10%–25%), molybdenum (0%–40%), and iron (0%–5%), that is hard and non-corroding; it is used for surgical instruments.

St Elmo's fire A luminous electrical discharge (see *brush discharge*) seen around the pointed parts of ships, aircraft, or church steeples, or at the top of trees, when there is a strong atmospheric *electric field* associated with thunderstorms. St Elmo was either the popular name of St Peter Gonzalez (1190–1246), the patron saint of seamen, or a corruption of St Erasmus (known as St Ermo), also a patron saint of seamen. St Elmo's fire was regarded by seamen as a sign of the saint's protection.

step-rocket See *rocket*.

steradian The *SI unit* of *solid angle*. The solid angle that encloses a surface on the *sphere* equal to the *square* of the radius. Symbol sr.

stere A metric unit of *volume*; 1 cubic *metre*.

stereochemistry *Chemistry* involving consideration of the arrangement in *space* of the *atoms* in a *molecule*. If a molecule is considered as a three-dimensional entity in space, possibilities of *stereoisomerism* or space *isomerism* arise; thus, a molecule consisting of four different groups or *atoms* attached to a central carbon atom can exist in two distinct space arrangements, one being a *mirror image* of the other. Such isomerism is associated with *optical activity*. See also *chirality*; *stereospecific*.

stereoisomerism *Isomerism* caused by possibilities of different arrangement in three-dimensional space of the *atoms* within a *molecule*, resulting in two isomers that are mirror images of each other. See also *cis-trans isomerism*; *optical isomerism*; *stereochemistry*.

stereoregular Having a regular arrangement in space of the atoms and groups within a *molecule*. See *stereoregular rubbers*.

stereoregular rubbers A group of synthetic *rubbers* manufactured by a solution *polymerization* process using special *catalysts* that control the stereoisomeric (see *stereoisomerism*) regularity of the products. These materials can therefore be made to resemble closely the structure of natural rubber. In cis-1,4-poly-*isoprene*, the structure of natural rubber is substantially duplicated, and this *elastomer* can be used for many of the purposes that were the exlusive preserve of natural rubber. A similar product is cis-1,4-poly*butadiene*, which is also used in place of natural rubber. See also *ethene-propene rubber*.

stereoscope An optical device by which two-dimensional pictures are given the appearance of depth and solidity. This is achieved by taking photographs from two slightly different positions to imitate *binocular* vision.

stereospecific 1. Having a particular arrangement in space of the atoms and groups within a *molecule*. See also *tactic polymer*. **2.** Denoting a reaction that produces such molecules.

steric hindrance (chem.) The hindering or retarding of a chemical reaction, as a result of the arrangement in space of the atoms of the reacting molecules.

steroids Derived *lipids* based on a four-fused-ring structure that include *sterols*, the *bile acids*, certain *hormones* (including the *sex hormones* and the corticosteroid hormones produced by the adrenal cortex) and *glucosides*, and *vitamin* D.

sterols Derived *alcohols* of the *steroid* group. *Cholesterol* and *ergosterol* are typical examples. Sterols are present in many living *organisms* in which they play an essential part.

stibine Antimony hydride. SbH_3. A poisonous *gas*.

stibnite Natural *antimony trisulphide*, Sb_2S_3. The principal *ore* of antimony.

stilb A unit of *luminance* equal to 1 *candela* per square centimetre.

stilbene 1,2-Diphenylethene. $C_6H_5CH:CHC_6H_5$. A colourless *insoluble* substance, m.p. 124°C., used in the manufacture of *dyes*.

stilboestrol $(HO.C_6H_4.CH:)_2$. A white crystalline *organic compound*, m.p. 171°C., used in medicine as an *oestrogen*.

still A *metal* or *glass* apparatus used for the *distillation* of *liquids*.

stimulated emission See *maser* and *laser*.

STM See *scanning tunnelling microscope*.

stochastic process A process that has some element of *probability* in its structure.

stoichiometric A *compound* is said to be stoichiometric when its component *elements* are present in the exact proportions represented by its chemical *formula*. A stoichiometric *mixture* is one that will yield on reaction a stoichiometric compound (e.g. two *molecules* of hydrogen and one molecule of oxygen constitute a stoichiometric mixture because they yield exactly two molecules of *water* on *combustion*). A non-stoichiometric compound is sometimes called a *Berthollide compound*.

stoichiometry The part of *chemistry* dealing with the composition of substances; more particularly with the determination of combining proportions or *chemical equivalents*.

stokes *C.g.s.* unit of *kinematic viscosity* equal to the *viscosity* of a *fluid* in *poises* divided by the *density* in grams per cubic centimetre. 1 centistokes = 10^{-6} $m^2 s^{-1}$. Named after Sir George Stokes (1819–1903).

Stokes' law A spherical ball, radius r, moving through a viscous medium, *viscosity* η, at a *velocity* v, will experience a frictional force, $F = 6\pi r\eta v$. From this law it can be deduced that a small *sphere* falling under the action of *gravity* through a *viscous* medium ultimately reaches a constant *velocity* equal to:

$$v = 2gr^2(d_1 - d_2)/9\eta,$$

where d_1 = *density* of the sphere and d_2 = density of the medium.

stop A *diaphragm* or disc with a central aperture used to restrict the light entering an optical system (e.g. a camera, telescope, etc.) or to reduce the size of a lens to cut down spherical *aberration*.

stopping power A measure of the ability of a substance to reduce the *kinetic energy* of a charged particle passing through it. The 'linear' stopping power is the energy lost per unit distance $-dE/dx$, often expressed in MeV cm^{-2}; the 'mass' stopping power is the linear stopping power divided by the *density* of the substance; the 'atomic' stopping power is the energy lost per atom per unit area of the substance perpendicular to the motion of the charged particle, i.e. $(1/n)dE/dx = (A/\rho N)dE/dx$, where n is the number of atoms in unit volume of the substance, N is the *Avogadro number*, and A the relative atomic mass of the substance. Stopping power is often expressed relative to such standard substances as air or aluminium.

storage battery See *accumulator*.

storage ring A modified form of *synchrotron* particle *accelerator*, in which particles can circulate for long periods, once accelerated to their maximum *energy*. Some storage rings only store particles without accelerating them but in many machines the two functions are combined. In some devices two beams of particles circulate in opposite directions. At the intersections of these two beams very high collision energies occur, enabling interactions to be studied. The 6 km diameter ring of *superconducting magnets* at Fermilab near Chicago, stores *protons* and antiprotons at energies up to 900 GeV, enabling a total collision energy of 1800 GeV to be reached.

store See *backing store*; *memory*.

s.t.p. Standard temperature and pressure. Formerly N.T.P. (normal temperature and pressure). A pressure of $1.013\ 25 \times 10^5$ *pascals* (760 mmHg) and a temperature of 273.15 K (0°C.). These are the standard conditions for comparing gas volumes.

straight chain A *hydrocarbon* molecule in which the carbon atoms are linked together in one long straight chain with no *side chains* attached.

strain (phys.) When a body is deformed by an applied *stress* the strain is the ratio of the dimensional change to the original or unstrained dimension. The strain may be a ratio of lengths, areas, or volumes. See also *shear stress*.

strain gauge, electrical A grid of fine resistance wire supported on a paper base, which is attached by a suitable adhesive to the surface under test, so that any strains set up in the latter are accurately transferred to the gauge wire. The change of electrical *resistance* of the gauge is proportional to the *strain*, so that methods of measuring resistance may be used for measuring strain. The gauge is suitable for measuring strains of the order of 10^{-4} to 10^{-2}. Some strain gauges use a *semiconductor* element in place of wire, while others measure a change in capacitance or inductance under stress of a suitable element.

strain hardening Work hardening. An increase in the hardness and *tensile strength* of a *metal*, due to cold working, that causes a permanent alteration (distortion) of its crystalline structure.

strange matter Matter in which the *nucleons* consist of up, down, and strange *quarks*, rather than the usual up and down quarks (see *elementary particles*). It has been suggested that some strange matter was created in the *big-bang* and that some remnants of this form of matter, called 'S-drops' still exist, although there is no evidence for their existence.

strangeness Certain *hadrons* (K-*mesons* and *hyperons*) *decay* about 10^{12} times more slowly than would be expected if they were decaying through *strong interactions*. These particles, which are called strange particles, have been arbitrarily assigned a *quantum number*, s, to account for this strangeness. This property is conserved in strong interactions, so the strange particles, which must lose their strangeness, can decay only by the much slower *weak interactions*. For ordinary particles (*nucleons*, *pions*, etc.) $s = 0$; each strange particle has a specific, *integral* value of s, which is not equal to 0. In the *quark* model, strange hadrons are postulated to contain the strange s quark (or its antiquark). See *strange matter*. Thus strangeness is one of the basic properties of *elementary particles*. See Appendix, Table 6.

stratigraphy The study of *rock strata*, their origin, composition, and sequence. It forms the basis of historical *geology*.

stratopause The boundary between the *stratosphere* and the *mesosphere*.

stratosphere A layer of the *atmosphere* beginning approximately 11 kilometres (7 miles) above the surface of the *Earth*. See Fig. 44, under *upper atmosphere*.

stratum A layer.

streamline A *streamline* is a line in a *fluid* such that the *tangent* to it at every point is in the direction of the *velocity* of the fluid particle at that point, at the instant under consideration. When the motion of the fluid is such that, at any instant, continuous streamlines can be drawn through the whole length of its course, the fluid is said to be in streamline flow. See *laminar flow*; *turbulent flow*.

streptomycin $C_{21}H_{39}N_7O_{12}$. An *antibiotic* produced by the *Streptomyces bacteria*. It is effective against several types of disease bacteria, including some against which *penicillin* is inactive.

stress (phys.) A *force* per unit area. When a stress is applied to a body (within its *elastic limit*) a corresponding *strain* is produced, and the ratio of stress to strain is a characteristic constant of the body (see *elastic modulus*). See also *compressive stress*; *shear stress*; *tensile stress*.

string theory A theory of *elementary particles* originally based on the concept of a particle as a *standing wave* in a string. It treats a particle, not as a point-like object that features in *quantum field theory*, but as a line or loop. By combining string theory with the idea of *supersymmetry*, the 'superstring theory' emerged. This theory may be more successful in producing a *unified field theory* involving all four *fundamental interactions* than quantum field theory, as it avoids the infinities that arise when the *gravitational interaction* is introduced into other field theories. Superstring theory makes use of multidimensional space (10 dimensions for *fermions* and 26 for *bosons*) in which superstrings have energies up to 10^{19} GeV. Because this level of energy is considerably higher than that achieved by any current *accelerator*, there is no direct evidence for superstrings.

stroboscope An instrument with the aid of which it is possible to view objects that are moving rapidly with a periodic motion (see *period*) and to see them as if they were at rest. For example, if a disc, rotating at n revolutions per second, is illuminated by a source flashing at the same *frequency*, then at any particular flash the eye will see the disc in exactly the same position as it was for the previous flash. The disc will therefore appear stationary. If the frequency of the motion is not quite equal to that of the flashing, the disc will appear to rotate slowly.

strong acid An *acid*, such as sulphuric acid, that is completely dissociated into *ions* in solution. Compare *weak acid*.

strong electrolytes See *electrolytic dissociation*.

strong interaction An interaction that occurs between *quarks*. It occurs only at very short range (about 10^{-15} metre) and is the force responsible for binding *nucleons* together in an atomic *nucleus*. The strong interaction is some 100 times stronger than the *electromagnetic interaction* at this short range. The force between nucleons and other hadrons (sometimes called an *exchange force*) can be visualized as the exchange of virtual *mesons* between the particles (see *virtual particle*) but the basic strong interaction between quarks is due to the exchange of intermediary particles called *gluons*. See also *elementary particles*.

strontia See *strontium oxide*. *Strontium hydroxide* is also sometimes known as strontia.

strontium Sr. Element. R.a.m. 87.62. At. No. 38. A *reactive metal* resembling calcium. R.d. 2.583, m.p. 770°C., b.p. 1390°C. It occurs as *celestine*, $SrSO_4$ and strontianite, $SrCO_3$. *Compounds* colour a *flame* crimson and they are used in fireworks. The *radioisotope* strontium-90 is present in the *fall-out* from nuclear explosions. It presents a health hazard as it has a relatively long *half-life* of 28 years and, owing to its chemical similarity to calcium, can become incorporated into bone. See *strontium unit*.

strontium carbonate $SrCO_3$. A white crystalline solid that occurs naturally as strontianite. It is a *phosphor* used to coat *cathode-ray tubes* and in sugar.

strontium hydroxide Strontia. $Sr(OH)_2$. A white *deliquescent* crystalline powder, m.p. 375°C., used in *sugar* refining as it combines with the sugar to form an *insoluble* saccharate.

strontium nitrate $Sr(NO_3)_2$. A colourless crystalline substance, m.p. 570°C., used in fireworks and flares to give a bright crimson colour.

strontium oxide Strontia. SrO. A grey *amorphous* powder, m.p. 2430°C., with similar properties to *calcium oxide*. It is used in the manufacture of strontium *salts*.

strontium unit SU. A measure of the *concentration* of strontium-90 in an *organic* medium (e.g. milk, bone, *soil*, etc.) relative to the concentration of calcium in the same medium. 1 SU = 10^{-12} *curie* of strontium-90 per gram of calcium.

structural formula A chemical *formula* that in addition to showing the *atoms* present in a *molecule*, also gives an indication of its structure. E.g. the structural formula of *benzene* (C_6H_6) is given by the *benzene ring*.

structural isomerism See *isomerism*.

strychnine $C_{21}H_{22}N_2O_2$. An *alkaloid* that occurs in the seeds of *Strychnos nux vomica*. It is a white crystalline substance, slightly *soluble* in *water*; m.p. 284°C. It has an intensely bitter taste and a powerful and very dangerous action on the nervous system. It is used in medicine in minute doses.

styrene Phenylethene. $C_6H_5.CH:CH_2$. A colourless *aromatic liquid*, b.p. 146°C., that polymerizes to a *thermoplastic* material (see *polystyrene*) and is used in the manufacture of synthetic *rubber*. See *styrene-butadiene rubber*.

styrene-butadiene rubber SBR. A widely used, general purpose synthetic *rubber*. A copolymer (see *polymerization*) of *butadiene* and about 35% of *styrene*, which is *vulcanized* in a similar manner to natural rubber. Properties are in general inferior to natural rubber, except for abrasion resistance, but passenger car tyres are made very largely from SBR. This *elastomer* is not suitable, however, for incorporation into heavy duty tyres.

sub- Prefix denoting under, below. In *chemistry* it was used to indicate either that the *element* mentioned is present in a lower proportion than usual, e.g. suboxide, or that the *compound* is *basic*, e.g. subacetate.

subatomic Consisting of particles smaller than, or forming a part of, the *atom*. See *atom, structure of*.

subcritical Said of a *nuclear reactor* in which the effective *multiplication constant* is less than unity, and in which the nuclear *chain reaction* is therefore not self-sustaining.

suberic acid Octanedioic acid. $HOOC(CH_2)_6COOH$. A white crystalline *dibasic acid*, m.p. 140°C., obtained from *castor oil* and used in the manufacture of *plastics* and *plasticizers*.

subgiant A *giant star* with a lower absolute *magnitude* than an ordinary giant.

sublate The product collected by ion *flotation*.

sublimate A *solid* obtained by the direct *condensation* of a vaporized solid without passing through the *liquid* state.

sublimation (chem.) The conversion of a *solid* direct into *vapour*, and subsequent *condensation*, without melting.

sub-shell A concept used in the *Bohr theory* of atomic structure. Each electron *shell* is divided into sub-shells, for which all the electrons have the same azimuthal *quantum number*. The shells are designated by the letters s,p,d,f. See also *orbital*.

subsonic Moving at, or relating to, a *speed* that is less than Mach 1. See *Mach number*.

substantive dyes See *direct dyes*.

substituent See *substitution product*.

substitution product A *compound* obtained by replacing an *atom* or group by another atom or group in a *molecule*. The new atom or group is known as the 'substituent'.

substrate 1. (bio.) A substance upon which a specific *enzyme* acts. **2.** (phys.) In *silicon chips*, the underlying piece of silicon in the surface layers of which the elements of an *integrated circuit* are constructed.

subtend (math.) Two points, *A* and *B*, are said to subtend the *angle ACB* at the point *C*.

subtractive process The process of producing *colours* by mixing three different *primary colours* of dyes or pigments together. The final colour is produced by the absorption of different wavelengths of light. Compare *additive process*.

successive approximations A method of solving a complex equation by substituting a likely numerical value for the variable in the left-hand side of the equation and making this as close as possible to the right-hand side, which is often zero, by taking successive values for the variable. With each successive value taken, the value of the l-h side should approximate more closely to that of the r-h side.

succinate A *salt* or *ester* of succinic acid (*butanedioic acid*).

succinic acid See *butanedioic acid*.

succinite See *amber*.

sucrase See *invertase*.

sucroclastic *Sugar*-splitting; applied to *enzymes* that have the power of hydrolysing complex *carbohydrates*. E.g. *invertase*.

sucrose Cane-sugar, beet sugar, saccharose. Common 'sugar' of the household. $C_{12}H_{22}O_{11}$. A white sweet crystalline *disaccharide* of *glucose* and *fructose*, m.p. $160°-186°C$. It is found in numerous plants, particularly the sugar cane, sugar beet, and maple-tree sap.

sugar 1. Any sweet *soluble monosaccharide* or *disaccharide*. **2.** *Sucrose*.

sugar of lead See *lead ethanoate*.

sulphane A compound that has the general formula H_2S_n. The simplest is *hydrogen sulphide*, H_2S. Others are H_2S_2, H_2S_3, H_2S_4, etc.

sulphanilic acid 4-aminobenzenesulphonic acid. $NH_2C_6H_4SO_3H.H_2O$. A grey crystalline *soluble* substance, m.p. $288°C$., used in the manufacture of *dyes*.

sulphate A *salt* or *ester* of *sulphuric(IV) acid*.

sulphate of ammonia See *ammonium sulphate*.

sulphation The formation of an insoluble layer of lead sulphate on the electrodes of a lead *accumulator*, when it is not in use and is left discharged for any length of time.

sulphide A *binary compound* of an *element* or group with sulphur; a *salt* of *hydrogen sulphide*, H_2S.

sulphite A *salt* or *ester* of *sulphurous acid*, H_2SO_3.

sulpho The *univalent* group $HO.SO_2 -$.

sulphonamide drugs Sulpha drugs. A group of *organic compounds*, containing the sulphonamide group $SO_2.NH_2$ or its *derivatives*; the group includes:
sulphanilamide ($NH_2C_6H_4.SO_2NH_2$),
sulphapyridine ($NH_2C_6H_4SO_2NHC_5H_4N$),
sulphathiazole ($NH_2C_6H_4SO_2NHC_3H_2NS$),
sulphadiazine ($NH_2C_6H_4SO_2NHC_4H_3N_2$),
and many others. They have been of great value in the treatment of many diseases caused by *bacteria*.

sulphonate A *salt* or *ester* of any *sulphonic acid*.

sulphonation The formation of a *sulphonic acid* by the addition of an $-SO_2OH$ group to a *benzene* molecule. It is done by heating with concentrated *sulphuric(VI) acid* or by treating with disulphuric(VI) acid.

sulphones *Organic compounds* having the general formula $R-SO_2-R'$, where R and R' are organic *radicals*.

sulphonic acids *Acids* (usually *organic*) containing the $HO.SO_2 -$ group; e.g. *benzenesulphonic acid*, $C_6H_5SO_2OH$.

sulphonyl The *bivalent* group $-SO_2 -$.

sulphoxides *Organic compounds* containing the group $=S=O$, called the sulphoxide group.

sulphur S. Element. R.a.m. 32.064. At. No. 16. A non-metallic *element* occurring in several *allotropic forms*. The stable form under ordinary conditions is rhombic or alpha-sulphur, a pale-yellow brittle crystalline *solid*, r.d. 2.086, m.p. 115.3°C., b.p. 444.6°C. Sulphur burns with a blue *flame* to give *sulphur dioxide*; it combines with many *metals* to form *sulphides*. Sulphur occurs as the element in many volcanic regions and as sulphides of many *metals*. It is extracted in vast quantities in Texas by the *Frasch process*. It is used in the manufacture of *sulphuric acid, carbon disulphide, dyes* and various chemicals. It is also used for *vulcanizing rubber*, in *fungicides*, and in medicine. Sulphur is an *essential element* that occurs in many *proteins*. See also *flowers of sulphur*.

sulphur dichloride dioxide Sulphuryl chloride. SO_2Cl_2. A colourless *liquid*, b.p. 69.1°C., used as a chlorinating agent.

sulphur dichloride oxide Thionyl chloride. $SOCl_2$. A colourless fuming liquid, b.p. 75.7°C., used as a chlorinating agent to replace –OH groups with Cl atoms.

sulphur dioxide Sulphur(IV) oxide. SO_2. A colourless *gas* with a choking penetrating smell; b.p. – 10°C.; *liquid* SO_2 is used in *bleaching*, fumigating, and as a *refrigerant*. It is also used in the manufacture of *sulphuric acid*, H_2SO_4 in the *contact process*.

sulphur dyes Dyes made by heating certain *organic* substances with sulphur and sulphides. They are usually polymeric and are insoluble in water, but when heated with *sodium sulphide* the large molecules break down to form a water-soluble leuco compound (see *vat dyes*), which dyes cellulosic fibres. The final dyeing is obtained by *oxidation*, as in the case of vat dyes. These dyes are very cheap but give dull hues; they are widely used for dyeing industrial fabrics.

sulphuretted hydrogen See *hydrogen sulphide*.

sulphuric acids Any of several acids. **1.** Sulphuric(VI) acid, tetraoxosulphuric(VI) acid, oil of vitriol. H_2SO_4. A colourless oily *liquid*, r.d. 1.84, m.p. 10.36°C. It is extremely corrosive, reacts violently with *water* with the evolution of heat, and chars *organic* matter. It is made by the *contact process* (and formerly by the *lead-chamber process*). It is used extensively in many processes (chemicals, fertilizers, detergents, paints, fibres, etc.) and in lead *accumulators*. **2.** Disulphuric(VI) acid, pyrosulphuric acid, Nordhausen sulphuric acid. $H_2S_2O_7$. A highly corrosive *hygroscopic* crystalline solid, m.p. 35°C. It is used in the *sulphonation* of *organic compounds*. **3.** Fuming sulphuric acid, oleum. Sulphuric(VI) acid containing an excess of *sulphur trioxide*. For example, 20% oleum contains 20% SO_3 and 80% H_2SO_4. It is extremely corrosive and contains some disulphuric(VI) acid. It is used for *nitration*. **4.** Permonosulphuric(VI) acid, peroxosulphuric(VI) acid, persulphuric acid, Caro's acid. H_2SO_5. A white crystalline substance that decomposes at 45°C. It is made by the action of *hydrogen peroxide* on concentrated sulphuric(VI) acid and is used as an *oxidizing agent*. **5.** Peroxodisulphuric(VI) acid, perdisulphuric acid. $H_2S_2O_8$. A white crystalline substance that decomposes at 65°C. It is made by the electrolysis of *sulphates* and is used in the manufacture of *hydrogen peroxide*. **6.** Sulphurous acid, trioxosulphuric(IV) acid, sulphuric(IV) acid. H_2SO_3. A weak acid that forms, together with H_2SO_4, when *sulphur dioxide* dissolves in water. It is known in the form of its *salts*, the *sulphites* and hydrogensulphites. **7.** See *thiosulphuric acid*.

sulphuric anhydride See *sulphur trioxide*.

sulphuric ether An obsolete name for *ethoxyethane*.

sulphurous acid See *sulphuric acids*.

sulphur point The *temperature* of equilibrium between *liquid* sulphur and its *vapour* at a *pressure* of one standard atmosphere; 444.6°C.

sulphur trioxide Sulphur(VI) oxide, sulphuric anhydride. SO_3. A white crystalline *solid*, that exists in three crystalline forms, m.p. 16.8°C. It combines with *water* to form *sulphuric acid*.

sulphuryl The *bivalent* group $-SO_2-$ in an *inorganic compound*.

sulphuryl chloride See *sulphur dichloride dioxide*.

Sun The incandescent approximately spherical heavenly body around which the *planets* rotate in elliptical *orbits* (see *solar system*). The Sun is a 'main sequence' *star* (see *Hertzsprung-Russell diagram*), being one of some 10^{11} stars that constitute our *Galaxy*. Mean distance from the *Earth* is approximately 149.6×10^6 kilometres, and the distance to nearest star is approximately 40×10^{12} km. The diameter of the Sun is about 1 392 000 km, its *mass* is approximately 2×10^{30} kilograms, and its average *relative density* 1.4. The visible surface of the Sun, called the *photosphere*, is at a *temperature* of about 6000°C.; its interior temperature is some 13 000 000°C. At this internal temperature *thermonuclear reactions* occur in which hydrogen is converted into helium, these reactions providing the Sun with its vast supply of *energy* (see *solar constant*). The Sun is composed of about 75% hydrogen, 24% helium, and only 1% of the heavier *elements*.

sunspots Large patches, which appear black by contrast with their surroundings, visible upon the surface (*photosphere*) of the *Sun*. Owing to the rotation of the Sun, they appear to move across its surface. Their appearance is spasmodic, but their number reaches a maximum approximately every eleven years. (See *eleven-year period*.) They are connected with such phenomena as magnetic storms and the *aurora borealis* and appear black because they result from a local drop in *temperature* to about 4000 K. See *solar flares*; *solar prominences*; *solar wind*.

super- Prefix denoting over, above.

superconductivity The electrical *resistance* of a *metal* or *alloy* is a function of *temperature*, decreasing as the temperature falls and tending to a constant low value at *absolute zero*. It is found that for certain metals and alloys (e.g. lead, vanadium, tin) the resistance changes abruptly, becoming vanishingly small at a temperature in the neighbourhood of a few degrees above absolute zero. This phenomenon is termed superconductivity, and the temperature at which it sets in is the transition temperature. A current induced by a changing *magnetic flux* in a ring of superconducting material will continue to circulate after the magnetic flux has been removed. (See also *cryotron*.) This effect has been used to produce large magnetic fields without the expenditure of appreciable quantities of *electrical energy* (except in maintaining the very low temperature). The explanation of superconductivity, known as the BCS theory (after its propounders J. Bardeen, L. N. Cooper, and J. R. Schrieffer); developed in 1957, it relies on the hypothesis that the current is carried in superconductors by bound pairs of *electrons*, called Cooper pairs. As the total *momentum* of a Cooper pair is not changed when one of its electrons interacts with the *lattice* of the superconducting crystal through which it is passing, the flow of electrons can continue indefinitely. In 1986 a different kind of superconductivity was discovered in certain substances; because it can occur at a *transition temperature* of 100 K it is known as 'high-temperature superconductivity'. It appears not to occur as a result of the BCS mechanism, although its mechanism has not yet been explained. However, high-temperature superconductivity has great advantages as it can operate at the temperature of liquid nitrogen (rather than liquid helium for BCS superconductors). An example of a high-temperature superconductor is $YBa_2Cu_3O_{1-7}$.

supercooling The *metastable* state of a *liquid* cooled below its *freezing point*. A supercooled liquid will usually freeze on the addition of a small particle of the *solid* substance, and often on the addition of any solid particle or even on shaking; the *temperature* then rises to the freezing point.

supercritical Said of a *nuclear reactor* in which the effective *multiplication factor* exceeds unity, and in which the nuclear *chain reaction* is therefore increasing with time.

superdense theory See *big-bang theory*.

superfluid A *fluid* that flows without *friction* and has an abnormally high thermal *conductivity*, e.g. helium-4 below 2.186 K. The temperature at which a fluid becomes superfluid is called the *lambda point*.

supergiant star A *star* of exceptionally high *luminosity*, low *density*, and a diameter some hundreds of times greater than the *Sun*. They lie above the *giant stars* on the *Hertzsprung-Russell diagram*.

supergravity A *unified field theory* based on *supersymmetry* that attempts to encompass all four *fundamental interactions* by abandoning point particles, as used in *quantum field theories*, replacing them with the extended objects that feature in *string theories*. The concept of supergravity has no experimental backing.

superheated steam *Steam* at a temperature in excess of the *phase* equilibrium temperature at the applied pressure, i.e. it is steam that has been heated to a temperature in excess of 100°C. at a pressure in excess of atmospheric pressure.

superheating Heating a *liquid* above its *boiling point* at an elevated pressure, when the liquid is in a *metastable* state. See *supercooling*.

superheterodyne Superhet. Abbreviation of 'supersonic heterodyne'. A method of *radio* reception in which the *frequency* of the *carrier wave* is changed in the receiver to a 'supersonic' *intermediate frequency* (i.e. a frequency above the audible limit for *sound*) by a *heterodyne* process. The intermediate frequency signal is amplified before *demodulation* and *amplification* by the audio-frequency amplifier. This enables distortion and *noise* to be reduced and *gain* and selectivity to be increased.

super high frequency SHF. *Radio frequencies* in the range 3000 to 30 000 *megahertz*.

supernatant Denoting a clear *liquid* that floats above a *precipitate*.

supernovae *Stars* that suffer an explosion becoming some 10^8 times brighter than the *Sun* during the process. They are relatively rare events, only six having been recorded within our *Galaxy*, although they have been observed fairly regularly in other *galaxies*. These explosions are believed to be caused when a star runs out of hydrogen and contracts under its own *gravitational field*. The contraction causes a sufficiently high *temperature* in the interior for *thermonuclear reactions* to occur, which produce heavy *elements*. The formation of heavy elements, with *atomic numbers* in excess of about 40, absorbs *energy* and the star collapses inwards, increasing its speed of rotation and ultimately flinging a large portion of its *matter* into *space*. It is believed that the *planets* of the *solar system* consist of matter thrown into space by a supernova, which was subsequently collected by the Sun's gravitational field. The residue of a supernova explosion is a *white dwarf* star. Compare *nova*.

superphosphate An artificial *fertilizer* consisting mainly of calcium hydrogenphosphate. See *calcium phosphate*.

superplasticity The property, exhibited by certain metallic *alloys*, of stretching several hundred per cent before failing, e.g. zinc in aluminium.

supersaturation The *metastable* state of a *solution* holding more dissolved *solute* than is required to *saturate* the solution. Condensation will occur if there are suitable surfaces or nuclei (e.g. dust particles).

supersonic Moving at, or relating to, a *speed* in excess of Mach 1. See *Mach number*.

supersonics See *ultrasonics*.

superstring theory See *string theory*.

supersymmetry A hypothetical *symmetry* in the behaviour of *elementary particles* that underlies most attempts to construct a *unified theory* of all four *fundamental interactions*. The symmetry occurs between *fermions* and *bosons* and gives rise to new bosonic partners for the existing fermions (the *quarks* and *leptons*), and fermionic partners for the known bosons (such particles as the *photon*, which mediate the interactions). There is as yet no experimental evidence for such a doubling of particles.

supplementary angles *Angles* together totalling 180°, or two right angles.

suppressor grid A grid, placed between the *screen grid* and the *anode* of a *thermionic valve*, to reduce the *secondary emission* of *electrons* between them.

surd An *irrational number*; a *root* that cannot be expressed as an exact number or fraction; e.g. $\sqrt{2}$.

surface acoustic wave SAW. An *ultrasonic* wave propagated along the surface of a solid at a *frequency* between a few megahertz and a few gigahertz. Electric signals can be converted to SAWs by means of *transducers* using *piezoelectric crystals*. SAW devices are used for signal processing and are particularly useful as filters.

surface-active agent See *surfactant*.

surface colour Certain reflecting surfaces, e.g. *metal* surfaces, exhibit selective reflection of *light* waves; i.e. they reflect some *wavelengths* (*colours*) more readily than others. When illuminated by *white light*, such surfaces reflect light deficient in certain wavelengths, and the body appears coloured. The body is then said to show surface colour, as opposed to *pigment colour*. Bodies showing surface colour when viewed by transmitted light appear to be of the *complementary colour* to that observed when viewed by reflected light. Substances that show pigment colour appear the same colour whether viewed by reflected or transmitted light.

surface tension γ. An open surface of a *liquid* is under a state of tension, causing a tendency for the portions of the surface to separate from each other; the surface thus shows properties similar to those of a stretched elastic film over the liquid. The tension is an effect of the *forces* of attraction existing between the *molecules* of a liquid. It is measured by the force per unit length (*newtons* per *metre*) acting in the surface at right angles to an element of any line drawn in the surface.

Alternatively it can be expressed as the *energy* required to create 1 square metre of new surface in $J\,m^{-2}$ (which is equivalent to $N\,m^{-1}$). A surface tension exists in any boundary surface of a liquid. Surface tension enables drops and bubbles to form in liquids and for *capillary action* to occur. It is also responsible for the absorption of liquids by porous solids and the ability of liquids to wet a surface. *Water* has a high surface tension due to the *hydrogen bonds* between molecules, which enables water to be taken in by plants against *gravity*.

surfactant Surface-active agent. A wide-ranging class of substances whose *molecules* contain structurally dissimilar groups with opposing solubility tendencies (amphipathic structure), such as water-soluble (*hydrophilic* or *polar*) and water-insoluble (*hydrophobic* or non-polar) groups. When dissolved in water, or in some other *solvent*, these substances are adsorbed at the interface between the solution and a *phase* in contact with it (air, another liquid, soiled material, etc.) and modify its properties (i.e. they exhibit surface activity). Depending on their nature, surfactant solutions have various functional properties, such as cleansing action (detergency), foaming, wetting, emulsifying, etc. They are classified, in accordance with the electric *charge* of their active groups, into three main classes: anionic, which include *soaps* (e.g. sodium stearate) whose active groups are *anions*; cationic, such as *quaternary ammonium compounds*, in which the active groups are positively charged quaternary ammonium *ions*; and non-ionic, exemplified by *condensation* products of higher *alcohols* with epoxyethane, in which polyoxyethene chains form the hydrophilic part. Most household detergents are based on anionic surfactants of various types (soaps, alkyl sulphates or sulphonates, etc.), usually with the addition of other agents to confer such specific properties as foaming, wetting, or emulsifying action.

susceptance *B*. The imaginary part of the *admittance* of a circuit. It is the reciprocal of the *reactance* and is measured in *siemens*.

susceptibility See *electric susceptibility*; *magnetic susceptibility*.

suspension (chem.) A two-phase system (see *phase*) consisting of very small *solid* or *liquid* particles distributed in a *fluid dispersion medium*.

suspensoid sol See *colloidal solutions*.

SVP See *saturated vapour pressure*.

sylvine Sylvite. Natural *potassium chloride*, KCl, usually containing *sodium chloride* as an impurity. It is an important source of potassium *compounds*.

symbiosis A relationship between two different types of *organism* that live together for their mutual benefit. E.g. the relationship between *cellulose*-digesting *bacteria* and the herbivores whose alimentary tract they inhabit.

symbol (chem.) A letter or letters representing an *element* or an *atom* of an element; e.g. Fe = iron or S = one atom of sulphur. See *formula*. The symbols of all the elements are given in the Appendix in Table 3.

symmetry 1. The correspondence of parts of a figure with reference to a *plane*, line, or point of symmetry. Thus, a *circle* is symmetrical about any diameter; a *sphere* is symmetrical about a plane of any great circle. **2.** The set of invariances of a system; a symmetry operation on a system is one that does not change it, e.g. reflections and rotations of molecules. More abstract symmetries occur, for example in *gauge theories*. See also *supersymmetry*.

synapse A junction between *neurones* by which nerve *impulses* are transferred within the nervous systems of animals. A synapse is usually formed between the *axon* of one neurone and the cell body or *dendrite* of another. See *neurotransmitter*.

synchrocyclotron A type of *cyclotron* that enables *relativistic velocities* to be achieved by modulating the *frequency* of the accelerating *electric field*, in synchronization with the increasing period of revolution of a group of the accelerating particles.

synchronous motor An *alternating current electric motor* whose speed of rotation is proportional to the *frequency* of its power supply.

synchronous orbit Geostationary orbit. The *orbit* of an artificial Earth *satellite* that has a *period* of 24 hours. The altitude corresponding to such an orbit is about 35 700 km; a satellite in a circular orbit parallel to the *equator* at this altitude would appear to be stationary in the sky. Communication satellites in stationary orbits are used for relaying *radio* signals between widely separated points on the Earth's surface.

synchrotron An accelerator of the cyclic type in which the *magnetic field* is modulated to keep the particles on the same circular path as their *energy* rises. With *proton* synchrotrons the *frequency* of the *electric field* must also increase with the rising velocity of the particles but in *electron* synchrotrons this frequency remains constant, as the injected electrons already have *relativistic speeds*.

synchrotron radiation High *energy electrons* within a *synchrotron* emit *light* as a consequence of their motion in a curved path around the machine's ring of *electromagnets*; this emission, known as synchrotron radiation, is inversely proportional to the *radius of curvature* of the orbit and to the fourth power of the *mass* of the particle. The term is also used to describe the emission of *radio frequency electromagnetic radiations* from interstellar gas clouds in *radio galaxies* (see *radio astronomy*) as this emission is believed to be due to electrons moving in curved paths at relativistic velocities in celestial *magnetic fields*.

syndiotactic polymer See *atactic polymer*.

syneresis The separation of *liquid* from a *gel*.

synodic month See *lunation*.

synodic period of a planet The period between two successive *conjunctions* with the *Sun*, as observed from the *Earth*.

synthesis (chem.) 'Putting together'; the formation of a *compound* from its *elements* or simpler compounds.

synthetic (chem.) Artificially prepared from the component *elements* or simpler materials; not obtained directly from natural sources.

Système International d'Unités See *SI units*.

syzygy A point of *opposition* or *conjunction* of a *planet*, or the *Moon*, with the *Sun*.

T

tachometer An instrument for measuring the rate of revolution of a revolving shaft.

tachyon A hypothetical particle that travels faster than the *speed* of light. To satisfy the *special theory of relativity* such a particle would have imaginary *energy* and *momentum* if it had a real *rest mass*, or imaginary rest mass if the energy was real. Its presence could be detected by the *Cerenkov radiation* it emits, but no such particle has yet been detected.

tacnode A point at which two branches of a curve touch each other and have a common *tangent*.

tactic polymer A *polymer* in which the groups attached to the polymer chain are regularly arranged, giving a *stereospecific* and a *stereoregular* structure. Compare *atactic polymer*.

talc *Hydrated* magnesium silicate, $3MgO.4SiO_2.H_2O$. It is used as a lubricant, a filler in paper, paint, etc., and in talcum powder.

tall oil A resinous substance obtained as a *by-product* in the manufacture of wood-pulp; it is used in *soaps* and *paints*.

tallow The rendered fat of animals, particularly cattle and sheep. It consists of various *glycerides*.

tandem generator An *accelerator* of the *electrostatic generator* type. The name is derived from the fact that it consists essentially of two *Van der Graaff generators* in series, thus enabling twice as much *energy* to be obtained for a given accelerating *potential* as could be obtained from a single machine. Negative *ions* are accelerated from ground potential, the *electrons* are then 'stripped' off and the positive particles accelerated back to ground potential. This type of machine can achieve particle energies up to 30 MeV.

tangent galvanometer A *galvanometer* consisting of a coil of wire (n turns of radius r) held in a vertical plane parallel to the Earth's *magnetic field*, H, with a small magnetic needle pivoted at the centre of the coil that is free to rotate in a horizontal plane. A direct *electric current*, I, flowing through the coil produces a magnetic field at right angles to that of the Earth. The needle takes up the direction of the *resultant* of these two fields: if θ is the *angle* of deflection of the needle from its equilibrium position parallel to the Earth's field, then the current will be given by: $I = Hr\tan\theta/2\pi n$. It is now used to measure the Earth's magnetic field, rather than a current.

tangent of an angle See *trigonometrical ratios*.

tangent to a curve A straight line touching the curve at a point. The tangent to a *circle* at any point is at right angles to the radius of the circle at that point.

tannic acid A white amorphous solid, extracted from gall nuts; it is a polymeric *ester*-type derivative of *gallic acid* and *glucose*. It is used in tanning, as a *mordant* in dyeing, and in ink manufacture.

tanning The conversion of raw animal hide into leather by the action of substances containing *tannins*, *tannic acid*, or other agents.

tannins A class of complex *organic compounds* of vegetable origin. Compounds consist of *mixtures* of *derivatives* of polyhydroxybenzoic acids; e.g. *tannic acid*.

tantalum Ta. Element. R.a.m. 180.9479. At. No. 73. A greyish-white *metal* that is very ductile and malleable. R.d. 16.67, m.p. 3000°C., b.p. 5500°C. It occurs together with niobium in a few rare *minerals* and is extracted by *reduction* of the *oxide* with carbon in an *electric-arc furnace* or by dissolving in *hydrofluoric acid* and reducing the *fluoride* with sodium. Ta-181 is stable and Ta-180 has a *half-life* of 10^7 years. It is used for electric lamp *filaments*, electronic components, in *alloys*, in cemented *carbides* for very hard tools, in surgical components (because of its low reactivity), and in electrolytic *rectifiers*.

tape recording See *magnetic tape*.

tar Various dark viscous *organic* materials; e.g. *coal-tar*.

tar sand See *oil sand*.

tartar See *argol*.

tartar emetic See *antimony potassium tartrate*.

tartaric acid 2,3-dihydroxybutanedioic acid. COOH.(CH.OH)$_2$.COOH. An *organic acid* existing in four stereoisomeric forms (see *stereoisomerism*). The common form, *d*-tartaric acid, obtained from *argol*, is a white *soluble* crystalline *solid*, m.p. 170°C. It is used in dyeing, calico-printing, and in making *baking-powder* and effervescent 'health salts'. *dl*-tartaric acid (*racemic acid*), m.p. 203–4°C., occurs in grapes.

tartrate A *salt* or *ester* of *tartaric acid*.

tauon Tau particle. A heavy negatively charged *lepton*. It has a very short lifetime (3×10^{-13} second) and a mass of approximately 1782 MeV (i.e. about 3500 times heavier than an electron). See *elementary particles* and Appendix, Table 6.

taurine NH$_2$(CH$_2$)$_2$SO$_3$H. A white crystalline substance, m.p. 328°C., obtained from the *bile* of mammals.

tautomerism Dynamic isomerism. The existence of a *compound* as a mixture of two *isomers* in equilibrium. The two forms are interconvertible and removal of one of the forms from the *mixture* results in the conversion of part of the other to restore the equilibrium; but each of the two forms may give rise to a stable series of *derivatives*. A substance exhibiting this property is called a 'tautomer'. See also *keto-enol tautomerism*.

taxon (plural **taxa**) A unit in biological *classification*: a group of *organisms* possessing similar characteristics. In evolutionary biology, an aim is that each taxon should be a clade, i.e. sharing common characteristics because all the organisms of that taxon are descended from a common ancestral population.

taxonomy The study of the *classification* of *organisms*.

tear gases Lachrymators. Substances that can be distributed in the form of a *vapour* or *smoke*, producing an irritating effect on the eyes.

technetium Masurium. Te. Element. At. No. 43. A radioactive metal, r.d. 11.496, m.p. 2200°C., b.p. 4600°C. The most stable *isotope*, technetium-98, has a *half-life* of 2.6×10^6 years. It is not found in nature but formed as a *fission product* of uranium.

Teflon* See *polytetrafluoroethene*; *fluorocarbons*.

tektites Small glass-like bodies whose chemical composition is unrelated to the geological formations in which they are found: they are believed to be associated with *meteorites* of extraterrestrial origin. Carbonaceous tektites contain traces of carbon *compounds*.

telecommunications The communication of signals, images, sounds, or other information by line or *radio* transmission.

telemeter Any apparatus for recording a physical event at a distance. Information is transmitted either by line (e.g. from the *core* of a *nuclear reactor*) or by radio (e.g. from an artificial *satellite* that transmits measurements made in *space* back to *Earth* by *radio*).

telephoto lens A combination of a *convex* and a *concave lens*, used to replace the ordinary lens of a *camera* in order to magnify the normal image. The size of the image obtained on the photographic *film* varies as the *focal length* of the lens. The telephoto lens system increases the effective focal length without the necessity of increasing the distance between the film and the lens.

telescope A device for viewing magnified images of distant objects. In the refracting telescope the *objective* is a large *convex lens* that produces a small bright real *image*; this is viewed through the *eyepiece*, which is another convex lens, serving to magnify the image. In the reflecting telescope a large *concave* mirror (see *speculum*) is used instead of the objective lens to produce the real image, which is then magnified by the eyepiece. For terrestrial needs, these types of telescope are unsuitable, since the images formed are inverted, but they are widely used in astronomical telescopes. For terrestrial purposes telescopes are equipped with a further lens or *prism* that causes the image to be seen erect. See also *Cassegrainian, coudé, Galilean, Gregorian, Herschelian, Maksutov, Newtonian, Schmidt*, and *radio telescopes*.

television The transmission of visible moving images by electrical means. In 'closed circuit' television the transmission is by line; in 'broadcast' television it is by *radio* waves. In either case *light* waves are converted into electrical impulses by a televison *camera* and reconverted into a picture on the screen of a *cathode-ray tube* in the receiver. In broadcast television the transmitter consists of equipment for broadcasting modulated *radio frequency electromagnetic radiations* representing a complete television signal, which includes *sound*, vision, and synchronizing signals. The receiver is based on the *superheterodyne* principle, the sound and vision signals being fed to separate *intermediate frequency* amplifiers, *detectors*, and output stages.

telluride A *binary compound* of an *element* or group with tellurium.

tellurium Te. Element. R.a.m. 127.6. At. No. 52. A silvery-white brittle non-metal, resembling sulphur in its chemical properties, r.d. 6.24, m.p. 450°C., b.p. 1000°C. It exists in several *allotropic forms*, and is used in *alloys*, for colouring *glass*, and in *semiconductors*.

temperament The distribution of the intervals between the notes of the musical scale. If the 12 keys are arranged so that the fundamental of each is the fifth of its predecessor, the interval between a low C and a high C seven octaves above it would be $(3/2)^{12} = 129.75$, which is not equal to the basic interval $2^7 = 128$. The difference $(129.75 - 128)$ is called the comma of Pythagoras. In the equal-temperament scale, advocated by J.S. Bach, the comma of Pythagoras is distributed equally between the 12 intervals of the scale over seven octaves. This makes a fifth $(128)^{1/12} = 1.4983$, instead of 1.5. See also *pitch of a note*.

temperature The property of a body or region of space determining the rate at which *heat* will be transferred to or from it. Temperature is a measure of the *kinetic energy* of the *molecules*, *atoms*, or *ions* of which *matter* is composed. The basic physical quantity, the *thermodynamic temperature*, is expressed in *kelvins*. These units are also used in the *International Practical Temperature Scale*. Other scales of temperature are the *Celsius* (*Centigrade*), *Fahrenheit*, and *Réaumur scales*.

tempering of steel Imparting a definite degree of hardness to *steel* by heating to a definite *temperature* (which is sometimes determined by the *colour* the steel assumes), holding it at that temperature for a specified time, and then *quenching*, i.e. cooling, in *oil* or *water*.

temporary hardness of water Hardness of *water* that is destroyed by *boiling*. See *hard water*.

temporary magnetism Induced magnetism. *Magnetism* that a body (e.g. soft iron) possesses only by virtue of being in a *magnetic field* and that largely disappears on removing the body from the field.

tenorite A naturally occurring *copper(II) oxide* that consists of small black scales. It is found in volcanic regions and in copper veins.

tensile stress An axial force per unit area that tends to extend a body to which it is applied. The ratio of the extension to its original length is called the *strain*. Compare *compressive stress*. See also *stress*.

tensimeter A manometer with sealed bulbs attached to each limb, used for measuring *vapour pressure*. If one of the bulbs contains a liquid of known vapour pressure (often water), the vapour pressure of the other liquid in the other bulb can be measured.

tensiometer 1. An apparatus for measuring the *surface tension* of a *liquid*. **2.** An apparatus for measuring the tension in a wire, fibre, or beam. **3.** An apparatus for measuring the moisture content of *soil*.

tensor A magnitude or set of *functions* by which the components of a system are transformed from one system of *coordinates* to another: a quantity expressing the ratio in which the length of a *vector* is increased.

tera- Prefix denoting one million million times; 10^{12}. Symbol T, e.g. Tm = 10^{12} *metres*.

teratogen A factor in the environment (e.g. *ionizing radiation* or a drug) that creates an abnormality in a *foetus*.

terbium Tb. Element. R.a.m. 158.9254. At. No. 65. A silvery metal, r.d. 8.267, m.p. 1360°C., b.p. 2500°C. The only stable *isotope* is terbium-159. It is used to dope certain *semiconductors*. See *lanthanides*.

terephthalic acid Benzene-1,4-dicarboxylic acid. $C_6H_4(COOH)_2$. A white insoluble crystalline substance, the *para-isomer* of phthalic acid, that sublimes (see *sublimation*) without melting above 300°C. It is used in the manufacture of *polyesters*; in particular, *polyethene terephthalate*.

terminal 1. (phys.) The point at which an electrical connection is made; the point, or the connecting device, at which *current* enters or leaves a piece of electric equipment. **2.** An input or output device connected to a *computer*; it is usually a *visual-display unit* connected to a keyboard and sometimes a printer. If it is capable of storing and processing data itself, rather than relying on a central processor, it is known as an 'intelligent terminal'.

terminal speed If a body free to move in a resisting medium is acted upon by a constant *force* (e.g. a body falling under the force of *gravity* through the atmosphere), the body accelerates until a certain terminal speed is reached, after which the *speed* remains constant. See *Stokes' Law*.

terminator The line on the surface of the *Moon*, or a *planet*, that separates the dark and light hemispheres.

ternary compound A chemical *compound* consisting of three *elements*. E.g. HNO_3 (*nitric acid*).

ternary fission A very rare form of *nuclear fission* as a result of which a heavy *nucleus* breaks up into three fragments of comparable *mass*. The term is also used for the more frequent case in which one of the three fragments (e.g. an *alpha-particle*) is much lighter than the others.

terpenes A class of *hydrocarbons* occurring in many fragrant *essential oils* of plants. They are colourless *liquids*, generally with a pleasant smell. Terpenes have the general formula $(C_5H_8)n$ and are made up of *isoprene* units. Monoterpenes (monocyclic terpenes) consist of two isoprene units, e.g. *pinene* $C_{10}H_{16}$. Sesquiterpenes have three units, $C_{15}H_{24}$, etc.

terpineol $C_{10}H_{17}OH$. Several *isomeric unsaturated alcohols* that occur in *essential oils*. α-terpineol, m.p. 35°C., b.p. 220°C., is used as a *solvent* and in perfumes.

Terramycin* Oxytetracycline. $C_{22}H_{30}N_2O_{11}$. An *antibiotic* powder obtained from *Streptomyces rimosus* bacteria, used to combat a wide variety of bacterial infections.

terrestrial guidance A method of missile or *rocket* guidance in which the missile steers itself with reference to the strength and direction of the *Earth's gravitational* or *magnetic field* (magnetic guidance).

terrestrial magnetism See *magnetism, terrestrial*.

terrestrial telescope. A *telescope* for use on land or sea, as opposed to an astronomical telescope.

tertiary colour A *colour* obtained by mixing two *secondary colours*. E.g. brown and grey.

tervalent Trivalent. Having a *valence* of three.

Terylene* Tradename for a *polyester* fibre derived from benzene-1,4-dicarboxylic acid (*terephthalic acid*), $C_6H_4(COOH)_2$ and *ethanediol* (ethylene glycol) by *condensation* and *polymerization*. It is widely used for making fabrics.

tesla The derived *SI unit* of *magnetic flux density*, defined as the density of one *weber* of *magnetic flux* per square *metre*. Symbol T. Named after Nikola Tesla (1870–1943).

Tesla coil A *transformer* for producing high *voltages* at high *frequencies*, consisting of a coil the primary circuit of which has a small number of turns but includes a spark gap and a fixed *capacitor*. The secondary winding has a large number of turns and the secondary circuit is *tuned*, by means of a variable capacitor, to resonate with the primary.

testosterone $C_{19}H_{28}O_2$. A male *sex hormone* (*androgen*), which in the pure form consists of a white *insoluble* crystalline substance, m.p. 155°C., whose function is to promote the development of male characteristics.

tetra- Prefix denoting four, fourfold.

tetraarsenic tetrasulphide Realgar. As_4S_4. A red *insoluble* poisonous powder, m.p. 307°C., used in the manufacture of fireworks. Before its tetrahedral structure was known it was called arsenic disulphide, As_2S_2.

tetrachloroethene Perchlorethylene. $Cl_2C{:}CCl_2$. A colourless non-flammable *liquid*, b.p. 121°C., used as a *solvent* and in dry cleaning.

tetrachloromethane Carbon tetrachloride. CCl_4. A heavy colourless *liquid* with a sweetish smell, b.p. 76.8°C. It was formerly used as a *solvent* and *fire extinguisher*, but has been replaced for those purposes as the moist compound is partially decomposed to yield the highly toxic gas *phosgene*.

tetracyclines A group of *antibiotics* derived from *Streptomyces* bacteria. They are effective against a wide range of bacterial infections.

tetrad An *element* having a *valence* of four.

tetraethyllead Lead tetraethyl(IV). $(C_2H_5)_4Pb$. A colourless oily *liquid*, used to reduce *knocking* in petrol engines. Because it can result in unacceptably high quantities of atmospheric lead its use has been restricted or prohibited in some countries. In the UK and some other countries, lead-free petrol is given a tax advantage to encourage its use.

tetraethyl pyrophosphate TEPP. $(C_2H_5)_4P_2O_7$. A colourless *hygroscopic liquid*, b.p. 155°C., used as an *insecticide* and rat poison.

tetrafluoroethene $CF_2{:}CF_2$. An unsaturated gaseous fluorocarbon, b.p. -76.3°C., that polymerizes (see *polymerization*) into a *thermoplastic* material with good electrical insulation properties. See *polytetrafluoroethene*.

tetrahedron A four-faced *solid* figure contained by four *triangles*; a *pyramid* with a triangular base.

tetrahydrofuran THF. C_4H_8O. A colourless *volatile* liquid, b.p. 67°C., widely used as a *solvent*.

tetranitromethane $C(NO_2)_4$. A colourless *volatile liquid*, b.p. 126°C., used as an *oxidant* in *rockets*.

tetraoxosulphuric(VI) acid See *sulphuric acids*.

tetravalent Quadrivalent. Having a *valence* of four.

tetrode A *thermionic valve* containing four *electrodes*; a *cathode*, an *anode* or plate, a *control grid*, and (between the two latter) a *screen grid*.

thalidomide $C_{13}H_{10}N_2O_4$. A white crystalline substance, formerly used as a *tranquillizer* but found to be the cause of deformed children when taken by pregnant women.

thallium Tl. Element. R.a.m. 204.383. At. No. 81. A white malleable *metal* resembling lead, r.d. 11.87, m.p. 303.5°C., b.p. 1460°C. It is used in *alloys*; its *salts* are used in *insecticides* and rat poisons.

thallium scan See *nuclear medicine*.

thebaine $C_{19}H_{21}NO_3$. A white *insoluble* substance, m.p. 193°C., present in *opium* in small quantities.

theine See *caffeine*.

theobromine $C_7H_8N_4O_2$. A white *insoluble* crystalline *alkaloid*, m.p. 337°C., that is *isomeric* with theophylline, m.p. 272°C. Both occur in tea and are used in medicine.

theodolite An instrument for the measurement of *angles*, used in surveying. It consists essentially of a *telescope* moving along a circular scale graduated in *degrees*.

theophylline See *theobromine*.

theorem A statement or proposition that is proved by logical reasoning from given facts and justifiable assumptions.

theory of games A mathematical treatment of competitive games with special reference to the strategic and tactical decisions that have to be made in situations involving conflicting interests in the light of specific odds and *probabilities*. The theory is extended for use in military and commercial situations.

therapeutics Healing; remedial treatment of diseases.

therm A practical unit of quantity of *heat*; 100 000 *British thermal units*, 25 200 000 *calories*, $1.055\,06 \times 10^8$ *joules*.

thermal analysis See *thermographic analysis*.

thermal barrier The limit to the *speed* with which an aircraft or *rocket* can travel in the Earth's *atmosphere* due to overheating caused by *friction* with the atmospheric *molecules*.

thermal capacity See *heat capacity*.

thermal conductivity See *conductivity, thermal*.

thermal cross-section A nuclear *cross-section* as measured with *thermal neutrons*.

thermal diffusion If a *temperature* gradient is maintained over a *volume* of *gas* containing *molecules* of different *masses*, the heavier molecules tend to diffuse down the temperature gradient, and the lighter molecules in the opposite direction. This forms the basis of a method of separating the different *isotopes* of an *element* in certain cases.

thermal diffusivity *a*. The quantity defined as $\lambda/\rho c_p$, where λ is the thermal *conductivity*, ρ is the *density*, and c_p is the *specific heat capacity* at constant pressure. It has the unit $m^2\,s^{-1}$.

thermal dissociation See *dissociation*.

thermal emission The emission of *electrons* from a solid surface as a result of raising its temperature (see also *thermionic emission*). Compare *cold emission*.

thermal equilibrium The state of a system in which there is no net flow of *heat* between its components.

thermal ionization The ionization of a substance A, by raising its *temperature* sufficiently for the reaction

$$A \rightarrow A^+ + e$$

to occur. The temperature required will depend on the *ionization potential* of the substance, the lower the ionization potential the lower the temperature required to cause ionization.

thermalize To bring *neutrons* into *thermal equilibrium* with their surroundings; to reduce the *energy* of neutrons with a *moderator*; to produce *thermal neutrons*.

thermal neutrons *Neutrons* of very slow speed and consequently of low *energy*. Their average *kinetic energy* is of the same order as the thermal energy of the *atoms* or *molecules* of the substance through which they are passing; i.e. about 0.025 eV, which is equivalent to a most probable *speed* of about 2200 metres per second at 20°C. Thermal neutrons are responsible for numerous types of *nuclear reactions*, including *nuclear fission*.

thermal pollution See *pollution*.

thermal power station 1. A power station in which electricity is generated by *combustion* of *coal*, *oil*, etc., as opposed to a nuclear power station, hydroelectric power station, etc. **2.** A nuclear power station in which electricity is generated by a *thermal reactor*, rather than a *fast reactor*.

thermal reactor A *nuclear reactor* in which most of the *nuclear fissions* are caused by *thermal neutrons*.

thermal spike The zone of high *temperature* briefly produced in a substance along the path of a high *energy* particle or *nuclear fission* fragment.

thermion An *ion* emitted by a hot body.

thermionic emission The emission of *electrons* from a heated *metal* usually in a *vacuum*, especially in the *electron gun* of *cathode-ray tubes* and formerly in *thermionic valves*. See *Richardson equation*.

thermionics The branch of *electronics* dealing with the emission of *electrons* from substances at high temperatures, particularly the study and design of *electron guns* and *thermionic valves*.

thermionic valve or tube A system of *electrodes* arranged in an evacuated *glass* or *metal* envelope. For special purposes a *gas* at low *pressure* may be introduced into the valve. The electrodes are: (1) a *cathode* that emits *electrons* when heated; (2) an *anode* or plate maintained at a positive *potential* with respect to the cathode; the electrons emitted by the latter are attracted to it. Most valves also contain a number of perforated electrodes or grids (see *control grid, screen grid, suppressor grid*) interposed between the cathode and anode, designed to control the flow of current through the valve. The cathode can be in the form of a *filament* heated by an *electric current* passing through it, or an electrode heated indirectly by a separate filament. See *diode, triode, tetrode, pentode*.

thermistor A *semiconductor*, the electrical *resistance* of which decreases rapidly with increase of *temperature*; e.g. the resistance may be of the order of 10^5 ohms at 20°C. and only 10 ohms at 100°C. It is used as a sensitive temperature-measuring device and to compensate for temperature variations of other components in a circuit.

thermite Thermit*. A *mixture* of aluminium powder and the *oxide* of a *metal*, e.g. *iron(III) oxide*. When ignited by magnesium ribbon, a *chemical reaction* begins in which the aluminium combines with the oxygen of the oxide, forming *aluminium oxide* and the metal:

$$2Al + Fe_2O_3 \rightarrow Al_2O_3 + 2Fe$$

A great quantity of *heat* is given out during the reaction, the reduced metal (see *reduction*) appearing in the molten state. The mixture is used for *welding* iron and *steel*, and in incendiary bombs; the principle is applied in the *extraction* of certain metals from their oxides (see *Goldschmidt process*).

thermobarograph An instrument for measuring and recording atmospheric *temperature* and *pressure*, consisting of a *thermograph* and a *barograph*.

thermochemistry The branch of *physical chemistry* dealing with the quantities of *heat* absorbed or evolved during *chemical reactions*. See *heat of reaction; Hess's law*.

thermocouple An instrument for the measurement of *temperature*. It consists of two wires of different *metals* joined at each end. One junction is at the point at which the temperature is to be measured and the other is kept at a lower fixed temperature. Owing to this difference of temperature of the junctions, a thermo-electric *EMF* is generated, the magnitude of which is related to the temperature difference. It causes an *electric current* to flow in the circuit (see *Seebeck effect*).

This current can be measured by means of a *galvanometer* in the circuit, or the thermoelectric EMF can be measured using a *potentiometer*.

thermodynamic energy *U*. See *internal energy*.

thermodynamics The study of the general laws governing processes that involve *heat* changes and the availability of *energy* to do *work*.

thermodynamics, laws of 1. The law of the conservation of *energy*. In a system of constant *mass*, energy can be neither created nor destroyed. The law is usually stated in the form:

$$\Delta U = Q + W,$$

where *U* is the *internal energy* of a system of constant mass, *Q* is the *heat* transferred to or from the system, and *W* is the *work* done on or by the system (taken as positive if work is done on the system). 2. Heat cannot be transferred by any continuous self-sustaining process from a colder to a hotter body. Or stated im terms of *entropy*; the entropy of a closed system increases with time. This is usually stated in the form:

$$\Delta U = T\Delta S - W,$$

where ΔS is the change in entropy of the system when its *thermodynamic temperature* is *T*. 3. See the *Nernst heat theorem*. The consequence of this law is that the *absolute zero* of temperature can never be attained.

One other law is fundamental to these three laws and is sometimes known as the zeroth law of thermodynamics. This states that if two bodies are each in thermal equilibrium with a third body, then the three bodies are in thermal equilibrium with each other.

thermodynamic temperature *T*. Although formerly referred to as a scale of *temperature* (Kelvin scale of temperature, or absolute scale of temperature), the concept of a temperature scale is now restricted to the *International Practical Temperature Scale*. The thermodynamic temperature is a basic physical quantity that depends on the concept of the *efficiency* of an ideal *heat engine* (see *Carnot's principle*), which is independent of the nature of the working substance. In practice, thermodynamic temperatures cannot be measured directly, they are inferred from measurements using a *gas thermometer* containing a nearly *perfect gas*. This is possible because the *internal energy*, *U*, of a *mole* of monatomic gas is given by $2RT/3$, where *R* is the universal *gas constant*. Originally, thermodynamic temperature was defined in terms of the *ice point* and *steam point* of *water* using a gas thermometer. However, in 1954 this was replaced by a definition using only one fixed point, the *triple point* of water, which was fixed as 273.16 *kelvins* exactly. The magnitude of the unit of thermodynamic temperature, the kelvin, is the same as the *degree* on the International Practical Scale of Temperature.

thermoelectric effect See *Seebeck effect*.

thermoelectricity An *electric current* produced by the direct conversion of *heat* energy into *electrical energy*. See *Peltier effect*; *Seebeck effect*; *Thomson effect*.

thermograph 1. A self-registering *thermometer*; an apparatus that records *temperature* variations during a period of time on a *graph*. 2. Thermogram. A record obtained by means of *thermography*.

thermographic analysis A group of methods of chemical analysis based on recording changes of mass (thermogravimetric analysis) due to decomposition, or of temperature (heating curves) due to *endothermic* or *exothermic* processes, when substances that undergo chemical changes on heating are heated at a definite rate.

thermography A technique used in medicine to detect cancers, especially of the breast. A tumour has an abnormally high blood supply and therefore the skin above it is at a slightly raised temperature. This is revealed by taking a photograph (*thermograph* or thermogram) of the infrared radiation emitted by an area of skin, using special film sensitive to infrared radiation.

thermoluminescence *Luminescence* resulting from a rise in the *temperature* of a body or substance. It occurs when *electrons*, trapped in crystal *defects*, are freed by heating the crystals.

As these defects are usually caused by *ionizing radiation*, the property is used as a method of dating archaeological remains, especially pottery. In 'thermoluminescent dating', the number of trapped electrons can be assumed to be related to the quantity of radiation to which the pottery has been subjected since it was fired. By assuming that this quantity is related to its age an estimate of age can be obtained by measuring the amount of light emitted by the pottery on heating and comparing it with the thermoluminescence of a similar material of known age.

thermoluminescent dating See *thermoluminescence*.

thermometer An instrument for the measurement of *temperature*. Any physical property of a substance that varies with temperature can be used to measure the latter; e.g. the *volume* of a *liquid* or *gas* maintained under a fixed *pressure*; the pressure of a gas at constant volume; the electrical *resistance* of a *conductor*; the *EMF* produced at a *thermocouple* junction, etc. The property chosen depends on the temperature range, the accuracy required, and the ease with which the instrument can be made and used. The common mercury thermometer depends upon the expansion of mercury with rise in temperature. The mercury is contained in a bulb attached to a narrow graduated sealed tube; the expansion of the mercury in the bulb causes a thin thread of it to rise in the tube. See also *gas thermometer*; *pyrometers*; *resistance thermometer*; *thermocouple*; *Beckmann thermometer*; *thermometer, clinical*; *thermometer, maximum and minimum*.

thermometer, clinical A mercury *thermometer* designed to measure the *temperature* of the human body, and graduated to cover a range of a few degrees on either side of the normal body temperature. A constriction in the tube near the bulb causes the mercury thread to break when the thermometer is taken away from the warm body, and the mercury in the bulb starts to contract. The thread thus remains in the tube to indicate the maximum temperature reached, until it is shaken down. For many purposes, clinical thermometers using a sensitive *thermocouple* and a digital display are now available.

thermometer, maximum and minimum A thermometer that records the highest and lowest *temperatures* reached during a period of time. It consists of a bulb filled with *alcohol*, which, by expansion, pushes a mercury thread along a fine tube, graduated in degrees. At each end of the mercury thread is a small *steel* 'index' that is pushed by the mercury; one is thus left at the farthest point reached by the mercury thread, corresponding to the maximum temperature, and the other at the lowest point.

thermomilliammeter An instrument for measuring small alternating *electric currents*. The current passes through a wire made of *constantan* or platinum, which is in contact with or very close to a *thermocouple*. The thermocouple is connected to a sensitive *milliammeter*, the heat of the constantan wire producing a thermoelectric current in the thermocouple; this current is recorded by the milliammeter. In a more sensitive instrument, the heater wire and thermocouple are arranged in an evacuated *quartz* envelope.

thermonuclear bomb See *nuclear weapons*.

thermonuclear reaction A *nuclear fusion* reaction in which the interacting particles or *nuclei* possess sufficient *kinetic energy*, as a result of their thermal agitation, to initiate and sustain the process. The hydrogen bomb (see *nuclear weapons*) makes use of thermonuclear reactions by employing a fission bomb to attain the required *temperature*, which is in excess of $40 \times 10^6 \,°C$. Controlled thermonuclear reactions attempt to make use of fusion reactions in *deuterium* and *tritium* gas at a temperature in the range 40×10^6 to $5 \times 10^9 \,°C$., for the purpose of generating *electrical energy*. The central problem in achieving this end is that of *containment*, i.e. separating the *plasma* (or high temperature ionized gas) from the walls of the containing vessel for long enough for the fusion energy released to enable the plasma to reach its ignition temperature (see *Lawson criterion*). In magnetic containment the plasma may be contained either by use of externally applied *magnetic fields*, or by the magnetic fields produced by currents flowing in the plasma itself (see *pinch effect*). The nature and instabilities of these magnetic fields are the subject of contemporary research. The machines in which these experiments are carried out may be classified according to whether the *magnetic lines of force* of the containing field are closed- or open-ended. The closed field group include *torus*-shaped machines called Tokamaks, while the open-ended machines include those using *magnetic mirrors* or rotating plasmas. The Tokamak configuration originated in the USSR but is used by other countries as the most hopeful type of machine to succeed in generating electricity. The Joint European Torus (JET) at Culham in the UK is a Tokamak device.

Another approach, pellet fusion, uses a tiny pellet of nuclear fuel in which fusion is initiated by means of a *laser* beam.

Methods of extracting the energy from a Tokamak or other type of machine fall into two categories. If most of the energy is in the form of energetic neutrons, as in the deuterium-tritium reaction:

$$D + T \rightarrow He + n + 17.6 \,MeV$$

the neutron energy would need to be absorbed by a *coolant* (e.g. liquid lithium) surrounding the reactor. The heat absorbed would then be transferred to a conventional steam-raising cycle. On the other hand if most of the energy is carried by charged particles, e.g. by:

$$D + D \rightarrow T + p + 4 \,MeV,$$

some of this energy might be used directly to drive an electric current, without a steam-raising cycle.

thermopile An instrument for detecting and measuring *heat* radiations. It consists of a number of rods of antimony and bismuth, connected alternately in series. When the hot blackened junctions are exposed to heat and the cold junctions are shielded, the thermoelectric current produced (see *thermocouple*) may be detected or measured by a sensitive *galvanometer*.

thermoplastic A substance that becomes plastic on being heated; a *plastic* material that can be repeatedly melted or softened by heat without change of properties.

446

thermosetting plastics *Plastics* that, having once been subjected to *heat* (and *pressure*), lose their plasticity.

thermosphere The region of the *upper atmosphere* in which the *temperature* increases with *altitude*. See Fig. 44, under *upper atmosphere*.

thermostat An instrument for maintaining a constant *temperature* by the use of a device that cuts off the supply of *heat* when the required temperature is exceeded and automatically restores the supply when the temperature falls below that required. It usually consists of a *bimetallic strip* so arranged that when it is heated (or cooled) the power supply contacts are opened (or closed).

theta pinch See *pinch effect*.

THF See *tetrahydrofuran*.

thiamine Aneurin. Vitamin B_1. See *vitamins*.

thiazines A group of *compounds* consisting of a six-membered ring, four of which are carbon *atoms,* one of which is a sulphur atom, and one a nitrogen atom.

thiazole $S.CH:N.CH:CH$. A colourless *volatile liquid*, b.p. 116.8°C., whose *molecule* consists of a five-membered ring. *Derivatives* are used in dyestuffs and in medicine.

thin-layer chromatography A form of *chromatography* in which the stationary phase consists of a thin layer of alumina *slurry* on a glass plate. After selective absorption of the mobile phase the plate is dried in an oven. The technique is very similar to that of *paper chromatography*.

thio- Prefix denoting sulphur, in the naming of chemical *compounds*.

thioacetamide CH_3CSNH_2. A colourless *soluble* crystalline substance, m.p. 115–16°C., used as a source of *hydrogen sulphide*.

thiocarbamide See *thiourea*.

thiocyanate A *salt* or *ester* of *thiocyanic acid*.

thiocyanic acid HSCN. An *unstable acid* that forms *salts* called *thiocyanates*.

thio ethers A group of compounds with the general formula RSR′, where R and R′ are hydrocarbon radicals.

Thiokols* *Rubber*-like *polymer* materials of the general formula $(RS_x)_n$, where R is an *organic bivalent radical*, and x is usually between 2 and 4. They are very resistant to the swelling action of *oils*, and undergo a form of *vulcanization* on being heated with certain metallic *oxides*.

thiolates Metallic salts of *thiols*, formerly known as "mercaptides"; they are sulphur analogues of *alcoholates*.

thiols A class of organic compounds of the general formula RSH, with sulphur attached directly to carbon; they are the sulphur analogues of alcohols, containing SH instead of OH groups. They were formerly called mercaptans but are now named with reference to the parent hydrocarbon, e.g. C_2H_5SH is *ethanethiol*.

thionin $C_{12}H_9N_3S$. A dark brown *thiazine derivative*, used as a *dye* in microscopy.

thionyl chloride See *sulphur dichloride oxide*.

thiophene C_4H_4S. A colourless liquid *heterocyclic* compound, b.p. 84.0°C., with a nauseating stench. It occurs in coal-tar; it is used as a solvent and in the manufacture of dyes, plastics, and pharmaceutical products.

thiosulphate A *salt* or *ester* of *thiosulphuric acid*.

thiosulphuric acid $H_2S_2O_3$. An *unstable acid* formed by replacing one oxygen *atom* of *sulphuric acid* (H_2SO_4) by one sulphur atom. It is known only in *solution* or in the form of its *salts* or *esters*, the *thiosulphates*.

thiourea Thiocarbamide. $NH_2CS.NH_2$. A colourless *organic compound*, m.p. 180°C., used in the manufacture of thiourea-aldehyde *plastics*.

thixotropy The rate of change of *viscosity* with time. Certain *liquids* possess the property of increasing in viscosity with the passage of time when the liquid is left undisturbed. On shaking, the viscosity returns to its original value. This property is made use of in nondrip paints, which are more viscous on the brush than when they are being worked on the wall. It is also used in some lubricants, which become thinner as the parts they are lubricating begin to move.

Thomson effect Kelvin effect. A *temperature* gradient along a conducting wire gives rise to an *electric potential* gradient along the wire. Named after Sir William Thomson (Lord Kelvin) (1824–1907).

Thomson scattering The scattering of *photons* of *electromagnetic radiations* by *electrons* according to classical theory, i.e. the energy lost by the photons is the radiation emitted by the electrons when they are accelerated by the transverse electric field of the radiation. Named after Sir Joseph John Thomson (1856–1940).

thoria See *thorium dioxide*.

thorides Natural *radioisotopes* that occur in the *radioactive series* containing thorium.

thorite A mineral consisting of thorium silicate, $ThSiO_4$; used as a source of thorium.

thorium Th. Element. R.a.m. 232.038. At. No. 90. A dark grey radioactive metal, r.d. 11.72, m.p. 1700°C., b.p. 4500°C. The most stable *isotope*, thorium-232, has a *half-life* of 1.4×10^{10} years. Compounds occur in *monazite* and *thorite*. It is used in alloys and as *nuclear fuel* in some *breeder reactors* (Th-232 captures slow *neutrons*, breeding U-235).

thorium dioxide Thoria. ThO_2. A white *insoluble* powder, m.p. 3050°C., used in *gas mantles*, *refractories* and special *glasses*.

thoron A gaseous *radioisotope* of radon, radon-220, produced by the *disintegration* of thorium. It has a *half-life* of 51.5 *seconds*.

threnardite A mineral form of *sodium sulphate*, Na_2SO_4.

threonine A colourless *soluble* crystalline *amino acid*, m.p. 230°C., that is essential to the diet of animals. See Appendix, Table 5.

threshold The lowest value of any stimulus, signal, or agency that will produce a specified effect. E.g. *threshold frequency*.

threshold freqency *Light* falling on a *metal* surface will give rise to the emission of *electrons* (see *photoelectric effect*) only if the *frequency* of the light is greater than a certain *threshold* value, which is characteristic of the metal used.

thrombin An *enzyme* formed in the *blood* of vertebrates that acts upon *fibrinogen* to form *fibrin*; it is therefore essential to the process of blood clotting. Thrombin is formed from a blood *protein*, prothrombin.

thrombocytes See *blood platelets*.

thrust The propulsive *force* produced by a *reaction propulsion* motor. See also *specific impulse*.

thulium Tm. Element. R.a.m. 168.9342. At. No. 69. A soft grey *metal*, r.d. 9.33, m.p. 1550°C., b.p. 1900°C. It has one natural *isotope*, thulium-169. See *lanthanides*.

thymine 5-Methyluracil. $C_5H_6N_2O_2$. One of the two *pyrimidine* bases occurring in the *nucleotides* of *deoxyribonucleic acid*, which plays a part in the formulation of the *genetic code*.

thymol 3-Hydroxy-1-isopropyl-4-methyl benzene. $C_{10}H_{14}O$. A white crystalline *phenol derivative*, m.p. 51.5°C., b.p. 233.5°C., that smells of thyme. It occurs in many *essential oils* and is used as a mild *antiseptic*.

thyratron A gas-filled *thermionic valve* (usually a *triode*) in which a *voltage* applied to the *control grid* initiates, but does not limit, the *anode* current. It was formerly widely used as an *electronic* switch but has now been largely replaced by the *thyristor*.

thyristor A *silicon-controlled rectifier* in which the *anode-cathode* current is controlled by a *gate*. It usually consists of three *p-n* junctions on a four-layer chip. It has largely replaced the *thyratron*, to which it has similar current-voltage characteristics.

thyroxin(e) $C_{15}H_{11}I_4NO_4$. An iodine-containing *amino-acid hormone* produced by the thyroid gland. The pure form is a white crystalline substance, m.p. 236°C., used in cases of thyroid deficiency.

tidal energy A *renewable energy source* that could produce an estimated 4×10^{18} joules per annum at known tidal sites throughout the world. The principle is that water collected at high tide behind a barrage is released at low tide to turn a *turbine* that, in turn, drives a *generator*. The Severn estuary in the UK, which has a tidal rise of nearly 9 metres could, it is estimated, provide 7% of the

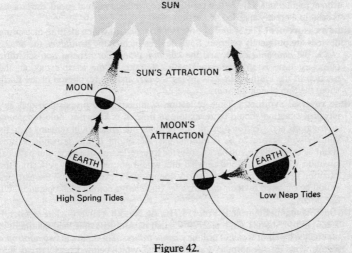

Figure 42.

electric power required in the UK. The capital cost, however, would be very high. French experience with a tidal power station on the river Rance has suggested an important future for this form of energy generation.

tides Movement of the seas caused by the attraction exerted upon the seas by the *Moon*, and to a lesser extent by the *Sun*. At full and new moon the tidal force of the Sun is added to that of the Moon, causing high spring tides; while at half-moons the forces are opposed, causing low neap tides. See Fig. 42. This shows how, on the side of the Earth nearest to the Moon, the seas are pulled away from the Earth, whereas on the other side, the Earth is pulled away from the sea. *Tidal energy* is a potentially large source of *power*.

timbre See *quality of sound*.

time base A voltage that varies in a strictly controlled manner with respect to time, often produced by a time-base generator using a *sawtooth waveform*. The time base is used to deflect the *electron beam* in a *cathode-ray tube*, making the luminous spot sweep the screen linearly, with a rapid return (flyback) to its starting position. A *cathode-ray oscilloscope* has a variable time base, which can be adjusted to suit the purpose for which it is used.

time dilation An effect predicted by the special theory of *relativity*. An observer measures the passage of a time t, on a clock travelling with him. Another observer travelling at a speed v relative to the first has an identical clock travelling with him. It will appear to the first observer that a time $t(1 - v^2/c^2)^{1/2}$ will have elapsed on the second observer's clock (c is the *speed of light*). The effect is only apparent at *relativistic velocities* and has been observed in the motions of some *muons*, which have a longer life time at relativistic velocities.

time exposure A long *photographic* exposure in which the *camera* shutter is operated manually or by some device not normally part of the camera.

time-lapse photography *Photography* in which a slow process, such as the growth of a plant, is photographed by a series of single exposures on cinematic film at regular intervals. When the film is projected at normal speed, the process is seen in a greatly speeded up version.

time measurement The *SI unit* of time is the *second*, to which all time-measuring devices are ultimately referred. Such devices include the *pendulum*, the *quartz clock*, the *ammonia clock*, and the *caesium clock*. For general non-scientific purposes, time is still measured with reference to the Earth's motion; the rotation of the Earth on its axis giving the *day* and the revolution of the Earth about the Sun giving the *year*.

time reversal symmetry Time reflection symmetry. The proposition that any physical situation should be reversible in time. It is known to hold for *strong interactions* and *electromagnetic interactions*, but some doubt remains as to its validity with respect to *weak interactions*. According to this principle, if time could be reversed (i.e. run backwards) the time reflection of a particular physical situation would correspond to what one would normally see by reflecting the situation in a space mirror (the *parity* operation, P), except that all the particles would be replaced by their *antiparticles* (the *charge conjugation* operation, C). See *parity*.

tin Sn. (Stannum.) Element. R.a.m. 118.710. At. No. 50. A silvery-white *metal*, r.d. 7.285, m.p. 231.9681°C., b.p. 2720°C., that is soft, malleable, and ductile. It is unaffected by air or *water* at ordinary *temperatures*. Tin occurs in two *allotropic* forms, white tin, the normal form of the metal, which below 13.2°C. passes into the powdery form known as grey tin. This causes *tin plague* but can be prevented by the addition of small amounts of antimony or bismuth. Tin is found in nature as *tin(IV) oxide*, SnO_2, *cassiterite* or *tinstone*. The metal is extracted by heating the *oxide* with powdered carbon in a *reverberatory furnace*. It is used for tin-plating and in many *alloys*.

tin ash See *tin(IV) oxide*.

tincal An impure form of *disodium tetraborate* (*borax*).

tin chlorides Either of two *compounds*. **1.** Tin(II) chloride, stannous chloride. $SnCl_2$. A white crystalline substance, m.p. 246°C., that forms a dihydrate known as tin salt, m.p. 37.7°C. It is used as a *reducing agent*, *mordant*, and a tinning agent. **2.** Tin(IV) chloride, stannic chloride. $SnCl_4$. A colourless fuming *liquid*, b.p. 114.1°C., used in the manufacture of mordants.

tincture An alcoholic extract or a *solution* in *alcohol*.

tin(IV) oxide Tin dioxide, stannic oxide, tin ash. SnO_2. A white *insoluble* powder, m.p. 1127°C., used in the manufacture of *glass* and polishes. It has two dihydrates, $SnO_2.2H_2O$, known also as α- and β-stannic acids. Tin(IV) oxide is *amphoteric* forming stannates with *alkalis*.

tin plague See *tin*.

tin plate Iron coated with a thin layer of tin, by dipping it into the molten *metal*.

tin salt See *tin chlorides*.

tinstone See *cassiterite*.

tin sulphides Either of two *compounds*. **1.** Tin(II) sulphide, stannous sulphide. SnS. A grey crystalline substance, m.p. 882°C. Above 265°C. it decomposes into tin(IV) sulphide and the *metal*. **2.** Tin(IV) sulphide, stannic sulphide, mosaic gold. SnS_2. A golden insoluble powder, used in the manufacture of gold paint.

tintometer Any of several instruments for comparing the colour of *solutions* with a series of standard solutions or stained glass slides. See also *Lovibond* tintometer*.

tints *Colours* that have the same *hue* but different *saturation*.

tissue culture The cultivation of fragments of the *tissues*, or of the *cells*, of an *organism* so that their growth and biochemical characteristics can be studied in controlled conditions *in vitro*. Tissue cultures are usually maintained in a sterile *physiological saline* and/or *culture medium*.

tissues A collection of similar *cells* and intercellular material, which forms the structural material of a plant or animal. They may be arranged to form *organs*.

titania See *titanium(IV) oxide*.

titanium Ti. Element R.a.m. 47.88, At. No. 22. A malleable and ductile *metal* resembling iron, r.d. 4.5, m.p. 1670°C., b.p. 3300°C. *Compounds* are fairly widely distributed, the principal ore being *rutile*. The metal is extracted by the *Kroll process*. Titanium is widely used where strong light alloys are required, e.g. aircraft, missiles, etc.

titanium(IV) oxide Titanium dioxide, titania. TiO_2. A white insoluble powder, m.p. 1850°C. It occurs in nature in several crystalline forms, including *anatase* and *rutile*. It is used as a white pigment in surface coatings and in the paper and textile industries, in ceramics, etc.

titration An operation forming the basis of *volumetric analysis*. The addition of measured amounts of a *solution* of one *reagent* (called the titrant) from a *burette*

TLR

to a definite amount of another reagent until the action between them is complete, i.e. till the second reagent is completely used up (see *end point*).

TLR Twin-lens reflex. See *reflex camera*.

TNT See *trinitrotoluene*.

tobacco mosaic virus TMV. A simple *virus* widely used in biochemical and biological studies, particularly concerning the transference of the *genetic code*. The virus particle consists of a single helix of *ribonucleic acid* containing some 6400 *nucleotides*, coated with about 2200 *molecules* of a single *protein*, each molecule of which comprises a *polypeptide* chain of 158 *amino acids* in a known sequence.

tocopherol Vitamin E. $C_{20}H_{50}H_2$. See *vitamins*.

Tokamak See *thermonuclear reaction*.

toluene See *methylbenzene*.

toluidine Methylphenylamine. $CH_3C_6H_4NH_2$. An aromatic amine that exists in three isomeric forms: 1,2-(*ortho*), 1,3-(*meta*), and 1,4-(*para*). The former two are liquids, b.p. 200–204°C., the third is a white crystalline solid, m.p. 45°C., b.p. 200°C.; all three are used as organic *intermediates*, especially in the manufacture of *dyes*.

tomography A technique for using *X-rays* to photograph one specific *plane* of the body, for diagnostic purposes. A computerized tomography (CT) scanner (formerly, computerized axial tomography, CAT scanner) is an X-ray machine that rotates through 180° around the horizontal patient, taking X-ray measurements every few degrees. A three-dimensional image is built up from these measurements by the scanner's own computer. The result provides much more information than a normal diagnostic X-ray, for about one fifth of the *dose*.

tone of sound See *quality of sound*.

tonne Metric ton; 1000 *kilograms*; 2204.62 lbs, 0.9842 ton.

tonometer 1. An instrument for measuring the *pitch* of a *sound*, usually consisting of a set of calibrated *tuning forks*. **2.** An instrument for measuring *vapour pressure*. **3.** An instrument for measuring blood pressure, or the pressure within an eyeball.

topaz A crystalline mineral, consisting of aluminium fluosilicate, with the formula $Al_2(OH,F)_2SiO_4$.

topology A branch of *geometry* concerned with the way in which figures are 'connected', rather than with their shape or size. Topology is thus concerned with the geometrical factors that remain unchanged when an object undergoes a continuous deformation (e.g. by bending, stretching, or twisting) without tearing or breaking.

toroidal Having the shape of a toroid or *torus*.

torque A *moment* of a force, or system of forces that tends to produce *angular acceleration*. See also *couple*.

torr A unit of *pressure* used in the field of high vacuum: it is equivalent to 1 mm of mercury or 133.322 *pascals*.

Torricellian vacuum The space, containing mercury *vapour*, that is produced at the top of a column of mercury when a long tube sealed at one end is filled with mercury and inverted in a trough of the metal. The mercury sinks in the tube until it is balanced by the atmospheric pressure (see *barometer*), the Torricellian

452

vacuum being the space above it. Named after Evangelista Torricelli (1608–47).

torsion 'Twisting' about an *axis*, produced by the action of two opposing forces (see *couple*).

torsion balance If a wire is acted upon by a *couple* the *axis* of which coincides with the wire, the wire twists through an angle determined by the applied couple and the *rigidity modulus* of the wire. The amount of twist produced can thus be used to measure an applied *force*. In the torsion balance, the force to be measured is applied at right angles to, and at the end of, an arm attached to the wire.

torus (phys.) A 'doughnut' or anchor-ring-shaped *solid* of circular or elliptical cross-section. If the cross-section is a circle of radius a, and the ring has a radius b, the volume of the torus is $2\pi^2a^2b$.

total internal reflection When *light* passes from one medium to another that is optically less dense, e.g. from *glass* to air (see *refraction* and *density, optical*), the *ray* is bent away from the *normal*. If the incident ray meets the surface at such an angle that the refracted ray must be bent away at an angle of more than 90°, the light cannot emerge at all, and is totally internally reflected.

totality The period in a total *eclipse* of the *Sun*, during which the bright surface of the Sun is totally obscured from view on *Earth* by the *Moon*.

tourmaline A class of natural crystalline *minerals*, consisting of *silicates* of various *metals* and containing boron. The *crystals* show some interesting *pyroelectric*, *piezoelectric* and optical *effects*. See *dichroism*.

toxic Poisonous.

toxicology The study of poisons.

toxin A poison, usually an intensely poisonous substance produced by certain *bacteria, fungi, algae*, and higher *organisms*.

trace element An *element* required in very small quantities by an *organism*. Such elements often form essential constituents of *enzymes, vitamins*, or *hormones*. The main trace elements are iron, manganese, zinc, copper, iodine, cobalt, selenium, molybdenum, chromium, and silicon.

tracer See *radioactive tracing*.

trajectory The path of a *projectile*.

tranquillizer A *drug* used to reduce tension and anxiety, without impairing alertness or causing drowsiness.

transactinides An *element* with an *atomic number* in excess of 103. They are elements 104, 105, and possibly 106. The discovery of 104 has been claimed by both the Soviet Union, as kurchatovium (Ku), and the USA, as rutherfordium (Rf), but the recommended name is unnilquadium (Unq). 105 has been called hahnium (Ha) and nielsbohrium (Nb). Transactinides are all radioactive with very short *half-lives*. See *unnil*.

transamination The transfer of an *amino group* from one *compound* (e.g. an *amino acid*) to another.

transcendental (math.) **1.** (Of a number or quantity) Not capable of being expressed as the *root* of an algebraic *equation* with *rational coefficients*, e.g. π or e. **2.** (Of a function) Not capable of being expressed by a finite number of algebraic operations, e.g. sin$x$, e^x. (See *exponential*.)

transconductance The *mutual conductance* between the *control grid* of a *thermionic valve* and its *anode*; it is usually expressed in *siemens*.

transcription The process in which a *molecule* of *ribonucleic acid* (RNA) is synthesized beside one of the strands of *deoxyribonucleic acid* (DNA). Ribonucleotides held by *hydrogen bonds* to the *complementary* nitrogenous *bases* of the DNA (see *base pairing*) are joined by RNA *polymerase*; the sequence of *nucleotides* in the DNA thus determines the sequence of nucleotides in the RNA, e.g. the sequence CGAT in DNA gives rise to GCUA in RNA.

transducer A device that receives signals (electrical, acoustical, or mechanical) from one or more media or transmission systems and supplies related signals (not necessarily of the same type as the input) to one or more other media or transmission systems. If the transducer derives *energy* from sources other than the input waves it is said to be 'active': if the input waves are the only source of energy it is said to be 'passive'.

transfer RNA See *ribonucleic acid.*

trans-form See *cis-trans isomerism.*

transformation, nuclear The change of one *nuclide* into another.

transformation constant See *disintegration constant.*

transformer A device by which an *alternating current* of one *voltage* is changed to another voltage, without alteration in *frequency*. A step-up transformer, which increases the voltage and diminishes the current, consists in principle of an iron core on which is wound a *primary coil* of a small number of turns of thick, insulated wire; and, forming a separate circuit, a secondary coil of a larger number of turns of thin, insulated wire. When the low-voltage current is passed through the primary coil, it induces a current in the secondary (see *induction*) by producing an alternating *magnetic field* in the iron core. The ratio of the voltage in the primary to that in the secondary is very nearly equal to the ratio of the number of turns in the primary to that in the secondary. The step-down transformer works on the same principle, with the coils reversed. In practice, the voltage ratio will not exactly equal the turns ratio as there will be a power loss due to *hysteresis*, heating of the coil, *eddy currents*, and incomplete magnetic linkage.

transgenic organisms See *genetic engineering.*

transient 1. (math.) A *function* whose value tends to *zero* as the independant *variable* tends to *infinity*. **2.** (phys.) A short-lived oscillation in a system caused by a sudden change of *voltage*, *current*, or load.

transistor A *semiconductor* device capable of amplification. A bipolar transistor consists of two *p-n semiconductor junctions* back to back forming either a *p-n-p* or *n-p-n* structure. In a *p-n-p* transistor the thin central *n*-region is called the *base*, one *p*-region is called the *emitter*, the other the *collector*. In an *n-p-n* transistor the *p*-region is the base. In order to obtain amplification an *n-p-n* transistor is included in a circuit that supplies a positive *voltage* to the collector (*n*-region) and a negative voltage to the emitter (the other *n*-region). The collector in this type of transistor therefore corresponds to the *anode* of a thermionic valve while the emitter corresponds to the *cathode*. The base (*p*-region) is also positively biased and is analogous to the *control grid*. With this arrangement the large number of *electrons* in the emitter region is attracted to the *p*-layer, which, if it is sufficiently thin, will allow the electrons to pass through it and be attracted into the positive collector. The magnitude of the collector current will depend on the extent of the positive bias on the *p*-layer base. By suitable design the device can be made to give a collector current some 20–100 times the base current. The advantages of a transistor over a valve are that it is less bulky and fragile, that it requires no heater current, and that the voltage at the collector need only be a few volts. A *p-n-p* transistor works in an exactly analogous manner to an *n-p-n* device, but the collector current consists mainly of *holes* instead of electrons. The device described here is a *junction transistor*, as this type has replaced the earlier point-contact transistor. See also *field-effect transistor*.

transition elements *Elements* that have chemical properties resembling those of their horizontal neighbours in the *periodic table*. These elements have incomplete inner electron *shells* and are characterized by their variable *valences*: they occur in the middle of the long periods of the periodic table.

transition, nuclear A change in the configuration of an atomic *nucleus*. It may involve a *transformation* (e.g. by *alpha-* or *beta-particle* emission) or a change in *energy level* by the emission of a *gamma-ray*.

transition temperature 1. Transition point. The *temperature* at which one form of a polymorphous substance (see *polymorphism*) changes into another; the temperature at which both forms can co-exist. **2.** See *superconductivity*.

translation 1. (bio.) The process in which the sequence of *nucleotides* in a *molecule* of messenger RNA (mRNA; see *ribonucleic acid*) determines the sequence of *amino acids* in a *polypeptide* being synthesized at a *ribosome*. Each amino acid is bound to an appropriate transfer RNA molecule, which bears a sequence of three bases (the anticodon) *complementary* to the *codon* for that amino acid. Amino acyl transfer RNA molecules are held in place on the mRNA by *hydrogen bonds* (see *base pairing*) and an *enzyme* on the ribosome uses *energy* (e.g. from a nucleotide triphosphate) to join the amino acids by *condensation*, forming *peptide* bonds. When the sequence of amino acids is complete, the mRNA and the polypeptide leave the ribosome, which can then join with another molecule of mRNA to make another polypeptide. **2.** (phys.) See *translatory motion*.

translatory motion Translation. A motion that involves a non-reciprocating movement of *matter* from one place to another.

translucent Permitting the passage of *light* in such a way that there is some *scattering* and *diffusion* so that an object cannot be seen clearly through the substance; e.g. frosted *glass*.

transmission coefficient See *transmittance*.

transmission electron microscope See *electron microscope*.

transmission of radio waves Radio waves travel from a transmitting aerial to a receiving aerial by *ground waves* (i.e. line-of-sight waves or waves reflected from the ground) or by *sky waves*, which are reflected by the *ionosphere* and enable transmission to be made around the Earth's curved surface. The *ultrahigh frequency* (UHF) and *very high frequency* (VHF) waves used in television broadcasting pass through the ionosphere without reflection. For long distance TV broadcasting, therefore, artificial *satellites* have to be used to reflect or rebroadcast the transmissions.

transmittance Transmission coefficient. T. When a *beam of light* (or other *elec-*

tromagnetic radiation) passes through a medium the radiation is absorbed to a greater or lesser extent (depending upon the medium and the *wavelength* of the radiation) and the intensity of the beam decreases. The ratio of the intensity after passing through unit distance of the medium to the original intensity is the transmittance. Compare *opacity*.

transmitter 1. The equipment required to broadcast *electromagnetic radiation* of *radio frequencies*. The transmitter consists of devices for producing the *carrier wave*, *modulating* and amplifying it, and feeding it to the *aerial* system. **2.** The part of a *telephone* system that converts *sound* waves into *electric currents*, or the part of a *telegraph* system that converts mechanical movements into electrical currents.

transmutation of elements Changing one chemical *element* into another. Once the aim of *alchemy*, it was subsequently held to be impossible; with the present knowledge of atomic structure it is seen that the process goes on continuously in *radioactive* elements. Artificial transmutation by suitable *nuclear reactions* is now commonplace in *nuclear physics*. See also *transition, nuclear; transformation, nuclear*.

transparent Permitting the passage of *light* in such a way that objects can be seen clearly through the substance. Compare *translucent*.

transpiration The loss of water vapour by plants into the *air*. It occurs mainly from leaves, the water being drawn from the earth by the roots of the plant.

transponder *Electronic* equipment designed to receive a specific signal and automatically transmit a reply.

transport number Transference number. The proportion of the total *electric current* passing through an *electrolyte* that is carried by a particular type of *ion*. The *anion* transport number plus the *cation* transport number equals unity.

transuranic elements *Elements* beyond uranium in the *periodic table*; i.e. elements of *atomic number* greater than 92. Such elements do not occur in nature, but may be obtained by suitable *nuclear reactions*; they are all *radioactive* and members of the *actinide* or *transactinide* groups. See Appendix, Table 8.

transverse Cross-wise; in a direction at right angles to the length of the body under consideration.

transverse waves Waves in which the vibration or displacement takes places in a *plane* at right angles to the direction of propagation of the wave; e.g. *electromagnetic radiation*. Compare *longitudinal waves*.

trapezium A *quadrilateral* having two of its sides parallel. The area of a trapezium having parallel sides a and b units in length, and vertical height h units is given by $h(a + b)/2$.

travelling wave See *standing wave*.

triangle A *plane* figure bounded by three straight lines. The three *angles* total $180°$. The area of any triangle is given by the following expressions: 1. Half the *product* of one of the sides and the perpendicular upon it from the opposite vertex ($\frac{1}{2} \times$ base $\times$ height). 2. Half the product of any two of the sides and the *sine* of the angle between them ($\frac{1}{2} bc$ sin$A$). 3. $[s(s - a)(s - b)(s - c)]^{1/2}$, where a, b, and c are the lengths of the sides, and s is half the sum of a, b, and c.

triangle of forces If three *forces* acting at the same point can be represented in magnitude and direction by the sides of a *triangle* taken in order, they will be in *equilibrium*.

triangle of velocities If a body has three component *velocities* that can be represented in magnitude and direction by the sides of a *triangle* taken in order, the body will remain at rest.

triatomic Having three *atoms* in the *molecule*, e.g. *ozone*, O_3; having three replaceable atoms or groups in the molecule.

triazine $C_3H_3N_3$. Three *isomeric compounds* having three nitrogen and three carbon *atoms* forming a six-membered ring. *Cyclonite* is a *derivative* of triazine.

triazole $C_2H_3N_3$. Four *isomeric compounds* having three nitrogen and two carbon *atoms* forming a five-membered ring.

tribasic acid An *acid* having three *atoms* of *acidic hydrogen* in the *molecule*, thus giving rise to three possible series of *salts*; e.g. *phosphoric(V) acid*, H_3PO_4, can give rise to trisodium phosphate(V), Na_3PO_4, disodium hydrogen phosphate(V), Na_2HPO_4, and sodium dihydrogen phosphate(V), NaH_2PO_4. See *sodium phosphates*.

triboelectricity See *electricity, frictional*.

tribology The study of *friction* and lubrication, i.e. the study of substances that prevent contact between two moving surfaces to reduce wear, overheating, and rust formation. Liquid *hydrocarbons* are the most widely used lubricants, but solid lubricants, such as *graphite* or molybdenum disulphide, are also used. Semisolid greases, made by mixing metallic *soaps* with hydrocarbon oils, are also widely used, especially on vertical surfaces.

triboluminescence The emission of *light* when certain *crystals* (e.g. *cane-sugar*) are crushed, as a result of *electric charges* generated by *friction*.

tribromoethanol CBr_3CH_2OH. A white crystalline powder, m.p. 79–82°C., used as a veterinary *anaesthetic*.

tribromomethane Bromoform. $CHBr_3$. A colourless *liquid*, m.p. 8.3°C., b.p. 149.5°C., used in organic synthesis.

tricarboxylic acid A *carboxylic acid* containing three *carboxyl* groups in its molecule, e.g. *citric acid*:
$$COOHCH_2C(OH)COOHCH_2COOH.$$
In systematic nomenclature, such compounds end in -trioic, e.g. citric acid is 2-hydroxypropane-1,2,3-trioic acid.

trichloroethanal Chloral. CCl_3CHO. A pungent colourless *liquid aldehyde*, b.p. 97.7°C., used in making *DDT*.

2,2,2,-trichloroethanediol Chloral hydrate. $CCl_3CH(OH)_2$. A white crystalline *solid*, m.p. 57°C., b.p. 97.8°C., used as a sedative and in the manufacture of *DDT*. It is made by adding *water* to *trichloroethanal*.

trichloroethene $CHCl:CCl_2$. A colourless *liquid*, b.p. 87°C.; widely used as industrial *solvent*, in dry cleaning, and as an *anaesthetic*.

trichloromethane Chloroform. $CHCl_3$. A *volatile* colourless *liquid* with a strong sweet odour, b.p. 61°C. It can be made by the *chlorination* of *methane* and was formerly used as an *anaesthetic*, but because it can cause liver damage has now been replaced by other halogenated *hydrocarbons*. It is used as a *solvent* and in organic synthesis.

trichloronitromethane Chloropicrin. CCl_3NO_2. An oily liquid, b.p. 112°C. A highly poisonous and chemically active substance, it is used in disinfectants and fungicides.

triethanolamine See *ethanolamines*.

triethylamine ($C_2H_5)_3$N. A colourless inflammable *liquid*, b.p. 89–90°C., used as a *solvent*.

triglycerides See *glycerides*.

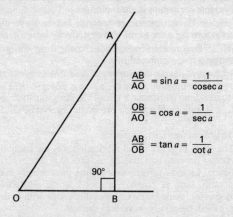

$$\frac{AB}{AO} = \sin a = \frac{1}{\csc a}$$

$$\frac{OB}{AO} = \cos a = \frac{1}{\sec a}$$

$$\frac{AB}{OB} = \tan a = \frac{1}{\cot a}$$

Figure 43.

trigonometrical ratios If a perpendicular *AB* is drawn from any point on arm *OA* of an *angle AOB* to the other arm, the following ratios are constant for the particular angle: *AB/AO*, sine (sin*AOB*); *OB/AO*, cosine (cos*AOB*); *AB/OB*, tangent (tan*AOB*); *AO/AB*, cosecant (cosec*AOB*); *AO/OB*, secant (sec*AOB*); and *OB/AB*, cotangent (cot*AOB*). See Fig. 43.

trigonometry A branch of mathematics using the fact that numerous problems may be solved by the calculation of unknown parts (i.e. sides and *angles*) of a *triangle* when three parts are known. The solution of such problems is greatly assisted by the use of the *trigonometrical ratios*.

trihydric Containing three *hydroxyl* groups in the *molecule*. See *triols*.

trihydroxybenzoic acid Gallic acid. $C_6H_2(OH)_3COOH$. A yellowish crystalline substance, used in *tanning* and the manufacture of inks.

triiodomethane See *iodoform*.

trillion 10^{18}, a million million million (British); 10^{12}, a million million (American). The use of this word is best avoided in scientific contexts.

trimer A substance composed of *molecules* that are formed from three molecules of a *monomer*.

trinitrobenzene TNB. $C_6H_3(NO_2)_3$. Three *isomeric* crystalline *compounds*, m.p. 121–127°C., used as high *explosives*, and having greater power than *TNT*.

2,4,6-trinitrophenol See *picric acid*.

trinitrotoluene TNT. Methyl-2,4,6-trinitrobenzene. $CH_3C_6H_2(NO_2)_3$. A pale yellow, crystalline *solid*, m.p. 82°C., made by the *nitration* of toluene (see *methylbenzene*). A widely used high *explosive*.

triode A *thermionic valve* containing three *electrodes*; and *anode* or plate, a *cathode*, and a *control grid*. It was the first electronic device to provide amplification, a role now taken over by the *transistor*.

-trioic See *tricarboxylic acid*.

triolein Olein. $(C_{17}H_{33}COO)_3.C_3H_5$. A *glyceride* of *oleic acid*, b.p. 235–240°C. It is a *liquid* oil that occurs in many natural *fats and oils*.

triols Trihydric *alcohols* derived from *aliphatic hydrocarbons* by the substitution of *hydroxyl groups* for three of the hydrogen *atoms* in the *molecule*.

triose A *sugar* containing three carbon atoms in the *molecule*.

trioxoboric(III) acid See *boric acid*.

trioxosulphuric(IV) acid See *sulphuric acids*.

tripalmitin Palmitin. $(C_{15}H_{31}COO)_3.C_3H_5$. A *glyceride* of *palmitic acid*, m.p. 65.5°C., b.p. 310–320°C. It is a fat-like substance that occurs in palm-oil and many other natural *fats and oils*.

triple bond Three *covalent bonds* linking two *atoms* in a chemical *compound*, e.g. *ethyne*, CH≡CH. See *chemical bond*.

triple point The point at which the gaseous, *liquid*, and *solid phases* of a substance are in *equilibrium*. For a given substance, the triple point occurs at a particular *temperature* and *pressure*, e.g. for water it occurs at 273.16 K and 611.2 Pa. This value is used in defining *thermodynamic temperature* and the *kelvin*.

trisaccharides A group of *sugars* the *molecules* of which consist of three *monosaccharides*.

trisodium phosphate(V) See *sodium phosphates*.

tristearin Stearin. $(C_{17}H_{35}COO)_3.C_3H_5$. A *glyceride* of stearic acid (*octadecanoic acid*), m.p. 53.5°C. It is a fat-like substance that occurs in natural *fats*; it is formed by the hydrogenation of *triolein*. See *hydrogenation of oils*.

tritiated compound A *compound* in which some hydrogen *atoms* have been replaced by *tritium*, so that it may be used in *radioactive tracing*. See *labelled compound*.

tritium T. ^3_1H. A *radioactive isotope* of hydrogen with *mass number* 3 and *atomic mass* 3.016. The *abundance* of tritium in natural hydrogen is only one *atom* in 10^{17}, and its half life is 12.5 years. It undergoes *beta decay* to helium-3. It can, however, be made artificially in *nuclear reactors* and *tritiated compounds* are used in *radioactive tracing* (see *labelled compound*).

triton The *nucleus* of a *tritium atom*.

trivalent Tervalent. Having a *valence* of three.

trochoid A curve formed by a point on the radius of a *circle* as the circle rolls along a straight line. If the point is on the circumference of the circle the curve is a *cycloid*.

trochotron A multi-*electrode thermionic valve* used as a *scaler*.

trona A natural crystalline *double salt* of *sodium carbonate* and *sodium hydrogencarbonate*, $Na_2CO_3.NaHCO_3.2H_2O$, found in dried lakes.

troostite 1. The constituent of *steel* produced when *martensite* is *tempered* below 450°C., consisting of *ferrite* and finely divided *cementite*. **2.** Troostitic pearlite. The constituent of steel produced by the decomposition of *austenite* when cooled at a slower rate than yields martensite and a faster rate than yields *sorbite*.

tropical year See *year*.

tropine $C_8H_{15}NO$. A white crystalline *hygroscopic soluble alkaloid*, m.p. 63°C.

tropopause The boundary between the *troposphere* and the *stratosphere*.

troposphere The lower part of the Earth's *atmosphere* in which *temperature* decreases with height, except for local areas of 'temperature inversion'. See Fig. 44, under *upper atmosphere*.

Trouton's rule The ratio of the *molar latent heat* of vaporization to the *boiling point* in *kelvin* is a constant for all substances. The rule is only approximate.

Troy weight 1 grain = 0.0648 gram.

    20 grains = 1 scruple.

    24 grains = 1 pennyweight.

    3 scruples = 1 drachm.

    8 drachms = 1 ounce Troy = 1.1 ounces avoirdupois.

trypsin An *enzyme* produced by the pancreas. In the process of digestion it catalyses the breakup of *peptide* bonds between specific pairs of *amino acids* in a *polypeptide*, producing peptide fragments, which are later degraded by less specific *proteases* into single amino acids.

tryptophan A colourless crystalline *amino acid*, m.p. 281–9°C., that is essential to the diet of animals and occurs in the seeds of some vegetables. See Appendix, Table 5.

tube of force A theoretical concept of a tube formed by the *lines of force* drawn out into space through every point on a small closed curve upon the surface of a charged *conductor*.

tuned circuit A *resonant circuit* in which the resonant frequency can be varied by adjusting its *capacitance* (often by means of a variable air capacitor) or its *inductance* (by means of a variable inductor). Tuning a *radio*, for example, involves changing the capacitance of a resonant circuit so that it resonates, in *amplitude modulation*, at the *frequency* of the *carrier wave* of the desired station.

tungstate A *salt* of *tungstic acid*.

tungsten W. Wolfram. Element. R.a.m. 183.85. At. No. 74. A grey hard ductile malleable metal that is resistant to *corrosion*, r.d. 19.25, m.p. 3422°C., b.p. 5700°C. It occurs as *wolframite*, $FeWO_4$, and *scheelite,* $CaWO_4$ and is obtained by converting the *ore* to the *oxide* and then reducing the latter. It is used in *alloys*, in cemented *carbides* for hard tools, and for electric lamp *filaments*.

tungsten carbide WC. A grey powder, m.p. 2780°C., obtained by direct combination of tungsten and carbon at 1600°C. It is almost as hard (9.5 on the *Mohs scale*) as *diamond* and is used in making abrasives and tools.

tungsten trioxide WO_3. A yellow *insoluble* powder, m.p. 1473°C., used in the manufacture of *tungstates*.

tungstic acid H_2WO_4. *Hydrated tungsten trioxide*. A white crystalline powder that loses a *molecule* of *water* at 100°C.; it is used in the manufacture of lamp *filaments*.

tuning, radio See *tuned circuit*.

tuning fork A two-pronged *metal* fork that, when struck, produces a pure *tone* of constant specified *pitch*. It is used in *acoustics* and for tuning musical instruments.

tunnel diode Esaki diode. A *semiconductor* device that has negative *resistance* over a part of its operating range. It consists of a *p-n* semiconductor junction in which both the *p*- and *n*- regions contain very large numbers of impurity *atoms*, thus producing a high *potential* barrier at the junction. If a small *voltage* is applied to the device, positive at the *p*-region, an *electron* current will flow (despite the high potential barrier) as a result of the *tunnel effect*. After a certain voltage has been reached this effect is reduced and the current declines with increasing voltage, thus exhibiting the negative resistance characteristic. At higher voltages the normal *majority carrier* current flows and the current again increases with voltage. It is used in switching circuits and where low-noise amplification is required up to *frequencies* of about 1000 *megahertz*.

tunnel effect The passage of an *electron* through a narrow *potential* barrier in a *semiconductor*, despite the fact that, according to classical *mechanics*, the electron does not possess sufficient *energy* to surmount the barrier. It is explained by *quantum mechanics* on the assumption that electrons are not completely localized in *space*, a part of the energy of the electron enabling it to 'tunnel' through the barrier. This effect is made use of in the *tunnel diode*.

turbine Any motor in which a shaft is steadily rotated by the impact or reaction of a current of *steam*, air, *water*, or other *fluid* upon blades of a wheel. In an 'impulse' turbine the fluid is directed from jets or nozzles on to the *rotor* blades. Some of the *kinetic energy* of the fluid jet is thus converted to rotational kinetic energy of the shaft to which the blades are attached. In a 'reaction' turbine the nozzles are attached to the rotor, the acceleration of the fluid jet producing a force of reaction on the rotor. Many turbines work on a combination of the reaction and impulse principles. Water turbines have been widely used to drive machinery situated beside fast flowing rivers, etc., and are now used in *hydroelectric power* stations to turn the *generators*. Steam turbines are used extensively in coal and oil-fired 'conventional' power stations and also in nuclear-powered steam-raising power stations. Thus most electricity is generated either by water or steam turbines. Steam turbines are also used to drive ships. See also *gas turbine*.

turbogenerator A steam *turbine* coupled to an electric *generator* for the production of *electric power*. It is the usual arrangement in a 'conventional' power station and in steam-raising *nuclear power* stations.

turbulent flow The type of *fluid* flow in which the motion at any point varies rapidly in direction and magnitude.

Turkey-red oil A mixture of sulphate *esters* obtained by treatment of *castor oil* with *sulphuric acid*. It is used in dyeing.

turpentine Oil of turpentine. A *liquid* extracted by *distillation* of the *resin* of pine trees. B.p. 155°–165°C. It is composed chiefly of *pinene* and is used as a *solvent*.

turquoise Natural basic aluminium phosphate, coloured blue or green by traces of copper.

Twaddell scale A scale for measuring the *relative density* of *liquids*. Degrees Twaddell = 200 (r.d. − 1); r.d. = 1 + degrees Twaddell/200. Named after W. Twaddell (19th century).

tweeter A *loudspeaker* designed to reproduce the higher *audiofrequency* sounds, i.e. 5 – 15 *kilohertz*.

Tyndall effect The *scattering of light* by particles of *matter* in the path of the light, thus making a visible 'beam'; a *ray* of light illuminates particles of dust floating

461

in the air of a room as a result of the Tyndall effect. Named after John Tyndall (1820–93).

type metal An *alloy* of 60% lead, 30% antimony, and 10% tin. Owing to the presence of antimony it expands on solidifying and thus gives a sharp cast.

tyrosine A white crystalline *amino acid*, m.p. 310°–320°C., obtained from most *proteins*. See Appendix, Table 5.

U

udometer Pluviometer. A rain gauge.

UHF See *ultra-high frequency*.

ultimate stress Tenacity. The load required to fracture a material divided by its original area of cross-section at the point of fracture. The ultimate stress is divided by the 'factor of safety', in order to obtain the 'working stress'.

ultracentrifuge A high speed *centrifuge* rotating at up to 75 000 rpm. It is used in the determination of the *relative molecular masses* of large *molecules* in high *polymers* and *proteins*. It is also used to separate large molecules having different relative molecular masses.

ultra-high frequencies UHF. *Radio frequencies* in the range 300 to 3000 *megahertz*.

ultramarine An artificial form of *lapis lazuli*, made by heating together *clay*, *sodium sulphate*, carbon, and sulphur.

ultramicroscope An instrument, making use of the *Tyndall effect* for showing the presence of particles that are too small to be seen with the ordinary optical *microscope*. A powerful *beam* of *light* is brought to a focus in the *fluid*, against a black background; suspended particles appear as bright specks by *scattering the light*.

ultrasonic frequency A *frequency* in excess of about 20 000 *hertz*.

ultrasonic generator A device for the production of pressure waves of *ultrasonic frequency*, usually using *piezoelectric* or *ferroelectric* materials or *magneto-striction* to function as *transducers*.

ultrasonics Supersonics. The study of pressure waves that are of the same nature as *sound* waves, but that have frequencies above the audible limit (i.e. *ultrasonic frequencies*). Ultrasonic scanning is now widely used as a diagnostic tool in medicine, especially for examining foetuses that could be damaged by *X-rays*. Ultrasonics is also used industrially to form *colloids*, clean surfaces, and detect flaws in *metals*.

ultraviolet microscope A *microscope* in which the object is illuminated by *ultraviolet radiation*. *Quartz* lenses are used and the image is recorded photographically. As ultraviolet radiation is of shorter *wavelength* than visible *light*, greater magnification can be obtained than with an optical microscope. The image obtained is either photographed or made visible by means of an *image converter*.

ultraviolet radiation UV radiation. *Electromagnetic radiation* in the *wavelength* range of approximately 4×10^{-7} to 5×10^{-9} *metre*; i.e. between visible *light* waves and *X-rays*. The longest ultraviolet waves have wavelengths just shorter than those of violet light, the shortest perceptible by the human eye. They affect photographic *films* and plates; their action on *ergosterol* in the human body produces *vitamin* D. It is usual to classify UV radiation as:

near UV $3.8 \times 10^{-7} - 3.0 \times 10^{-7}$ m
far UV $3.0 \times 10^{-7} - 2.0 \times 10^{-7}$ m

extreme UV below 2.0×10^{-7} m.

Radiation from the *Sun* is rich in such rays but most of it is absorbed by the *ozone layer* in the *upper atmosphere*. Ultraviolet radiation is produced artificially by the *mercury-vapour lamp*.

umbra A region of complete *shadow*. See Fig. 38 under *shadow*.

uncertainty principle Indeterminacy principle. It is impossible to determine with accuracy both the position and the *momentum* of a particle (e.g. an *electron*) simultaneously. The more accurately the position (x) is known, the less accurately can the momentum (p) be determined. If the indeterminacy in the x-component of the momentum is Δp_x, then $\Delta p_x \Delta x \geq h/4\pi$, where h is the *Planck constant*. A similar relationship exists between energy (E) and time (t), i.e.

$$\Delta E \Delta t \geq h/4\pi$$

The principle, which was first stated by Werner Heisenberg (1901–76), arises from the dual particle wave nature of *matter*. It appears to undermine the 'common-sense' view of cause and effect, at least on the atomic scale. How can two consecutive observations of the same particle be distinguished from two observations of different particles, if a particle cannot be located exactly? If a particle cannot be identified without uncertainty, how can one be sure what will happen to it in the future, or if the law of cause and effect is being obeyed? See *De Broglie wavelength*.

unfilled aperture A method of constructing a *radio telescope* in which two *aerials* of different shapes are combined into one *radio interferometer* in such a way that only two perpendicular arms of the aerial system are built, giving the effect of two large apertures. The two arms may be spaced at varying distances apart, or they may be superimposed upon one another as in the 'Mills Cross' radio telescope. Unfilled aperture telescopes are suitable for use at long *wavelengths*.

ungula A part of a *cylinder* or *cone* that is cut off by a *plane* not parallel to its base.

uniaxial crystal A double-refracting *crystal* possessing only one *optic axis*.

unicellular (Of an *organism*.) Consisting of only one *cell* (e.g. *bacteria*, protozoa, etc.).

unified field theory A theory that attempts to describe the *electromagnetic*, *gravitational*, *strong*, and *weak interactions* in one set of *equations*. No such satisfactory theory has yet been devised but there has been some progress in the unification of the weak and electromagnetic interactions in the *electroweak theory*. See also *grand unified theory*.

unit A quantity or dimension adopted as a standard of measurement. See *metric system*; *SI units*.

unitary symmetry SU3. A method of classifying *elementary particles* according to their properties in a similar manner to the classification of atomic properties in the *periodic table*. SU3 has successfully predicted the existence of particles that have subsequently been detected experimentally, e.g. omega-minus. The concept of SU3 has been extended to a larger symmetry group, called SU4, which leads to the concept known as *charm*.

unit cell The unit of which a crystal *lattice* is constructed. For example, the *body-centred* and face-centred lattices are forms of a cubic unit cell.

univalent (chem.) Monovalent. Having a *valence* of one.

universal gas constant See *gas constant*.

universal motor An *electric motor* that will run on either *d.c.* or *a.c.* Current is fed to the *stator* and reaches the *rotor* by means of a *commutator*. In a series-wound motor, the two are *in series*, but in a shunt-wound motor they are *in parallel*. Small universal motors are widely used to power domestic appliances.

Universe The total of all the *matter*, *energy*, and *space* that man is capable of experiencing, or whose existence he can deduce or has grounds for postulating. The universe is currently best described in terms of a four-dimensional curved *space-time continuum* (see *relativity*); it contains some 10^{41} *kilograms* of matter, collected in some 10^9 *galaxies*. See *heat death of the Universe*; *steady-state theory*; *superdense theory*.

unnil A system of naming *transactinide elements* that has been proposed to avoid controversy over eponyms. Names consist of un- (Latin *unus*, one) and -nil- (Latin *nil*, zero), plus -unium, -bium, -trium, -quadium, -pentium, or -hexium (1–6 in Latin) to spell out the *atomic numbers*, i.e.

   element 101, unnilunium; mendelevium.
   element 102, unnilbium; nobelium.
   element 103, unniltrium; lawtencium.
   element 104, unnilquadium; kurchatovium, rutherfordium.
   element 105, unnilpentium; hahnium, nielsbohrium.
   element 106, unnilhexium.

unsaturated compound (chem.) A *compound* having some of the *atoms* in its *molecule* linked by more than one *valence bond* (see *double bond* and *triple bond*); a compound that can form *addition compounds*.

unstable (chem.) Easily decomposed.

unstable equilibrium See *stable equilibrium*.

upper atmosphere The upper *atmosphere* of the *Earth* is usually taken to include its gaseous envelope from 30 *kilometres* upwards (i.e. the part of the atmosphere that is inaccessible to direct observations by balloons). Information is obtained from *space probes* and artificial Earth *satellites*. See Fig 44.

Up to about 100 km the composition of the upper atmosphere is similar to that at ground level (see *atmosphere*). Above this height the *dissociation* of oxygen into *atoms* is almost complete, and at above 150 km the nitrogen separates out owing to its greater *mass* so that monatomic oxygen predominates. There is considerable *ionization* in the upper atmosphere as a result of solar *ultraviolet radiation* and *X-rays*. *Densities* and *temperatures* in the upper atmosphere are shown in Fig. 44. See *ionosphere*.

uracil Pyrimidinedione. $C_4H_4N_2O_2$. A white crystalline *pyrimidine* base, m.p. 338°C., that occurs in one of the *bases* in *ribonucleic acid*.

uraninite See *pitchblende*.

uranium U. Naturally occurring *radioactive element*. R.a.m. 238.0289, At. No. 92. A hard white *metal*, r.d. 19.05, m.p. 1135°C., b.p. 4000°C. The natural element consists of 99.28% uranium-238 (*half-life* 4.5×10^9 years), 0.71% uranium-235 (half-life 7.1×10^8 years), and 0.006% uranium-234 (half-life 2.5×10^5 years). Uranium-235 is capable of sustaining a nuclear *chain reaction* and is of greater importance in *nuclear reactors* and *nuclear weapons*. The principal *ore* is *pitchblende*.

uranium dioxide Uranium(IV) oxide. UO_2. A black *insoluble* crystalline *radioactive* substance, m.p. 2500°C., used as a *fuel* in advanced *gas-cooled reactors*.

uranium-lead dating Various methods of *dating rocks* that depend on the decay

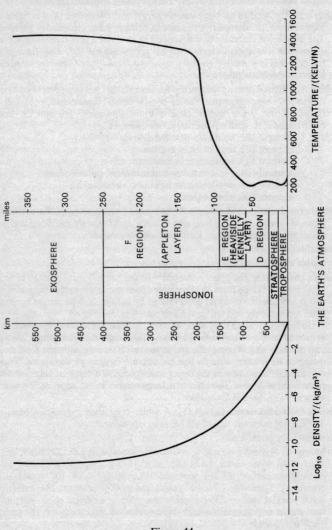

Figure 44.

of one of the *isotopes* of uranium to lead. One method relies on measuring the ratio of radiogenic lead (Pb-206, Pb-207, and Pb-208) present in a sample of rock to the nonradiogenic lead (Pb-204). Another method relies on measuring the ratio of the helium present in a rock sample to the amount of uranium. The decay U-238 → Pb-206 releases eight alpha particles (helium nuclei). The methods are reliable for ages in the range $10^7 - 10^9$ years.

uranium trioxide Uranyl(VI) oxide. UO_3. A red *insoluble radioactive* powder, which decomposes on heating.

Uranus (astr.) A *planet* possessing 15 *satellites*, with its *orbit* lying between those of *Saturn* and *Neptune*. Mean distance from the *Sun*, 2869.6 million kilometres. *Sidereal period* ('year') 84 years. *Mass* approximately 14.52 times that of the *Earth*, diameter 51 800 kilometres. Surface temperature, about $-240°C$. It has a system of some 20 rings, which have been observed by Voyager probes.

uranyl The *bivalent* group, $=UO_2$, which forms *salts* with *acids*.

urea Carbamide. $CO(NH_2)_2$. A white crystalline *soluble organic compound*, m.p. 132°C., that occurs in urine. It was the first organic compound to be prepared artificially and is used as a *fertilizer*, in medicine, and in *urea-formaldehyde resins*.

urea-formaldehyde resins *Thermosetting resins* with good *oil* resistant properties, produced by the condensation *polymerization* of *urea* and formaldehyde (*methanal*).

urease An *enzyme* capable of splitting *urea* into *ammonia* and *water*.

ureido The *univalent* group, NH_2CONH-, derived from *urea*.

urethane resins Polyurethanes. A class of *polymers* that are chemically related to *urethanes*, generally made by condensation of *isocyanates* with *polyhydric* compounds. They form valuable materials for a number of purposes, including the manufacture of coatings and foam plastics.

urethan(e)s *Esters* of carbamic acid, NH_2COOH. The name is usually applied to ethyl carbamate, $NH_2COOC_2H_5$.

ureylene group The *divalent* group, $-NHCONH-$, derived from *urea*.

uric acid $C_5H_4N_3O_4$. An *organic acid*, belonging to the *purine* group; a colourless crystalline *solid* that is slightly *soluble* in *water*. It occurs in the urine of some animals (e.g. birds) as a breakdown product of *amino acids* and *nucleic acids*. Sodium and potassium *salts* of the acid are deposited in the joints in cases of gout.

UV See *ultraviolet radiation*.

V

vacancy Schottky defect. An irregularity that occurs in a *crystal lattice* when a site normally occupied by an *atom* or *ion* is unoccupied. See *defect*; *Frenkel defect*.

vaccine A preparation containing disease-causing *viruses* or other *microorganisms* (either killed or of attenuated virulence) that is introduced into humans or other animals to stimulate the formation of *antibodies*. In this way immunity (partial or complete) to subsequent infection by this type of microorganism is conferred.

vacuum A *space* in which there are no *molecules* or *atoms*. A perfect vacuum is unobtainable, since every material that surrounds a space has a definite *vapour pressure*. The term is generally taken to mean a space containing air or other *gas* at very low *pressure*. A low (or soft) vacuum is one in which the pressure is above 10^{-2} *pascal*, while in a high (or hard) vacuum it is $10^{-2} - 10^{-7}$ Pa. 'Ultra-high' vacua (i.e. vacua in which the pressure does not exceed 10^{-7} Pa) occur naturally at heights of more than 800 kilometres above the *Earth's* surface, and by special techniques pressures of 10^{-13} Pa can be achieved in the laboratory (see *vacuum pump*).

vacuum distillation The process of *distillation* carried out at a reduced *pressure*. The reduction in pressure is accompanied by a depression in the *boiling point* of the substance to be distilled, thus lower *temperatures* can be employed. This process therefore enables substances to be distilled, which at normal pressures would decompose.

vacuum evaporation A technique for covering a solid surface with a thin layer of a substance. The substance is heated in a *vacuum*, the atoms escaping from its surface being allowed to condense on the surface to be coated.

vacuum gauge See *pressure gauge*.

vacuum pump Any device used to produce a low pressure. The common type of rotary oil pump can produce pressures down to 10^{-1} *pascal*. A *condensation* (diffusion) *pump* will produce a vacuum down to about 10^{-7} Pa and an *ion pump* will go down to 10^{-9} Pa. With a *cryogenic pump* in conjunction with a condensation pump pressures down to 10^{-13} Pa can be reached. See also *sorption pump*.

vacuum tube See *thermionic valve*; *discharge in gases*.

valence Valency. The combining power of an *atom*; the number of hydrogen atoms that an atom will combine with or replace. E.g. the valence of oxygen in water, H_2O, is 2. Hydrogen always has a valence of 1. In *ionic compounds* the valence equals the ionic charge, e.g. in NaCl, Na^+ and Cl^- each have a valence of 1. In covalent compounds, the valence is equal to the number of bonds formed, e.g. in CH_4, C has a valence of 4. See *benzene ring*; *chemical bond*; *hydrogen bond*; *resonance* (chem.).

valence band The range of energies (see *energy bands*) in a *semiconductor* corresponding to states that can be occupied by the *valence electrons* binding the *crystal* together. Electrons missing from the valence band give rise to *holes*.

valence bond The link holding *atoms* together in a *molecule*. In the case of two *univalent* atoms joined together, a single valence bond holds them together; it is possible for an atom to satisfy two or three valence bonds of another atom, giving rise to a *double* or *triple bond*. See *chemical bond*.

valence electron An outer *electron* of an *atom* that takes part in formation of a *valence bond*.

valeric acid See *pentanoic acid*.

valine A white crystalline *soluble amino acid* that occurs in most *proteins*. It is used in medicine and *culture media*. See Appendix, Table 5.

valve See *thermionic valve*.

vanadate A *salt* or *ester* of *vanadic acid*.

vanadic acid HVO_3. A yellow *insoluble* crystalline substance. Other *acids* are formed by the addition of *water molecules*, e.g. H_3VO_4.

vanadium V. Element. R.a.m. 50.9415. At. No. 23. A very hard white *metal*, r.d. 6.09, m.p. 1920°C., b.p. 3400°C. It occurs in a few rather rare *minerals*, such as *carnotite* and *patronite*. It is used in vanadium *steels*.

vanadium pentoxide Vanadium(V) oxide. V_2O_5. A yellowish crystalline substance, m.p. 690°C., used as a catalyst in oxidation processes in gases.

vanadyl The *divalent* group $=VO_2$, which forms *salts* with *acids*.

Van Allen radiation belts Two belts of charged particles trapped within the Earth's *magnetic field*, which were discovered by J. Van Allen (born 1914) in 1958 from the results of artificial *satellite* and *space probe* experiments. The inner belt, ranging from 1000 to 5600 km above the Earth's surface, is believed to consist of secondary charged particles emitted by the Earth's *atmosphere* as a consequence of the impact of *cosmic rays*. The outer belt lies between 15 000 and 25 000 km above the Earth, and contains electrons that originate from the *Sun* (see *solar wind*).

Van de Graaff generator An *electrostatic generator* for producing high voltages for use with particle accelerators and for other purposes. It consists of a large metal dome fed with charged particles by an endless insulated belt. Voltages of several megavolts can be achieved. The apparatus works with either an electron source or a positive-ion source. The voltage obtained is proportional to the diameter of the sphere. To obtain 1 MV requires a sphere of 2 metres diameter. However, this can be reduced by enclosing the equipment in nitrogen at $10-20$ *atmospheres*. In some later machines the belt has alternate metal and insulator beads or links, which enables a higher current to be taken. Named after R. J. Van de Graaff (1901–67).

Van der Waals' equation of state For a *mole* of a substance in the gaseous or liquid phases:

$$(p + a/V^2)(V - b) = RT,$$

where p = *pressure*, V = *volume*, T = *thermodynamic temperature*, R = the *gas constant*; a/V^2 is a correction for the mutual attraction of the *molecules* (see *Van der Waals' forces*), and b is a correction for the actual volume of the molecules themselves. The equation represents the behaviour of ordinary *gases* more correctly than the *perfect gas* equation $pV = RT$. Named after J. D. Van der Waals (1837–1923).

Van der Waals' force An attractive *force* existing between *atoms* or *molecules* of all substances. The force arises as a result of *electrons* in neighbouring atoms or molecules (see *atom, structure of*) moving in sympathy with one another. This

469

force is responsible for the term a/V^2 in *Van der Waals' equation of state*. In many substances this force is small compared with the other inter-atomic attractive and repulsive forces present.

vanillin $CH_3O(OH)C_6H_3CHO$. A white *insoluble* crystalline substance, m.p. 80–81°C., obtained from vanilla beans synthesized from *lignin*; it is used as a flavour and in perfumes.

Van't Hoff's law The *osmotic pressure* of a dilute *solution* is equal to the *pressure* that the *solute* would exert in the gaseous state, if it occupied a *volume* equal to the volume of the solution, at the same *temperature*. Named after Jacobus Van't Hoff (1852–1911).

vapour A substance in the gaseous state that can be liquefied by increasing the *pressure* without altering the *temperature*. A *gas* below its *critical temperature*.

vapour density A measure of the *density* of a *gas* or *vapour*; usually given relative to oxygen or hydrogen. The latter is the ratio of the *mass* of a certain *volume* of the gas to the *mass* of an equal volume of hydrogen, measured under the same conditions of *temperature* and *pressure*. Numerically this ratio is equal to half the *relative molecular mass* of the gas.

vapour pressure All *liquids* and *solids* give off *vapour*, consisting of *molecules* of the substance. If the substance is in an enclosed space, the *pressure* of the vapour will reach a maximum that depends only upon the nature of the substance and the *temperature*; the vapour is then saturated and its pressure is the *saturated vapour pressure*.

varec Kelp. The *ash* of seaweed, from which iodine is extracted.

variable (math.) **1.** A symbol or term that assumes, or to which may be assigned, different numerical values. An 'independent variable' is a variable in a *function* that determines the value of other variables. A 'dependent variable' has its value determined by other variables. E.g. in $y = 5x^2 + 2$, x is the independent variable and y is the dependent variable. **2.** Not *constant*.

variance 1. (statistics) The *square* of the mean *deviation*. **2.** (chem.) The number of *degrees of freedom* that a system can have.

variate (statistics) A *variable* that can have any of a *set* of values according to specified *probabilities*.

variation (bio.) The differences between individuals of animal and plant species. Differences acquired during an individual's lifetime, resulting from environmental conditions are not passed on to successive generations. Inherited variations assist a species in surviving environmental changes, since the wider the variation the more likely it is that some individuals will have the characteristics required to survive; these characteristics will be passed on to their offspring.

variation (math.) If a quantity y is some *function* of another quantity x, i.e. if $y =$ f(x), then, as x varies, y varies in a manner determined by the function. If f(x) = ax (where a is a constant), then y is said to vary directly as x, or to be directly proportional to x, $y = ax$. If f(x) = a/x, y is said to vary inversely as x, or to be inversely proportional to x; $y = a/x$.

variometer A *variable inductance* consisting of two coils in *series*, arranged so that one coil can rotate within the other. It is also used as a means of measuring inductance.

varistor A *resistor* that does not follow *Ohm's law*. It often consists of two *semiconductor diodes* in *parallel*, with opposite polarity.

varve dating An absolute *dating* technique used in *geochronology*, based on the thin layers of sedimentary clay called varves. These layers are alternately light and dark, corresponding to winter and summer deposits. They occur commonly in Scandinavia; by counting varves an absolute time scale (up to some 20 000 years) can be established for *fossils*.

Vaseline* See *petrolatum*.

vasopressin Antidiuretic hormone. A *hormone* produced in the hypothalamus of the brain and transported in the bloodstream to the pituitary gland, from which it is released. It stimulates the absorption of water by the kidneys, controlling the concentration of body fluids.

vat dyes A class of *insoluble dyes* that are applied by first reducing them to leuco-compounds, which are *soluble* in *alkalis*. The *solution* is applied to the material, and the insoluble dye is regenerated in the fibres by *oxidation*. *Indigo* and many synthetic dyes belong to this class.

VDU See *visual-display unit*.

vector 1. Any physical quantity that requires a direction to be stated in order to define it completely. For example, *displacement* and *velocity* are vector quantities; the corresponding *scalar* quantities are *speed* and distance. **2.** (bio.) An animal, often an insect, that transmits a pathogenic *microorganism*, usually from another animal to man. **3.** (genetics) A self-replicating molecule of *deoxyribonucleic acid* (DNA), such as a virus, that enables useful *genes* to be inserted into the *genome* of a host cell in *genetic engineering*.

vectors, parallelogram of If a particle is under the action of two like *vector* quantities, which are represented by the two sides of a *parallelogram* drawn from a point, the *resultant* of the two vectors is represented in magnitude and direction by the diagonal of the parallelogram drawn through the point. See *parallelogram of forces*; *parallelogram of velocities*.

vectors, triangle of For a particle under the action of three coplanar *vectors*, such that each side of the triangle represents one of these vectors, the vectors will be in *equilibrium* if the triangle is complete, with no gaps between the sides. See *triangle of forces*; *triangle of velocities*.

vegetable oils Oils obtained from the leaves, fruit, or seeds of plants; they consist of *esters* of *fatty acids* and *glycerol*. See also *fats and oils*.

velocity v. The *vector* of *speed*; i.e. the speed of a body in a specified direction; the rate of *displacement (s)* in a given direction gives a velocity, $v = ds/dt$. It is measured in metres per second.

velocity, relative The *velocity* of one body relative to another is the rate at which the first body is changing its position with respect to the second. If two bodies A and B are moving in the same direction, the velocity of A relative to B is $v_A - v_B$; if the bodies are moving in opposite directions, it is $v_A + v_B$ (provided that v_A and v_B are small compared to the *speed of light*).

velocity modulation The *modulation* of the *velocity* of a stream of *electrons* by alternately accelerating and decelerating them. See also *klystron*.

velocity of light See *light, speed of*.

velocity ratio of a machine Distance ratio. The ratio of the distance through which the point of application of the applied *force* moves, to the distance through which the point of application of the resistance moves in the same time.

For an 'ideal' machine, which requires no *energy* to move its component parts, the velocity ratio is equal to the *mechanical advantage*.

Venetian white A mixture of *lead carbonate* (white lead) and *barium sulphate* (blanc fixe) in equal parts. It is used in *paints*.

Venn diagram See *sets*.

Venturi tube A device for measuring the rate of flow of a *fluid*; it consists of an open-ended tube flared at each end, so that the fluid *velocity* in the narrow central portion is higher than at the flared ends. The fluid velocity can be calculated from the difference in *pressure* between the centre and the ends. Named after G. B. Venturi (1746–1822).

Venus (astr.) A *planet* with its *orbit* between those of *Mercury* and the *Earth*. Mean distance from the *Sun*, 108.21 million kilometres. *Sidereal period* ('year'), 224.701 days. *Mass*, approximately 0.815 that of the Earth, diameter 12 100 kilometres. The *atmosphere* of the planet consists of 98% *carbon dioxide* and 2% nitrogen. The Mariner *space probe* indicated that its surface temperature is about 475°C. with an atmospheric pressure 90 times that of the Earth and that it is covered by a dense cloud layer with freezing temperatures high up in the atmosphere. The planet rotates slowly on its axis, with a period of 243 days, in an opposite direction to the Earth's rotation.

verdigris A green deposit formed upon copper; it consists of *basic* copper(II) carbonate or *sulphate*, i.e. $CuCO_3.Cu(OH)_2$ and $CuSO_4.Cu(OH)_2.H_2O$, in variable proportions. The basic chloride is also present in some atmospheres, i.e. $CuCl_2.Cu(OH)_2$.

vermicide A substance used to kill worms.

vermiculite A group of low-grade *micas* that expand and exfoliate on heating to a light water-absorbent material. They are used in the exfoliated form as *heat* and *sound* insulating materials, and in special (potting) *soils*.

vermifuge A substance used for expelling (or killing) intestinal worms.

vermilion A scarlet form of *mercury(II) sulphide*, HgS; it is used as a *pigment*.

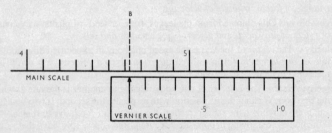

Figure 45.

vernier A device for measuring subdivisions of a scale. For a scale graduated in (say) centimetres and tenths, a vernier consists of a scale that slides alongside the main scale, and on which a length of nine-tenths of a cm is subdivided into ten equal parts. Each vernier division is thus 0.09 cm. If it is desired to measure a length AB, the main scale is placed with its zero mark at A, and the vernier scale is slid till its zero mark is at B. By noting which division on the vernier scale is exactly in line with a division on the main scale, the second decimal place of the length AB is obtained. Thus, if B falls between 4.6 and 4.7 cm on the main scale, and the fourth division on the vernier scale is just in line with a main scale division line, the length AB is 4.64 cm. See Fig. 45. Named after Pierre Vernier (1580–1637).

vernier motor (engine) A small *rocket* motor used to correct the flight path or *velocity* of a missile or *spacecraft*.

versed sine One minus the cosine of an angle (see *trigonometrical ratios*).

vertex 1. (math.) The point on a geometrical figure furthest from the *base*. **2.** (astr.) The point on the *celestial sphere* towards which, or from which, a *star* appears to move.

very high frequencies VHF. *Radio frequencies* in the range 30 to 300 *megahertz*.

very low frequencies VLF. *Radio frequencies* below 30 *kilohertz*.

vesicant Blister-producing.

VHF See *very high frequency*.

vibration, plane of see *polarization of light*.

vicinal When two similar *substituents* are added to a carbon compound, the positions of the substituents (or the molecule itself) are referred to as 'vicinal' (or *vic-*) if the substituents have attached to adjacent carbons and 'gem' if they have attached to the same carbon atom.

video-frequency signal The signal that transmits the picture and synchronizing information in a *television* system.

vinasse The residual *liquid* obtained after *fermentation* and *distillation* of beetroot molasses. It is used as a source of *potassium carbonate*.

vinegar A *liquid* containing 3%–6% *ethanoic acid*, obtained by the *oxidation* of *ethanol* by the action of *bacteria* on wine, beer, or fermented wort.

vinyl The *unsaturated univalent* group $CH_2:CH–$.

vinyl acetate See *ethenyl ethanoate*.

vinyl chloride See *chloroethene*.

vinylene The *bivalent* group $–CH:CH–$.

vinyl ether See *divinyl ether*.

vinylidene The *unsaturated bivalent* group $CH_2:C=$.

vinylidene chloride $CH_2:CCl_2$. A colourless inflammable *liquid*, b.p. 32°C., that *polymerizes* to form *polyvinylidene chloride*.

virgin neutrons *Neutrons*, produced by any means, before they have experienced a collision.

virial equation A *gas law* that attempts to account for the behaviour of a real gas. It usually takes the form:
$$pV = RT + Bp + Cp^2 + Dp^3 \ldots,$$
where B,C,D are empirical constants known as the virial coefficients.

virology The branch of *microbiology* concerned with the study of *viruses* and the diseases they cause.

virtual image See *image, virtual*.

virtual particle In *classical physics* a *force* between bodies not in contact (e.g. electrostatic repulsion) is represented by a *field*. In *quantum mechanics* this force may be represented by an exchange of particles between the interacting bodies. The exchanged particle is not in a 'real' state, as its emission and reabsorption violate the law of *conservation of mass and energy*; however, this is allowed within the context of the *uncertainty principle*, as it occurs within a very short time. Such a particle is described as a virtual particle. E.g. electrically charged particles may be visualized as interacting as the result of the exchange of virtual *photons*. A virtual particle that is responsible for a force can, by the addition of *energy* to the system, be converted into a real particle.

virtual work If a body, acted upon by a system of *forces*, is imagined to undergo a small displacement, then in general the forces will do *work*, termed the virtual work of the forces. If the body is in *equilibrium*, the total virtual work done is zero. This principle of virtual work is used to determine the positions of equilibrium of a body or a system of bodies under the action of given forces, and to determine relations between the forces acting on such a system in a given equilibrium position.

virus A disease-producing particle, too small to be seen by an optical *microscope* but visible with an *electron microscope*. Viruses are only capable of multiplication within a living *cell*, each type of virus requiring a specific host cell. The simplest virus consist of a single helical strand of *ribonucleic acid* coated with *protein molecules* (see *tobacco mosaic virus*). The active principle of these viruses resides in the RNA as it is only this part of the particle that enters the cell. Other viruses are considerably more complex and contain DNA rather than RNA; they may be up to 0.2 *micrometre* in diameter. Viruses are considered to be on the borderline between the animate and the inanimate. See also *bacteriophage*; *retrovirus*.

viscometer An instrument for the measurement of *viscosity*. In the Ostwald viscometer the time taken for a liquid to fall to a specified level through a *capillary* tube, from a bulb above, is measured. In another type the *speed* at which a ball falls through a liquid is measured.

viscose A thick treacly brownish *liquid*, consisting mainly of a *solution* of cellulose *xanthate* in dilute *sodium hydroxide*. It is made from *cellulose* by the action of sodium hydroxide and *carbon disulphide*. It is used for the production of viscose *rayon* and of cellulose film, of the type used for transparent wrappings.

viscose rayon See *rayon*.

viscosity η. The property of a *fluid* that enables it to resist flowing when it is subject to a shear stress. If different layers of a fluid are moving with different *velocities*, *viscous forces* come into play, tending to slow down the faster-moving layers and to increase the velocity of the slower-moving layers. For two parallel layers in the direction of flow, a short distance apart, this viscous force is proportional to the velocity gradient between the layers (provided the velocity gradient is sufficiently small for the flow not to be *turbulent*; see *Newtonian fluid*), i.e. $F = \eta A.\mathrm{d}v/\mathrm{d}x$, where A is the area of the adjacent layers and η is a constant of proportionality called the coefficient of viscosity of the fluid. Viscosity is measured in *pascal seconds* (*SI units*) or *poise* (*c.g.s. units*). 1 centipoise $= 10^{-3}$ Pa s. See also *kinematic viscosity*.

viscous Having high *viscosity*; a viscous *liquid* drags in a treacle-like manner.

visible spectrum The range of *electromagnetic radiations* that are visible to man. See *spectrum colours*.

vision See *photopic vision; scotopic vision*.

visual binary A *binary star* system that can be resolved into two *stars* with an optical *telescope*. See also *spectroscopic binary*.

visual-display unit VDU. A *computer peripheral* device whose output is a *cathode-ray tube* for displaying text or diagrams. It may have an input device consisting of a keyboard or it may be a *light pen*.

visual purple See *rhodopsin*.

vitalistic theory The view that life, and all consequent biological phenomena, are due to a 'vital force'.

vitamins Accessory food factors. A group of organic substances, about 14 in number, occurring in various foods, which are necessary for a normal diet. Absence or shortage leads to various deficiency diseases. Before the chemical nature of any of the vitamins was known, they were named by the letters of the alphabet.

Vitamin A, retinol, $C_{20}H_{29}OH$, cannot be synthesized by humans, although the *carotenes* that occur in green plants are converted to vitamin A in the intestinal wall and the liver. Deficiency causes 'night-blindness' (see *rhodopsin*) and ultimately more serious eye troubles; the resistance of the mucous membranes to infection also decreases.

Vitamin B, originally regarded as a single substance, is now known to consist of nine *compounds* termed the vitamin B complex.

Vitamin B_1, thiamine, aneurin, $C_{12}H_{17}ON_4SCl$, is important in *carbohydrate metabolism*. Deficiency causes beriberi. It occurs in yeast, wheatgerm, beans, peas, and other vegetables.

Vitamin B_2, riboflavin, lactoflavin, $C_{17}H_{20}N_4O_6$, is an important component of various *coenzymes* involved in the metabolism of most foods. Deficiency causes inflammation of the skin, especially of the tongue and mouth. It occurs in green vegetables, yeast, liver, and milk.

Vitamin B_6, pyridoxine, $C_8H_{11}NO_3$, is a component of the coenzymes involved in the metabolism of *amino acids*. Deficiency causes retarded growth, skin infections, convulsions, etc. It occurs in cereals, milk, yeast, and liver.

Vitamin B_{12}, cyanocobalamin, $C_{63}H_{90}O_{14}N_{14}PCo$, functions as a *coenzyme* in the synthesis of *deoxyribonucleic acid* (DNA) and the *oxidation* of *fatty acids*. It also functions, in conjunction with *folic acid*, in the production of red blood cells. Deficiency causes pernicious anaemia. Liver is the best source.

Other vitamins in the B complex are: *biotin*; *choline*; *inositol*; *lipoic acid*; *nicotinic acid*; *pantothenic acid*.

Vitamin C, ascorbic acid, $C_6H_8O_6$, is needed to maintain connective tissues. Deficiency causes scurvy. Many organisms can synthesize it from *glucose*, but primates (including man) must acquire it from their diet. It occurs in citrus fruits and vegetables.

Vitamin D, a fat-soluble steroid that occurs as vitamin D_2, ergocalciferol, calciferol, $C_{28}H_{43}OH$, which is formed from *ergosterol* by the action of ultra-violet radiation, and vitamin D_3, cholecalciferol, which is produced by the action of sunlight on a cholesterol derivative in the skin. Vitamin D deficiency causes rickets and other bone diseases as it is important in enabling the body to make use of calcium for the production of bones. The major source of vitamin D is fish-liver oils.

Vitamin E, tocopherol, $C_{20}H_{50}H_2$, is needed to prevent the *oxidation* of *unsaturated fatty acids* in *cell membranes* and so to preserve the structure of these membranes. Deficiency can lead to a range of problems including infertility, liver damage, and muscular diseases. It occurs in cereals and green vegetables. Vitamin K is a fat-soluble substance that occurs as phytomenadione (in plants) and menaquinone (in animals). It is required for the production of prothrombin in the liver (which is itself needed for its blood-clotting properties). It does not involve deficiency diseases as it can be synthesized by bacteria in the large intestine. It also occurs widely in green vegetables and meat.

Vitreosil* A translucent form of *silica*, SiO_2, prepared from sand. It is used for making laboratory apparatus that is required to withstand large and sudden changes in *temperature* as it does not crack owing to its very low expansion.

vitreous Pertaining to, composed of, or resembling *glass*.

vitreous silica See *quartz*.

vitriol Concentrated *sulphuric acid*, H_2SO_4, oil of vitriol; *copper(II) sulphate*, $CuSO_4.5H_2O$, was formerly called blue vitriol; *ferrous sulphate*, $FeSO_4.7H_2O$, was called green vitriol; *zinc sulphate*, $ZnSO_4.7H_2O$, was sometimes known as white vitriol.

VLF See *very low frequency*.

volatile Passing readily into *vapour*; having a high *vapour pressure*.

volcanic rock See *igneous rock*.

volcano A gap in the *Earth's* crust from which hot *gases, liquid magma*, and *ash* can escape. The conical hill or mountain above it is formed from an accumulation of the *lava* and ash. Chains of volcanos are formed at the margins of lithospheric plates (see *plate tectonics*), where one plate is pushed underneath the next plate.

volt The derived *SI unit* of electric *potential* defined as the difference of potential between two points on a conducting wire carrying a constant *current* of one *ampere* when the *power* dissipated between these points is one *watt*. It is also the unit of *potential difference* and *electromotive force*. 1 volt $= 10^8$ *electromagnetic units*. Symbol V (= W/A). Named after Alessandro Volta (1745 – 1827).

voltage The *potential, potential difference*, or *electromotive force* of a supply of electricity, measured in *volts*.

voltage divider Potential divider, potentiometer. A *resistor* or series of resistors connected across a source of *voltage* (V) and tapped at a point to give a fraction (v) of the total voltage. In Fig. 46:

$$v/V = r_2/(r_1 + r_2).$$

voltage doubler An *electronic circuit* that delivers a *direct current voltage* approximately twice the peak *alternating current* voltage it feeds on. It usually consists of two *semiconductor diode rectifiers* whose outputs are connected in *series*.

voltage drop The *voltage* between two points on a *conductor* resulting from a flow of current between them. In a *d.c. circuit* the voltage drop is equal to the product of the current and the *resistance* between the two points. In an *a.c.* circuit, the resistance is replaced by the *impedance*.

voltaic cell See *cell* (phys.).

voltaic pile The earliest electric *battery*, devised by Volta. It consists of a number of *cells* joined in series, each consisting of a sheet of zinc and copper separated by a piece of cloth moistened with dilute *sulphuric acid*.

voltameter Coulometer. An electrolytic cell in which a *metal*, generally silver or copper, is deposited by *electrolysis* of a *salt* of the metal upon the *cathode*. From the increase in mass (m) of the cathode and a knowledge of the *electrochemical equivalent* (z) of the metal, the quantity of electric charge (Q) that has passed through the circuit can be found from the relationship $Q = m/z$.

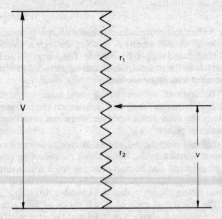

Figure 46.

volt-ampere VA. The unit of electrical *power* in an *a.c. circuit*, equal to the product of the *root-mean-square* values of the voltage and the current.

voltmeter An instrument for measuring the *potential difference* between two points. In principle it consists of an arrangement similar to an *ammeter* with a high *resistance* in series incorporated in the instrument, the scale being cali- brated in *volts*. When the instrument is connected in parallel between the points at which the p.d. is being measured, very little current flows through it, and a correct reading of the *voltage* is obtained. A *moving-coil* instrument can only be used with *d.c.* unless a *rectifier* is used. A *moving-iron* instrument can be used with either *a.c.* or d.c. Digital voltmeters are electronic instruments that give a reading in digits.

volume 1. *V*. The measure of bulk or space occupied by a body. **2.** The loudness of sound, especially of the sound produced electronically by a *loudspeaker*.

volumetric analysis A group of methods of *quantitative chemical analysis* involv- ing the measurement of *volumes* of the reacting substances. The amount of a substance present is determined by finding the volume of a *solution* of another substance, of known *concentration*, that is required to react with it. The added volume is measured by adding the reacting solution from a *burette* (see *titra- tion*); the *end point* of the reaction is often shown by a suitable *indicator*.

vulcanite A hard insulating material made by the action of *rubber* on sulphur.

vulcanized rubber The product obtained by heating *rubber* with sulphur.

vulgar fraction Common fraction. A fraction expressed in terms of a *numerator* and a *denominator*, e.g. $\frac{1}{3}$.

W

Wacker process A chemical process used to make *ethanal* by oxidizing *ethene* in air. The mixture of air and ethene is bubbled through a solution containing palladium(II) chloride and copper(II) chloride. The copper(II) and palladium(II) ions act as catalysts in the process, which can be extended to produce *ethanoic acid* and is also used with other *alkenes*.

wall of a cell The layer of *polysaccharides* and other substances (*cellulose, hemicellulose, pectins, lignin,* etc.) that lies outside the *plasma membrane* of plant *cells*, giving the cells a characteristic shape and contributing to the mechanical properties of plants.

Wankel rotary engine Epitrochoidal engine. A type of *internal-combustion engine* employing a 4-stroke cycle, but without reciprocating parts. It consists essentially of an elliptical combustion chamber fitted with valveless inlet and outlet ports, and a conventional *sparking-plug*. An epicyclically-driven roughly triangular-shaped piston rotates within this chamber dividing it into three gastight sections, the volume of each of which varies as the piston rotates. The explosive mixture sucked in through the inlet port is compressed by the rotating piston and exploded by the sparking-plug. The explosion provides the power to rotate the piston and sweep the exhaust gases round to the outlet port. The small number of moving parts and the absence of vibration are the chief advantages of this type of engine although the problem of making an effective seal between the piston edges and the walls of the chamber have made it unsatisfactory in use. Named after Felix Wankel (1902–88).

warfarin $C_{19}H_{16}O_4$. A colourless crystalline substance, m.p. 161°C., used as a rat poison and an anticoagulant in medicine.

washing-soda Crystalline *sodium carbonate*, $Na_2CO_3.10H_2O$.

water H_2O. The normal *oxide* of hydrogen. Natural water (river, spring, rain, etc.) is never quite pure but contains dissolved substances. Pure water is a colourless, odourless *liquid*, m.p. 0°C., b.p. 100°C., which has a maximum *density* at 3.98°C. of 999.973 kg m^{-3}. In the gaseous phase, water (or steam) molecules exist as single entities with an angle of 105° between the two OH bonds that form the H-O-H structure. Liquid water consists of associated *polar molecules*, $(H_2O)_n$, with *hydrogen bonds* between the molecules (see illustration at *hydrogen bond*).

water equivalent (phys.) See *heat capacity*.

water, expansion of *Water* on cooling reaches its maximum *density* at 3.98°C. when its density is 999.973 kg m^{-3}; it then expands as its temperature falls to 0°C., the density at 0° being 999.841 kg m^{-3}; on freezing, it expands still further, giving *ice* with a density of 916.8 kg m^{-3} at 0°C. This accounts for the bursting of water-pipes in frosts, and the fact that ice floats on water.

water gas A *fuel gas* obtained by the action of *steam* on glowing hot *coke*; the gas formed consists of *carbon monoxide* and hydrogen. The formation of water gas is accompanied by absorption of *heat* (an *endothermic* reaction); thus the coke is rapidly cooled and has to be reheated at intervals by a blast of hot air, which causes partial *combustion* and makes the coke incandescent again.

water glass See *sodium silicate*.

water of crystallization A definite molecular proportion of water chemically combined with certain substances in the crystalline state; e.g. the *crystals* of *copper(II) sulphate* contain 5 *molecules* of *water* with every molecule of copper(II) sulphate, $CuSO_4.5H_2O$. The water molecules may form bonds with the *ions* of the *salt* or they may occupy positions in the crystal *lattice*. In the case of copper(II) sulphate, for example, four of the water molecules form *coordinate bonds* with the copper ions. These bonds break at about 100°C., leaving a *monohydrate* in which one water molecule is held by a hydrogen bond to the sulphate ion. This bond breaks at a temperature of 250°C., when the substance becomes *anhydrous*.

water softening The removal of the causes of hardness of water (see *hard water*). It generally depends on the *precipitation* or removal from *solution* of the *metals* the *salts* of which cause the hardness.

water turbine See *turbine*.

water vapour *Water* in the gaseous or *vapour* state; it is present in the *atmosphere* in varying amounts. See *humidity*.

watt The derived *SI* unit of *power*, equal to one *joule* per *second*. The *energy* expended per second by an unvarying *electric current* of 1 *ampere* flowing through a *conductor* the ends of which are maintained at a *potential difference* of 1 *volt*. It is equivalent to 10^7 *ergs* per second. The power in watts is given by the product of the current in amperes and the potential difference in volts (see also *volt-ampere*). 1000 watts = 1 *kilowatt*; 745.7 watts = 1 *horsepower*. Symbol W (= J/s). Named after James Watt (1736–1819).

wattage *Power* measured in *watts*.

wattmeter An instrument for the direct measurement of the *power*, in *watts*, of an electrical *circuit*. It usually consists of an *electrodynamometer*.

watt-second A unit of *work* or *energy* equivalent to one *joule*. See also *kilowatt-hour*.

wave A periodic disturbance in a medium or in *space* that involves the elastic displacement of material particles or a periodic change in some physical quantity, such as *temperature*, *pressure*, *electric potential*, *electromagnetic field* strength, etc. See *wave motion*.

waveband A range of *wavelengths* in the *electromagnetic spectrum* that is characterized by some property of the radiation or the transmitting system.

wave energy A *renewable energy source* that relies on the wave motion in the sea to generate *energy*. The most likely method would be to use a string of floats (called Salter ducks), which rise and fall with the waves, and in so doing turn a generator. It is estimated that off the coast of the UK there are sufficient suitable sites to generate 120 GW of power, although not all of this could be turned into electrical energy. Moreover, the design and maintenance of the generators and transmission lines present formidable problems.

wave equation The *equation* that gives mathematical expression to *wave motion*:
$$\nabla^2 \psi = 1/c^2 . \partial^2 \psi / \partial t^2$$
where ∇^2 is the *Laplace operator*, ψ is the *wave function*, c is the *speed of light*, and t is the time at any instant. See also *Schrödinger's wave equation*.

479

wave form The shape of a *wave*, illustrated graphically by plotting the values of the periodic quantity against time.

wave front The locus of adjacent points in the path of a *wave motion* that possess the same *phase*.

wave function In *wave mechanics*, *orbital electrons* are not treated as particles moving in precisely defined *orbits*, but as 3-dimensional *standing wave* systems represented by a wave function, ψ, the magnitude of which represents the varying *amplitudes* of the wave system at various points around the *nucleus*. The volume containing all the points at which ψ has an appreciable magnitude is called the *orbital* of the electron. Thus, according to wave mechanics, the precise position and *velocity* of an electron (which cannot be defined without error, see *uncertainty principle*) is replaced by a *probability* that an electron, visualized as a particle, will be present in a volume element dV. The *probability* of finding it at a certain point is zero. See also *wave equation*.

wave guide A hollow *metal conductor* through which *microwaves* may be propagated with little *attenuation*. Having either rectangular or circular cross-sections, they are used extensively in *radar*.

wavelength λ. The distance between successive points of equal *phase* of a *wave*; e.g. the wavelength of the waves on water could be measured as the distance from crest to crest. The wavelength is equal to the *speed* of the *wave motion* divided by its *frequency*. For *electromagnetic radiation* $\lambda = c/f$, where c is the *speed of light* and f is the frequency.

wave mechanics A development of *quantum mechanics*. Every particle is considered to be associated with a periodic *wave*, whose *frequency* and *amplitude* are determined by rules (see *de Broglie wavelength*) derived partly by analogy with the propagation of *light* waves, partly by ad hoc hypothesis from known quantum conditions, and partly from necessary conditions of continuity. These waves, however, are not conceived as having any real physical existence, the term 'wave' being used only by analogy as a description of the mathematical relations employed, since in all but the simplest cases the waves would have to be imagined in a 'hyperspace' of very many dimensions. Wave mechanics is based on *Schrödinger's wave equation* relating the energy of a system to its *wave function*, only certain values for which are allowed (see *eigenfunction*).

wavemeter An instrument for measuring the *wavelength* of a *radio frequency electromagnetic radiation*. Up to about 100 MHz a *tuned circuit* is used, but above this *frequency* a cavity resonator in a *waveguide* is employed.

wave motion The propagation of a periodic disturbance carrying *energy*. At any point along the path of a wave motion, a periodic displacement or vibration about a mean position takes place. This may take the form of a displacement of air *molecules* (e.g. *sound* waves in air), of water molecules (waves on water), a displacement of elements of a string or wire, displacement of electric and magnetic *vectors* (*electromagnetic waves*), etc. The locus of these displacements at any instant is called the *wave*. The wave motion moves forward a distance equal to its *wavelength* in the time taken for the displacement at any point to undergo a complete *cycle* about its mean position. Thus the speed (c) with which the wave motion is propagated is the product of the *wavelength* (λ) and the *frequency* (f), i.e. $c = \lambda f$. This is known as the 'phase speed'. In some wave motions the phase speed varies with the wavelength. As a result, a non-sinusoidal wave appears to travel at a speed different to that of the phase speed. This is given by:

$$U = c - \lambda.dc/d\lambda$$

where U is known as the 'group speed'. See *longitudinal waves*; *transverse waves*.

wave number $\sigma = 1/\lambda$. The number of *waves* in unit length. It is the reciprocal of *wavelength*.

wave theory of light The theory that *light* is propagated as a *wave motion* (see *electromagnetic radiation*). Formerly the existence of a medium, the *ether*, was postulated for the transmission of light waves. This hypothesis has been rejected as unnecessary, and the classical wave theory has been modified to include the dual particle (*photon*) wave concept, which is required to explain all the observed phenomena. See *complementarity*.

wax True waxes (e.g. *beeswax*) are simple *lipids* consisting of *esters* of higher *fatty acids* that are found in *fats and oils*, with *monohydric alcohols*. The term is often loosely applied to *solid*, non-greasy, *insoluble* substances that soften or melt at fairly low *temperatures*, e.g. *paraffin wax*.

weak acid An *acid*, such as *ethanoic acid*, that is only partly dissociated in solution. Compare *strong acid*.

weak electrolytes See *electrolytic dissociation*.

weak interactions A fundamental interaction between *elementary particles* that is some 10^{12} times weaker than *strong interactions*. *Beta decay* is a form of weak interaction as are the initial reactions between *protons* that fuel the Sun's conversion of hydrogen to helium. According to the *electroweak theory*, such interactions are the result of an exchange of *virtual particles* called W and Z *particles*.

weber The derived *SI unit* of *magnetic flux* defined as the flux that, linking a circuit of one turn, produces in it an *EMF* of one *volt* as it reduces to zero at a uniform rate in one *second*. Symbol Wb (= Vs). 1 weber = 10^{8} *maxwells*. Named after Wilhelm Weber (1804–91).

weight The *force* of attraction of the *Earth* on a given *mass* is the weight of that mass. Being a force, weight is correctly measured in units of force, such as the *newton*. The weight of a mass m, is equal to mg, where g is the *acceleration of free fall*. Thus the weight of a body depends on its geographical position (because of the variation in the value of g). The weight of a body is sometimes loosely expressed in units of mass, though this is not correct scientifically.

weight, British units of Avoirdupois weights.

$437\frac{1}{2}$ grains = 1 ounce
7000 grains = 16 ounces = 1 *pound* = 0.453 592 *kilograms*
14 pounds = 1 stone
2 stone = 1 quarter
4 quarters = 1 hundredweight
2240 pounds = 20 cwt = 1 ton
2000 pounds = 1 short ton

See Appendix, Table 1.

weight, metric units of

1000 milligrams = 1 gram = 15.432 grains
1000 grams = 1 kilogram = 2.204 62 lb

1000 kilograms = 1 tonne = 0.984 207 ton
See Appendix, Table 1.

weightlessness A condition in which a body is infinitely far from any other body so that it experiences no *gravitational force*. This theoretical concept can be simulated in space when the Earth's gravitational force on an orbiting body is equal to the *centripetal* force required by its orbital motion, so that the body is in free fall. Free fall can also be simulated for a short period by an aircraft flying in a parabolic flight path.

welding Joining of two *metal* surfaces by raising their *temperature* sufficiently to melt and fuse them together.

Weston cell Cadmium cell. A *primary cell* used as a standard of *EMF*. It produces 1.0186 *volts* at 20°C. It consists of a mercury *anode* covered with mercury(I) sulphate and a cadmium *amalgam cathode* coated with cadmium sulphate *crystals*. The *electrolyte* is a *saturated solution* of cadmium sulphate.

wet and dry bulb hygrometer An instrument for determining the *relative humidity* of the *atmosphere*. It consists of a pair of *thermometers* side by side, the bulb of one being surrounded by moistened muslin. This one will indicate a lower temperature than the other, on account of loss of *heat* by *evaporation*; the difference in the readings will depend upon the relative humidity, which can be found by reference to special tables calculated for the purpose.

wetting agent A substance that lowers the *surface tension* of a *liquid*.

whale oil Animal *fat* obtained from the fatty layer of blubber of true whales. After extraction it is divided into various fractions and used for *soap* manufacture and other purposes; on *hydrogenation* a hard tasteless edible fat is obtained. As whaling is now restricted, substitutes are usually used for whale oil.

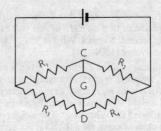

Figure 47.

Wheatstone bridge A divided electrical *circuit* used for the measurement of *resistances*. When no current flows from $C = D$, as indicated by the absence of deflection on the *galvanometer G*, $R_1/R_2 = R_3/R_4$, where R_1, etc., are resistances. See Fig. 47. This principle is applied in the metre bridge. A wire, *AB*, of uniform resistance and generally 1 metre in length, corresponds to R_3 and R_4 in the Wheatstone bridge diagram; for R_1 a standard resistance is used, while R_2 is the resistance to be measured. By a sliding contact a point of no deflection in the galvanometer is found along *AB*, the resistances R_3 and R_4 being proportional to the lengths cut off. Named after Sir Charles Wheatstone (1802–75).

whistler An *atmospheric* whistle of descending *pitch* that can be picked up under certain circumstances by a *radio* receiver. It is caused by *electromagnetic radiations*, produced by *lightning* flashes, which follow the *lines of force* of the Earth's *magnetic field* and are reflected back to Earth by the *ionosphere*.

white arsenic See *arsenic(III) oxide*.

white blood cells See *leucocytes*.

white bronze *Bronze* that contains a high proportion of tin.

white dwarf A class of small highly dense *stars* of low *luminosity*. They are the remnants of stars that have consumed nearly all their available hydrogen and have suffered a gravitational collapse. Owing to their small size they have high surface *temperatures* and therefore appear white. See *supernovae*.

white lead See *lead carbonate*.

white light *Light* that can be resolved into a continuous *spectrum* of *wavelengths* (i.e. *colours*) in the correct balance; e.g. the light from an incandescent 'white-hot' *solid*.

white spirit A *mixture* mainly of *alkanes* of boiling range $150°-200°C$. It is used as a *solvent* and in the *paint* and varnish industry, usually as a substitute for *turpentine*.

white vitriol See *zinc sulphate*.

wide-angle lens A *camera lens* with a wide angle of view (up to $100°$) and a short *focal length*.

Wiedemann-Franz law The ratio of the thermal *conductivity* to the electrical conductivity is the same for all *metals* at a given *temperature*. This ratio is proportional to the *thermodynamic temperature*. Most pure metals obey the law with reasonable accuracy at ordinary temperatures.

Wien displacement The product of the *thermodynamic temperature* and the wavelength at which maximum emission occurs from a *black body* is constant. A graph of wavelength against temperature has a maximum, which is displaced towards shorter wavelengths as the temperature increases. Named after Wilhelm Wien (1864–1928).

Wigner effect The effect produced when the *atoms* in a *crystal* are displaced as a result of *irradiation*. If *graphite*, for example, is bombarded with *neutrons*, the shape of the crystal *lattice* is altered and the material suffers a change of physical dimensions. See also *Wigner energy*. Named after Eugene Paul Wigner (born 1902).

Wigner energy *Energy* stored within a crystalline substance as a result of the *Wigner effect*. In a *nuclear reactor* in which *graphite* is used as the *moderator*, some of the energy lost by the *neutrons* is stored in the graphite; this is known as the Wigner energy.

Wigner nuclides Pairs of *isobars* of odd *mass number* in which the *atomic number* and *neutron* number differ by one, e.g. ^3_1H and ^3_2He.

Wilson cloud chamber See *cloud chamber*. Named after C. T. R. Wilson (1869–1961).

WIMP Weakly Interacting Massive Particle. See *missing mass*.

Wimshurst machine An early laboratory apparatus for generating static *electric charge*. It consists of two insulating discs to which radial metal strips are attached. The discs are rotated in opposite directions. The charge, produced by friction, is collected by metal combs. Named after J. Wimshurst (1836–1903).

wind A large-scale movement of air, generally caused by a *convection* effect in the *atmosphere*.

wind energy A *renewable energy source* that makes use of wind turbines (see *aerogenerators*) to generate electrical energy. Some 12 aerogenerators are currently feeding *power* into the UK national *grid*, mainly making use of the Atlantic winds on the west coast of the country. It is estimated that if all the suitable sites were to be used, some 20% of the UK's electricity could be generated in this way.

window 1. A wavelength band to which a particular medium is transparent. The atmosphere, for example, has a radio window in the range 8 mm–20 m. **2.** A period during which an event may take place. For example, a spacecraft may be launched during a launch window to achieve a desired rendezvous.

wire chamber A device used to record electronically the tracks of ionizing particles. The basic multiwire proportional chamber consists of three layers of parallel wires in a suitable gas. An *electric field* is set up between the layers, with the outer layers acting as *cathodes*, while the central layer forms the *anode*. When a particle ionizes the gas, the *electrons* released travel to the anode wires, producing small *avalanches* of additional electrons as they accelerate in the high *electric field* near the wires. These electrons are detected as pulses of *electric charge* on the wires nearest the path of the particle. In a different variation, called the 'drift chamber', the electrons produced by the ionizing particle are allowed to move at a constant velocity in a region of uniform electric field so that their time of arrival at the anode wires depends on the position at which they were produced.

wireless See *radio*.

witherite See *barium carbonate*.

wolfram W. See *tungsten*.

wolframite 'Wolfram'. Natural iron tungstate, $FeWO_4$.

Wollaston prism A *prism* for obtaining plane-polarized light (see *polarization of light*). Constructed of *quartz*, this prism, like the *Rochon prism*, may be used for *ultraviolet radiation*. Named after W. H. Wollaston (1766–1828).

wood naphtha See *methanol*.

Wood's metal An *alloy* of 50% bismuth, 25% lead, 12.5% tin, 12.5% cadmium. It has a low m.p. of 71°C.

wood sugar See *xylose*.

woofer A *loudspeaker* designed to reproduce the lower *audiofrequency sounds*.

word The smallest number of *bits* of information that a particular *computer* can conveniently process as a single unit; usually 12 to 64 bits.

word processor A form of computerized typewriter that consists of a keyboard, a *microcomputer* designed to function as a word processor, a memory store, a *VDU*, and a word processing packaged *program*. The program enables the user to amend and correct without retyping, to search for specified items, to store in memory, to check spelling, to merge documents, etc.

work (phys.) The work done by a *force f* when it moves its point of application through a distance s is equal to $fs \cos \theta$, where θ is the *angle* between the line of action of the force and the displacement. The derived *SI unit* of work is the *joule*; other units include *erg*, *foot-pound*, *foot-poundal*.

work function 1. At the *absolute zero* of temperature, the *free electrons* present in a *metal* are distributed amongst a large number of discrete *energy* states E_1, E_2, etc., up to a state of maximum energy E. At higher temperatures a small proportion of the electrons have energies greater than E. The work function of a metal is the energy that must be supplied to free electrons possessing energy E, to enable them to escape from the metal. It is usually expressed in joules although it is sometimes expressed as a potential difference in volts. **2.** Helmholtz *free energy*.

work hardening See *strain hardening*.

working stress See *ultimate stress*.

wort See *brewing*.

W particle W *boson*. The intermediary *elementary particle* with one unit of *electric charge* that is exchanged as a *virtual particle* in some *weak interactions*. The W particle, like its neutral relation the *Z particle*, is very heavy, with a mass of 80 GeV.

wrought iron The purest commercial form of iron; iron nearly free from carbon. It is very tough and fibrous and can be welded.

Wurtz reaction A method of preparing an *alkane* from a haloalkane by refluxing it with sodium in *ether*, i.e.
$$2RX + 2Na \rightarrow 2NaX + RR,$$
where X is a halogen atom and R is a univalent organic group, e.g.
$$CH_3Cl + 2Na \rightarrow 2NaCl + C_2H_6.$$
Named after C-A Wurtz (1817–84).

X

xanthates *Salts* or *esters* of the series of xanthic acids that have the general formula ROCSSH. Cellulose xanthate is the important intermediate product in the manufacture of *viscose*.

xanthene Dibenzo-1,4-pyran. $C_6H_4O.CH_2C_6H_4$. A yellow crystalline *heterocyclic* compound, m.p. 100.5°C., which forms the basis of the xanthene *dyes*.

xanthine 2,6-dioxypurine. $C_5H_4N_4O_2$. A yellow *soluble heterocyclic* compound, found in urine, *blood*, and certain animal *tissues*.

xanthone $C_6H_4.CO.O.C_6H_4$. A yellow *insoluble* crystalline *ketone*, m.p. 174°C., that occurs in several natural yellow *pigments*.

xenon Xe. Element. R.a.m. 131.29. At. No. 54. An *inert gas* occurring in exceedingly minute amounts in the air (about 0.006 parts per million by volume); m.p. −111.9°C., b.p. −108.1°C. It is used in filling certain types of *fluorescent tubes* and light bulbs. Several compounds, e.g. XeF_2 and XeO_3, are known.

xerography A method of photographic copying in which an electrostatic image is formed on a surface coated with selenium when it is exposed to an optical image. A dark powder (consisting of *graphite* and a *thermoplastic resin*), oppositely charged to the electrostatic image, is dusted on to the surface after exposure so that particles adhere to the charged regions; the image thus formed is then transferred to a sheet of charged paper and fixed by heating.

X-radiation *Electromagnetic radiation* consisting of *X-rays*.

X-ray astronomy See *X-ray sources*.

X-ray crystallography The study of crystalline substances by observation of the *diffraction* patterns that occur when a *beam* of *X-rays* is passed through a *crystal*. It is principally as a result of the use of X-ray crystallography that the structure of certain *proteins* (e.g. *haemoglobin*) and *nucleic acids* has been analysed.

X-ray diffraction See *X-ray crystallography*.

X-rays Röntgen rays. *Electromagnetic radiations* of the same type as *light*, but of much shorter *wavelength*, in the range of 5×10^{-9} *metre* to 6×10^{-12} metre approximately. They are produced when a stream of *electrons* strikes a material object. The *atoms* of all the *elements* emit a characteristic *X-ray spectrum* when bombarded by electrons, as a result of the inner orbital electrons being displaced by the bombarding electrons. An outer electron then falls into the inner shell to replace the displaced electron, losing potential energy ΔE; the *frequency* of the emitted X-rays is $\Delta E/h$, where h is the *Planck constant*. X-rays affect a photographic plate in a way similar to light. The absorption of the rays by *matter* depends upon the *atomic number* and the concentration of the atoms of the material. The lower the r.a.m. and density, the more *transparent* is the material to X-rays. Thus, bones are more opaque than the surrounding flesh; this makes it possible to take an X-ray photograph (*radiograph*) of the bones of a living person.

X-ray sources Astronomical sources that emit *X-rays*: they were discovered by instruments carried outside the *Earth's atmosphere* by *space probes*. Owing to the absorption of X-rays by the Earth's atmosphere they cannot be seen by land-based telescopes, although some high-energy X-rays can be detected by telescopes on balloons in the upper atmosphere. X-ray astronomy is now largely carried out by means of *rockets* and artificial *satellites*.

X-ray spectrum Each *element*, when bombarded by *electrons*, emits *X-rays* of several characteristic *frequencies*, depending on the *atomic number*; a photograph of the *line spectrum* corresponding to various elements may thus be obtained from the X-rays emitted. There is also a continuous spectrum caused by *Bremsstrahlung*.

X-ray tube An evacuated tube for producing *X-rays*; it contains an *electron gun* and a heavy *metal* target forming part of a massive *anode*. The metal emits X-rays when it is bombarded by high-energy *electrons*. The *spectrum* of the radiation depends on the *voltage* between the *cathode* and the anode, the *temperature* of the cathode, and the metal of the target.

xylan A complex *polysaccharide* that occurs closely associated with *cellulose* in plants.

xylene Xylol, dimethylbenzene. $C_6H_4(CH_3)_2$. A *liquid* resembling *toluene* that occurs in *coal-tar*. It exists in three *isomeric* forms, a *mixture* of which boils at 137°–140°C. It is used in the manufacture of *dyes*.

xylidine Dimethylaniline. $(CH_3)_2C_6H_3NH_2$. An aromatic amine that exists in six isomeric forms, of which five are liquids above 20°C.; b.p. in the range 216°–230°C. It is used in the manufacture of *dyes*.

xylol See *xylene*.

xylose Wood sugar. $C_5H_{10}O_5$. A colourless crystalline *pentose*, m.p. 144°C., found in *xylan*.

xylyl The *univalent* group $CH_3C_6H_4CH_2-$.

xylylene The *bivalent* group $-H_2CC_6H_4CH_2-$.

Y

Yagi aerial A directional *aerial* consisting of one or two *dipoles*, a parallel reflector, and a series of directors in front of the dipole, all so arranged that *radiation* is focused on to the dipole. It is used in *television* and *radio astronomy*. Named after Hidetsuga Yagi (1886–1976).

yard British unit of length. The Imperial standard yard used to be defined as the distance, at 62°F., between the central traverse lines on two gold plugs in a certain *bronze* bar. The yard was redefined by the 1963 Weights and Measures Act as 0.9144 *metre*.

year A measure of time, commonly understood to be the time taken by the *Earth* to complete its *orbit* round the *Sun*. The civil year has an average value of 365.2425 mean *solar days*; 3 successive years consisting of 365 days, the fourth or leap year of 366. Century years do not count as leap years unless divisible by 400. The tropical, astronomical, or solar year, the average interval between two successive returns of the Sun to the first point of Aries, is 365.2422 mean solar days; the sidereal year, the interval in which the Sun appears to perform a complete revolution with reference to the fixed *stars*, is 365.2564 mean solar days. The anomolistic year, the average period of revolution of the Earth round the Sun from *perihelion* to perihelion, is 365.2596 mean solar days.

yeasts *Unicellular fungi* producing *zymase*, which converts *sugars* into *alcohol* and *carbon dioxide*. Yeast is used in *brewing* for the production of alcohol, and in baking because the carbon dioxide produced causes the dough to 'rise'. It is also used in more modern aspects of *biotechnology*.

yield point If a wire or rod of a material, such as *steel*, is subjected to a slowly increasing tension, the elongation produced is at first proportional to the tension (*Hooke's law*). If the tension is increased beyond the *limit of proportionality* and the *elastic limit*, a point is reached at which a sudden increase in elongation occurs with only a small increase in tension; this is the yield point.

YIG Yttrium iron garnet. A synthetic *ferrite* used in *microwave* technology.

yocto- Prefix denoting one million million million millionth; 10^{-24}. Symbol y, e.g. $ys = 10^{-24}$ *second*.

yotta- Prefix denoting one million million million million times; 10^{24}. Symbol Y, e.g. $Ym = 10^{24}$ *metres*.

Young's modulus The *elastic modulus* applied to a stretched wire or to a rod under tension or compression; the ratio of the *stress* on a cross-section of the wire or rod to the longitudinal *strain*. Named after Thomas Young (1773–1829).

yperite See *mustard gas*.

ytterbium Yb. Element. R.a.m. 173.04. At. No. 70. A silvery metal, r.d. 6.97, m.p. 824°C., b.p. 1500°C. The element is used in certain *steels*. See *lanthanides*.

yttrium Y. Element. R.a.m. 88.9059. At. No. 39. A greyish metal, r.d. 4.475, m.p. 1510°C., b.p. 3300°C. The natural element is Y-89. It is used in some *alloys* for *superconductors* and the oxide, Y_2O_3, is used in *television phosphors*. See also *YIG*.

Z

Zeeman effect When a substance that emits a *line spectrum* is placed in a strong *magnetic field*, the single lines are split up into groups of closely spaced lines. The 'normal Zeeman effect' can be explained by classical theory as a result of the effect on the *velocity* of the orbital electrons of the applied field. From the separation of the lines in these groups information on atomic structure can be deduced. The 'anomalous Zeeman effect' is a further splitting of these lines, which is described as anomalous as it cannot be explained by classical theory. It is predicted by *quantum* mechanics as a result of *electron spin*. Named after Pieter Zeeman (1865–1943).

Zener current The current in a *semiconductor*, consisting of *electrons* that have escaped from the *valence band* into the *conduction band* under the influence of a strong *electric field*. Named after C. M. Zener (b. 1905).

Zener diode A *semiconductor diode* consisting of a *p-n* junction. It is a *rectifier* until the Zener breakdown voltage is reached, when the device becomes conducting as a result of the *Zener current*. They are used in voltage-limiting circuits.

zenith (astr.) The highest point; the point on the *celestial sphere* directly overhead. See Fig. 2 under *azimuth*, which also illustrates the 'zenith angle'. Compare *nadir*.

zeolites A large class of *hydrated aluminosilicates*, both natural and synthetic, used for *ion exchange* in water softening and as adsorbents (e.g. in *sorption pumps*). The *water molecules* are held in lattice cavities, and can be released by heating. The cavities can then be filled by other molecules of the appropriate size. Zeolites are sometimes known as molecular sieves.

zepto- Prefix denoting one thousand million million millionth; 10^{-21}. Symbol z, e.g. zs = 10^{-21} *second*.

zero Nought; the starting-point of any scale of measurement.

zero point energy The *energy* possessed by the *atoms* or *molecules* of a substance at the *absolute zero* of temperature. This arises because according to the *uncertainty principle* a particle cannot be said to be at rest at exactly the centrepoint of its oscillation.

zeroth law of thermodynamics See *thermodynamics, laws of*.

zerovalent Having zero *valence*.

zeta pinch See *pinch effect*.

zeta-potential See *electrokinetic potential*.

zetta- Prefix denoting one thousand million million million times; 10^{21}. Symbol Z, e.g. Zm = 10^{21} *metres*.

Ziegler catalysts *Catalysts* capable of initiating the *polymerization* of *ethene* and *propene* at normal *temperatures* and *pressures*, e.g. titanium trichloride and aluminium alkyl. Named after Carl Ziegler (1897–1973).

zinc Zn. Element. R.a.m. 65.39. At. No. 30. A hard bluish-white *metal*, r.d. 7.14, m.p. 419°C., b.p. 910°C. It occurs as *calamine*, $ZnCO_3$, *zincite*, ZnO, and *zinc blende*, ZnS. The metal is extracted by roasting the ore to form the *oxide*, which is then reduced with carbon and the resulting zinc distilled. It can also be extracted by electrolysis of the sulphate, obtained by dissolving the oxide in *sulphuric acid*. It is used in *alloys*, especially *brass*, and in *galvanized iron*.

zincate A *salt* containing the *ion* ZnO_2^{2-}.

zinc blende Sphalerite. Natural *zinc sulphide*, ZnS. An important *ore* of zinc.

zinc carbonate *Calamine*. $ZnCO_3$. A white *insoluble* crystalline substance, used in medicine in the treatment of skin diseases.

zinc chloride Butter of zinc. $ZnCl_2$. A white *deliquescent soluble* substance, m.p. 283°C., used as an *antiseptic*, a wood preservative, and as a *flux*.

zinc-copper couple Metallic zinc coated with a thin film of copper by immersing zinc in *copper sulphate solution*. It evolves hydrogen with hot *water*.

zincite Natural *zinc oxide*, ZnO. An important *ore* of zinc.

zinc oxide ZnO. A white *amorphous* powder, m.p. 1975°C., widely used as a *pigment* (Chinese white), in *glass* manufacture, in cosmetics, and in medicine. It occurs naturally as *zincite*.

zinc phosphide Zn_3P_2. A grey *insoluble* crystalline substance, used as a rat poison.

zinc silicate Several silicates of zinc exist. Natural zinc silicate, hemimorphite, is $2ZnO.SiO_2.H_2O$ (see also *calamine*). Zinc *meta*silicate, $ZnSiO_3$, has m.p. 1437°C. Zinc *ortho*silicate, Zn_2SiO_4, has m.p. 1509°C.

zinc sulphate White vitriol. $ZnSO_4.7H_2O$. A white *soluble* crystalline powder, m.p. 100°C., used as a *mordant*, in zinc plating, in the manufacture of *paper*, and in medicine.

zinc sulphide ZnS. A white or yellowish *insoluble* crystalline substance that occurs naturally as *zinc blende*. It is used as a *pigment*.

zircon Zirconium silicate. $ZrSiO_4$. A colourless or yellowish *insoluble* substance, m.p. 2550°C. It is used as a gemstone when *transparent* and a *refractory* when coloured.

zirconia See *zirconium dioxide*.

zirconium Zr. Element. R.a.m. 91.224. At. No. 40. A rare greyish *metal*, r.d. 6.507, m.p. 1850°C., b.p. 4500°C. It occurs as *zircon*, which is the main source. Chlorination of zircon gives $ZrCl_4$, which is reduced by the *Kroll process*, after purification. It is used in *alloys*, abrasives, and flame proofing *compounds*. It is a *neutron* absorber and is used in *nuclear reactors*.

zirconium dioxide Zirconium(IV) oxide, zirconia. ZrO_2. A white crystalline *insoluble* substance, m.p. 2715°C., used as a *pigment* and a *refractory*. It is also used as an *electrolyte* in *fuel cells*. The *hydrated* form, $ZrO_2.xH_2O$, also known as 'zirconium hydroxide' and 'zirconic acid', is a white *amorphous* powder.

zirconyl The *univalent* group ZrO–.

zodiac The zone of the *celestial sphere* that contains the paths of the Sun, the Moon, and the planets. It is bounded by two circles, which are equidistant from the *ecliptic* and about 18° apart. It is divided into the 12 signs of the zodiac, which were named by the ancient Greeks after the 12 constellations that occupied these signs some 2000 years ago. As a result of the *precession of the equinoxes* the constellations no longer coincide with these signs.

zodiacal light A faint luminous patch seen in the sky, on the western horizon after sunset or on the eastern horizon before sunrise, believed to be due to the *scattering* of sunlight by dust particles revolving round the *Sun*.

zone of sphere A portion of the surface of a *sphere* cut off by two parallel *planes*. Its area is given by $2\pi rd$, where r is the radius of the sphere and d the distance between the two planes.

zone refining A purification method, applied mainly to *alloys*, *metals*, and *semiconductors*, based on the principle that the solubility of an impurity B in a main component A in the solid state may differ from the solubility of B in A in the liquid state. When a narrow molten zone is made to pass (e.g. by movement of a heater outside a tube containing a long bar of the material) along a bar of impure A, the distribution of B between the solid and liquid material alters so that the impurity B tends to segregate towards one end of the bar, with pure material at the other end.

zones, fresnel See *half-period zones*.

zones of audibility An intense *sound*, e.g. due to an *explosion*, can usually be heard or detected at all points in a large area around the source of the sound, and also in distant zones of audibility separated from that area by regions in which the sound cannot be detected. Sound waves can reach these zones by reflection down from the *upper atmosphere*.

zoogeography The study of the geographical distribution of animals throughout the world.

zoology The scientific study of animals.

zoom lens A cinematic, television, or still *camera lens* whose *focal length* can be adjusted continuously to vary the *magnification* without loss of focus.

Z particle Z *boson*. The intermediary *elementary particle* with no *electric charge* that is exchanged as a *virtual particle* in some *weak interactions*. The Z particle, like its charged relation the *W particle*, is very heavy, with a mass of 91 GeV. The discovery of the Z particle in 1983 provided confirmation of the *electroweak theory*.

zwitterion An *ion* carrying both a positive and negative *electric charge*.

zygote A fertilized *ovum*; the product of the union of two *gametes*.

zymase An *enzyme* present in *yeast* that acts on *sugar* with the formation of *alcohol* and *carbon dioxide*. See *fermentation*.

zymology Enzymology. The study of *fermentation* and the action of *enzymes*.

zymotic Relating to, or caused by, *fermentation*.

APPENDIX

APPENDIX

TABLE 1

6-FIGURE CONVERSION FACTORS
SI. CGS AND FPS UNITS

Length

	m	cm	in	ft	yd
1 metre	1	100	39.3701	3.28084	1.09361
1 centimetre	0.01	1	0.393701	0.0328084	0.0109361
1 inch	0.0254	2.54	1	0.0833333	0.0277778
1 foot	0.3048	30.48	12	1	0.333333
1 yard	0.9144	91.44	36	3	1

	km	mi	n.mi
1 kilometre	1	0.621371	0.539957
1 mile	1.60934	1	0.868976
1 nautical mile	1.85200	1.15078	1

1 light year $= 9.46070 \times 10^{15}$ metres $= 5.87848 \times 10^{12}$ miles.

1 Astronomical Unit $= 1.495 \times 10^{11}$ metres.

1 parsec $= 3.0857 \times 10^{16}$ metres $= 3.2616$ light years.

TABLE 1—cont.

6-FIGURE CONVERSION FACTORS

SI, CGS AND FPS UNITS

Area

	m²	cm²	in²	ft²	acre
1 square metre	1	10⁴	1550	10.7639	2.47105×10^{-4}
1 square centimetre	10^{-4}	1	0.155	1.07639×10^{-3}	247.105
1 square inch	6.4516×10^{-4}	6.4516	1	6.94444×10^{-3}	2.06612×10^{-4}
1 square foot	9.2903×10^{-2}	929.03	144	1	640

	m²	km²	yd²	mi²	acre
1 square metre	1	10^{-6}	1.19599	3.86019×10^{-7}	2.47105×10^{-4}
1 square kilometre	10^{6}	1	1.19599×10^{6}	0.386019	247.105
1 square yard	0.836127	8.36127×10^{-7}	1	3.22831×10^{-7}	2.06612×10^{-4}
1 square mile	2.58999×10^{6}	2.58999	3.0976×10^{6}	1	640
1 acre	4.04686×10^{3}	4.04686×10^{-3}	4840	1.5625×10^{-3}	1

1 are = 100 square metres.
1 hectare = 10 000 square metres = 2.47105 acres.

TABLE 1. Conversion factors—*cont.*

Volume

	m³	cm³	in³	ft³	gal
1 cubic metre	1	10^6	6.10236×10^4	35.3146	219.969
1 cubic centimetre	10^{-6}	1	0.0610236	3.53146×10^{-5}	2.19969×10^{-4}
1 cubic inch	1.63871×10^{-5}	16.3871	1	5.78704×10^{-4}	3.60464×10^{-3}
1 cubic foot	0.0283168	28316.8	1728	1	6.22882
1 gallon (UK)	4.54609×10^{-3}	4546.09	277.42	0.160544	1

1 gallon (US) = 0.832 68 gallon (UK).
1 cubic yard = 0.764 555 cubic metre.
The *litre* is now recognized as a special name for a cubic decimetre, but is not used to express high precision measurements.

Velocity

	m/sec	km/hr	mi/hr	ft/sec
1 metre per second	1	3.6	2.23694	3.28084
1 kilometre per hour	0.277778	1	0.621371	0.911346
1 mile per hour	0.44704	1.609344	1	1.46667
1 foot per second	0.3048	1.09728	0.681817	1

1 knot = 1 nautical mile per hour = 0.514 444 metre per second.

497

TABLE 1-cont.

6-FIGURE CONVERSION FACTORS
SI, CGS AND FPS UNITS

Mass	kg	g	lb	long ton
1 kilogram	1	1000	2.20462	9.84207×10^{-4}
1 gram	10^{-3}	1	2.20462×10^{-3}	9.84207×10^{-7}
1 pound	0.453592	453.592	1	4.46429×10^{-4}
1 long ton	1016.047	1.016047×10^{6}	2240	1

1 slug = 14.5939 kg = 32.174 lbs.

Density	kg/m³	g/cm³	lb/ft³	lb/in³
1 kilogram per cubic metre	1	10^{-3}	0.062428	3.61273×10^{-5}
1 gram per cubic centimetre	1000	1	62.428	3.61273×10^{-2}
1 pound per cubic foot	16.0185	0.0160185	1	5.78704×10^{-4}
1 pound per cubic inch	2.76799×10^{4}	27.6799	1728	1

1 lb/gal (UK) = 0.099 7763 kg/dm³.

Force	N	kg	dyne	poundal	lb
1 newton	1	0.101972	10^{5}	7.23300	0.224809
1 kilogram force	9.80665	1	9.80665×10^{5}	70.9316	2.20462
1 dyne	10^{-5}	1.01972×10^{-6}	1	7.23300×10^{-5}	2.24809×10^{-6}
1 poundal	0.138255	1.40981×10^{-2}	1.38255×10^{4}	1	0.031081
1 pound force	4.44822	0.453592	4.44823×10^{5}	32.174	1

TABLE 1. Conversion factors—*cont.*

Pressure

	N/m² (Pa)	kg/cm²	lb/in²	atmos
1 newton per square metre (pascal)	1	1.01972×10^{-5}	1.45038×10^{-4}	9.86923×10^{-6}
1 kilogram per square centimetre	980.665×10^2	1	14.2234	0.967841
1 pound per square inch	6.89476×10^3	0.0703068	1	0.068046
1 atmosphere	1.01325×10^5	1.03323	14.6959	1

1 pascal = 1 newton per square metre = 10 dynes per square centimetre.
1 bar = 10^5 newtons per square metre = 0.986 923 atmosphere.
1 torr = 133.322 newtons per square metre = 1/760 atmosphere.
1 atmosphere = 760 mm Hg = 29.92 in Hg = 33.90 ft water (all at 0°C.).

Work and Energy

	J	cal_{IT}	kWhr	btu_{IT}
1 joule	1	0.238846	2.77778×10^{-7}	9.47813×10^{-4}
1 calorie (IT)	4.1868	1	1.16300×10^{-6}	3.96831×10^{-3}
1 kilowatt hour	3.6×10^6	8.59845×10^5	1	3412.14
1 British Thermal Unit (IT)	1055.06	251.997	2.93071×10^{-4}	1

1 joule = 1 newton metre = 1 watt second = 10^7 ergs = 0.737 561 ft lb.
1 electron volt = $1.602\ 10 \times 10^{-19}$ joule.

TABLE 2. FUNDAMENTAL CONSTANTS

Constant	Symbol	Value in SI Units
electronic charge	e	$1.602\,177\,33 \times 10^{-19}$ C
electronic rest mass	m_e	$9.109\,3897 \times 10^{-31}$ kg
electronic radius	r_e	$2.817\,940\,92 \times 10^{-15}$ m
proton rest mass	m_p	$1.672\,6231 \times 10^{-27}$ kg
neutron rest mass	m_n	$1.674\,929 \times 10^{-27}$ kg
Planck constant	h	$6.626\,076 \times 10^{-34}$ J s
speed of light	c	$2.997\,924\,58 \times 10^{8}$ m s^{-1}
Avogadro constant	L, N_A	$6.022\,1367 \times 10^{23}$ mol^{-1}
Loschmidt constant	N_L	$2.686\,763 \times 10^{25}$ m^{-3}
gas constant	R	$8.314\,510$ J K^{-1} mol^{-1}
Boltzmann constant	$k = \dfrac{R}{N_A}$	$1.380\,658 \times 10^{-23}$ J K^{-1}
Faraday constant	F	$9.648\,4531 \times 10^{4}$ C mol^{-1}
Stefan-Boltzmann constant	σ	$5.670\,51 \times 10^{-8}$ W m^{-2} K^4
gravitational constant	G	$6.672\,59 \times 10^{-11}$ N m^2 kg^{-2}
acceleration of free fall	g	$9.806\,65$ m s^{-2}
magnetic constant	μ_0	$4\pi \times 10^{-7}$ H m^{-1}
electric constant	ε_0	$8.854\,187\,817 \times 10^{-12}$ F m^{-1}

[*R.a.m. values in brackets denote mass number of the most stable known isotope*]

Element	Symbol	At. No.	R.a.m.
Actinium	Ac	89	[227]
Aluminium	Al	13	26.981 54
Americium	Am	95	[243]
Antimony	Sb	51	121.75
Argon	Ar	18	39.948
Arsenic	As	33	74.9216
Astatine	At	85	[210]
Barium	Ba	56	137.33
Berkelium	Bk	97	[247]
Beryllium	Be	4	9.012 18
Bismuth	Bi	83	208.9804
Boron	B	5	10.811
Bromine	Br	35	79.904
Cadmium	Cd	48	112.41
Caesium	Cs	55	132.9054
Calcium	Ca	20	40.078
Californium	Cf	98	[251]
Carbon	C	6	12.011
Cerium	Ce	58	140.12
Chlorine	Cl	17	35.453
Chromium	Cr	24	51.9961
Cobalt	Co	27	58.9332
Copper	Cu	29	63.546
Curium	Cm	96	[247]
Dysprosium	Dy	66	162.50
Einsteinium	Es	99	[252]
Erbium	Er	68	167.26
Europium	Eu	63	151.96
Fermium	Fm	100	[257]
Fluorine	F	9	18.998 403
Francium	Fr	87	[223]
Gadolinium	Gd	64	157.25
Gallium	Ga	31	69.723
Germanium	Ge	32	72.59
Gold	Au	79	196.9665
Hafnium	Hf	72	178.49
Helium	He	2	4.002 602
Holmium	Ho	67	164.9304

TABLE 3. TABLE OF ELEMENTS, etc.–*cont.*

Element	Symbol	At. No.	R.a.m.
Hydrogen	H	1	1.007 94
Indium	In	49	114.82
Iodine	I	53	126.9045
Iridium	Ir	77	192.22
Iron	Fe	26	55.847
Krypton	Kr	36	83.80
Lanthanum	La	57	138.9055
Lawrencium	Lr	103	[260]
Lead	Pb	82	207.2
Lithium	Li	3	6.941
Lutetium	Lu	71	174.967
Magnesium	Mg	12	24.305
Manganese	Mn	25	54.938
Mendelevium	Md	101	[258]
Mercury	Hg	80	200.59
Molybdenum	Mo	42	95.94
Neodymium	Nd	60	144.24
Neon	Ne	10	20.179
Neptunium	Np	93	[237]
Nickel	Ni	28	58.69
Niobium	Nb	41	92.9064
Nitrogen	N	7	14.0067
Nobelium	No	102	[259]
Osmium	Os	76	190.2
Oxygen	O	8	15.9994
Palladium	Pd	46	106.42
Phosphorus	P	15	30.973 76
Platinum	Pt	78	195.08
Plutonium	Pu	94	[244]
Polonium	Po	84	[209]
Potassium	K	19	39.0983
Praseodymium	Pr	59	140.9077
Promethium	Pm	61	[145]
Protactinium	Pa	91	[231]
Radium	Ra	88	[226]
Radon	Rn	86	[222]
Rhenium	Re	75	186.207
Rhodium	Rh	45	102.9055
Rubidium	Rb	37	85.4678
Ruthenium	Ru	44	101.07
Samarium	Sm	62	150.36
Scandium	Sc	21	44.955 91
Selenium	Se	34	78.96
Silicon	Si	14	28.0855

Element	Symbol	At. No.	R.a.m.
Silver	Ag	47	107.868
Sodium	Na	11	22.989 77
Strontium	Sr	38	87.62
Sulphur	S	16	32.066
Tantalum	Ta	73	180.9479
Technetium	Tc	43	[98]
Tellurium	Te	52	127.60
Terbium	Tb	65	158.9254
Thallium	Tl	81	204.383
Thorium	Th	90	232.038
Thulium	Tm	69	168.9342
Tin	Sn	50	118.710
Titanium	Ti	22	47.88
Tungsten	W	74	183.85
Uranium	U	92	238.0289
Vanadium	V	23	50.9415
Xenon	Xe	54	131.29
Ytterbium	Yb	70	173.04
Yttrium	Y	39	88.9059
Zinc	Zn	30	65.39
Zirconium	Zr	40	91.224

TABLE 4. THE SOLAR SYSTEM

Planet	Equatorial Diameter (kilometres)	Mass (Earth masses)*	Mean Distance from Sun (millions of kilometres)	Sidereal period
Mercury	4878	0.054	57.91	87.969 days
Venus	12 100	0.8150	108.21	224.701 days
Earth	12 756	1.000	149.60	365.256 days
Mars	6762	0.107	227.94	686.980 days
Jupiter	142 700	317.89	778.34	11.86 years
Saturn	120 800	95.14	1430	29.46 years
Uranus	51 800	14.52	2869.6	84.01 years
Neptune	49 400	17.46	4496.7	164.8 years
Pluto	3500	0.1 (approx.)	5900	248.4 years
Sun	1 392 000	332 958	149.60†	—
Moon	3476	0.0123	0.3844†	27.32

* The Mass of the Earth is 5.976×10^{24} kilogram.
† Distance to Earth.

503

TABLE 5. TABLE OF AMINO ACIDS

Name	Formula	Molecular weight
Glycine	$CH_2(NH_2).COOH$	75.1
Alanine	$CH_3CH.(NH_2).COOH$	89.1
Phenylalanine	$C_6H_5CH_2CH.(NH_2).COOH$	165.2
Tyrosine	$C_6H_4OH.CH_2CH.(NH_2).COOH$	181.2
Valine	$(CH_3)_2CH.CH.(NH_2).COOH$	117.1
Leucine	$(CH_3)_2CH.CH_2CH.(NH_2).COOH$	131.2
Isoleucine	$(CH_3).CH_2CH(CH_3)CH.(NH_2).COOH$	131.2
Serine	$CH_2OH.CH.(NH_2).COOH$	105.1
Threonine	$CH_3CHOH.CH.(NH_2).COOH$	119.1
Cysteine	$SH.CH_2CH.(NH_2).COOH$	121.1
Cystine	$[HOOC.CH(NH_2)CH_2S]_2$	240.3
Methionine	$CH_3.S.(CH_2)_2CH.(NH_2).COOH$	149.2
Asparagine	$NH_2CO.CH_2CH.(NH_2).COOH$	132.1
Glutamine	$NH_2CH.(CH_2)_2(CO.NH_2).COOH$	146.1
Lysine	$NH_2(CH_2)_4CH.(NH_2).COOH$	146.2
Arginine	$NH_2C(:NH).NH(CH_2)_3CH.(NH_2).COOH$	174.2
Aspartic	$COOH.CH_2CH.(NH_2).COOH$	133.1
Glutamic	$COOH.(CH_2)_2CH.(NH_2).COOH$	147.1
Histidine	$C_3H_3N_2.CH_2CH.(NH_2).COOH$	155.2
Tryptophan	$C_6H_4.NH.C_2H.CH_2CH.(NH_2).COOH$	204.2
Proline	$NH.(CH_2)_3CH.COOH$	115.1

TABLE 6. ELEMENTARY PARTICLES

	Particle	Quark content	Mass MeV/c^2	Lifetime/s
gauge bosons	γ		0	stable
	$W^{\pm}$		80000	
	Z^0		91000	
leptons	ν		0	stable
	e		0.511	stable
	μ		105.7	2.2×10^{-6}
	τ		1784.1	3.0×10^{-13}
mesons	$\pi^{\pm}$	$u\bar{d}, \bar{u}d$	139.6	2.6×10^{-8}
	π°	$u\bar{u}, d\bar{d}$	105.7	8.4×10^{-17}
	$K^{\pm}$	$u\bar{s}, s\bar{u}$	493.6	1.2×10^{-8}
	K°	$d\bar{s}$	497.7	
	K°_s		497.7	8.9×10^{-11}
	K°_L		497.7	5.2×10^{-8}
	η°	$u\bar{u}, d\bar{d}, s\bar{s}$	548.8	
	$D^{\pm}$	$c\bar{d}, d\bar{c}$	1869	1.1×10^{-12}
	D°	$c\bar{u}$	1865	4×10^{-13}
	$D^{\pm}_s$	$c\bar{s}, s\bar{c}$	1969	4×10^{-13}
	$B^{\pm}$	$u\bar{b}, b\bar{u}$	5278	1×10^{-12}
	B°	$d\bar{b}$	5279	1×10^{-12}
baryons	p	uud	938.3	stable
	n	udd	939.6	896
	Λ°	uds	1115.6	2.6×10^{-10}
	Σ^+	uus	1189.4	8.0×10^{-10}
	Σ°	uds	1192.5	7.4×10^{-20}
	Σ^-	dds	1197.4	1.5×10^{-10}
	Ξ°	uss	1314.9	2.9×10^{-10}
	Ξ^-	dss	1321.3	1.7×10^{-10}
	Ω^-	sss	1672.5	1.3×10^{-10}
	Λ^+_c	udc	2285	2×10^{-13}

TABLE 7. ELECTRON CONFIGURATIONS AND IONIZATION POTENTIALS OF THE COMMONER ELEMENTS

Element	Atomic Number	Electron Configuration Shell						Ionization Potentials (electron-volts)				
		1 K	2 L	3 M	4 N	5 O	6 P	I	II	III	IV	V
H	1	1						13.59				
He	2	2						24.48	54.40			
C	6	2	4					11.26	24.38	47.87	64.48	392.0
N	7	2	5					14.53	29.59	47.43	77.45	97.86
O	8	2	6					13.61	35.11	54.89	77.39	113.9
F	9	2	7					7.87	16.18	30.64	56.80	114.2
Ne	10	2	8					21.56	41.07	63.50	97.02	126.3
Na	11	2	8	1				5.14	47.29	71.71	98.88	138.4
Mg	12	2	8	2				7.64	15.03	80.14	109.29	141.2
Al	13	2	8	3				5.98	18.82	28.44	119.96	153.8
Si	14	2	8	4				8.15	16.34	33.49	45.13	166.7
P	15	2	8	5				10.48	19.72	30.16	51.35	65.0
S	16	2	8	6				10.36	23.40	35.0	47.29	72.5
Cl	17	2	8	7				13.01	23.80	39.9	53.50	67.8
Ar	18	2	8	8				15.75	27.62	40.9	59.8	75.0

TABLE 7. Electron configurations, etc.—cont.

Element	Atomic Number	Electron Configuration Shell						Ionization Potentials (electron-volts)				
		1 K	2 L	3 M	4 N	5 O	6 P	I	II	III	IV	V
K	19	2	8	8	1			4.34	31.81	46.0	60.9	82.6
Ca	20	2	8	8	2			6.11	11.87	51.2	67.0	84.4
Fe	26	2	8	14	2			7.87	16.18	30.6	56.8	—
Cu	29	2	8	18	1			7.72	20.30	36.8	—	—
Zn	30	2	8	18	2			9.39	17.96	39.7	—	—
Br	35	2	8	18	7			11.84	21.60	35.9	47.3	59.7
Kr	36	2	8	18	8			13.99	24.50	36.9	43.5	63.0
Ag	47	2	8	18	18	1		7.57	21.5	34.8	—	—
Sn	50	2	8	18	18	4		7.34	14.63	30.5	40.7	72.3
I	53	2	8	18	18	7		10.45	19.13	—	—	—
Xe	54	2	8	18	18	8		12.13	21.2	31.3	42.0	53.0
Cs	55	2	8	18	18	8	1	3.89	25.1	35.0	—	—
Ba	56	2	8	18	18	8	2	5.21	10.0	35.5	—	—
Hg	80	2	8	18	32	18	2	10.43	18.75	34.2	49.5	—

TABLE 8. PERIODIC TABLE OF THE ELEMENTS

1A	2A	3B	4B	5B	6B	7B	8			1B	2B	3A	4A	5A	6A	7A	0
1 H																	2 He
3 Li	4 Be											5 B	6 C	7 N	8 O	9 F	10 Ne
11 Na	12 Mg											13 Al	14 Si	15 P	16 S	17 Cl	18 Ar
19 K	20 Ca	21 Sc	22 Ti	23 V	24 Cr	25 Mn	26 Fe	27 Co	28 Ni	29 Cu	30 Zn	31 Ga	32 Ge	33 As	34 Se	35 Br	36 Kr
37 Rb	38 Sr	39 Y	40 Zr	41 Nb	42 Mo	43 Tc	44 Ru	45 Rh	46 Pd	47 Ag	48 Cd	49 In	50 Sn	51 Sb	52 Te	53 I	54 Xe
55 Cs	56 Ba	57* La	72 Hf	73 Ta	74 W	75 Re	76 Os	77 Ir	78 Pt	79 Au	80 Hg	81 Tl	82 Pb	83 Bi	84 Po	85 At	86 Rn
87 Fr	88 Ra	89† Ac															

← TRANSITION ELEMENTS →

*Lanthanides	57 La	58 Ce	59 Pr	60 Nd	61 Pm	62 Sm	63 Eu	64 Gd	65 Tb	66 Dy	67 Ho	68 Er	69 Tm	70 Yb	71 Lu
†Actinides	89 Ac	90 Th	91 Pa	92 U	93 Np	94 Pu	95 Am	96 Cm	97 Bk	98 Cf	99 Es	100 Fm	101 Md	102 No	103 Lr

508

TABLE 9. DIFFERENTIAL COEFFICIENTS AND INTEGRALS

y	$\dfrac{\mathrm{d}y}{\mathrm{d}x}$	$\int y.\mathrm{d}x$
x^n	nx^{n-1}	$\dfrac{1}{n+1}.x^{n+1}$
$\dfrac{1}{x}$	$\dfrac{-1}{x^2}$	$\log_e x$
e^{ax}	$a\mathrm{e}^{ax}$	$\dfrac{1}{a}.\mathrm{e}^{ax}$
$\log_e x$	$\dfrac{1}{x}$	$x(\log_e x - 1)$
$\log_q x$	$\dfrac{1}{x}.\log_a \mathrm{e}$	$x.\log_a \dfrac{x}{\mathrm{e}}$
$\cos ax$	$-a.\sin ax$	$\dfrac{1}{a}.\sin ax$
$\sin ax$	$a.\cos ax$	$-\dfrac{1}{a}.\cos ax$
$\tan ax$	$a.\sec^2 ax$	$-\dfrac{1}{a}.\log_e \cos ax$
$\cot x$	$-\mathrm{cosec}^2 x$	$\log_e \sin x$
$\sec x$	$\tan x.\sec x$	$\log_e(\sec x + \tan x)$
$\mathrm{cosec}\ x$	$-\cot x.\mathrm{cosec}\ x$	$\log_e(\mathrm{cosec}\ x - \cot x)$
$\sin^{-1}\dfrac{x}{a}$	$\dfrac{1}{(a^2-x^2)^{1/2}}$	$x.\sin^{-1}\dfrac{x}{a}+(a^2-x^2)^{1/2}$
$\cos^{-1}\dfrac{x}{a}$	$\dfrac{-1}{(a^2-x^2)^{1/2}}$	$x.\cos^{-1}\dfrac{x}{a}-(a^2-x^2)^{1/2}$
$\tan^{-1}\dfrac{x}{a}$	$\dfrac{a}{a^2+x^2}$	$x.\tan^{-1}\dfrac{x}{a}-a\log_e(a^2+x^2)^{1/2}$

TABLE 10. SPECTRUM OF ELECTROMAGNETIC RADIATIONS

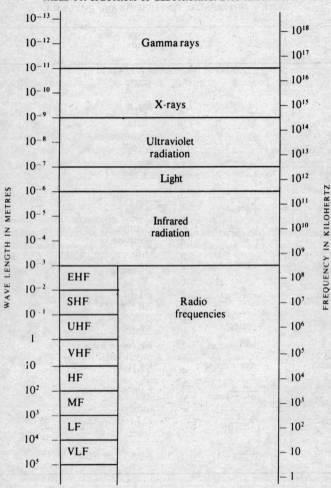

TABLE 11. THE GREEK ALPHABET

Letters		Name
A	α	alpha
B	β	beta
Γ	γ	gamma
Δ	δ	delta
E	ϵ	epsilon
Z	ζ	zeta
H	η	eta
Θ	θ	theta
I	ι	iota
K	κ	kappa
Λ	λ	lambda
M	μ	mu
N	ν	nu
Ξ	ξ	xi
O	o	omicron
Π	π	pi
P	ρ	rho
Σ	σ	sigma
T	τ	tau
Υ	υ	upsilon
Φ	ϕ	phi
X	χ	chi
Ψ	ψ	psi
Ω	ω	omega

READ MORE IN PENGUIN

In every corner of the world, on every subject under the sun, Penguin represents quality and variety – the very best in publishing today.

For complete information about books available from Penguin – including Puffins, Penguin Classics and Arkana – and how to order them, write to us at the appropriate address below. Please note that for copyright reasons the selection of books varies from country to country.

In the United Kingdom: Please write to *Dept. JC, Penguin Books Ltd, FREEPOST, West Drayton, Middlesex UB7 OBR*

If you have any difficulty in obtaining a title, please send your order with the correct money, plus ten per cent for postage and packaging, to *PO Box No. 11, West Drayton, Middlesex UB7 OBR*

In the United States: Please write to *Penguin USA Inc., 375 Hudson Street, New York, NY 10014*

In Canada: Please write to *Penguin Books Canada Ltd, 10 Alcorn Avenue, Suite 300, Toronto, Ontario M4V 3B2*

In Australia: Please write to *Penguin Books Australia Ltd, 487 Maroondah Highway, Ringwood, Victoria 3134*

In New Zealand: Please write to *Penguin Books (NZ) Ltd,182–190 Wairau Road, Private Bag, Takapuna, Auckland 9*

In India: Please write to *Penguin Books India Pvt Ltd, 706 Eros Apartments, 56 Nehru Place, New Delhi 110 019*

In the Netherlands: Please write to *Penguin Books Netherlands B.V., Keizersgracht 231 NL–1016 DV Amsterdam*

In Germany: Please write to *Penguin Books Deutschland GmbH, Friedrichstrasse 10–12, W–6000 Frankfurt/Main 1*

In Spain: Please write to *Penguin Books S. A., C. San Bernardo 117–6° E–28015 Madrid*

In Italy: Please write to *Penguin Italia s.r.l., Via Felice Casati 20, I–20124 Milano*

In France: Please write to *Penguin France S. A., 17 rue Lejeune, F–31000 Toulouse*

In Japan: Please write to *Penguin Books Japan, Ishikiribashi Building, 2–5–4, Suido, Bunkyo-ku, Tokyo 112*

In Greece: Please write to *Penguin Hellas Ltd, Dimocritou 3, GR–106 71 Athens*

In South Africa: Please write to *Longman Penguin Southern Africa (Pty) Ltd, Private Bag X08, Bertsham 2013*

READ MORE IN PENGUIN

SCIENCE AND MATHEMATICS

QED   Richard Feynman
The Strange Theory of Light and Matter

'Physics Nobelist Feynman simply cannot help being original. In this quirky, fascinating book, he explains to laymen the quantum theory of light – a theory to which he made decisive contributions' – *New Yorker*

Does God Play Dice?   Ian Stewart
The New Mathematics of Chaos

To cope with the truth of a chaotic world, pioneering mathematicians have developed chaos theory. *Does God Play Dice?* makes accessible the basic principles and many practical applications of one of the most extraordinary – and mind-bending – breakthroughs in recent years.

Bully for Brontosaurus   Stephen Jay Gould

'He fossicks through history, here and there picking up a bone, an imprint, a fossil dropping and, from these, tries to reconstruct the past afresh in all its messy ambiguity. It's the droppings that provide the freshness: he's as likely to quote from Mark Twain or Joe DiMaggio as from Lamarck or Lavoisier' – *Guardian*

The Blind Watchmaker   Richard Dawkins

'An enchantingly witty and persuasive neo-Darwinist attack on the anti-evolutionists, pleasurably intelligible to the scientifically illiterate' – Hermione Lee in the *Observer* Books of the Year

The Making of the Atomic Bomb   Richard Rhodes

'Rhodes handles his rich trove of material with the skill of a master novelist ... his portraits of the leading figures are three-dimensional and penetrating ... the sheer momentum of the narrative is breathtaking ... a book to read and to read again' – Walter C. Patterson in the *Guardian*

Asimov's New Guide to Science   Isaac Asimov

A classic work brought up to date – far and away the best one-volume survey of all the physical and biological sciences.

READ MORE IN PENGUIN

SCIENCE AND MATHEMATICS

The Panda's Thumb   Stephen Jay Gould

More reflections on natural history from the author of *Ever Since Darwin*. 'A quirky and provocative exploration of the nature of evolution … wonderfully entertaining' – *Sunday Telegraph*

Einstein's Universe   Nigel Calder

'A valuable contribution to the demystification of relativity' – *Nature*. 'A must' – *Irish Times*. 'Consistently illuminating' – *Evening Standard*

Gödel, Escher, Bach: An Eternal Golden Braid
Douglas F. Hofstadter

'Every few decades an unknown author brings out a book of such depth, clarity, range, wit, beauty and originality that it is recognized at once as a major literary event' – Martin Gardner. 'Leaves you feeling you have had a first-class workout in the best mental gymnasium in town' – *New Statesman*

The Double Helix   James D. Watson

Watson's vivid and outspoken account of how he and Crick discovered the structure of DNA (and won themselves a Nobel Prize) – one of the greatest scientific achievements of the century.

The Quantum World   J. C. Polkinghorne

Quantum mechanics has revolutionized our views about the structure of the physical world – yet after more than fifty years it remains controversial. This 'delightful book' (*The Times Educational Supplement*) succeeds superbly in rendering an important and complex debate both clear and fascinating.

Mathematical Circus   Martin Gardner

A mind-bending collection of puzzles and paradoxes, games and diversions from the undisputed master of recreational mathematics.

READ MORE IN PENGUIN

SCIENCE AND MATHEMATICS

The Dying Universe Paul Davies

In this enthralling book the author of *God and the New Physics* tells how, from the instant of its fiery origin in a big bang, the universe has been running down. With clarity and panache Paul Davies introduces the reader to a mind-boggling array of cosmic exotica to help chart the cosmic apocalypse.

The Newtonian Casino Thomas A. Bass

'The story's appeal lies in its romantic obsessions ... Post-hippie computer freaks develop a system to beat the System, and take on Las Vegas to heroic and thrilling effect' – *The Times*

Wonderful Life Stephen Jay Gould

'He weaves together three extraordinary themes – one palaeontological, one human, one theoretical and historical – as he discusses the discovery of the Burgess Shale, with its amazing, wonderfully preserved fossils – a time-capsule of the early Cambrian seas' – *Mail on Sunday*

The New Scientist Guide to Chaos edited by Nina Hall

In this collection of incisive reports, acknowledged experts such as Ian Stewart, Robert May and Benoit Mandelbrot draw on the latest research to explain the roots of chaos in modern mathematics and physics.

Innumeracy John Allen Paulos

'An engaging compilation of anecdotes and observations about those circumstances in which a very simple piece of mathematical insight can save an awful lot of futility' – Ian Stewart in *The Times Educational Supplement*

Fractals Hans Lauwerier

The extraordinary visual beauty of fractal images and their applications in chaos theory have made these endlessly repeating geometric figures widely familiar. This invaluable new book makes clear the basic mathematics of fractals; it will also teach people with computers how to make fractals themselves.

READ MORE IN PENGUIN

REFERENCE

Medicines: A Guide for Everybody Peter Parish

Now in its seventh edition and completely revised and updated, this bestselling guide is written in ordinary language for the ordinary reader yet will prove indispensable to anyone involved in health care – nurses, pharmacists, opticians, social workers and doctors.

Media Law Geoffrey Robertson, QC, and Andrew Nichol

Crisp and authoritative surveys explain the up-to-date position on defamation, obscenity, official secrecy, copyright and confidentiality, contempt of court, the protection of privacy and much more.

The Slang Thesaurus

Do you make the public bar sound like a gentleman's club? The miraculous *Slang Thesaurus* will liven up your language in no time. You won't Adam and Eve it! A mine of funny, witty, acid and vulgar synonyms for the words you use every day.

The Penguin Dictionary of Troublesome Words Bill Bryson

Why should you avoid discussing the *weather conditions*? Can a married woman be celibate? Why is it eccentric to talk about the aroma of a cowshed? A straightforward guide to the pitfalls and hotly disputed issues in standard written English.

The Penguin Dictionary of Musical Performers Arthur Jacobs

In this invaluable companion volume to *The Penguin Dictionary of Music* Arthur Jacobs has brought together the names of over 2,500 performers. Music is written by composers, yet it is the interpreters who bring it to life; in this comprehensive book they are at last given their due.

The Penguin Dictionary of Physical Geography John Whittow

'Dr Whittow and Penguin Reference Books have put serious students of the subject in their debt, by combining the terminology of the traditional geomorphology with that of the quantitative revolution and defining both in one large and comprehensive dictionary of physical geography ... clear and succinct' – *The Times Educational Supplement*

READ MORE IN PENGUIN

DICTIONARIES

Abbreviations
Archaeology
Architecture
Art and Artists
Biology
Botany
Building
Business
Chemistry
Civil Engineering
Computers
Curious and Interesting Numbers
Curious and Interesting Words
Design and Designers
Economics
Electronics
English and European History
English Idioms
French
Geography
Historical Slang
Human Geography
Information Technology

Literary Terms and Literary Theory
Mathematics
Modern History 1789–1945
Modern Quotations
Music
Musical Performers
Physical Geography
Physics
Politics
Proverbs
Psychology
Quotations
Religions
Rhyming Dictionary
Saints
Science
Sociology
Spanish
Surnames
Telecommunications
Troublesome Words
Twentieth-Century History